T0320765

QUANTUM MONTE CARLO METHODS

Featuring detailed explanations of the major algorithms used in quantum Monte Carlo simulations, this is the first textbook of its kind to provide a pedagogical overview of the field and its applications. The book provides a comprehensive introduction to the Monte Carlo method, its use, and its foundations, and examines algorithms for the simulation of quantum many-body lattice problems at finite and zero temperature. These algorithms include continuous-time loop and cluster algorithms for quantum spins, determinant methods for simulating Fermions, power methods for computing ground and excited states, and the variational Monte Carlo method. Also discussed are continuous-time algorithms for quantum impurity models and their use within dynamical mean-field theory, along with algorithms for analytically continuing imaginary-time quantum Monte Carlo data. The parallelization of Monte Carlo simulations is also addressed. This is an essential resource for graduate students, teachers, and researchers interested in quantum Monte Carlo.

J. E. GUBERNATIS works at the Los Alamos National Laboratory. He is a Fellow of the American Physical Society (APS) and served as a Chair of the APS' Division of Computational Physics. He represented the United States on the Commission of Computational Physics of International Union of Pure and Applied Physics (IUPAP) for nine years and chaired the Commission for three years.

N. KAWASHIMA is a professor at the University of Tokyo. He is a member of the Society of Cognitive Science and has been a Steering Committee member for the public use of the supercomputer at the Institute for Solid State Physics (ISSP) for the last 15 years. He received the Ryogo Kubo Memorial Prize for his contributions to the development of loop and cluster algorithms in 2002.

P. WERNER is a professor at the University of Fribourg. In 2010, he received the IUPAP Young Scientist Prize in computational physics for the development and implementation of quantum Monte Carlo methods for impurity models.

QUANTUM MONTE CARLO METHODS

METHODS

Algorithms for Lattice Models

J. E. GUBERNATIS

Los Alamos National Laboratory

N. KAWASHIMA

University of Tokyo

P. WERNER

University of Fribourg

CAMBRIDGE
UNIVERSITY PRESS

CAMBRIDGE
UNIVERSITY PRESS

University Printing House, Cambridge CB2 8BS, United Kingdom

Cambridge University Press is part of the University of Cambridge.

It furthers the University's mission by disseminating knowledge in the pursuit of
education, learning and research at the highest international levels of excellence.

www.cambridge.org
Information on this title: www.cambridge.org/9781107006423

© Cambridge University Press 2016

First published 2016

A catalogue record for this publication is available from the British Library

Library of Congress Cataloging in Publication data
Gubernatis, J. E., author.
Quantum Monte Carlo methods : algorithms for lattice models / J.E. Gubernatis (Los Alamos National
Laboratory), N. Kawashima (University of Tokyo), P. Werner (University of Fribourg).
pages cm
Includes bibliographical references and index.
ISBN 978-1-107-00642-3 (Hardback : alk. paper)
1. Monte Carlo method.   2. Many-body problem.   I. Kawashima, N. (Naoki), author.
II. Werner, P., 1975– author.   III. Title.
QC174.85.M64G83 2016
530.1201'518282–dc23   2015026699

ISBN 978-1-107-00642-3 Hardback

Contents

Preface

Fast computers enable the solution of quantum many-body problems by Monte Carlo methods. As computing power increased dramatically over the years, similarly impressive advances occurred at the level of the algorithms, so that we are now in a position to perform accurate simulations of large systems of interacting quantum spins, Bosons, and (to a lesser extent) Fermions. The purpose of this book is to present and explain the quantum Monte Carlo algorithms being used today to simulate the ground states and thermodynamic equilibrium states of quantum models defined on a lattice. Our intent is not to review all relevant algorithms – there are too many variants to do so comprehensively – but rather to focus on a core set of important algorithms, explaining what they are and how and why they work.

Our focus on lattice models, such as Heisenberg and Hubbard models, has at least two implications. The first is obviously that we are not considering models in the continuum where extensive use of quantum Monte Carlo methods traditionally has focused on producing highly accurate ab initio calculations of the ground states of nuclei, atoms, molecules, and solids. Quantum Monte Carlo algorithms for simulating the ground states of continuum and lattice models, however, are very similar. In fact, the lattice algorithms are in many cases derived from the continuum methods. With fewer degrees of freedom, lattice models are compact and insightful representations of the physics in the continuum.

The second implication is a focus on both zero and finite temperature algorithms.[1] On a lattice, it is natural to study phase transitions. In particular, the recent dramatic advances in quantum Monte Carlo lattice methods for the simulation of quantum spin models were prompted by a need for more efficient and effective ways to study finite-temperature transitions. While quantum Monte Carlo is profitably used to study zero temperature phase transitions (quantum critical phenomena),

---

[1] Temperature zero is a finite temperature, but common usage separates $T = 0$ (zero temperature) from $T > 0$ (finite temperature) when classifying algorithms. Throughout, we adopt the common usage.

some ground state algorithms have no finite temperature analogs and vice versa. In many respects, the lattice is where the current algorithmic action is.

The book is divided into four parts. The first part is a self-contained, more advanced than average, discussion of the Monte Carlo method, its use, and its foundations. With the basics in place, this part then steps toward the more recent worm and loop/cluster Monte Carlo algorithms for simple classical models, and finally for simple quantum models. Our intent is to be as tutorial as possible and impart a good taste for what quantum Monte Carlo is like. In this introduction, we only briefly mention ground state simulations. The foundations for ground state simulations require less introduction, and we wanted to keep Part I reasonably sized so it can be used as teaching material for a course on computational classical and quantum many-body physics.

Parts II and III present the main quantum Monte Carlo algorithms. Part II discusses finite-temperature methods for quantum spin and Fermion systems. The quantum spin chapter, plus its associated appendices, present a more extensive and sophisticated treatment of the worm and loop/cluster algorithms introduced in Part I. The Fermion chapter details the most important methods for the finite-temperature simulation of lattice Fermion models. Many of the formal techniques developed in this chapter are used in the discussion of the zero temperature Fermion methods in Part III. Besides well-established algorithms for Fermionic lattice models, we also discuss the more recent continuous-time Monte Carlo technique for quantum impurity models. This method is the dynamo driving today's lattice calculations based on the dynamical mean-field approximation, which in turn is being coupled to ab initio calculations of the electronic properties of solids. The chapter on impurity models hence connects the lattice and continuum modeling and simulations of many-electron physics.

Part III discusses the two main zero temperature methods, the variational Monte Carlo and the power method. The power method is a more universal term for what is often called the Green's function Monte Carlo method. Part III also includes a special chapter on the use of the power method for the simulation of Fermion systems. It is in this chapter that the "sign problem" is discussed. This problem is the biggest inhibitor to quantum Monte Carlo reaching its full potential.

The final part does not discuss quantum Monte Carlo algorithms but rather topics that accompany their use. First, we present a widely used method to analytically continue simulation data from imaginary time to real time, so dynamical properties can be extracted from finite-temperature simulations. Finally, we address the parallelization of Monte Carlo simulations. While Monte Carlo calculations per se are naturally parallel (the code, with different random number seeds, can be run on independent processors and the results combined when all calculations have finished), this chapter is about a relatively recent trend, namely, the complexity of simulations is making the sharing of specific computational tasks among several processors desirable or mandatory, and about what is lying in the future, namely, running even more complex simulations on computers with orders of magnitude more processors.

We happily thank Tom Booth, Yan Chen, Kenji Harada, Akiko Kato, Yasuyuki Kato, Tsuyoshi Okubo, Gerado Ortiz, Brenda Rubenstein, Richard Scalettar, Devinder Sivia, Synge Todo, and Cyrus Umrigar for helpful comments and suggestions about various parts of the manuscript. One of the coauthors (NK) thanks Yan Chen, in particular, for offering a quiet place for writing a part of the book. Another coauthor (PW) would like to thank Matthias Troyer, Emanuel Gull, and Andrew Millis for fruitful discussions and collaborations on quantum Monte Carlo and dynamical mean-field related topics. The third coauthor (JG) thanks his wife, Michele, for her proofreading of a manuscript that to say the least was not her "cup of tea." Of course, for the errors that remain we take full responsibility. We also thank our editor, Simon Capelin. His polite, but yearly, inquiry about interest in writing such a textbook caused one of the coauthors eventually to "cave in" and to start writing this book. Simon's patience in waiting for something that took longer than just a few years is also appreciated.

To the researchers, students, and teachers using this book: we hope that it helps you appreciate and understand the essence of quantum Monte Carlo algorithms. We also hope that it inspires you to develop even better ways to do quantum simulations. Without doubt, this important research tool has limitations needing mitigation to realize its full potential. Realizing this potential, however, will contribute to our understanding of quantum many-body physics, which in the long run is our attractor to this research area and our motivation to explore and develop new algorithms.

J. E. Gubernatis

N. Kawashima

P. Werner

Part I

Monte Carlo basics

1

Introduction

A quantum Monte Carlo method is simply a Monte Carlo method applied to a quantum problem. What distinguishes a quantum Monte Carlo method from a classical one is the initial effort necessary to represent the quantum problem in a form that is suitable for Monte Carlo simulation. It is in making this transformation that the quantum nature of the problem asserts itself not only through such obvious issues as the noncommutivity of the physical variables and the need to symmetrize or antisymmetrize the wave function, but also through less obvious issues such as the sign problem. Almost always, the transformation replaces the quantum degrees of freedom by classical ones, and it is to these classical degrees of freedom that the Monte Carlo method is actually applied. Succeeding chapters present and explain many of the quantum Monte Carlo methods being successfully used on a variety of quantum problems. In Chapters 1 and 2 we focus on discussing what the Monte Carlo method is and why it is useful.

1.1 The Monte Carlo method

The Monte Carlo method is not a specific technique but a general strategy for solving problems too complex to solve analytically or too intensive numerically to solve deterministically. Often a specific strategy incorporates several different Monte Carlo techniques. In what is likely the first journal article to use the phrase "Monte Carlo," Metropolis and Ulam (1949) discuss this strategy. To paraphrase them,

The Monte Carlo method is an iterative stochastic procedure, consistent with a defining relation for some function, which allows an estimate of the function without completely determining it.

This is quite different from the colloquialism, "a method that uses random numbers." Let us examine the definition piece by piece. A key point will emerge.

Ulam and Metropolis were presenting the motivation and a general description of a statistical approach to the study of differential and integro-differential equations. These equations were their "defining relation for some function." The "function" was the solution of these equations. This function is of course unknown a priori. Metropolis, Rosenbluth, Rosenbluth, Teller, and Teller (1953) a few years later would propose a statistical approach to the study of equilibrium statistical mechanics. The defining relation there was a thermodynamic average of a physical quantity over the Boltzmann distribution. The function was the physical quantity, and the unknown its average. The general description of the Monte Carlo method, given by Ulam and Metropolis and paraphrased by us, covers statistical mechanics applications plus many more applications than the original ones for differential and integro-differential equations.

The method is an "iterative stochastic procedure," meaning the procedure is applied over and over. Ostensibly, one purpose is to produce a large number of measurements on which a statistical analysis is made and validated by appealing to the law of large numbers. Certainly, this application is one use of the Monte Carlo method Ulam and Metropolis had in mind. For many other applications, the meaning of this phrase is more subtle. The movement from one step to the next depends on a stochastic procedure, creating a chain of events on which a statistical study is made. Usually this chain is a Markov chain.

Random numbers enter through the back door by saying it is a stochastic procedure. Besides being amazed at the speed with which the first computers did arithmetic, Ulam and Metropolis were also amazed that this arithmetic could generate "random numbers" uniformly distributed over the interval from zero to one, thereby enabling various schemes to sample from almost any distribution.

The defining character of the Monte Carlo strategy, however, comes from the rest of the statement: the Monte Carlo method estimates some function "without completely determining it." For Ulam and Metropolis this meant not having to obtain point-by-point values of functions obeying the differential and integro-differential equations. More broadly, presaging applications to problems in statistical mechanics, we paraphrased them by saying the Monte Carlo method makes estimates without needing to determine the function completely. Because it does not completely determine the function, the Monte Carlo method can thus provide only an estimate. As we will discuss, this estimate can have a high degree of certainty.

The Monte Carlo method is not the only method that does not need to determine a function completely. Deterministic methods also often do not need to do this. For example, a standard numerical problem is estimating the integral of a function of many variables. Deterministic algorithms use discretization schemes that estimate the integral by evaluating the integrand only at N points in each dimension. The computational effort of these methods hence scales as N^D where D is the number of dimensions (variables). Thus the effort for high dimensional problems scales exponentially with dimension. As we will see, many Monte Carlo methods break "the curse of dimensionality" by needing to evaluate the function at a number of points that shows only a weak dependence on dimensionality. This is the key point.

The Monte Carlo method is a powerful but relatively simple idea. The seed of the idea came to Ulam over his frustration with the card game of Solitaire. At any intermediate point in the game, he was frustrated by the difficulty in calculating whether he would win or lose, but he realized that if he played the game often enough, he could reliably estimate the probability of winning. The speed with which the first electronic computer could do calculations prompted him to grow this idea into a new way to solve mathematical problems. Sharing his thoughts with von Neumann led them to develop a stochastic method to simulate the transport of neutrons through fissile material. This transport was naturally stochastic: defined by mean-free paths, the neutron has a specific probability of traveling a certain distance without a collision, and on collision, a scattering cross-section specifies probabilities for the collision to be elastic, inelastic, fissile, or absorptive. The first Monte Carlo method was a stochastic procedure, expressible on a computer, that mimicked what occurs in nature. It iteratively and stochasticly solved a Boltzmann-like transport problem without ever directly appealing to a Boltzmann equation.

The Metropolis-Ulam paper was written a couple of years after the first application of the method. It noted the general utility of stochastic procedures for solving problems, even those without stochastic analogs. Metropolis takes credit for the name "Monte Carlo" (Metropolis, 1985, 1987). It was not chosen to connect the method with the games of chance in this famous casino area of Monaco, but was Metropolis's way of ribbing Ulam, who had an uncle who occasionally had a "need" to go to the casino. The Monte Carlo method was seen as a procedure to which theorists would occasionally need to resort. It was perhaps unforeseen how frequently we have to resort to it.

1.2 Quantum Monte Carlo

Metropolis and Ulam (1949) credit Fermi with foreseeing that the Monte Carlo method would be useful for solving quantum problems. As they relate, Fermi noted that this new way of computing allows us to obtain the ground state of the time-independent Schrödinger equation

$$-\nabla^2 \psi (x, y, z) = [E - V(x, y, z)] \, \psi (x, y, z)$$

by introducing an imaginary-time dependence via the transformation

$$u (x, y, z, \tau) = \psi (x, y, z) \, e^{-E\tau},$$

so $u(x, y, z, \tau)$ obeys the diffusion-like equation

$$\frac{\partial u}{\partial \tau} = \nabla^2 u - Vu.$$

Already known was that this equation had a Monte Carlo expression as a collection of weighted particles, in which each independently performs a random walk while at the same time having its weight subjected to multiplication by the value of V at the point (x, y, z). In this type of Monte Carlo simulation the total weight of particles decays exponentially in (imaginary) time at a rate controlled by the eigenvalue E, and the spatial distribution of particles provides an estimate of $\psi(x, y, z)$, the solution of the time-independent Schrödinger equation. As we will show later, this eigenpair corresponds to the ground state solution. Indeed, Fermi's transformation is the starting point for all ground state quantum Monte Carlo methods. Computers, however, had to advance before this application of the Monte Carlo method became feasible for nontrivial quantum problems.

Fermi's observation is equivalent to transforming the time-dependent Schrödinger equation

$$i\frac{\partial \psi}{\partial t} = -\nabla^2 \psi + V\psi \tag{1.1}$$

from real time to imaginary time by the change of variables $it = \tau$:

$$\frac{\partial \psi}{\partial \tau} = \nabla^2 \psi - V\psi. \tag{1.2}$$

As Feynman later noted (Feynman and Hibbs, 1965), the same change of variables (with periodic boundary conditions in imaginary time) transforms the real-time path-integral formulation of quantum mechanics into finite-temperature quantum statistical mechanics. Finite-temperature quantum Monte Carlo simulations are typically performed in imaginary time. We will flesh out this observation later. We have a more modest, immediate task – fleshing out the Monte Carlo method and the need to resort to it.

1.3 Classical Monte Carlo

Because it is a convenient model to explain why we sometimes need Monte Carlo simulations, we start by discussing the one-dimensional Ising model in an external magnetic field. It is not a quantum model, but it is likely the simplest possible model with many interacting degrees of freedom.

The energy for the model is given by

$$E = -\sum_i (Js_i s_{i+1} + Hs_i), \tag{1.3}$$

where $J > 0$ is the (ferromagnetic) exchange integral, i identifies a lattice site, the on-site Ising variable s_i takes the values ± 1, and H is the external magnetic field. The lattice has N sites, and we assume periodic boundary conditions; that is, we assume $s_i = s_{i+N}$.

Two questions are typically asked of an interacting many-body problem: as N approaches the thermodynamic limit, what are its zero temperature properties? And what are its finite temperature properties? For the Ising model in an external magnetic field, the zero temperature state, that is, ground state, has an energy of $-N(J \pm H)$ with all Ising variables being $+1$ or -1, depending on the sign of H. In zero field, the ground state is doubly degenerate with the energy $-NJ$ being the lowest for spins all positive and all negative. Knowing the energy and the state is sufficient to determine all ground state properties. We thus focus on the finite temperature properties.

For finite temperatures, the partition function Z for the one-dimensional Ising model is known (Thompson, 1972):

$$Z = \lambda_+^N + \lambda_-^N, \qquad (1.4)$$

where

$$\lambda_\pm = e^{J/kT} \cosh(H/kT) \pm \left[e^{2J/kT} \sinh^2(H/kT) + e^{-2J/kT} \right]^{1/2}, \qquad (1.5)$$

and hence so is the free energy, $F = -kT \ln Z$. From the free energy all the equilibrium thermodynamics of the model follows by differentiation.

Typically, the finite-temperature property of interest is the equation of state, with emphasis on the behavior of the system near phase transitions. For a model of interacting spins, the appearance of magnetic long-range order is the candidate transition. Statistical mechanics teaches us that to study phase transitions we need to extrapolate the system to the thermodynamic limit and look for singular behavior in the free energy or in derived thermodynamic functions.

It is easy to take the thermodynamic limit of the present model. From (1.4) and (1.5), we find that the free energy per site is

$$F/N = -kT \ln \lambda_+,$$

and the magnetization per site is

$$M/N = -\frac{\partial}{\partial H}(F/N) = \sinh(H/kT)\left[\sinh^2(H/kT) + e^{-4J/kT}\right]^{-1/2}.$$

In zero field, we see that the magnetization is zero for all nonzero temperatures. This precludes the model from spontaneously magnetizing, a property that distinguishes the one-dimensional Ising model from Ising models in higher dimensions.

Being able to write an exact expression for the partition function is equivalent to saying the model is solved. Equations (1.4) and (1.5) were obtained by the transfer matrix technique (Thompson, 1972). While this technique also yields the exact solution of other models, for example, the zero-field two-dimensional Ising model, statistical mechanics is most often tasked with producing good approximations or good numerical simulations. We now address the basic challenges for the efficient Monte Carlo simulation of an Ising model.

For Monte Carlo methods, the most convenient approach to thermodynamics is the computation of averages from the Boltzmann probabilities for the possible configurations of the microscopic variables. If X represents a physical quantity of interest that is a function of the Ising variables, we can compute the temperature-dependent average (thermal expectation value) of X from

$$\langle X(T) \rangle = \frac{\sum_C X(C) e^{-E(C)/kT}}{\sum_C e^{-E(C)/kT}} = \sum_C X(C) \frac{e^{-E(C)/kT}}{Z}. \tag{1.6}$$

Here, C represents a configuration of the Ising variables; that is, $C = (s_1, s_2, \ldots, s_N)$, $X(C)$ and $E(C)$ are the values of X and the energy E for this configuration, and the summation is over the set of all possible configurations. The exponential $e^{-E(C)/kT}$ is the *Boltzmann factor* for configuration C, and the partition function $Z = \sum_C e^{-E(C)/kT}$ is its normalization constant. We are able thus to rewrite (1.6) as

$$\langle X(T) \rangle = \sum_C X(C) p(C), \tag{1.7}$$

where

$$p(C) = \frac{e^{-E(C)/kT}}{Z} \tag{1.8}$$

is the probability of configuration C and $\sum_C p(C) = 1$.

In an application, X might represent the energy

$$\langle E(T) \rangle = \sum_C E(C) p(C),$$

the magnetization

$$\langle M(T) \rangle = \sum_C \left(\sum_i s_i(C) \right) p(C),$$

or the spin-spin correlation function

$$\langle s_i s_j(T) \rangle = \sum_C s_i(C) s_j(C) p(C).$$

The critical tasks in computing an average this way are generating all configurations C and computing for each the probability $p(C)$. To do the latter, we have to evaluate the partition function, which requires the computation of all the Boltzmann factors. For our N-site model, the number of possible configurations is 2^N, so the complexity of this task grows exponentially fast, quickly making the computation of the partition function not practical. A way of providing a reliable solution to a problem without completely solving the problem is needed. We need the Monte Carlo method.

A Monte Carlo method for problems like the Ising model does not generate all configurations but instead only selects M representative configurations with the correct probability. It replaces (1.7) by

$$\langle X \rangle_{\text{MC}} = \frac{1}{M} \sum_{i=1}^{M} X(C_i). \tag{1.9}$$

If all the measurements of $X(C_i)$ are statistically independent, then the Monte Carlo method also provides an estimate of the statistical error σ_X associated with this average,

$$\sigma_X \approx \sqrt{\frac{1}{M-1} \left(\frac{1}{M} \sum_{i=1}^{M} X^2(C_i) - \left(\frac{1}{M} \sum_{i=1}^{M} X(C_i) \right)^2 \right)}. \tag{1.10}$$

Ideally, $M \ll 2^N$; that is, the average is estimated without completely solving the problem.

We now begin the process of defining specific Monte Carlo techniques that allow us to generate the configurations C_i appearing in (1.9). One basic question we need to answer is, how does the Monte Carlo method select the configurations with the correct probability if we cannot in a reasonable amount of computer time generate all the configurations necessary to compute the partition function?[1] As we shall see there are ways to select the configurations with the proper probability from an unnormalized probability density. The most famous way is called the *Metropolis algorithm*, the work by Metropolis, Rosenbluth, Rosenbluth, Teller, and Teller we mentioned earlier. We discuss this algorithm in Section 2.5.1. After discussing this and multiple other Monte Carlo topics, we return to the Ising model in Chapters 4 and 5. In the latter chapter, we quantize it, and then develop our first quantum Monte Carlo algorithms for simulating quantum spin models.

---

[1] Several modern Monte Carlo methods developed for finite-temperature classical statistical mechanics enable the estimation of the logarithm of the partition function. The value of this estimate, however, is a product of the simulation and is not something needed to execute the simulation.

Overall, we focus on Markov chain Monte Carlo algorithms for quantum many-body models on a lattice. Not all Monte Carlo algorithms are Markov chain algorithms. Some just throw darts in a blindfolded way. The concept of a Markov chain is defined and discussed in Chapter 2. It produces the stochastic chain of events on which we perform a statistical analysis. Lattice models are models such as the Heisenberg model for interacting quantum spins and the Hubbard model for interacting electrons where spins and electrons exist only on lattice sites. Zero-temperature algorithms for interacting electrons and Bosons on a lattice are very similar to zero-temperature algorithms used for interacting electrons and Bosons in the continuum. For lattice models a greater variety of finite-temperature algorithms exists.

Suggested reading

N. Metropolis and S. Ulam, "The Monte Carlo method," *J. Am. Stat. Assoc.* **44**, 335 (1949).

N. Metropolis, "The beginning of the Monte Carlo method," *Los Alamos Science*, Special Issue, 1987.

R. Eckhardt, "Stan Ulam, John von Neumann, and the Monte Carlo method," *Los Alamos Science*, Special Edition, 1987.

J. E. Gubernatis, "The heritage," in *The Monte Carlo Method in the Physical Sciences: Celebrating the 50th Anniversary of the Metropolis Algorithm*, American Institute of Physics Conference Proceedings 690, ed. J. E. Gubernatis (Melville, NY: American Institute of Physics, 2003).

2

Monte Carlo basics

The Monte Carlo method is built on probability and statistics. Here, we introduce the few basic concepts of probability and statistics necessary to understand the concept of sampling from a probability distribution. We then discuss several useful techniques to do this sampling from distribution functions that depend on just a few random variables. Functions depending on a large number of random variables require a different technique. For these functions we discuss the use of Markov chain sampling.

2.1 Some probability concepts

As a stochastic procedure, a Monte Carlo simulation generates a set of random events. Sometimes the order in which the events are generated matters; other times, it does not. The theory of Monte Carlo sampling, the procedure by which we access the random events, is based on probability theory. We begin by discussing a few essential basic concepts from probability theory.

In probability theory, the set of all possible *outcomes* $\{\chi_1, \chi_2, \ldots, \chi_n, \ldots\}$ of a real or imagined experiment defines a *sample space*. If our experiment was, for example, the roll of a pair of dice, there would be 36 outcomes in the sample space. An *event* is a set of one or more outcomes of this space that satisfies some criterion. The *probability* of an event A is a number assigned to the event in a way consistent with three axioms: (1) $0 \leq P(A) \leq 1$; (2) if the event includes all possible outcomes in the sample space, then $P(A) = 1$; and (3) if an event A breaks into events B and C that share no common outcome, then $P(A) = P(B) + P(C)$. In our roll of the dice experiment, if the event A were those outcomes whose sum equals 3, it would encompass two outcomes, and $P(A) = \frac{1}{18}$. If the event A were those outcomes whose sums are odd, $P(A) = \frac{1}{2}$.

A consequence of the axioms is that the probability maps events to the interval $[0, 1]$ in such a way that the sum over a set $\{A_1, A_2, \dots\}$ of mutually exclusive events that covers the sample space equals 1. Two events A_i and A_j are *mutually exclusive*, that is, share no common outcome, if and only if the occurrence of A_i implies that A_j does not occur and vice versa. Because the sum is 1, at least one event is possible.

A *random variable X* maps an outcome to a real number. The events

$$A = \{X(\chi) \leq x\}, \quad B = \{x_1 < X(\chi) < x_2\}, \quad \text{and} \quad C = \{X(\chi) = x_0\}$$

are the set of outcomes χ such that $X(\chi)$ is less than or equal to x, between x_1 and x_2, and equal to x_0, respectively. Normally, we work at the level of random variables; that is, our experiments generate events with numbers assigned. In the following, we write the probability $P(X(\chi) = x)$ as $P(x)$ or $P(X)$; in other words, the random variable and its value are used interchangeably. Also, we often do not distinguish an outcome (an elemental event) from an event.

The *cumulative distribution function* of a random variable X, $F_X(x)$, is defined by

$$F_X(x) = P(X \leq x). \tag{2.1}$$

More generally,

$$P(a < X \leq b) = F_X(b) - F_X(a). \tag{2.2}$$

$F_X(x)$ is a positive, monotonic, nondecreasing function of x defined over the entire real axis with the properties $F_X(-\infty) = 0$ and $F_X(\infty) = 1$. If $F_X(x)$ is everywhere differentiable, the random variable is continuous. Then a probability density $f_X(x)$ exists and satisfies

$$f_X(x) = \frac{dF_X(x)}{dx}.$$

This expression says that $f_X(x)dx$ is the probability that x is in the interval $[x, x + dx]$. By convention, $F_X(x)$ is continuous from the right; that is, $F_X(x) = \lim_{\epsilon \to 0} F_X(x + \epsilon)$. If $F_X(x)$ is step-wise continuous, with jumps $f_1, f_2, \dots$ at $x_1, x_2, \dots$, then the random variable is discrete. If discrete, we often write $f_X(x_i)$ as simply f_i. We do not explicitly consider distributions of mixed random variable types. Doing so is relatively straightforward.

Often it is useful to change from one random variable to another. Let us consider, for example, $Y = y(X)$ where $y(x)$ is a nondecreasing function of x. If we know $F_X(x)$ and $f_X(x)$, what are $F_Y(y)$ and $f_Y(y)$? Because $y(X) \leq y(x)$ when $X \leq x$ and vice versa, it follows that

$$P(y(X) = Y \leq y(x)) = P(X \leq x),$$

that is, $F_Y(y) = F_X(x)$. By differentiation, we obtain

$$f_Y(y)\frac{dy}{dx} = f_X(x).$$

Because $y(x)$ is nondecreasing, the derivative of y with respect to x is positive and both sides of the equation are positive.

If the change of variables is nonincreasing, it is easy to show that

$$F_Y(y) = 1 - F_X(x)$$

and

$$f_Y(y)\frac{dy}{dx} = -f_X(x).$$

More generally, we write

$$f_Y(y)\left|\frac{dy}{dx}\right| = f_X(x),$$

or for the inverse mapping $f_Y(y) = f_X(x)\left|\frac{dx}{dy}\right|$.

If we have two random variables X and Y, we represent the probability of both occurring as $P(X, Y)$. Of course, $P(Y, X) = P(X, Y)$. This bivariate function is called the *joint probability* of X and Y and is related to $P(X)$ and $P(Y)$ by

$$P(X) = \sum_Y P(X, Y) \quad \text{and} \quad P(Y) = \sum_X P(X, Y). \tag{2.3}$$

If $P(X, Y) = P(X)P(Y)$, the random variables are said to be *statistically indepen-dent*.

The *conditional probability* $P(X|Y)$ of X given that Y occurred is related to the joint probability by

$$P(X|Y) = \frac{P(X, Y)}{P(Y)}. \tag{2.4}$$

Similarly, the conditional probability of Y given that X occurred is

$$P(Y|X) = \frac{P(X, Y)}{P(X)}. \tag{2.5}$$

By comparing these two equations we find that

$$P(X|Y)P(Y) = P(Y|X)P(X), \tag{2.6}$$

and consequently

$$P(X|Y) = \frac{P(Y|X)P(X)}{P(Y)}, \tag{2.7}$$

which is called *Bayes's theorem*. We note that

$$\sum_X P(X|Y) = \sum_X P(X, Y)/P(Y) = 1. \tag{2.8}$$

Statistical independence implies $P(X|Y) = P(X)$, provided $P(Y) > 0$.

A bivariate cumulative distribution is defined by

$$F_{XY}(x, y) = P(X \le x, Y \le y).$$

The analog to (2.2) is

$$P(a < X \le b, c < Y \le d) = F_{XY}(b, d) - F_{XY}(a, d) - F_{XY}(b, c) + F_{XY}(a, c)$$

and the associated densities are

$$f_{XY}(x, y) = \frac{\partial^2 F_{XY}(x, y)}{\partial x \partial y}. \tag{2.9}$$

Distributions and densities for more than two random variables are defined analogously.

If we change both random variables to new random variables U and V,

$$U = g(X, Y), \qquad V = h(X, Y),$$

and if the inverse mappings are

$$X = p(U, V), \qquad Y = q(U, V),$$

then their densities are related by

$$f_{UV}(u, v) = f_{XY}(x, y) \left| \frac{\partial(p, q)}{\partial(u, v)} \right|.$$

Note the absolute value of the Jacobian of the transformation.

It is often useful to integrate out one or more random variables from a multivariate distribution. The result is a new distribution called the *marginal probability density*. Let us say $f_{XY}(x, y)$ is the joint distribution of two random variables X and Y and suppose we are interested in the probability that $a < X < b$. This event occurs only when $a < X < b$ and $-\infty < Y < \infty$. For the continuum and discrete cases

$$P(a < X < b, -\infty < Y < \infty) = \begin{cases} \int\limits_a^b \int\limits_{-\infty}^{\infty} f_{XY}(x, y) dy dx, \\ \sum\limits_{a<x<b} \sum\limits_{y} f_{XY}(x, y). \end{cases}$$

The result of either the integration or summation is a function of x alone,

$$f_X(x) = \begin{cases} \int\limits_{-\infty}^{\infty} f_{XY}(x,y)dy, \\ \sum\limits_{y} f_{XY}(x,y). \end{cases}$$

This function $f_X(x)$ is the *marginal probability distribution* of X.

With the concept of a marginal distribution and the multivariate definition of a conditional probability distribution function,

$$f(x_1,\ldots,x_k|x_{k+1},\ldots,x_n) = \frac{f(x_1,\ldots,x_n)}{f(x_{k+1},\ldots,x_n)},$$

we can derive a number of interesting relations. One is the *Chapman-Kolmogoroff equation*

$$\int f(x_3|x_1,x_2)f(x_2|x_1)dx_2 = f(x_3|x_1).$$

To derive it we use the definition of a conditional density and write the left-hand side of the above as

$$\int \frac{f(x_1,x_2,x_3)}{f(x_1,x_2)} \frac{f(x_1,x_2)}{f(x_1)} dx_2.$$

Reusing this definition and invoking the one for a marginal distribution, we find

$$\int f(x_2,x_3|x_1)dx_2 = f(x_3|x_1).$$

We can use a similar reasoning to prove expressions such as

$$\int \int f(x_4,x_3,x_2|x_1)f(x_3,x_2|x_1)dx_3dx_2 = f(x_4|x_1).$$

We also note that repeated application of the definition of the multivariate conditional probability produces the *conditional probability chain rule*

$$f(x_1,\ldots,x_n) = f(x_n|x_{n-1},\ldots,x_1)\cdots f(x_2|x_1)f(x_1).$$

2.2 Random sampling

The most fundamental Monte Carlo concept we need is that of *random sampling*. Random sampling, or *sampling* for short, bridges probability theory and statistics. Implicit in the concept of probability is the existence of some experiment that produces a realization of a random variable from a set of possible outcomes.

If executing this experiment N times produces the value x of the random variable X N_x times, then the ratio N_x/N is an empirical measure of $P(x)$. To sample a probability distribution means that we generate events with a frequency proportional to it.

An alternative approach to random sampling is to consider N experiments simultaneously (an *ensemble*). Associated with the i-th experiment is the random variable X_i, which is identical to the random variable X and by assumption has the same distribution as X; that is, $P_i(X_i) = P(X)$. The joint distribution of these N random variables is $P(X_1, X_2, \ldots, X_N)$. The statistical independence of the random variables means

$$P(X_1, X_2, \ldots, X_N) = P_1(X_1)P_2(X_2) \cdots P_N(X_N).$$

The ensemble of N experiments thus produces a set of outcomes $\chi = \{\chi_1, \chi_2, \ldots, \chi_N\}$ such that the i-th random variable takes the values $X_i(\chi) = X(\chi_i)$. The combination of independence and identical distributions is a hallmark of random samples.

There are various procedures for sampling a probability distribution, and they almost always assume the existence of a *random number generator*, which is some numerical procedure producing a random variable uniformly distributed over the interval $[0, 1]$.[1] In other words, it samples $u(x)dx = dx$ over this interval.[2] The numbers produced are actually quasi-random as opposed to being truly random: if a generator is sampled long enough, the numbers repeat themselves. For most applications the length of useful sequences is sufficient.

In Fig. 2.1 we show schematically the type of histogram a typical random number generator produces. It is flat to within statistical fluctuations. If the random variables are uniformly distributed over $[0, 1]$, then pairs of them should be uniformly distributed over the unit square; triplets, over the unit cube, etc. Generally, the problem with a random number generator is not an inability to produce a reasonably flat histogram, but its ability to produce independent random numbers. Correlations can exist between pairs or triplets or higher order tuples. Volumes have been written about what is a good random number generator and how it is constructed. We do not discuss these issues but simply assume that we have access to a good one. Today, vendor-supplied generators accompanying modern compilers and others found in recent textbooks are adequate. Serious, large-scale simulations might require other choices.

---

[1] In practice, random number generators return numbers from the interval $(0, 1]$, $[0, 1)$, or $(0, 1)$. Which is the case can matter if we need, for example, the logarithm of the random number. We ignore this practical issue in our discussions.

[2] Throughout we use $u(x)$ to represent a uniform distribution of the random variable over the interval $[0, 1]$ and ζ to represent a sample drawn from this distribution.

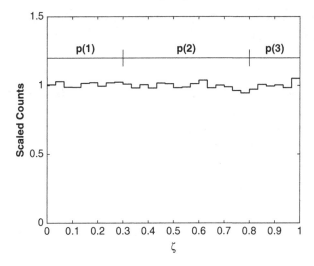

Figure 2.1 Schematic histogram of a uniform distribution returned by a random number generator. The generator was sampled 50,000 times and the results accumulated in 30 bins. The number of counts in a bin was divided by $50,000/30$, the expected number of counts per bin. Segments of a discrete probability are overlaid on the uniform distribution.

Direct sampling and *Markov chain methods* are two classes of sampling methods that are relevant for quantum Monte Carlo algorithms. We now discuss for each class various techniques for sampling discrete and continuous distribution functions.

2.3 Direct sampling methods

2.3.1 Discrete distributions

The simplest case of a direct sampling method is the standard procedure for sampling from a discrete probability function f_i for N events. To motivate it, we assume that $N = 3$ and that $f_1 = 0.3, f_2 = 0.5$, and $f_3 = 0.2$. Next, we lay out these probabilities over the uniform distribution as shown in Fig. 2.1. They segment the $[0, 1]$ interval into three regions. We see from this figure that f_1 overlays approximately 30% of the events generated by the random number generator, f_2 50%, and f_3 the remaining 20%. Thus, if we draw a uniformly distributed random number ζ from our generator, we observe that it properly samples event 1 if it is less or equal to f_1, samples event 2 if it is greater than f_1 but less than or equal to $f_1 + f_2$, and samples event 3 if it is greater than $f_1 + f_2$ but less than or equal to $f_1 + f_2 + f_3 = 1$. This observation translates into a simple strategy for sampling from a discrete distribution by using its cumulative probability function F_i. We detail this strategy in Algorithm 1.

Algorithm 1 Sample a discrete distribution function via its cumulative distribution.

Input: Vector f of probabilities.

  $F(0) \leftarrow 0$;

  for $i = 1$ to N **do**

    $F(i) \leftarrow f(i) + F(i - 1)$;   ▷ Create cumulative distribution function

  end for

  Generate a uniform random number $\zeta \in [0, 1]$;

  $k \leftarrow 0$;

  repeat

    $k \leftarrow k + 1$;

  until $\zeta \leq F(k)$.

  return k.

More formally, for a uniform random variable we have $u(x) = f_X(x) = 1$ when $0 \leq x \leq 1$, so its cumulative distribution function is $F_X(x) = x$. We also have

$$P\left(0 \leq x_1 < \zeta \leq x_2 \leq 1\right) = F_X(x_2) - F_X(x_1) = x_2 - x_1,$$

which says that the probability that ζ lies in an interval $[x_1, x_2]$ of $[0, 1]$ is proportional to the length of the interval. Now if we have a set of discrete events with probability f_i, $i = 1, \ldots, n$, and we wish to sample one at random, we can divide $[0, 1]$ into segments of length f_i. The interval in which ζ lies then selects the event. We can accomplish the selection of event k by finding the k that satisfies

$$k = \min_n \left\{ \sum_{i=1}^{n} f_i \geq \zeta \right\}. \tag{2.10}$$

If the discrete probability has just a few elements, a simple linear search algorithm, as described in Algorithm 1, suffices. For a larger number of elements, a binary search becomes important for efficiency. If the vector of probabilities changes from sampling to sampling, Algorithm 1 is easily modified to eliminate the need to create and store the cumulative distribution; see Algorithm 2. If the vector of probabilities is very large and changes infrequently, if at all, other sampling algorithms such as the cut and the alias methods might be preferred (see Appendix A for Walker's alias method).

Algorithm 1 provides a way to sample an Ising configuration C_i from the Boltzmann probability $p(C)$ (1.8). After the $p(C_i)$ are constructed, we can use (2.10) to sample the C_i. As we already noted (Section 1.3), the construction of all the $p(C_i)$ becomes impractical when the lattice size becomes large. Sampling configurations from Boltzmann densities for large lattices is usually done efficiently by means of a Markov chain, a method we discuss in the next section.

Algorithm 2 Sample a discrete distribution function.

Input: Vector f of probabilities.

Generate a uniform random number $\zeta \in [0, 1]$;

$k \leftarrow 0$;

repeat

$\quad k \leftarrow k + 1$;

$\quad \zeta \leftarrow \zeta - f(k)$;

until $\zeta < 0$

return k.

Some physical problems have very specific discrete distributions, such as the Poisson and binomial distributions, that require sampling. The cumulative probability method (2.10) can be used to sample from them. For these, and many other discrete distributions, exploiting specifics is often more efficient (Everett and Cashwell, 1983; Fishman, 1996). For example, while we could sample the Poisson probability (Fig. 2.2),

$$f_i = \frac{t^i}{i!} e^{-t}, \quad i = 0, 1, 2, \ldots \text{ and } t > 0 \tag{2.11}$$

via

$$k = \min_n \left\{ \sum_{i=0}^{n} \frac{t^i}{i!} \geq e^t \zeta \right\}, \tag{2.12}$$

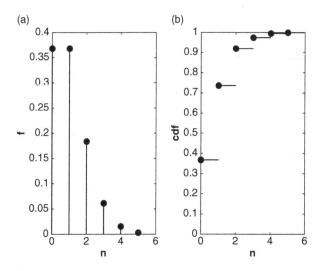

Figure 2.2 The (a) discrete and (b) cumulative probabilities for the Poisson distribution (2.11). Here $t = 1$.

a more efficient way is

$$k = -1 + \min_n\{\zeta_1\zeta_2\cdots\zeta_n < e^{-t}\}, \tag{2.13}$$

where the ζ_i are successive draws from the random number generator. We leave the derivation of the latter procedure to the Exercises. It is also straightforward to show that the Poisson distribution (2.11) has a mean equal to t and a variance equal to t.

Equation (2.13) defines a way to determine the integer k. This integer, however, is a function of t. In many applications of a Poisson distribution, t defines the length of some interval, for example, $(0, t)$, and what is of interest is not only the number of events $k(t)$ occurring in this interval but also an ascending sequence of points $0 < t_1 < t_2 \cdots < t_k < t$ associated with these events. A stochastic process generating these $k(t)$ points is called a Poisson process. We discuss this process in Section 5.2.3.

2.3.2 Continuous distributions

As we shift from sampling discrete distributions to continuous ones, we also shift our attention from probability functions to probability densities $f(x)$. We can sample $f(x)$ by analogy with the procedure we described for a discrete distribution. After drawing our random number, we solve

$$\zeta = \int_{-\infty}^{x} f(y)dy. \tag{2.14}$$

for x. The content of this expression is $f(x)dx = d\zeta$; that is, the probability of x on $[x, x+dx]$ corresponds to the probability of the random number on $[\zeta, \zeta+d\zeta]$. This is the fundamental sampling concept for a continuous distribution.

Intuition likely suffices to justify the correctness of this procedure and the one for the discrete distribution. Mathematically, for both, we need to start with the observation that a function of a random variable is another random variable and the range of a cumulative distribution is $[0, 1]$, the same as the domain of our uniform random variable ζ. Next, we consider the properties of the generalized inverse of a nondecreasing function $F(x)$, for example, a cumulative distribution.

A *generalized inverse* (Fig. 2.3) of a nondecreasing function is defined as

$$F^-(y) = \inf\{x|F(x) \geq y\}.$$

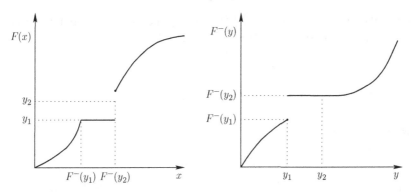

Figure 2.3 Graphical representation of the definition of a generalized inverse.

In general this inverse satisfies

$$F(F^-(y)) \ge y \quad \text{and} \quad F^-(F(x)) \le x.$$

Therefore, the set $\{(x, y)\}$ of points for which $F^-(y) \le x$ is the same as the set for which $F(x) \ge y$. Equating y with ζ and $F(x)$ with the cumulative distribution of the random variable X, we note that if y is in $[0, 1]$, then x is in $[F^-(0), F^-(1)]$. Then using the definition of a cumulative distribution (2.1), we can state

$$P(F^-(\zeta) \le x) = P(\zeta \le F(x)) = F(x).$$

Thus, to generate a random variable X that has a distribution $F(X)$ it is sufficient to generate a uniform random variable ζ and then make the transformation $x = F^-(\zeta)$.

As it stands, (2.14) is seldom easy to solve for x. If $F(x)$ is an explicit functional representation of the integral of $f(x)$, the problem, as just argued, is reduced to solving $\zeta = F(x)$ for x. This too can be difficult, but if the analytic inverse of $F(x)$ is known, then

$$x = F^{-1}(\zeta). \tag{2.15}$$

The simplest application of (2.14) is sampling from a uniform density over the interval $[a, b]$

$$\zeta = \int_a^x \frac{1}{b-a} dx = \frac{x-a}{b-a}.$$

Solving for x, we find

$$x = a + (b - a)\,\zeta,$$

a result we could easily have guessed.

A more useful application of (2.15) is sampling from the exponential density defined over the interval $[0, \infty)$. We can integrate it to obtain

$$\zeta = \int_0^x \lambda e^{-\lambda y} dy = 1 - e^{-\lambda x},$$

and then solve for x,

$$x = -\ln(1 - \zeta)/\lambda.$$

If ζ is a random variable on $(0, 1)$, then so is $1 - \zeta$. Thus, the result is $x = -\ln \zeta/\lambda$.

We can sometimes sample other densities by a *change of variables*. Let us use this technique and a trick to derive the *Box-Muller method* for sampling from a Gaussian distribution

$$N(x) = \frac{1}{\sqrt{2\pi}} \exp\left(-\tfrac{1}{2}x^2\right), \quad -\infty < x < \infty.$$

The trick is the observation that it is easier to sample from the product of two Gaussian distributions

$$N(x_1)N(x_2) = \frac{1}{2\pi} \exp\left[-\tfrac{1}{2}\left(x_1^2 + x_2^2\right)\right].$$

By transforming to polar coordinates

$$x_1 = r\cos\theta, \quad x_2 = r\sin\theta,$$

we can write

$$N(x_1)N(x_2)dx_1 dx_2 = \left[\exp\left(-\tfrac{1}{2}r^2\right) r dr\right]\left[\tfrac{1}{2\pi}d\theta\right].$$

There are now two probability densities we need to sample, and hence we need two random numbers. For the r-dependent distribution we have

$$\zeta_1 = \int_0^r \exp\left(-\tfrac{1}{2}s^2\right) s\, ds = 1 - \exp\left(-\tfrac{1}{2}r^2\right),$$

from which it follows that $r = [-2\log\zeta_1]^{1/2}$. For the θ-distribution we need to sample a uniform distribution over $[0, 2\pi]$ yielding

$$\theta = 2\pi\zeta_2.$$

From the two random numbers, two independent Gaussian random variables result:

$$x_1 = \left[-2\log\zeta_1\right]^{1/2}\cos(2\pi\zeta_2), \quad x_2 = \left[-2\log\zeta_1\right]^{1/2}\sin(2\pi\zeta_2). \tag{2.16}$$

In Appendix B we discuss the *rejection method*, a general-purpose technique for sampling from continuous distributions of a few random variables. An important feature of this technique is that the distribution being sampled need not be normalized. In the following section, we begin our discussion of Markov chains and common methods for generating them. Using these chains is almost the universal way we sample from discrete and continuous distributions of very high dimension.

2.4 Markov chain Monte Carlo

Before we define a Markov chain, we first make some notational adjustments and several remarks. In our discussion of Markov chains, and our subsequent discussion of Monte Carlo methods more generally, our random variable X is best regarded as a "random vector" $X = (Y_1, Y_2, \ldots, Y_n, \ldots)$ whose components Y_i are random variables. This understanding simplifies the notation for defining procedures that sample multivariate distributions by eliminating long strings of subscripts. The Y_i are independent random variables in the sense that we can vary the values of each irrespective of the values of the others. As the Y_i vary, they cause X to vary over the sample space of the problem. The Y_i the Markov chain generates, however, are not necessarily statistically independent from one sequence to another.

Let us recall our discussion of the Ising model in Chapter 1. There, we considered each Ising spin as a random variable that takes a value of plus or minus one. A *configuration* or a *state* corresponds to one outcome $(s_1, s_2, \ldots, s_N)$ in the sample space of the 2^N outcomes. The energy $E(C)$ of the Ising model is a random variable mapping a configuration to a real number. In this example, the random variable X represents a configuration C, and Y_i a single spin such as s_i. It is convenient to express the energy (and other observables) in terms of the individual values of the Ising variables, and thus to regard C as the collection $(s_1, s_2, \ldots, s_N)$. Furthermore, in a Markov chain, the transition from one configuration to the next is most easily implemented by changing one component of the configuration at a time.

The definition and properties of a Markov chain apply to both classical and quantum Monte Carlo algorithms. In a quantum Monte Carlo simulation, we may have to distinguish between the state of the system and the Monte Carlo configuration. In quantum mechanics, we work with the Hamiltonian operator H and the state vector $|\psi\rangle$. To deal with them numerically, we refer to some basis. Given some set $\{|C\rangle\}$ of "configuration states," we can evaluate matrix elements $\langle C'|H|C\rangle$ and wave functions $\psi(C) = \langle C|\psi\rangle$, which are classical objects and therefore amenable to the type of sampling we next discuss.

In previous sections, we distinguished between random variables that were discrete (having a finite number of values) and continuous (having an infinite number of values). To continue doing so becomes cumbersome. In subsequent sections, the notation is most proper for discrete random variables, but in most cases it immediately generalizes to the continuous case. A few results we quote or prove are, however, most obviously correct only for finite, discrete-time Markov chains.

2.4.1 Markov chains

A *Markov chain* is a procedure to generate a sequence $x_1, x_2, \ldots, x_j, \ldots$ of the values of a random variable. Such a sequence is said to be Markovian if all the conditional probabilities for the associated random variables X_j satisfy

$$P(X_j|X_{j-1}, \ldots, X_1) = P(X_j|X_{j-1}),$$

that is, the probability of X_j depends only on the random variable X_{j-1} immediately preceding it and not on the others. The order of the random variables is important.

For simplicity we consider only discrete sequences of random variables with a finite number of values. Let us define $P_{ij} = P(X_i|X_j)$ and $p_i = P(X_i)$. A Markov chain is defined by an initial probability p_i^0 and a *transition probability* matrix P_{ij} normalized for each j by $\sum_i P_{ij} = 1$. The chain is typically generated recursively,

$$p_i^{(k+1)} = \sum_j P_{ij} p_j^{(k)}, \quad k = 0, 1, \ldots \tag{2.17}$$

with $p_i^{(0)} = p_i^0$. If we sum both sides of this equation over i and use the normalizations $\sum_i P_{ij} = 1$ and $\sum_i p_i^0 = 1$, we find that $\sum_i p_i^{(k)} = 1$ so that the iteration conserves probability.

What is remarkable about (2.17) is that under very general conditions on the P_{ij}, and independent of the starting point p_i^0, the iteration eventually reaches a stationary state p_i that satisfies the *stationary condition*

$$p_i = \sum_j P_{ij} p_j, \tag{2.18}$$

that is, it produces the eigenvector p_i of the matrix P_{ij} whose eigenvalue is 1 and whose vector components satisfy $\sum_i p_i = 1$.

In general, p_i is unknown. As we will see, in ground state quantum Monte Carlo simulations, finding p_i is the objective. In other applications, such as an Ising model simulation, producing a specific distribution is necessary. To do so requires specific choices of the transition probability. Metropolis et al. (1953) proposed that we will obtain a specific distribution p_i if the transition probability obeys

$$P_{ij}p_j = P_{ji}p_i. \tag{2.19}$$

This is the *detailed balance condition*, which equates the probability of being in state j and going to i with the probability of being in i and going to j.[3] If the right-hand side of (2.19) is substituted into the right-hand side of (2.18), then $p_i = \sum_j P_{ji}p_i = p_i$, where we used the normalization $\sum_j P_{ji} = 1$. Hence, P_{ij} and p_i are consistent in the stationary state. As we soon discuss, Metropolis et al. also gave a general definition for such a P_{ij} and an algorithm for sampling from it. The algorithm is called the *Metropolis algorithm* and has the remarkable property that the function to be sampled need not be normalized.

2.4.2 Stochastic matrices

We are now poised to define Monte Carlo algorithms that allow us to answer the two basic questions asked of a system of interacting particles, namely, what are its ground state properties? And what are its finite-temperature properties? To find the ground state, we need to construct a Markov chain whose transition probability projects to it. It is this type of Monte Carlo calculation that Fermi was suggesting (Chapter 1). To compute thermodynamic quantities we need a transition probability that satisfies detailed balance for the Boltzmann distribution. We defer the discussion of the ground state algorithms to Chapters 9, 10, and 11. Before we start discussing several detailed balance algorithms for equilibrium thermodynamics, we first justify the remarkable properties of a Markov chain and discuss the general conditions for their validity.

The property of a Markov chain that a unique stationary distribution satisfying (2.18) always exists follows from the transition probability P_{ij} being an ergodic stochastic matrix. A *stochastic matrix* is a matrix with no negative elements (a nonnegative matrix) and each column sum equal to 1, that is, $\sum_i P_{ij} = 1$.[4] *Irreducibility* means that a series of permutations of rows and columns *must not* transform P_{ij} into the form

$$\begin{pmatrix} A & B \\ 0 & C \end{pmatrix},$$

where A and C are square matrices of any possible order. If it does, then the matrix is called *reducible*. Similarly, *aperiodicity* means that there is no permutation that transforms P_{ij} into the form

---

[3] Equation (2.19) is a sufficient but not a necessary condition for stationarity (see also the discussion in Sec. 2.6).

[4] A stochastic matrix is often defined as having the row (instead of the column) sums equal to 1. We choose the column-wise definition because it is consistent with the natural order of matrix-vector multiplication.

$$\begin{pmatrix} 0 & A_1 & 0 & \cdots & & 0 \\ \vdots & 0 & A_2 & \ddots & & \vdots \\ \vdots & \vdots & 0 & \ddots & & 0 \\ \vdots & \vdots & \vdots & \ddots & A_{n-1} & \\ A_n & 0 & 0 & \cdots & & 0 \end{pmatrix}.$$

An irreducible and aperiodic stochastic matrix is called an ergodic matrix.[5]

In general, a stochastic matrix is not symmetric; that is, $P_{ij} \neq P_{ji}$. Consequently, it has unequal left (x^α) and right (y^α) eigenvectors that share the same eigenvalues

$$\sum_i x_i^\alpha P_{ij} = \lambda_\alpha x_j^\alpha, \quad \sum_j P_{ij} y_j^\alpha = \lambda_\alpha y_i^\alpha.$$

If the eigenvalues are distinct, $\lambda_\alpha \neq \lambda_\beta$, then the right and left eigenvectors, x^α and y^β, are linearly independent and satisfy $[x^\alpha]^T \cdot y^\beta = 0$.

While the proof lies outside the scope of this book, one can show that an irreducible, stochastic matrix has a nondegenerate eigenvalue equal to 1 and the corresponding right eigenvector's components are all positive (Meyer, 2000, Chapter 8). These results follow from the application of the *Perron-Frobenius theorem*.[6] This theorem says that if an irreducible matrix (not necessarily stochastic) is non negative, then it has a nondegenerate eigenvalue equal to its spectral radius.[7] Accordingly, this eigenvalue is real and positive. The theorem states further that the components of the associated right eigenvector are also positive, and this eigenvector is unique (up to an overall scaling). If the matrix is stochastic, then we can show that the absolute value of the eigenvalue cannot exceed unity, and therefore at least one of the dominating eigenvalues is unity, and other dominating eigenvalues, if any, are complex numbers with absolute value 1.

To further nail down the dominating eigenvalue, we must require an additional condition, namely, the aperiodicity. We can show that, if the matrix P is aperiodic as well as irreducible, there exists a finite number n of matrix multiplications so that P^n is a positive matrix; that is, $[P^n]_{ij} > 0$ for any pair of i and j. When a non-negative matrix satisfies this condition, it is called *primitive*, and the sampling is called *ergodic*.

---

[5] In general, an irreducible and aperiodic non negative matrix is called a *primitive* matrix. An ergodic matrix is a primitive stochastic matrix.

[6] The analogous theorem for continuous operators is called the Jentzsch-Hopf theorem (Camp and Fisher, 1972; van Hove, 1950).

[7] The spectral radius is $\rho(A) = \max |\lambda|$, where λ is a member of the set of eigenvalues of A.

If a stochastic matrix is positive, then from any given state the chain can transition to any other state in a single step. Note that an irreducible, stochastic matrix generally has zero elements (and may have more than one eigenvalue of unit magnitude). This means that some states are inaccessible from a given state by one step in the Markov chain.[8] Further, one can show (Meyer, 2000, Chapter 8) that for a positive stochastic matrix, there is only one eigenvector with an eigenvalue equal to 1. It is straightforward to show that these properties also apply to the primitive stochastic matrix. Namely, if a stochastic matrix is primitive, the right eigenvector of P with the eigenvalue of 1 is unique and positive. In addition, when P is chosen so that it satisfies the stationary condition (2.18) with a given target distribution p, the right dominating eigenvector equals p. It means that after sufficiently many iterations in the Markov-chain Monte Carlo simulation, the probability distribution converges to the target distribution.

Now, let us take some (unnormalized) vector ψ and express it as a linear combination of the right eigenvectors y^α, that is, $\psi = \sum_\alpha a_\alpha y^\alpha$, with $a_1 = 1$, and order the eigenvalues as $1 > |\lambda_2| \geq |\lambda_3| \geq \ldots$ Repeated multiplication of the matrix P with this vector yields

$$P^k \psi = y^1 + \sum_{\alpha \geq 2} \lambda_\alpha^k y^\alpha.$$

As k becomes large, the eigenvectors with subdominant eigenvalues project out, leaving just the right eigenvector of the nondegenerate unit eigenvalue. Setting $y_i = p_i > 0$ leads us to (2.18). We note that this projected state does not depend on the starting state p_i^0. This exercise reveals another important feature of Markov chain Monte Carlo: *before we start taking samples, we have to iterate for a while to "equilibrate" the system.* This contrasts with the direct sampling methods previously considered.

While an algorithm might be formally ergodic, in an actual simulation it might behave as if it is not. For example, there might exist several high probability regions of phase space connected by low probability paths. Moving between these regions of high probability is a rare event, costly in computation time. As a result, the simulation might lock into one of the regions, implying false convergence and producing configurations uncharacteristic of the entire phase space. In such a situation, a proper sampling requires long simulation times, and the data analysis (Chapter 3) needs special care to assure statistical independence among measurements.

---

[8] An irreducible matrix must have at least one nonzero off-diagonal element in each column. In general, the product of a stochastic matrix with itself may also have zero elements, again leaving some states inaccessible.

2.5 Detailed balance algorithms

We now discuss two classes of algorithms that satisfy detailed balance and hence allow the sampling from a prespecified probability distribution. The first class encompasses the Metropolis and related algorithms; the second class, the *heat-bath algorithm* and related algorithms. For both classes, the new state is usually sampled by means of relatively local configuration changes. In contrast to the Metropolis algorithm, the heat-bath algorithm has no explicit rejection. In the broader literature, this algorithm is called the *Gibbs sampler*, and in the statistics literature in particular, a litany of Gibbs samplers exists. Although the two classes employ different methodologies, the heat-bath algorithm is actually a special case of what is called the *Metropolis-Hastings algorithm*.

2.5.1 Metropolis algorithm

The Metropolis algorithm (Metropolis et al., 1953) is ingenious with an elegance that derives from its simplicity. It was proposed more as a method that "seems reasonable" than one with any obviously intended connections to Markov chain mathematics. The inventors did recognize it as a computational tool that in principle could solve any problem in classical equilibrium statistical mechanics. We quote from the introductory paragraph of Metropolis et al. (1953):

The purpose of this paper is to describe a general method, suitable for fast electronic computing machines, of calculating the properties of any substance which may be considered as composed of interacting individual molecules. Classical statistics is assumed ...

Indeed, the Metropolis algorithm has endured for decades, spreading well beyond statistical mechanics into all areas of analysis using stochastic processes. We now present the algorithm and show that it defines a stochastic transition matrix that satisfies the detailed balance condition. We then explain the Metropolis et al. procedure for sampling from this matrix and discuss its use in statistical mechanics.

To sample a probability distribution p_i asymptotically, Metropolis et al. proposed the following form of the transition probability matrix

$$P_{ij} = T_{ij}A_{ij}, \tag{2.20}$$

where the T_{ij} are the elements of a symmetric matrix of *trial transition (proposal) probabilities* satisfying

$$T_{ij} \geq 0, \quad T_{ij} = T_{ji}, \quad \text{and} \quad \sum_i T_{ij} = 1, \tag{2.21}$$

and the A_{ij} are elements of an *acceptance matrix*,

$$A_{ij} = \begin{cases} 1 & \text{if } p_i/p_j \geq 1, \\ p_i/p_j & \text{if } p_i/p_j < 1, \end{cases} \tag{2.22}$$

often written as

$$A_{ij} = \min\{1, p_i/p_j\}. \tag{2.23}$$

Strictly speaking the definitions (2.22) and (2.23) are valid only for $i \neq j$. This condition is typically an ingrained part of the algorithm's implementation: what is proposed, by using T_{ij}, is by design a change, so $i \neq j$. However, for the purpose of demonstrating that the transition probability matrix defined by the algorithm is a stochastic matrix, let us state the transition probability matrix as

$$P_{ij} = \begin{cases} \begin{cases} T_{ij} & \text{if } p_i/p_j \geq 1 \\ T_{ij}p_i/p_j & \text{if } p_i/p_j < 1 \end{cases} & i \neq j, \\ T_{jj} + \sum_{\{k \mid p_k < p_j\}} T_{kj}(1 - p_k/p_j) & i = j. \end{cases}$$

The expression for P_{ij} defines the rejection component of the Metropolis algorithm. If the proposed move is rejected, then the current state is added to the Markov chain.

With the transition probability matrix now completely defined, its stochastic nature is easily demonstrated.[9] We need only to show that $\sum_i P_{ij} = 1$, because it is obvious from its definition that $P_{ij} \geq 0$. The steps of the proof are

$$\sum_i P_{ij} = P_{jj} + \sum_{\{i \mid p_i > p_j\}} T_{ij} + \sum_{\{i \mid p_i < p_j\}} T_{ij}p_i/p_j$$

$$= T_{jj} + \sum_{\{k \mid p_k < p_j\}} T_{kj}(1 - p_k/p_j) + \sum_{\{i \mid p_i > p_j\}} T_{ij} + \sum_{\{i \mid p_i < p_j\}} T_{ij}p_i/p_j$$

$$= T_{jj} + \sum_{i \neq j} T_{ij}$$

$$= 1.$$

To show detailed balance (2.19), we first note that it is obvious for $i = j$. Then, for $i \neq j$, by assuming $p_i < p_j$ without loss of generality (due to the arbitrariness of labeling),

$$P_{ij} = T_{ij}p_i/p_j = T_{ji}p_i/p_j = P_{ji}p_i/p_j,$$

from which it follows that $P_{ij}p_j = P_{ji}p_i$. In obtaining this result, we used the symmetry of T, (2.20), and (2.23). Thus, the demonstration that detailed balance holds for all i and j is complete.

---

[9] The irreducibility of the matrix follows from nonzero diagonal elements.

Algorithm 3 Metropolis algorithm.

Input: Proposal probability T, limiting distribution p, configuration j.

    Sample an i from T_{ij} ;

    Generate a uniform random number $\zeta \in [0, 1]$;

    if $\zeta \leq p_i/p_j$ **then**

      $j \leftarrow i$;

    end if

    return j.

To sample an i given a j is exceptionally simple (see Algorithm 3), but possibly not obvious in afterthought. In this procedure, Metropolis et al. use T_{ij} to create a proposal for the next configuration i in the Markov chain. This new configuration could in principle be any allowable one. Once it is proposed, they then use the A_{ij} element of the acceptance matrix to decide between i and j.

What does this sampling procedure mean in practice? From (2.23) we see that the transition probabilities depend only on the ratio of the weight of the proposed configuration to the current one and hence are independent of their normalization constant. In classical statistical mechanics, this ratio is a ratio of Boltzmann factors and equals $\exp(-\Delta E/kT)$ where ΔE is the energy difference between the two configurations. The Markov chain is constructed by visiting each lattice site or particle, deterministically or stochastically, and proposing a new configuration by changing the state of the random variable on the site or moving the particle. The changes are typically local. Rarely does a Metropolis Monte Carlo simulation change all, or even more than just a few, of the variables defining a configuration at once. The algorithm says "accept the proposed local change" if the new configuration has a lower energy. The algorithm also says "sometimes accept a change that increases the energy, but only with probability $\exp(-\Delta E/kT)$."

For the Ising model, a typical T_{ij} selects a specific lattice site and proposes a change in the spin state. The random variable at this site, that is, the Ising spin, has two possible values, the present one and the flipped one. The Metropolis transition probability selects which of the two configurations is added to the Markov chain. *It is important to note that the Metropolis algorithm makes repeating the configuration in the Markov chain one or more times a definite and necessary possibility.*

Is the Metropolis algorithm ergodic? The Metropolis algorithm transfers the burden of ergodicity to T_{ij}. If in a finite number of steps T_{ij} enables all of the phase space to be reached, then the Metropolis algorithm is ergodic. For the Ising example, T_{ij} allows each site to be visited, and at each visit any value of the random variable can be selected. Thus, all spin configurations may be realized.

2.5.2 Generalized Metropolis algorithms

Generalizations of the Metropolis algorithm exist. Let us view them in terms of the general prescription that

$$P_{ij} = T_{ij}A_{ij}, \quad \text{and} \quad A_{ij} = \min\left\{1, \mathcal{R}_{ij}\right\}. \tag{2.24}$$

Here we no longer require that $T_{ij} = T_{ji}$. First we note that there is a general class of algorithms associated with the acceptance ratio

$$\mathcal{R}_{ij} = \frac{S_{ij}}{1 + \dfrac{T_{ij}p_j}{T_{ji}p_i}}, \tag{2.25}$$

where S_{ij} is any non negative symmetric matrix[10] that ensures $0 \le A_{ij} \le 1$ for all i and j. Two choices for S_{ij} are common. One is

$$S_{ij} = 1, \tag{2.26}$$

leading to

$$\mathcal{R}_{ij} = \frac{T_{ji}p_i}{T_{ij}p_j + T_{ji}p_i}, \tag{2.27}$$

which is called the Metropolis-Barker algorithm (Barker, 1965). The second, due to Hastings (1970), is

$$S_{ij} = \begin{cases} 1 + \dfrac{T_{ji}p_i}{T_{ij}p_j}, & \text{if} \quad T_{ij}p_j \ge T_{ji}p_i, \\ 1 + \dfrac{T_{ij}p_j}{T_{ji}p_i}, & \text{if} \quad T_{ji}p_i > T_{ij}p_j, \end{cases} \tag{2.28}$$

leading to the Metropolis-Hastings algorithm based on

$$\mathcal{R}_{ij} = \frac{T_{ji}p_i}{T_{ij}p_j}. \tag{2.29}$$

If T is symmetric, the Metropolis-Hastings algorithm reduces to the original Metropolis algorithm. We remark that Barker's and Hastings's algorithms share with the original Metropolis algorithm the property of not needing the normalization of the stationary distribution. We leave it to the Exercises to show that all three algorithms satisfy the detailed balance condition.

Not all algorithms that satisfy detailed balance are of the Metropolis form (2.24). Another general class has an acceptance matrix of the form $A_{ij} = S_{ij}/p_j T_{ij}$. Again, S_{ij} is any non negative symmetric matrix that ensures $0 \le A_{ij} \le 1$ for all i and j.

---

[10] A non negative matrix has all elements greater than or equal to zero.

In this case the transition probability of the Markov chain becomes independent of the proposal probability, but dependent on the normalization of the stationary distribution. Many more generalizations of the original Metropolis et al. insights exist. The challenge is not in creating them but in deciding which is the most efficient. Fortunately, there are a few guidelines worth noting.

First we comment on the difference between computational efficiency and statistical efficiency. *Computational efficiency* is measured by the average computer time per step needed to execute the algorithm. Here, we address *statistical efficiency*. From the discussion in Section 2.4.2 we saw that the statistical efficiency, in the sense of the convergence of the Markov chain, is controlled by the magnitude of the second-largest eigenvalue of the transition probability matrix: the smaller the better. Because of the size of the matrix,[11] quantitative knowledge of this eigenvalue is typically unavailable. Recently, Monte Carlo methods were developed to determine this eigenvalue (Rubenstein et al., 2010). The available results support the analytic work by Peskun (1973).

The mathematical analysis of Peskun, and subsequently by others (Liu, 2001) established a theorem that if the off-diagonal elements of one transition matrix are greater than or equal to those of another, the variances of any expectation value produced by the Markov chain generated with the off-diagonally dominant transition matrix are smaller than those produced by the other chain. The general rule proposed by Peskun is that statistical efficiency increases as the off-diagonal matrix elements of P_{ij} dominate the diagonal ones. This makes sense. Off-diagonal dominance decreases the probability of a state being repeated in the chain and thus reduces statistical correlations among successive estimates of an observable. For *discrete* spaces Peskun also proved that the Metropolis-Hastings algorithm is statistically more efficient than the Metropolis-Barker algorithm.

Very generally, the detailed balance condition and the Metropolis-Hastings algorithm provide flexible tools for designing algorithms to sample from a specified probability density. Various alternatives to this algorithm are possible and may control the probability flow more optimally. Dramatic increases in computational efficiency have occurred by designing algorithms that are *not* of the Metropolis (or heat-bath) type. The rejection character of the algorithm, repeating the current configuration one or more times, is the principal source of statistical inefficiency as it inhibits the generation of statistically independent measurements. Breaking away from the Metropolis algorithm is, however, not easy. Successful examples are the cluster and loop algorithms, which are discussed in Chapters 4, 5, and 6.

---

[11] The elements of the transition probability matrix are computed on the fly. Except for trivial problems or small physical systems, computer memory is insufficient to store the entire matrix.

The important innovation in these algorithms is the step from local configuration changes to global ones.

2.5.3 Heat-bath algorithm

The *heat-bath algorithm* (Creutz, 1980) involves local moves, as does the Metropolis algorithm, but in contrast to the Metropolis algorithm, the proposed change in the value of the local variable x_i is independent of its current value and is always accepted. It may get repeated in the Markov sequence many times, but this is not the result of a rejection – rather, it is the result of a selection.

The strategy is to sample x_i from a univariate density that is conditional on the values of a small number of variables. In other words, we sample x_i from some $P(x_i|x_{i_1}, \ldots, x_{i-1}, x_{i+1}, \ldots, x_{i_m})$ where $m \ll N$. If the local environment of x_i were replaced with a heat bath, defined by the conditional variables, then we would be sampling from the expected equilibrium distribution for that environment.

We can achieve the same effect with the Metropolis algorithm by using the standard proposal probability and repeatedly updating the configuration at the same site. Eventually, we start sampling from the local equilibrium probability. Thus, the heat-bath algorithm achieves in one step something that might take the Metropolis algorithm many steps. Because of this, the heat-bath algorithm is generally believed to be more efficient than the Metropolis algorithm. However, the repetition of the state in the chain because of selection can have the same effect as the rejection in the Metropolis case. The situation with respect to which algorithm is more efficient is in reality more complex (Rubenstein et al., 2010).

To illustrate the algorithm, we again invoke the Ising model as the example. We select some lattice site i and then fix the values of the spins at the other sites. The local energy of the spin at the selected site is $E(s_i) = -Js_i \sum_{j \neq i} s_j$, leading to the local Boltzmann factor $\exp[-E(s_i)/kT]$. Taking into account the two spin orientations, we can write the local probability for the spin $s_i = \pm 1$ as

$$P(s_i) = \frac{\exp[-E(s_i)/kT]}{\exp[-E(s_i)/kT] + \exp[-E(-s_i)/kT]}. \tag{2.30}$$

Sampling from this probability places an s_i with a value of $+1$ or -1 at site i, thereby generating a new configuration from the old one without regard to the previous value of the spin at the selected site.

To present more details and justification, we proceed as follows. If the probability function for a collection of random variables is $P(s_1, s_2, \ldots, s_N)$, the heat-bath algorithm generates a new configuration from the old one by sampling the conditional probability for each variable while the others are fixed (see Algorithm 4).

Algorithm 4 Heat-bath algorithm.

Input: A configuration $(s_1, s_2, \ldots, s_N)$.

   for $i = 1$ to N **do**

      Sample s_i' from $P(s_i'|s_1, \ldots, s_{i-1}, s_{i+1}, \ldots, s_N)$;

      $s_i \leftarrow s_i'$;

   end for

   return the updated $(s_1, s_2, \ldots, s_N)$.

The definition of a multivariate conditional probability

$$P(s_i|s_1, \ldots, s_{i-1}, s_{i+1}, \ldots, s_N) = \frac{P(s_1, \ldots, s_N)}{P(s_1, \ldots, s_{i-1}, s_{i+1}, \ldots, s_N)}, \tag{2.31}$$

plus that of the marginalization

$$P(s_1, \ldots, s_{i-1}, s_{i+1}, \ldots, s_N) = \sum_{s_i} P(s_1, \ldots, s_N),$$

give the probabilities we need. For the zero-field, one-dimensional Ising model with nearest neighbor interactions, we can express these probabilities in a simple formula. The conditional probability becomes

$$P(s_i|s_{i-1}, s_{i+1}) = \frac{e^{J(s_{i-1}s_i + s_i s_{i+1})/kT}}{\sum_{\tilde{s}_i = \pm 1} e^{J(s_{i-1}\tilde{s}_i + \tilde{s}_i s_{i+1})/kT}}$$

$$= \frac{e^{J(s_{i-1} + s_{i+1})s_i/kT}}{e^{J(s_{i-1} + s_{i+1})/kT} + e^{-J(s_{i-1} + s_{i+1})/kT}},$$

which is a more detailed expression of (2.30). Note that we started with all the random variables but finished with an expression dependent on only a few. The reduction is a consequence of the Ising model's short-ranged exchange interaction.

The heat-bath algorithm is a special case of the Metropolis-Hastings algorithm (2.28), with an acceptance probability equal to one. To see this, let us at the i-th step of the heat-bath algorithm form the Metropolis-Hastings acceptance ratio

$$\mathcal{R}_{s_i', s_i} = \frac{P(s_1, \ldots, s_i', \ldots s_N)P(s_i|s_1, \ldots, s_{i-1}, s_{i+1} \ldots s_N)}{P(s_1, \ldots, s_i, \ldots s_N)P(s_i'|s_1, \ldots, s_{i-1}, s_{i+1} \ldots s_N)}$$

$$= \frac{P(s_i'|s_1, \ldots, s_{i-1}, s_{i+1} \ldots s_N)P(s_i|s_1, \ldots, s_{i-1}, s_{i+1} \ldots s_N)}{P(s_i|s_1, \ldots, s_{i-1}, s_{i+1} \ldots s_N)P(s_i'|s_1, \ldots, s_{i-1}, s_{i+1} \ldots s_N)} = 1,$$

where we invoked (2.31) twice, once for $P(s_1, \ldots, s_i', \ldots s_N)$ and the second time for $P(s_1, \ldots, s_i, \ldots s_N)$. Thus each step of the heat-bath algorithm corresponds to a Metropolis-Hastings sampling with $A_{ij} = 1$.

2.6 Rosenbluth's theorem

As previously noted, detailed balance and the Metropolis algorithm were originally proposed more as something that seemed reasonable than as something that followed from the mathematics. About four years after the Metropolis et al. publication, Wood and Parker (1957) connected the algorithm with a Markov process, and this often repeated but seldom cited analysis defines the standard justification of the original proposal. Shortly after the Metropolis et al. paper, Rosenbluth (1953) wrote an unpublished, and only recently noticed, report (Gubernatis, 2003) proving the validity of the algorithm. His proof focuses on the conservation of probability. In fact, he proved that under the conditions of ergodicity and detailed balance, probability flows through phase space in such a way that the average squared-deviation of the probability density from the canonical ensemble not only becomes zero but does so monotonically. His insightful proof points to a special character of the convergence, absent from the standard Markov chain proof.

Let us offer here a related but simpler proof for the convergence of the Markov chain and the H-theorem. While to prove them in a general setting we need the Perron-Frobenius theorem, which itself requires many lines to prove, the assumption that the stationary condition (2.18) is satisfied with the target distribution simplifies the task. The fact that most of the Monte Carlo algorithms satisfy an even stronger condition, namely, the detailed-balance condition, justifies this assumption. As we will see, we require only the stationary condition and ergodicity (Section 2.4.2) in proving the convergence of the Markov process. We need the detailed balance condition (2.19) in establishing the H-theorem, or generalized Rosenbluth's theorem.

We consider two ensembles $\mathcal{E}_p$ and $\mathcal{E}_q$ of configurations i with probability densities p_i and q_i. We define the distance between these ensembles as

$$\|\mathcal{E}_p - \mathcal{E}_q\| = \sum_i |p_i - q_i|.$$

Now we consider the case where p_i and p_i' are related by one step in the Markov chain $p_i' = \sum_j P_{ij} p_j$. We already have noted that this operation conserves probability if P is a stochastic matrix. The distance between the ensemble $\mathcal{E}'$ and the target ensemble $\mathcal{E}^{\mathrm{eq}}$, which satisfies the stationarity condition, is

$$\|\mathcal{E}' - \mathcal{E}^{\mathrm{eq}}\| = \sum_i \left| \sum_j P_{ij} \left(p_j - p_j^{\mathrm{eq}} \right) \right|$$

$$\leq \sum_i \sum_j P_{ij} \left| p_j - p_j^{\mathrm{eq}} \right| = \sum_j \left| p_j - p_j^{\mathrm{eq}} \right|. \tag{2.32}$$

Note that we have used the stationarity condition $\sum_j P_{ij} p_j^{eq} = p_i^{eq}$ in the first step and the conservation of probability $\sum_i P_{ij} = 1$ in the last step.[12] On the right-hand side, we recognize $\|\mathcal{E} - \mathcal{E}^{eq}\|$, so

$$\|\mathcal{E}' - \mathcal{E}^{eq}\| \leq \|\mathcal{E} - \mathcal{E}^{eq}\|.$$

Thus, each step of the algorithm reduces the distance between the current ensemble and the equilibrium ensemble. Because the distance is trivially bounded from below by zero, it must converge to some value. The remaining question is whether this value is zero or not. We can answer this question by examining (2.32) more closely.

First, we note that (2.32) must hold even if we replace P_{ij} in (2.32) by $[P^n]_{ij}$ and let $\mathcal{E}'$ be the ensemble after n steps. Let n be large enough so that $[P^n]_{ij} > 0$ for all combinations of i and j. Choosing such an n is possible due to the ergodicity. Now, after convergence is reached, the equality must hold in (2.32). For it to hold, $p_j - p_j^{eq}$ must have the same sign (or be zero) for all j reachable from the same i by n steps. Since $[P^n]_{ij} > 0$ for all j, it means $p_j \leq p_j^{eq}$ for all j or $p_j \geq p_j^{eq}$ for all j. But because of the normalizations $\sum_j p_j = \sum_j p_j^{eq} = 1$, each inequality must reduce to $p_j = p_j^{eq}$. Thus, p_j must converge to p_j^{eq}. This observation completes the proof of the theorem that an ergodic Markov process with a stationary condition (or detailed balance condition) must converge to the target distribution. It also follows that the eigenvalue with the largest magnitude is unity and that the corresponding eigenvector is unique and equal to p_i^{eq}, because otherwise the process would not always converge to p_i^{eq} regardless of the initial distribution.

We can also show that if a Markov process satisfies detailed balance, the free energy decreases monotonically and eventually converges to its equilibrium value. We call this type of H-theorem for Monte Carlo simulations *Rosenbluth's theorem*. This monotonicity is a feature characteristic of a more general convex function than the averaged squared-deviation from the equilibrium density (Renyi, 1960; Kawashima, 2007). The assertion is that

$$p_i' = \sum_j P_{ij} p_j \tag{2.33}$$

together with the condition that $P_{ij} p_j^{eq} = P_{ji} p_i^{eq}$ causes

$$\Phi = \sum_i p_i^{eq} f\left(\frac{p_i}{p_i^{eq}}\right)$$

---

[12] The fact that we have not used the detailed balance condition is important for proving the convergence of Monte Carlo algorithms that do not satisfy the detailed balance condition, such as the directed-loop algorithm, although we do not discuss many such cases in the present book.

to converge monotonically if $f(x)$ is a convex function. The most important example of this family of observables can be obtained by setting $f(x) = x \ln x$. The result is the Kullback-Leibler information:

$$\Phi = I_{\text{KL}}(p \parallel p^{\text{eq}}) \equiv \sum_i p_i \ln \left(\frac{p_i}{p_i^{\text{eq}}} \right).$$

For physicists, a more familiar interpretation of this quantity is the excess free energy, which measures how much larger the current free energy is than its equilibrium value. Namely, when $p_i^{\text{eq}} \propto e^{-E_i/kT}$, then $I_{\text{KL}} = (F - F^{\text{eq}})/(k_B T)$ with $F = \langle E \rangle - TS$ and $F^{\text{eq}} = \langle E \rangle^{\text{eq}} - TS^{\text{eq}}$, where S is the von Neuman entropy, $S \equiv -\sum p \log p$. If we set $f(x) = x^n$, we obtain the n-th Renyi information:

$$\Phi = I_R^{(n)}(p \parallel p^{\text{eq}}) \equiv \sum_i \frac{p_i^n}{(p_i^{\text{eq}})^{n-1}} = \left\langle \left(\frac{p_i}{p_i^{\text{eq}}} \right)^{n-1} \right\rangle.$$

The proof is as follows. At a particular step in the iteration, we have

$$\Phi' = \sum_i p_i^{\text{eq}} f \left(\frac{p_i'}{p_i^{\text{eq}}} \right).$$

Using (2.33), we rewrite this equation as

$$\Phi' = \sum_i p_i^{\text{eq}} f \left(\frac{1}{p_i^{\text{eq}}} \sum_j P_{ij} p_j \right) = \sum_i p_i^{\text{eq}} f \left(\sum_j P_{ij} \frac{p_j^{\text{eq}}}{p_i^{\text{eq}}} \frac{p_j}{p_j^{\text{eq}}} \right),$$

which on the use of the detailed balance condition becomes

$$\Phi' = \sum_i p_i^{\text{eq}} f \left(\sum_j P_{ji} \frac{p_j}{p_j^{\text{eq}}} \right).$$

Now we employ Jensen's inequality for the convex function f, that is, $f(\sum_i a_i x_i) \le \sum_i a_i f(x_i)$ for any probability distribution a_i, to obtain

$$\Phi' \le \sum_{ij} P_{ji} p_i^{\text{eq}} f \left(\frac{p_j}{p_j^{\text{eq}}} \right).$$

Using the detailed balance condition and $\sum_i P_{ij} = 1$, we find

$$\Phi' \le \sum_j p_j^{\text{eq}} f \left(\frac{p_j}{p_j^{\text{eq}}} \right) = \Phi.$$

This quick derivation represents a proof quite different from the one presented by Rosenbluth. His proof was in the continuum and he argued on the basis of the need to conserve the number of particles $\rho(r)dr$ in a differential phase space volume dr as they move through phase space.

2.7 Entropy content

The *entropy* of a distribution,

$$S = -\sum_{i=1}^{n} p_i \ln p_i,$$

provides information about its global character. For example, if all the p_i are equal, the entropy for an n-state distribution takes its maximum value of $\ln n$, and we say that the distribution contains the minimal amount of information. In other words, one event is as likely as any other, and we have the maximum uncertainty about which event the chain would generate next. If the probability of a particular event is one, which means the probability of all others must be zero, then the value of the entropy takes its minimum value of zero. The event is a sure thing with no uncertainty. Once the Markov chain reaches stationarity, that is, samples a fixed distribution, its information entropy content, as defined by the distribution it generates, becomes constant. We refer to this constant as the entropy content of the Markov process. (In fact, the information that is actually minimized at the stationary point is the Kullback-Leibler information, which is proportional to the excess free energy; see Section 2.6.)

If the chain takes r steps beyond some point of stationarity, it generates a sequence of events $K = \{k_1, k_2, \ldots, k_r\}$ with probability $P(K) = P_{k_r, k_{r-1}} \cdots P_{k_3, k_2} P_{k_2, k_1} p_{k_1}$. Let us consider the set of all n^r sequences. Then, one can prove (Khinchin, 1957) that given $\varepsilon > 0$ and $\eta > 0$, no matter how small, for a sufficiently large r, the set divides into two subsets with one subset having the property that the probability $P(K)$ of any sequence in it satisfies

$$e^{-r(R+\eta)} < P(K) < e^{-r(R-\eta)}$$

and the other subset having the property that the sum of the probabilities of all sequences in it is less than ε. The expression

$$R = -\sum_i \sum_j p_j P_{ij} \ln P_{ij}$$

is a measure of the amount of information obtained when the Markov chain moves one step ahead.

We note that there are in total n^r possible r-term sequences. If we were to arrange these sequences in order of decreasing probability $P(K)$, then another theorem (Khinchin, 1957) says that if we sum these probabilities until the sum just exceeds the positive number λ that satisfies $0 < \lambda < 1$, then the number $N_r(\lambda)$ of sequences used satisfies

$$\lim_{r \to \infty} \frac{\ln N_r(\lambda)}{r} = R,$$

independent of the value of λ.

While the proof of these two theorems is outside the scope of this book, they make the important point that while the number of possible sequences of events equals $n^r = \exp(r \ln n)$, the number the Markov chain selects is approximately $\exp(rR)$. In other words, e^R is the effective number of candidate states that we can choose for the next step in the Markov chain. Since the maximum value of the entropy is $\ln n$, we almost certainly have $R < \ln n$. For example, in the case of the single-spin update for the Ising model, we choose one of N spins at random and choose between two possible choices (spin up or spin down) for the value of the spin in the next Monte Carlo step. Since there are $N \times 2$ choices for the next spin configuration, R is bounded from above by $\ln(2N)$, and it is natural to assume that R is of the order of $\ln N$. Considering $n = 2^N$ in this example, R is not only smaller than $\ln n$, it is much smaller than $\ln n$ for large systems. As is clear from this example, in typical situations, the latter theorem by Khinchin indicates that a negligibly small fraction of the total number of sequences accounts for almost all the probability; that is, for this small fraction $\sum_K P(K) \approx 1$. Therein lies the power of Markov chain sampling.

We can also show that the entropy decreases as the chain steps away from the initial state. The behavior of the entropy, while similar, does differ from the actual relaxation behavior of the chain to stationarity, which depends on the magnitude of the second largest eigenvalue of the transition probability matrix. The entropy content of the chain does provide a complementary perspective on the behavior of Markov chains and why they can be an effective sampler of distributions of a large number of random variables.

Suggested reading

M. H. Kalos and P. A. Whitlock, *Monte Carlo Methods*, vol. 1: *The Basics* (New York: Wiley-Interscience, 1986), chapters 1–3 and appendix.

W. H. Press, S. A. Teukolsky, W. T. Vettering, and B. P. Flannery, *Numerical Recipes* (Cambridge University, 1992), chapter 7.

J. S. Liu, *Monte Carlo Strategies in Scientific Computing* (New York: Springer, 2001), chapters 5, 7, and 13.

Exercises

2.1   Use the method of mathematical induction to prove the validity of (2.13); that is, show the result is true for $n = 1$, assume it is true for n, and then show it is true for $n + 1$.

2.2   If $0 \leq p \leq 0$ and $q = 1 - p$, the binomial distribution

$$p(k) = \binom{n}{k} p^s q^{n-k} \qquad (2.34)$$

describes the probability of k successes in n trials. Show that one way to sample k is

$$k = \min_{n} \{\zeta_1 \zeta_2 \cdots \zeta_n \leq p\}.$$

2.3   If $0 \leq p \leq 0$ and $q = 1 - p$, show that

$$\binom{n}{k} p^k q^{n-k} = \frac{(np)^k}{k} \left(1 - \frac{np}{n}\right)^n Q_n,$$

where

$$Q_n = \frac{\prod\limits_{r=2}^{k} \left(1 - \frac{r-1}{n}\right)}{(1 - p)^k}.$$

Now let $p = \lambda/n$ for $n > \lambda$ and show that as $n \to \infty$, $Q_n \to 1$ and

$$\binom{n}{k} p^k q^{n-k} \to \frac{\lambda^k}{k!} e^{-\lambda} \quad \text{as } n \to \infty.$$

This result shows that for large n and small p the binomial distribution is approximately the same as the Poisson distribution, provided np is constant.

2.4   Propose a method to sample x from the density

$$p(x) = \sum_{i=1}^{N} f_i(x), \quad \text{where } f_i(x) \geq 0$$

if the f_i are easily normalized and if they are not.

2.5   Consider a two-state Markov chain defined by the stochastic matrix

$$P = \begin{pmatrix} \alpha & 1 - \beta \\ 1 - \alpha & \beta \end{pmatrix} \quad \text{for } 0 < \alpha, \beta < 1.$$

1. Find its eigenvalues and verify that the dominant eigenvalue $\lambda_1 = 1$.
2. Find its left- and right-hand eigenvectors x_i and y_i for each λ_i and verify that the components of the left-hand eigenvector of the dominant eigenvalues are all equal and that $x_i^T y_j = 0$ if $i \neq j$.

3. Use these eigenvectors to find P^n analytically.
4. Show that as $n \to \infty$, $P^n p^0$ is independent of the vector p_0 if its components sum to unity.
5. Show that the limiting vector of $P^n p^0$ is the right-hand eigenvector associated with λ_1.

2.6 Demonstrate that the Metropolis-Barker (2.26) and the Metropolis-Hastings (2.28) algorithms satisfy detailed balance. Do the same for the heat-bath algorithm.

2.7 A typical simulation uses just one sampling method, say, either the Metropolis or the heat-bath method. For the heat-bath method, the conditional probabilities may not always be as easy to sample as is the case for the Ising model.

1. Discuss the algorithmic issues and opportunities of using the Metropolis algorithm within the heat-bath algorithm to do this sampling.
2. Propose a scenario where it might be advantageous to use the heat-bath algorithm within the Metropolis algorithm.

2.8 If the proposal probability T_{ij} is independent of the current state and equals $\pi(i)$, the transition probability for the Metropolis-Hastings algorithm is

$$
P_{ij} = \begin{cases} \pi_i \min\left(0, w_i/w_j\right) & \text{if } i \neq j, \\ \pi_j + \sum_k \pi_k \max\left(0, 1 - w_k/w_j\right) & \text{if } i = j, \end{cases}
$$

where p_i is the target distribution and $w_i = p_i/\pi_i$.

1. Show that this transition probability satisfies detailed balance.
2. If the states are labeled so that $w_1 \geq w_2 \geq \cdots \geq w_n$, show that

$$
P = \begin{pmatrix} \pi_1 + \lambda_1 & \pi_1 & \cdots & \pi_1 & \pi_1 \\ p_2/w_1 & \pi_2 + \lambda_2 & \cdots & \pi_2 & \pi_2 \\ \vdots & \vdots & \ddots & \vdots & \vdots \\ p_{n-1}/w_1 & p_{n-1}/w_2 & \cdots & \pi_{n-1} + \lambda_{n-1} & \pi_{n-1} \\ p_n/w_1 & p_n/w_2 & \cdots & p_n/w_{n-1} & \pi_n \end{pmatrix}
$$

where $\lambda_k = \sum_k (\pi_i - p_i/w_k)$. Use your knowledge of the value of the first eigenvalue and the components of eigenvectors of a transition matrix to show that $\lambda_2 = 1 - 1/w_1$. What are the corresponding eigenvectors?
3. Argue that λ_k is the probability of rejection if the next step is the current state at k. Liu (2001) has shown that for this transition probability the λ_k are the eigenvalues of P.

2.9   Starting with

$$S(A) = -\sum_i p(A_i) \ln p(A_i)$$

and

$$S(A, B) = -\sum_{ij} p(A_i, B_j) \ln p(A_i, B_j),$$

prove $S(A, B) = S(A) + S(A|B)$.

3

Data analysis

An important part of any Monte Carlo simulation is the statistical analysis of the data it generates. Central to this analysis is computing the average values of the relevant physical observables and estimating the statistical error of these averages. Computing the averages is easy. It mainly requires ensuring the simulation is sampling configurations from the equilibrium distribution before tallying the values of the observables. Estimating the error is more difficult. The increased difficulty is not in evaluating its mathematical definition, the so-called standard deviation, but rather it is in ensuring the measurements used in this formula are statistically independent. The correlations among the data points depend on the values of a model's parameters, the observable, and the sampling algorithm being used. We now discuss the statistical basis of common procedures for analyzing the results of a Monte Carlo simulation.

3.1 Equilibrating the sampling

We know from Section 2.4 that the Markov chain requires a certain number of steps before it begins to sample from its limiting distribution. This distribution is also called the *equilibrium distribution* even for cases where we are not simulating a problem in equilibrium statistical mechanics. Verifying that equilibrium has been reached is a generic task of all Markov chain Monte Carlo simulations. Once in equilibrium, we then want to use the chain samples to estimate the values of various physical quantities and their associated statistical errors.

There are physics-related issues that can make equilibration particularly difficult, many of which affect simulations in the vicinity of phase transitions. Although we might know what phases to expect, we generally do not know the precise location of phase boundaries. Identifying the phases and determining their locations in parameter space are common tasks for which Monte Carlo simulations are employed.

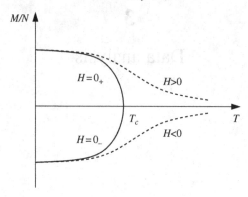

Figure 3.1 Sketch of the magnetization of the two-dimensional Ising model as a function of T for zero field (solid lines) and nonzero field (dashed lines). T_c is the critical temperature of the zero field case. The magnetization per site (M/N) is the order parameter.

The two-dimensional Ising model $E = -J \sum_{\langle ij \rangle} s_i s_j + H \sum_i s_i$, with $\langle ij \rangle$ denoting nearest neighbor sites, is a convenient example for illustrating the difficulty in equilibrating near a phase transition and the other main points of this chapter.

The magnetization of the two-dimensional Ising model as a function of temperature T is sketched in Fig. 3.1. In zero field, as the temperature is lowered, a second-order (continuous) phase transition occurs at the critical temperature $T_c = 2J/\log(1 + \sqrt{2}) \approx 2.269J$, where the system transitions from a disordered to an ordered phase. In a nonzero field, at fixed $T < T_c$, a first-order transition occurs as we switch the sign of the field.

In the vicinity of either type of transition, Monte Carlo simulations, especially those based on local update Metropolis or heat-bath algorithms, experience difficulties in equilibrating. These algorithms move through phase space with small changes in configurations associated with small changes in energy. A first-order transition is marked by an energy barrier created by the surface tension between the two phases that is hard to surmount in a single or a chained couple of moves. The problem here is getting stuck in one part of phase space for a long time. A second-order phase transition is marked by critical slowing down. This phenomenon, which is related to a diverging correlation length, affects a Monte Carlo simulation if the updating procedure changes the configuration only locally. The problem here is knowing when the system has relaxed.

Before we can begin a proper analysis of the data we must first determine the number of Monte Carlo steps necessary to equilibrate. This number is generally called the *equilibration time*. Typically, we measure this time by the number of *sweeps*. One sweep is the result of a procedure that attempts at least one local Monte Carlo update of all the random variables. In the case of the Ising model,

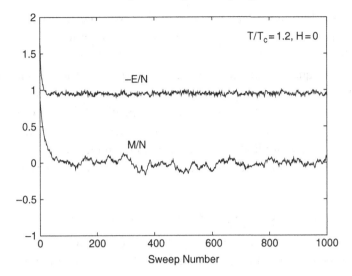

Figure 3.2 Relaxation to equilibrium as a function of the number of Monte Carlo sweeps for the magnetization per site M/N and the negative of the energy per site $-E/N$. The results are for the two-dimensional zero-field Ising model simulated by the heat-bath algorithm. $T = 1.2T_c$ and $H = 0$. The lattice size is 100×100.

for example, it would involve a Monte Carlo update of the Ising spin at each lattice site.

A simple way to estimate the number of equilibration steps is to plot the measured values as a function of the Monte Carlo sweep number and then conservatively estimate the number at which they begin to fluctuate around a mean value. We illustrate this technique in Fig. 3.2 where we present plots for the magnetization per site (M/N) and the negative of the energy per site $(-E/N)$ for the two-dimensional zero field Ising model. We used the heat-bath algorithm and performed the simulation at $T = 1.2T_c$, that is, above the critical temperature in the disordered phase. The lattice size is 100×100, and our initial configuration was the state in which the Ising variable at each site was "up." From the figure it is clear that different physical quantities relax at different rates. The energy per site appears equilibrated within 50 sweeps. The equilibration time for the magnetization is longer and more ambiguous. A value between 500 and 1000 seems reasonable. Taking the larger value is wise.

While not presented here, it is useful to test the sensitivity of the estimate to changes in the random number seed. In many cases, it is important to try initial configurations with different symmetries. Also important is performing simulations with a number of sweeps much longer than the initially estimated equilibration time, plotting only a subset of values if convenient and appropriate, to check whether the equilibrium plateau is stable, that is, whether it does not drift up or down or whether it does not eventually drop or rise precipitously.

Obviously, successive measurements shown in Fig. 3.2 are correlated. A proper definition of the equilibration time requires the calculation, for each physical quantity measured, of the *autocorrelation time*, which tells us over how many Monte Carlo sweeps the correlations persist. We give a precise definition of this quantity in Section 3.3. The *equilibration time* may then be defined as the maximum of the autocorrelation times for all observables of interest. In practice, we should choose the number of equilibration steps several times larger than this equilibration time.

3.2 Calculating averages and estimating errors

With the system equilibrated, we want to make measurements and estimate their statistical errors. Here we introduce the statistical basis for the common procedures.

The Monte Carlo sampling produces a sequence of configurations C_i from which we calculate a sequence of values $x_i = X(C_i)$ of some physical observable X. A function of a random variable is also a random variable. Hence, for each observable, we produce a different sequence of the values of a random variable. The properties of a random variable X follow from its distribution function $f_X(x)$. In Monte Carlo simulations we are mainly interested in two properties, the *mean*

$$\langle x \rangle = \int dx \, x f_X(x),$$ (3.1)

which is also called the *average* or the *expectation value*, and the *variance*

$$\sigma_X^2 = \int dx (x - \langle x \rangle)^2 f_X(x) = \langle x^2 \rangle - \langle x \rangle^2,$$ (3.2)

which is also called the *dispersion*.

With the mean and variance known, *Tchebycheff's inequality* allows us to calculate the likelihood that the value of the random variable lies in a certain interval. This inequality says that *independent of the form of $f_X(x)$*

$$P\left(\langle x \rangle - k\sigma_X < x < \langle x \rangle + k\sigma_X\right) \geq 1 - \frac{1}{k^2}.$$ (3.3)

This theorem is relatively easy to prove: From the definition (3.2) of the variance, it follows that

$$\sigma_X^2 \geq \int_{-\infty}^{\langle x \rangle - k\sigma} (x - \langle x \rangle)^2 f_X(x) dx + \int_{\langle x \rangle + k\sigma}^{\infty} (x - \langle x \rangle)^2 f_X(x) dx$$

$$\geq (k\sigma)^2 \int_{-\infty}^{\langle x \rangle - k\sigma} f_X(x) dx + (k\sigma)^2 \int_{\langle x \rangle + k\sigma}^{\infty} f_X(x) dx.$$

Hence, upon setting $\sigma^2 = \sigma_X^2$, we obtain $P(|x - \langle x \rangle| \geq k\sigma_X) \leq k^{-2}$, or equivalently (3.3).

Even in a detailed balance simulation, where we know a priori the distribution function of the configuration, we do not know a priori the mean and variance for any physical observable we are measuring relative to this distribution, so we cannot yet use Tchebycheff's inequality.

To obtain this information we begin with a sequence $x_1, x_2, \ldots, x_M$ of values of the random variable X, which we assume is uncorrelated. Their arithmetic mean

$$\bar{x} = \frac{1}{M} \sum_{i=1}^{M} x_i \tag{3.4}$$

is called the *sample mean*. The expectation value of the sample mean is

$$\langle \bar{x} \rangle = \frac{1}{M} \sum_{i=1}^{M} \langle x_i \rangle = \frac{1}{M} \sum_{i=1}^{M} \langle x \rangle = \langle x \rangle$$

and its variance is

$$\sigma_{\bar{X}}^2 = \langle \bar{x}^2 \rangle - \langle \bar{x} \rangle^2 = \frac{1}{M^2} \left\langle \sum_{i,j} x_i x_j \right\rangle - \langle x \rangle^2$$

$$= \frac{1}{M^2} \left\langle \sum_i x_i^2 + \sum_{i \neq j} x_i x_j \right\rangle - \langle x \rangle^2. \tag{3.5}$$

The absence of correlations among the x_i implies $\langle x_i x_j \rangle = \langle x_i \rangle \langle x_j \rangle = \langle x \rangle^2$, so the above becomes

$$\sigma_{\bar{X}}^2 = \frac{1}{M} \left\langle \frac{1}{M} \sum_{i=1}^{M} x_i^2 \right\rangle - \frac{1}{M} \langle x \rangle^2 = \frac{\sigma_X^2}{M}. \tag{3.6}$$

Thus, we have shown that the sample mean is the same as the mean of the random variable, but its variance is reduced by the number of terms in the sequence. Tchebycheff's inequality implies the confidence interval

$$P\left(\langle x \rangle - k\frac{\sigma_X}{\sqrt{M}} < \bar{x} < \langle x \rangle + k\frac{\sigma_X}{\sqrt{M}} \right) \geq 1 - \frac{1}{k^2}, \tag{3.7}$$

or equivalently that

$$P\left(\bar{x} - k\frac{\sigma_X}{\sqrt{M}} < \langle x \rangle < \bar{x} + k\frac{\sigma_X}{\sqrt{M}} \right) \geq 1 - \frac{1}{k^2}.$$

We see that as M becomes large, $\bar{x} \rightarrow \langle x \rangle$.

Our discussion assumed we know the variance σ_X^2. We do not. How do we estimate it? Corresponding to the sample mean is a *sample variance*,

$$s_{\bar{x}}^2 = \frac{1}{M-1} \sum_{i=1}^{M} (x_i - \bar{x})^2. \tag{3.8}$$

$s_{\bar{x}}$ is called the *standard deviation*. We previously showed that $\langle \bar{x} \rangle = \langle x \rangle$ and $\sigma_{\bar{x}}^2 = \sigma_X^2/M$. We now show that

$$\langle s_{\bar{x}}^2 \rangle = \sigma_X^2. \tag{3.9}$$

First we write $(x_i - \langle x \rangle)^2 = (x_i - \bar{x})^2 + 2(x_i - \bar{x})(\bar{x} - \langle x \rangle) + (\bar{x} - \langle x \rangle)^2$. Summing both sides of this equation from 1 to M, observing that $\sum_{i=1}^{M}(x_i - \bar{x}) = 0$, and taking the expectation value, we find that

$$M\sigma_X^2 = \left\langle \sum_{i=1}^{M} (x_i - \bar{x})^2 \right\rangle + M\sigma_{\bar{x}}^2 = \left\langle \sum_{i=1}^{M} (x_i - \bar{x})^2 \right\rangle + \sigma_X^2,$$

from which (3.9) follows.

The *central limit theorem* is a stronger theorem than Tchebycheff's inequality. It enables us to define tighter confidence intervals and to estimate a minimal number of measurements to obtain this confidence. Under very general conditions, mainly that the variance of the distribution is finite, this theorem says that as $M \to \infty$ the distribution of the random variable $\bar{x} = \sum_i x_i/M$ approaches a normal distribution

$$f_{\bar{X}}(\bar{x}) = \frac{1}{\sqrt{2\pi\sigma_{\bar{X}}^2}} e^{-(\langle x \rangle - \bar{x})^2/2\sigma_{\bar{X}}^2}.$$

With the theorem telling us that the form of the limiting distribution is Gaussian, we are able to use the well-known properties of this function to state tighter confidence intervals,

$$P\left(\bar{x} - \sigma_{\bar{X}} < \langle x \rangle < \bar{x} + \sigma_{\bar{X}}\right) = 0.65, \tag{3.10}$$

$$P\left(\bar{x} - 2\sigma_{\bar{X}} < \langle x \rangle < \bar{x} + 2\sigma_{\bar{X}}\right) = 0.95, \tag{3.11}$$

$$P\left(\bar{x} - 3\sigma_{\bar{X}} < \langle x \rangle < \bar{x} + 3\sigma_{\bar{X}}\right) = 0.99, \tag{3.12}$$

which are known as the "one-sigma," "two-sigma," and "three-sigma" confidence intervals.

The analysis of this section was based on the assumption that the Monte Carlo configurations are uncorrelated. In practical simulations, this is not the case, and we therefore must shift our attention to promoting statistical independence so we can use the expressions for the sample variance and the confidence intervals (3.10)–(3.12) to estimate error bars.

3.3 Correlated measurements and autocorrelation times

If we retrace the analysis of the last section, we can easily convince ourselves that correlations in a sequence of measured values affect only our estimation of the variance but do not affect our estimation of the mean. Let us return to our calculation of the variance of $\bar{x}$, (3.5), and proceed without the assumption of statistical independence:

$$\sigma_{\bar{x}}^2 = \langle \bar{x}^2 \rangle - \langle \bar{x} \rangle^2 = \frac{1}{M^2} \left\langle \sum_{ij} (x_i x_j - \langle x_i \rangle \langle x_j \rangle) \right\rangle$$

$$= \frac{1}{M^2} \left\langle \sum_i (x_i^2 - \langle x_i \rangle^2) + 2 \sum_{j>i} (x_i x_j - \langle x_i \rangle \langle x_j \rangle) \right\rangle. \qquad (3.13)$$

Defining the correlation function

$$\chi_{|i-j|} \equiv \chi_{ij} \equiv \langle x_i x_j \rangle - \langle x_i \rangle \langle x_j \rangle$$

we first remark that it is a function of only $|i-j|$ and $\chi_0 = \langle x^2 \rangle - \langle x \rangle^2 = \sigma_X^2$. Thus, we can rewrite (3.13) as

$$\sigma_{\bar{x}}^2 = \frac{1}{M} \left[\sigma_X^2 + 2 \sum_{k=1}^{M-1} \left(1 - \frac{k}{M} \right) \chi_k \right],$$

which in turn we rewrite as

$$\sigma_{\bar{x}}^2 = \frac{\sigma_X^2}{M} (1 + 2\tau_X), \qquad (3.14)$$

where

$$\tau_X = \sum_{k=1}^{M-1} \left(1 - \frac{k}{M} \right) \frac{\chi_k}{\chi_0}. \qquad (3.15)$$

Inspection shows that $\tau_X \geq 0$ and equals zero only if the data are uncorrelated, that is, only if $\chi_k = 0$ for $k \neq 0$. If so, (3.14) becomes our previous result (3.6). If correlations are present, (3.14) says that for a fixed number of measurements M, the variance of $\bar{x}$ increases. Correlations reduce the effective amount of statistical information, since we generated only $M/(1 + 2\tau_X)$ uncorrelated samples. If we were to ignore the correlations, (3.14) implies that our error estimates would be too small. This discrepancy underscores the need for care in estimating the variance.

The quantity τ_X defined in (3.15) is called the *autocorrelation time* for the observable X. Computing σ_X^2 and τ_X is one way of estimating $\sigma_{\bar{x}}^2$. The problem is finding an unbiased estimator of τ_X. The discussion in the following section suggests another procedure, based on promoting the statistical independence of measurements, that goes by the names of binning, batching, bunching, blocking, etc. We adopt the terminology of *blocking*.

3.4 Blocking analysis

The basis for the blocking method is the following. Let us first define the function $\text{var}(X) = \sigma_X^2$, which returns the variance of a random variable X. We note that

$$\text{var}(aX) = a^2\,\text{var}(X), \tag{3.16}$$

where a is some constant and

$$\text{var}(X + Y) = \text{var}(X) + \text{var}(Y), \tag{3.17}$$

if X and Y are independent. The sample mean is a random variable so we write its variance as

$$\text{var}(\bar{X}) = \text{var}\left(\frac{1}{M}\sum_{i=1}^{M}x_i\right) = \text{var}\left(\frac{1}{N_{\text{blocks}}}\sum_{j=1}^{N_{\text{blocks}}}\left(\frac{1}{m}\sum_{i=(j-1)m+1}^{jm}x_i\right)\right),$$

where in the last step we have split the M measurements into N_{blocks} blocks of length $m = M/N_{\text{blocks}}$. We assume here that the block length is sufficiently long that the *block averages*

$$\bar{x}_j = \frac{1}{m}\sum_{i=(j-1)m+1}^{jm}x_i$$

are statistically independent. In this case, (3.16) and (3.17) imply

$$\text{var}(\bar{X}) = \frac{1}{N_{\text{blocks}}^2}\sum_{j=1}^{N_{\text{blocks}}}\text{var}(\bar{x}_j).$$

Next we assume that for sufficiently long blocks the block averages have a common variance. Thus,

$$\text{var}(\bar{X}) = \frac{1}{N_{\text{blocks}}}\text{var}(\bar{X}_{\text{block}}), \tag{3.18}$$

where $\bar{X}_{\text{block}}$ is the random variable representing the block averages. Explicitly, we use the estimate

$$\text{var}(\bar{X}_{\text{block}}) \approx s_{\bar{X}_{\text{block}}}^2 = \frac{1}{N_{\text{blocks}} - 1}\sum_{j=1}^{N_{\text{blocks}}}(\bar{x}_j - \bar{x})^2. \tag{3.19}$$

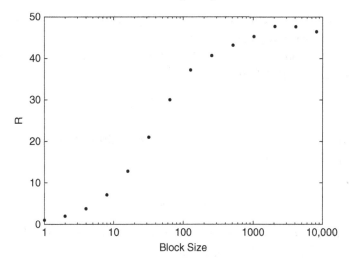

Figure 3.3 Estimating statistical independence by blocking. The simulation method, model, and its parameters are the same as in Fig. 3.2, and the observable is the magnetization per site. The size of the data stream was 1,000,000 with the first 500 eliminated. We started with a block size of $m = 1$ and successively doubled it.

The "one-sigma" error bar is given by $\sqrt{\mathrm{var}(\bar{X})}$. To determine $\mathrm{var}(\bar{X})$, we compute

$$R_X \equiv \frac{m\,\mathrm{var}(\bar{X}_{\mathrm{block}})}{\mathrm{var}(X)},$$

which *in the limit of independent blocks* relates $\mathrm{var}(\bar{X})$ to the "uncorrelated" estimate $\mathrm{var}(X)/M$:[1]

$$\mathrm{var}(\bar{X}) = R_X \frac{\mathrm{var}(X)}{M}. \tag{3.20}$$

We plot R_X for increasing block sizes m, as shown in Fig. 3.3 for the magnetization per site (the full data stream's length is $M = 10^6$). As the block size increases, the block averages become statistically independent, and $\mathrm{var}(\bar{X}_{\mathrm{block}})$ becomes inversely propositional to m. When this happens, R_X saturates and we can then use (3.18) to obtain $\mathrm{var}(\bar{X})$. In the example of Fig. 3.3, saturation occurs around $m^{\mathrm{plateau}} = 1024$ and the saturation value is $R_X^{\mathrm{plateau}} \approx 50$. This means that by neglecting correlations between successive measurements, we would have underestimated the one-sigma error bar by a factor $\sqrt{50} \approx 7$. If no plateau is reached before the block size m

---

[1] Using (3.18) one finds $R\frac{\mathrm{var}(X)}{M} = \frac{m}{M}\mathrm{var}(\bar{X}_{\mathrm{blocks}}) = \frac{\mathrm{var}(\bar{X}_{\mathrm{blocks}})}{N_{\mathrm{blocks}}} = \mathrm{var}(\bar{X})$.

approaches the largest useful value of about $\mathcal{O}(M/100)$, then we need much more data to compute a meaningful error bar.

From the definitions (3.14) and (3.20) it follows that the autocorrelation time τ is related to the value R_X^{plateau} by

$$\tau_X = \frac{1}{2}\left(R_X^{\text{plateau}} - 1\right). \tag{3.21}$$

The autocorrelation times for different observables differ. The largest value determines the equilibration time.

3.5 Data sufficiency

The central limit theorem does not tell us how many independent blocks we need, so the distribution of our data becomes Gaussian. It turns out N_{blocks} need not be that large. When one of the authors was writing his first Monte Carlo code, he asked a senior colleague who had been performing Monte Carlo simulations for several decades, "How many uncorrelated values do I need?" He replied, "32," and walked away. This brusque reply is usually about right.

While the field of statistics has many procedures for defining the probability that a set of data is distributed as a Gaussian, a convenient, simple, and useful alternative is overlaying a histogram of the data with a Gaussian centered at $\bar{x}$ and half-width $\sigma_{\bar{x}}$ to see if they look alike. We illustrate this in Fig. 3.4 for the cases of $N_{\text{blocks}} = 32$ and 64. In both cases, the agreement is actually satisfactory, even though the "fit" might look poor. In each case we have 20 bins in our histogram. Throwing 32 or 64 values into these bins produces histograms with bin counts subject to large fluctuations relative to the number of data in a bin.[2]

Deviations from Gaussian behavior tend to appear as a histogram leaning too left or too right. This is called *skewness*. Or the histogram appears too narrow or too squat. This is called *kurtosis*. Multiple peaks can also occur. A Gaussian has a skewness and a kurtosis of 0.[3] As a supplementary test quantitative measures of both quantities,

$$\text{skewness} = \frac{\sum_{i=1}^{N_{\text{blocks}}} (\bar{x}_i - \bar{x})^3}{(N_{\text{blocks}} - 1)\, s_{\bar{X}}^3}, \quad \text{kurtosis} = \frac{\sum_{i=1}^{N_{\text{blocks}}} (\bar{x}_i - \bar{x})^4}{(N_{\text{blocks}} - 1)\, s_{\bar{X}}^4} - 3$$

---

[2] We constructed our histograms in the following manner: A Gaussian, centered around 0 with a variance of 1, has an effective width of 6 ranging from -3 to 3 along the x-axis. For the case of data with an arbitrary mean and variance, we chose our bin width Δ to be $\Delta = 6\sigma_{\bar{x}}^2/20$ and positioned our bins so that one was centered at $\bar{x}$. Instead of a normalized Gaussian $G(\bar{x}, \sigma_{\bar{x}})$, we plotted $N_{\text{blocks}} G(\bar{x}, \sigma_{\bar{x}})$, that is, the expected count number in the interval $(x, x + \Delta)$.

[3] The mathematical definition of kurtosis yields 3 for a Gaussian. For convenience, one often subtracts 3 from the definition, so that Gaussian behavior corresponds to both the skewness and kurtosis approaching 0.

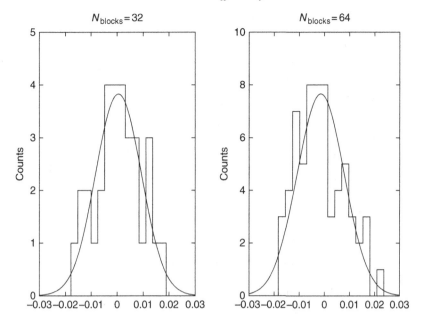

Figure 3.4 Histograms and expected Gaussian counts. The simulation method, model, and its parameters are the same as in Fig. 3.2.

are easily computed for comparison to the Gaussian values. Still another test is a goodness-to-fit test between the Gaussian and the histogram.

A *control chart* is a simple, quick, visual check of the consistency of the proposed values of equilibration steps, block size, and sweeps. Plotted in Fig. 3.5 are the block averages of the energy as a function of block number. The solid horizontal line is the average magnetization per site; the dashed horizontal lines lie one and two sigmas above and below it. We used the proverbial 32 blocks. If we underestimated the number of equilibration steps badly, we would see a drift in the data. This appears not to be the case. If the number of sweeps and the block size are about right, we expect to see about 21 (0.65 times 32) markers between $\bar{x} - \sigma_{\bar{x}}$ and $\bar{x} + \sigma_{\bar{x}}$, about 30 or 31 (0.95 times 32), between $\bar{x} - 2\sigma_{\bar{x}}$ and $\bar{x} + 2\sigma_{\bar{x}}$, and hence maybe 1 or 2 outside this range. Our control chart is consistent with these expectations. Because these confidence intervals are statistically based, fluctuations might produce deviations from the expected values. This type of chart provides a good test for whether the data require a closer analysis.

It is also useful to count in this chart the number of "runs up" and "runs down," that is, the number of successive values that increase or decrease. If the data were Gaussian distributed, the result would look like "white noise" with small runs up and down. If the runs are consistently large, then the block averages may be correlated and further analysis is warranted.

Data analysis

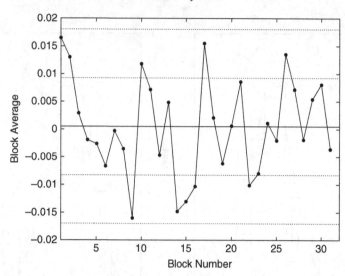

Figure 3.5 Control chart for the $N_{\text{blocks}} = 32$ case of Fig. 3.4.

While the data analysis discussed above provides an estimate of the error, a given application might require the error bars to be small. To make them smaller we generate more blocks. The error decreases as $1/\sqrt{N_{\text{blocks}}}$, while the computation time increases proportional to N_{blocks}.

The field of statistics has numerous additional tests and theorems that are sometimes helpful for troublesome simulations. They lie outside the scope of this book. The methods discussed in this chapter are generally sufficient for analyzing Monte Carlo data and stating with confidence the estimated means and their statistical errors. There is another issue that should be kept in mind: Unless the simulation is properly equilibrated and the sampling is ergodic, these estimates might be meaningless. As noted in Section 2.4.2, a simulation can get stuck in a limited region of phase space. The mean values and errors of the data sampled then reflect only this part of phase space and might be completely unrelated to the values obtained if the sampling visited the entire phase with the proper frequency.

We began this chapter by discussing the equilibration of the sampling, noting certain "sticking" mechanisms, which might affect the simulations of both first-order and continuous phase transitions. We also illustrated (Fig. 3.2) that two different observables generally equilibrate at different rates. Somewhat in passing, we noted the need to experiment with the simulation to test the sensitivity of the results to changes in initial configurations, their symmetries, etc. These studies and others are just as important for the uncertainty quantification of the simulation as the relatively routine statistical analysis just discussed. Different problems generally require different types of experimentation. There are no sure-fire procedures to ensure equilibration and ergodic sampling. The experience of the researcher and the use of effective algorithms are important factors.

3.6 Error propagation

Sometimes we need to estimate the variance of a function of two or more random variables. An example of such a quantity of interest is the *Binder ratio*, defined as

$$B_X = \frac{\langle X^4 \rangle}{\langle X^2 \rangle^2}. \tag{3.22}$$

With the random variable X being the magnetization, this ratio is useful in the finite-size scaling analysis of phase boundaries of the Ising model and other spin models. Because the values of the numerator and denominator are computed with the same $\{x_i = X(C_i)\}$, they are statistically correlated. In computing the error associated with B_X, how do we account for these correlations?

In making multivariate variance estimates, several generalizations of the concept of variance are useful. The *covariance* of two random variables X and Y is

$$\text{cov}(X, Y) = \langle (x - \langle x \rangle)(y - \langle y \rangle) \rangle, \tag{3.23}$$

and the *correlation coefficient* of these two variables is

$$\rho(X, Y) = \frac{\text{cov}(X, Y)}{\sigma_X \sigma_Y}.$$

The numerical estimate of $\text{cov}(X, Y)$ is

$$\text{cov}(X, Y) \approx \frac{1}{M - 1} \sum_{i=1}^{M} (x_i - \bar{x})(y_i - \bar{y}), \tag{3.24}$$

and the numerical estimate of $\rho(X, Y)$ is

$$\rho(X, Y) \approx \frac{\sum_{i=1}^{M} (x_i - \bar{x})(y_i - \bar{y})}{\sqrt{\sum_{i=1}^{M} (x_i - \bar{x})^2 \sum_{j=1}^{M} (y_j - \bar{y})^2}}. \tag{3.25}$$

The correlation coefficient satisfies $|\rho(X, Y)| \leq 1$.

Let us first discuss how to estimate $g(\langle x \rangle)$ for an arbitrary function g, and the associated error. One possibility to estimate $g(\langle x \rangle)$ is to compute

$$\overline{g(x)} \equiv \frac{1}{M} \sum_{i=1}^{M} g(x_i), \tag{3.26}$$

and another is

$$g(\bar{x}) \equiv g\left(\frac{1}{M}\sum_{i=1}^{M}x_i\right). \tag{3.27}$$

When M is large and g is nonlinear, the latter estimator is better. Expanding (3.26) with respect to $\Delta x \equiv x - \langle x \rangle$ and then taking the expectation value generates

$$\left\langle \overline{g(x)} \right\rangle = \left\langle g(\langle x \rangle) + g'|_{\langle x \rangle}\Delta x + \frac{1}{2}g''|_{\langle x \rangle}(\Delta x)^2 + \cdots \right\rangle$$

$$= g(\langle x \rangle) + \frac{1}{2}g''|_{\langle x \rangle}\mathrm{var}(X) + \cdots.$$

Similarly, by expanding (3.27) with respect to $\Delta\bar{x} \equiv \bar{x} - \langle x \rangle$, we obtain

$$\langle g(\bar{x}) \rangle = \left\langle g(\langle x \rangle) + g'|_{\langle x \rangle}\Delta\bar{x} + \frac{1}{2}g''|_{\langle x \rangle}(\Delta\bar{x})^2 + \cdots \right\rangle$$

$$= g(\langle x \rangle) + \frac{1}{2}g''|_{\langle x \rangle}\mathrm{var}(\bar{X}) + \cdots, \tag{3.28}$$

where in the second term we have $\mathrm{var}(\bar{X}) = \mathrm{var}(X)/M$. We note that neither (3.26) nor (3.27) is an unbiased estimator for $g(\langle x \rangle)$. An estimator is *unbiased* if its expected value equals its true value. The above estimators have a systematic error proportional to the curvature of the function $g(x)$ at $x = \langle x \rangle$. However, for the second estimator (3.27), the systematic error is proportional to $1/M$. This bias is negligible because it is much smaller than the statistical error, which is proportional to $1/\sqrt{M}$.

In fact, let us calculate the statistical error of the estimator (3.27). The expectation value of the squared error is

$$\left\langle (\Delta g(\bar{x}))^2 \right\rangle = \left\langle (g(\bar{x}) - g(\langle x \rangle))^2 \right\rangle \approx \left\langle \left(g'|_{\bar{x}}\Delta\bar{x}\right)^2 \right\rangle = \left(g'|_{\bar{x}}\right)^2 \mathrm{var}(\bar{X}). \tag{3.29}$$

The numerical variance estimate is given by (3.18) and (3.19), and we evaluate g' at $\bar{x}$ instead of $\langle x \rangle$. This approximation should have a small effect on the estimate of the error.

Similarly, in the case of a bivariate function, we use the estimate $g(\langle x \rangle, \langle y \rangle) \approx g(\bar{x}, \bar{y})$ and compute the variance as

$$\left\langle (\Delta g(\bar{x}, \bar{y}))^2 \right\rangle = \left\langle (g(\bar{x}, \bar{y}) - g(\langle x \rangle, \langle y \rangle))^2 \right\rangle$$

$$\approx \left\langle \left(\partial_x g|_{\langle x \rangle, \langle y \rangle}\Delta\bar{x} + \partial_y g|_{\langle x \rangle, \langle y \rangle}\Delta\bar{y}\right)^2 \right\rangle$$

$$\approx \left(\partial_x g|_{\bar{x}, \bar{y}}\right)^2 \mathrm{var}(\bar{X}) + \left(\partial_y g|_{\bar{x}, \bar{y}}\right)^2 \mathrm{var}(\bar{Y}) + 2\partial_x g|_{\bar{x}, \bar{y}}\,\partial_y g|_{\bar{x}, \bar{y}}\,\mathrm{cov}(\bar{X}, \bar{Y}),$$

where again we evaluate the derivatives at $x = \bar{x}$ and $y = \bar{y}$. For example, if $g(\langle x \rangle, \langle y \rangle) = \langle x \rangle / \langle y \rangle$, we find

$$\frac{\langle (\Delta(\bar{x}/\bar{y}))^2 \rangle}{(\bar{x}/\bar{y})^2} \approx \frac{\text{var}(\bar{X})}{\bar{x}^2} + \frac{\text{var}(\bar{Y})}{\bar{y}^2} - 2\frac{\text{cov}(\bar{X}, \bar{Y})}{\bar{x}\bar{y}}. \tag{3.30}$$

This particular expression is useful for error estimations, for example, when the simulation has a small to moderate sign problem (Section 5.4). In this case, the numerator is the sign-weighted average and the denominator is the average sign.

3.7 Jackknife analysis

It can be tedious to compute the whole correlation matrix and derivatives of a function to estimate the function of a mean and its variance. The jackknife method is a data resampling method which often provides a good alternative. Its use is simple in the sense that the estimates are generalizations of the standard definitions of a sample mean (3.4) and a sample variance (3.8).

Let us define the "deleted average" $x_{[i]}$ as the sample average in which we ignore the value x_i,

$$x_{[i]} = \frac{1}{M - 1} \sum_{j \neq i} x_j = \frac{M\bar{x} - x_i}{M - 1}.$$

It is a simple matter to show that the sample average of the deleted averages $x_{[i]}$ equals the sample average $\bar{x}$,

$$\overline{x_{[\,]}} = \bar{x}.$$

It is also a simple matter to show that

$$s_{[\,]}^2 = \frac{M - 1}{M} \sum_{i=1}^{M} \left(x_{[i]} - \overline{x_{[\,]}} \right)^2$$

equals the sample variance (3.8).

In practice, the jackknife method is applied to the block averages that we introduced in the discussion of the blocking analysis. The procedure is as follows: We split the M measurements, which should be large, into N_{blocks} blocks of length m, with m larger than the autocorrelation time τ. Then we compute the averages

$$x_{[i]} = \frac{1}{N_{\text{blocks}} - 1} \sum_{j \neq i} \bar{x}_j, \quad i = 1, \ldots, N_{\text{blocks}}, \tag{3.31}$$

where we use all the blocks, *except* block number i, and $\bar{x}_j$ denotes the average over block j. We next define $g_{[i]} = g(x_{[i]})$. The simple jackknife estimate of $g(\langle x \rangle)$ is the average $\overline{g_{[\,]}}$ of these $g_{[i]}$:

$$g(\langle x \rangle) \approx \overline{g_{[]}} \equiv \frac{1}{N_{\text{blocks}}} \sum_{i=1}^{N_{\text{blocks}}} g_{[i]}. \qquad (3.32)$$

Using (3.19) and (3.18), we find that the expectation value of the Taylor series expansion of

$$g_{[i]} = g\left(\frac{1}{N_{\text{blocks}} - 1} \sum_{j \neq i} \bar{x}_j \right) = g\left(\bar{x} + \frac{1}{N_{\text{blocks}} - 1} (\bar{x} - \bar{x}_i) \right)$$

equals

$$\langle g_{[i]} \rangle = \langle g(\bar{x}) \rangle + \frac{1}{2(N_{\text{blocks}} - 1)} g''|_{\bar{x}} \frac{\text{var}(\bar{X}_{\text{block}})}{N_{\text{blocks}}} + \cdots$$

$$= \langle g(\bar{x}) \rangle + \frac{1}{2(N_{\text{blocks}} - 1)} g''|_{\bar{x}} \, \text{var}(\bar{X}) + \cdots . \qquad (3.33)$$

This equation says that the bias in the jackknife estimate is reduced compared with that in (3.28) by a factor $1/(N_{\text{blocks}} - 1)$. By combining (3.28) and (3.33), we obtain an estimator without g''-bias:[4]

$$g(\langle x \rangle) \approx N_{\text{blocks}} g(\bar{x}) - (N_{\text{blocks}} - 1)\overline{g_{[]}}. \qquad (3.34)$$

We now show that the error estimate on $g(\langle x \rangle)$ is

$$\langle (\Delta g(\bar{x}))^2 \rangle \approx \frac{N_{\text{blocks}} - 1}{N_{\text{blocks}}} \sum_{i=1}^{N_{\text{blocks}}} (g_{[i]} - g(\bar{x}))^2. \qquad (3.35)$$

Since $g_{[i]} - g(\bar{x}) \approx g'|_{\bar{x}}(\bar{x} - \bar{x}_i)/(N_{\text{blocks}} - 1)$, it follows from (3.18) and (3.19) that

$$\frac{N_{\text{blocks}} - 1}{N_{\text{blocks}}} \sum_{i=1}^{N_{\text{blocks}}} (g_{[i]} - g(\bar{x}))^2 \approx \frac{(g'|_{\bar{x}})^2}{N_{\text{blocks}}(N_{\text{blocks}} - 1)} \sum_{i=1}^{N_{\text{blocks}}} (\bar{x} - \bar{x}_i)^2$$

$$= (g'|_{\bar{x}})^2 \frac{\text{var}(\bar{X}_{\text{blocks}})}{N_{\text{blocks}}} = (g'|_{\bar{x}})^2 \, \text{var}(\bar{X}),$$

which is consistent with (3.29).

The jackknife procedure is easily extended to arbitrary functions f of one or several observables, such as (3.22). If $f(\{\bar{x}_j\})$ is the estimate based on the expectation

---

[4] The expectation value of the right-hand side is $N_{\text{blocks}} \langle g(\bar{x}) \rangle - (N_{\text{blocks}} - 1) \langle \overline{g_{[]}} \rangle = \langle g(\bar{x}) \rangle - \frac{1}{2} g''|_{\bar{x}} \, \text{var}(\bar{X}) + \cdots =$
$g(\langle x \rangle) + \frac{1}{2} g''|_{\bar{x}} \, \text{var}(\bar{X}) - \frac{1}{2} g''|_{\bar{x}} \, \text{var}(\bar{X}) + \cdots .$

values of the observables in the block ensemble $\{\bar{x}_j\}$, we define $f_{[i]} = f(\{\bar{x}_{j\neq i}\})$. The jackknife estimate of the mean and variance is then

$$\overline{f_{[]}} = \frac{1}{N_{\text{blocks}}} \sum_{i=1}^{N_{\text{blocks}}} f_{[i]}, \tag{3.36}$$

$$\langle (\Delta f)^2 \rangle = \frac{N_{\text{blocks}} - 1}{N_{\text{blocks}}} \sum_{i=1}^{N_{\text{blocks}}} (f_{[i]} - \overline{f_{[]}})^2. \tag{3.37}$$

Let us close this section with some remarks about the limitations of the jackknife method. By rewriting the definition (3.31) of the deleted averages as $x_{[i]} = \bar{x} + \frac{1}{N_{\text{blocks}}-1}(\bar{x} - \bar{x}_i)$, it becomes obvious that the members of the jackknife ensemble are distributed in a small sample space, centered around the average of the original ensemble, whose linear size is $N_{\text{blocks}} - 1$ times smaller than that of the original ensemble. Therefore, the linear size of this space is approximately $\sqrt{N_{\text{blocks}}}$ times smaller than the statistical error, which is $\mathcal{O}(1/\sqrt{N_{\text{blocks}}})$. As a result, we need to multiply the variance of the jackknife ensemble in (3.35) by a factor of order N_{blocks} to obtain the estimator for the squared statistical error in $g(\bar{x})$. While the jackknife ensemble picks up the local curvature of g correctly and works well when the gradient of the function does not change much in the region of size $1/\sqrt{N_{\text{blocks}}}$ centered around the correct expectation value, it may fail when the nonlinearity of the function g is strong (Fig. 3.6). In the latter case, the small sample space covered by the jackknife ensemble may not include the correct expectation value $\langle x \rangle$.

3.8 Bootstrap analysis

We can overcome the limitations of the jackknife analysis by employing a resampling method that mimics the broader distribution of the sample average $\bar{x}$. The *bootstrap method* is a popular choice for this purpose. A member of this ensemble, which we denote by x_i' ($i = 1, 2, \ldots, N_{\text{bootstrap}}$), is generated by randomly choosing N_{blocks} samples $\bar{x}_j$ from the original ensemble of block averages, without trying to avoid picking the same one multiple times. As a result, some members of the original ensemble are not chosen at all, while some are chosen once, some twice, and so on. Then x_i' is defined as the average of these N_{blocks} values. Thus, formally, $x_i' = \frac{1}{N_{\text{blocks}}} \sum_{k=1}^{N_{\text{blocks}}} \bar{x}_{\rho(i,k)}$, where $\rho(i,k) = 1, 2, \ldots, N_{\text{blocks}}$ is a uniform random integer such that $\rho(i,k)$ and $\rho(j,k')$ are statistically independent if $(i,k) \neq (j,k')$.

Because of this averaging, the standard deviation of the bootstrap ensemble is of the same order as the standard deviation of the original ensemble multiplied by $1/\sqrt{N_{\text{blocks}}}$, i.e., $\text{var}(\{x_i'\}) \sim \text{var}(\{\bar{x}_i\})/N_{\text{blocks}}$. Hence, the variance of the bootstrap

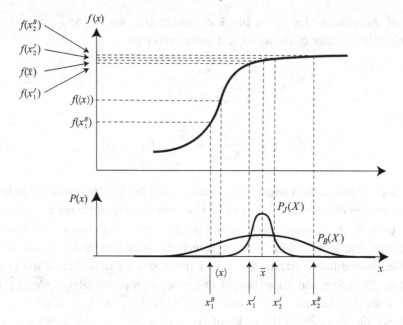

Figure 3.6　Estimation of a strongly nonlinear function by jackknife and bootstrap analyses. The bootstrap samples are distributed, roughly, over the region $x_1^B < x < x_2^B$. As a result, the confidence interval for $f^* \equiv f(\langle x \rangle)$ is $f(x_1^B) < f^* < f(x_2^B)$. The jackknife samples are distributed over a much narrower interval $x_1^J < x < x_2^J$. The confidence interval for f^* of the jackknife analysis is then $f(x_1^J) < f^* < f(x_2^J)$ expanded by the factor $\sqrt{N_{\text{blocks}}}$, which is about 4 in this example. Obviously, the factor 4 is not sufficient to cover the correct value of f^*.

ensemble is comparable to the variance of the sample average: $\text{var}(\{x_i'\}) \sim \text{var}(\bar{x})$. As a result, the correct value $\langle x \rangle$ should be contained with high probability in the sample space covered by the bootstrap ensemble, and the correct value of $g(\langle x \rangle)$ should fall within the image of this region. We illustrate this in Fig. 3.6.

In case the non-linearity is not so strong, a little more precise ... arithmetic yields that the estimate of the statistical error in $g(\bar{x})$ is obtained by the variance of the bootstrap samples as

$$\sigma^2 \equiv \left\langle (g(\bar{x}) - g(\langle x \rangle))^2 \right\rangle \approx \frac{N_{\text{blocks}}}{N_{\text{blocks}} - 1} \left\langle \left(g(x_i') - g(\bar{x}) \right)^2 \right\rangle$$

when $N_{\text{bootstrap}} \gg 1$. In contrast to the jackknife analysis, this estimator should produce a sensible result even in the case where the nonlinearity of the function g is strong.

3.9 Monte Carlo computer program

Algorithm 5 shows the core structure of a Monte Carlo program, which is very simple. This structure assumes that the data analysis is a post-simulation process. Besides the continuing dramatic increases in processor speeds, the cost and access speeds of external memory have also improved dramatically over the past few years, making this separation of tasks feasible and advantageous. Not too many years ago it would not always have been possible. Some of the methods mentioned for the analysis of the data, such as using a control chart and the jackknife method, in fact require a post-simulation analysis.

Algorithm 5 requires as input the parameters of the physical model. It first defines or reads the simulation parameters, initializes arrays and variables needed by the program, defines the initial Monte Carlo configuration, and perhaps even sets the seed for the random number generator. The most important building block of the program is a routine that performs an update of the entire configuration. For example, in a simulation of the classical Ising model based on local updates, this function would loop over all lattice sites and propose an update for each spin. We call this a "Monte Carlo sweep." The second most important procedure is the one that computes observables, such as the energy or the magnetization, for a given Monte Carlo configuration. Which quantities are computed depends on the problem. As the measurements are made, they are accumulated in an appropriate data container. This data collection is the result of the simulation and is saved at the end of the simulation. Depending on the amount of data and the time needed to save the data relative to the execution cost of a sweep, the data might instead

---

Algorithm 5 Structure of a Monte Carlo program.

Input: Model and control parameters (N_{equil} = number of equilibration sweeps, N_{sweep} = number of sweeps, N_{skip} = number of updates between measurements), plus accumulators for storing measured quantities.

Initialize simulation ;

for $i = 1$ to N_{equil} **do**

    Perform Monte Carlo sweep ;

end for

for $i = 1$ to N_{sweep} **do**

    Perform a Monte Carlo sweep ;

    if i is multiple of N_{skip} **then**

        Perform measurements and accumulate the results ;

    end if

end for

return the accumulators.

---

be written immediately after the measurement, thus eliminating the final output step. To avoid a bias from the choice of the initial configuration, the first N_{equil} sweeps (with N_{equil} larger than the thermalization time) are not measured. N_{sweep} is the number of sweeps performed after equilibration. The cost of computing the measurements relative to the cost of a sweep often makes it advisable to measure and save the measurements from only every N_{skip} configurations, with N_{skip} of the order of the shortest autocorrelation time.

For post-simulation data processing, another computer program loads the data from the disk and computes means and variances. For the blocking analysis, a loop over the block sizes $m_l = 2^l$ is performed, and the averages of the measured observables are computed for each block. The variance of these block-averaged measurements divided by the number of blocks yields the square of the error estimate for iteration l. The increase of the error with increasing l is monitored, and if saturation occurs, the saturated value gives a reliable estimate of the square of a one-sigma error bar. If no saturation occurs up to the largest possible block size (which is about two orders of magnitude smaller than the length of the data stream), then a reliable error estimate is not possible. In this case, much more simulation data are needed.

Some simulations require the computation of a lot of observables. While locating the phase boundary of the Ising model requires the computation of just a few simple quantities, such as the energy and the order parameter, the discovery of a proposed novel phase in a more exotic model might require the computation of the spatial dependence of multiple correlation functions for a wide range of model and control parameters. In such cases, the storage cost may become an issue, and the data file management requires pre-simulation thought.

Different people have different preferences about doing all or at least part of the data analysis on the fly instead of doing it all after the simulation. On a practical note, outputting a stream of on-the-fly block averages of select measurements is a convenient real-time indicator of the state of the simulation. Having such indicators is especially helpful in the code development phase, that is, in the development of the procedure that updates the configurations. In the code development phase, the ability to rerun a simulation with the same sequence of random numbers is also essential. Even if all the measurements are done on the fly, it is advisable to output to external storage a few streams of select measurements for inspection in case something weird appears to be happening. As a stochastic process, a Monte Carlo simulation is susceptible to rare events occurring. Having these streams of information on hand to help distinguish among rare events, coding errors, and algorithm instabilities might help to avoid rerunning the simulation (using the same random number seed) to replicate the problem and might help to resolve the issue more rapidly.

Suggested reading

M. H. Kalos and P. A. Whitlock, *Monte Carlo Methods*, vol. 1: *The Basics* (New York: Wiley-Interscience, 1986), chapter 2.

A. Papoulis, *Probability and Statistics* (Englewood Cliffs, NJ: Prentice Hall, 1990), chapters 4, 7, and 9.

W. H. Press, S. A. Teukolsky, W. T. Vettering, and B. P. Flannery, *Numerical Recipes* (Cambridge University Press, 1992), chapter 14.

Exercises

3.1 Write a short computer program to simulate the two-dimensional Ising model in a magnetic field, using either the heat-bath or Metropolis algorithm. Examples of FORTRAN and C programs of varying degrees of complexity can be found by searching the Web or by looking in Parisi (1988), Landau and Binder (2000), or Newman and Barkema (1999). Add the computation of the specific heat to the set of measurements.

3.2 For a 32×32 lattice, repeat the analysis in Figs. 3.2 through 3.5.

3.3 Repeat Exercise 3.2 for a series of temperatures approaching T_c.

3.4 Repeat Exercise 3.2 for a temperature equal to T_c for several large lattice sizes.

3.5 Repeat Exercise 3.2 for a series of magnetic fields increasing from zero.

3.6 If var(X) is the function that returns the variance (3.2) of the random variable X and if cov(X, Y) returns the covariance (3.23) of the random variables X and Y, prove that $\text{cov}(X, Y)^2 \leq \text{var}(X)\text{var}(Y)$ and hence that $-1 \leq \rho(X, Y) \leq 1$.

3.7 The *conditional mean* of a function $g(x, y)$ with respect to the variable x relative to a distribution $p(x, y)$ is

$$E[g(x, y)|x] = \int g(x, y)p(x, y)dy.$$

Show

1. $E[g(x, y)|x] = \int g(x, y)p(x, y)dy.$
2. $E[E[g(x, y)|x]] = E[g(x, y)].$
3. $E[g_1(x)g_2(y)|x] = g_1(x)E[g_2(y)|x].$
4. $E[g_1(x)g_2(y)] = E[g_1(x)E[g_2(y)|x]].$

3.8 A *conditional variance* is defined as

$$\text{var}[g(x, y)|x] = E\left[(g(x, y) - E[g(x, y)|x])^2\right]$$
$$= E\left[g^2(x, y)|x\right] - E^2[g(x, y)|x].$$

Show that

$$E\left[\text{var}\left[g(x, y|x)\right]\right] + \text{var}\left[E\left[g(x, y)\,|x\right]\right]$$
$$= E\left[g^2(x, y)\right] - E^2\left[g(x, y)\right] = \text{var}\left[g(x, y)\right].$$

3.9   For a target distribution $p(z)$, the Monte Carlo method estimates $E[g(z)]$ via

$$I_1 = \frac{1}{M} \sum_{i=1}^{M} g\,(z_i),$$

where the z_i are samples drawn for $p(z)$. If we split up z as (x, y) and assume we know or can calculate easily $E[g(x, y)|x]$, then an alternative estimate of $E[g(z)]$ is

$$I_2 = \frac{1}{M} \sum_{i=1}^{M} E\left[g(z)|x_i\right].$$

Use the results of the two previous exercises to argue that

1. Both estimates I_1 and I_2 are unbiased and give the same expectation value.
2. If the computational costs are comparable, then the second estimate I_2 is preferred because its variance is potentially smaller.

These results illustrate a Monte Carlo rule of thumb: If part of the sampling is replaced by an exact result, the variance is often reduced. Numerous variance reduction methods are based on this rule.

3.10 If $\{f_i(X_j)\}$ is a set of functions of a set $\{X_j\}$ of random variables X_j, show that

$$\text{var}(f_i) = \sum_{j} \left(\frac{\partial f_i}{\partial x_j}\right)^2 \text{var}(X_j)$$

$$+ \sum_{j} \sum_{j \neq k} \left(\frac{\partial f_i}{\partial x_j}\right) \left(\frac{\partial f_i}{\partial x_k}\right) \text{cov}(X_j, X_k),$$

$$\text{cov}(f_i, f_j) = \sum_{kl} \left(\frac{\partial f_i}{\partial x_k}\right) \left(\frac{\partial f_j}{\partial x_k}\right) \text{var}(X_k)$$

$$+ \sum_{k} \sum_{k \neq l} \left(\frac{\partial f_i}{\partial x_k}\right) \left(\frac{\partial f_j}{\partial x_l}\right) \text{cov}(X_k, X_l).$$

The functions $\text{var}(X)$ and $\text{cov}(X, Y)$ are defined in Exercise 3.6.

3.11 In a one-dimensional random walk along a line, the walker at the i-th position makes a displacement of $X_i = \pm d$ from its current position with equal probability. After n steps, its displacement is $X(n) = X_1 + X_2 + X_n = \cdots + X_n$.

1. Show that $\langle X(n) \rangle = 0$ and $\langle X(n)^2 \rangle = nd^d$.
2. Find $\text{cov}(X(n), X(m))$ and $\rho(X(n), Y(n))$.
3. Show that the $X(m)$ and $X(n)$ become completely correlated as $m/n \to 1$ and completely uncorrelated as $m/n \to 0$.

4

Monte Carlo for classical many-body problems

Our discussion of Markov chains, with the exception of mentioning the Metropolis and heat-bath algorithms, has so far been very general with little contact with issues and opportunities related to specific applications. In this chapter, we recall that our target is many-body problems defined on a lattice and introduce several frameworks exploiting what is special about Markov processes for these types of problems. We consider here classical many-body problems, using the Ising model as the representative. Our discussion will be extended to various quantum many-body problems and algorithms in subsequent chapters.

4.1 Many-body phase space

The numerical difficulty of studying many-body problems on a lattice arises from the fact that their phase space Ω is a direct product of many phase spaces of local degrees of freedom. Generally, a local phase space is associated with each lattice site. If n is the size of this phase space and N is the number of lattice sites, then the number of states $|\Omega|$ available to the whole system is n^N. In other words, the number of states in the phase space grows exponentially fast with the physical size of the system. For example, in the Ising model, the Ising spin s_i on each site can take one of the two values ± 1, and hence the number of states in the total phase space is 2^N.

In Chapter 1, we noted that this exponential scaling thwarts deterministic solutions and is a reason why we use the Monte Carlo method. The exponentially large number of states implies that the enumeration of all states, which requires a computational effort proportional to the size of the phase space, is not an option for a problem with a large number of sites. As discussed in Section 2.7, the Monte Carlo method generally samples from a compressed phase space, avoiding the need to solve the entire problem. However, even with this advantage, can we reduce the computational effort to a manageable level?

An obvious requirement is that we can equilibrate the simulation in a reasonable amount of computer time. We know that the Markov process converges to a stationary state sampling some distribution (Section 2.4), and for detailed balance algorithms, Rosenbluth's theorem (Section 2.6) guarantees monotonic convergence. However, as we noted before, these theorems do not tell us how rapid the convergence is. While we know that the convergence rate is controlled by the value of the second largest eigenvalue λ_2 of the transition probability matrix, in general, we do not know this eigenvalue.

We do know from the properties of stochastic matrices that $|\lambda_2| < 1$. The difference between it and unity is a finite margin of the order $\mathcal{O}(|\Omega|^{-1})$ or larger (Section 2.4.2). This margin characterizes the slowest relaxation mode and sets an upper bound to the computational auto-correlation time discussed in Sections 3.3 and 3.4,

$$\tau_{\text{comp}} = \frac{-1}{\log|\lambda_2|} \le \mathcal{O}(|\Omega|),$$

which implies that importance sampling, as embodied, for example, in the Metropolis and heat-bath algorithms, is at least as good as random sampling. However, if the upper bound is fulfilled, the simple enumeration of all states is just as fast, and it has the advantage of being exact.

Fortunately, in many important many-body simulations, τ_{comp} is of the order of unity, independent of the size of the system. Even in the vicinity of a critical point, where Monte Carlo simulations usually have longer relaxation times than in noncritical cases, τ_{comp} depends on the system size polynomially rather than exponentially. Hence, Monte Carlo simulation has become a standard technique for solving many-body problems.

But why can τ_{comp} be independent of the system size? While a mathematically rigorous answer is beyond the scope of this book, we can obtain a rough idea why by reminding ourselves that the transition probability matrix of a many-body problem is not an arbitrary Markov matrix. To see what we mean by this last statement, let us divide the physical system into many identical subsystems, the size of each being independent of the size of the whole system, and also assume that they are large enough that their properties are approximately independent of each other. Then, the distribution function of the whole system is the product of the distribution functions of the subsystems, and the relaxation rate of the whole system is therefore controlled by the relaxation rates of the subsystems. Since by assumption all subsystems are identical, the Markov chain auto-correlation time of the whole system is that of any one of the subsystems with the latter being independent of system size. This situation applies to most Monte Carlo simulations of many-body problems. There usually exists a characteristic length scale beyond which the spatial correlations do not extend. If we are simulating continuous phase transitions, the divergence of the correlation length as we approach a critical point is the reason why the auto-correlation time becomes unusually large in many Monte Carlo simulations. The cause is the physics of the problem and not per se the physical size of the many-body phase space.

4.2 Local updates

We adopt the Ising model as a working example and explore three different Monte Carlo algorithms for simulating its properties. As we remarked in Chapter 1, we can use Monte Carlo simulations to compute both ground state and finite-temperature properties. As we also remarked, for the Ising model the ground state properties are not as interesting as the finite-temperature ones. We defer the discussion of ground state Monte Carlo methods to other chapters (Chapters 9, 10, and 11). To simulate finite-temperature properties we have to sample from the Boltzmann distribution, which means it is sufficient that the algorithms satisfy the detailed balance condition. The Metropolis algorithm sets the standard here. We start by revisiting this method and discuss what is general about it as opposed to dwelling on its specific use for the Ising model.

Our goal is sampling from a distribution whose partition function[1]

$$Z = \sum_{C \in \Omega} W(C) \qquad (4.1)$$

is defined by the non negative weight

$$W(C) = \prod_{\langle ij \rangle} w(s_i, s_j), \qquad w(s_i, s_j) = e^{K s_i s_j}, \qquad (4.2)$$

where the product is over all pairs of neighboring sites and $K \equiv J/kT$. We seek a procedure for updating a configuration C to another configuration C', where $C \equiv (s_1, s_2, \ldots, s_N)$ denotes a configuration of Ising spin variables for a lattice of N sites.

A simple way of accomplishing this sampling is to use the Metropolis algorithm with single spin-flip updates. It proposes and then accepts or rejects the flipping of the spin on a given site, one site at a time. Its Markov chain transition matrix $P(C'|C)$, which satisfies the detailed balance condition, is a product of a proposal and an acceptance probability. To be specific, we pick a random site with probability $1/N$ and propose to flip the spin on that site. If ΔE is the change in total energy associated with this spin flip, then the spin flip is accepted with a probability $\min(1, \exp(-\Delta E/kT))$. Otherwise we keep (and measure) the old configuration. This simple algorithm is sketched in Algorithm 6 and illustrated in Fig. 4.1.

---

[1] The goal is computing the derivatives of $\log Z$ with respect to various control parameters, not the calculation of Z per se.

Algorithm 6 Single spin-flip Metropolis algorithm for the Ising model.

Input: A configuration of N spins.

   Randomly select one of the N spins ;

   Compute the energy change ΔE associated with the spin flip ;

   Flip the spin with probability $\min(1, \exp(-\Delta E/kT))$;

   return the updated configuration.

Figure 4.1 A cycle in the single spin-flip update of the Ising model.

4.3 Two-step selection

The single spin-flip Metropolis algorithm can be viewed as a two-step sampling process with the following steps:

1. Select a given spin with some probability that is independent of the current configuration,
2. Select the new state of this spin with the appropriate probability.

The transition matrix for this procedure has the form

$$P(C'|C) = \sum_{i=1}^{N} p_i q_i(C'|C), \tag{4.3}$$

where p_i is the probability that the spin s_i is chosen in the first step, and q_i is the probability that the configuration C is changed to C' in the second step. We can show (Exercise 4.1) that if $q_i(C'|C)$ satisfies the detailed balance condition, then so does $P(C'|C)$.[2] Because of this property, we can select the spins deterministically; that is, we can implement a sweep over the lattice sites j, and for the individual updates use $p_i = \delta_{i,j}$, instead of selecting the spins at random ($p_i = 1/N$).

In the two-step procedure, instead of choosing the next state directly from the vast space of all states, we first restrict the set of states from which we select the

---

[2] It is rather obvious that we can generalize the single-spin algorithm to a *single unit algorithm* where a unit u consisting of more than one spin is flipped. More specifically, if we express all degrees of freedom as a sum of units, then in (4.3) we can replace i by u and the summation over i by the summation over all units. Now $q_u(C'|C) = 0$ if C' differs from C on some i outside of u. Although it involves flipping several spins at once, the procedure still has two steps.

final state. More specifically, if we are currently in the state $C = (1, 1, 1, 1, \ldots)$ and select the second spin in this state as the candidate for flipping, then we must choose the next state from a set of two states: C itself and $C' = (1, -1, 1, 1, \ldots)$. Thus, in the selection of the second spin, we are actually selecting from a two-state subset of the phase space. We can define this type of two-state subspace for all initial and final spin configurations. In the example of the single-spin update of the Ising model, every single state is covered by N elements of the family of such subspaces since there are N choices for the spin to be updated.

In an abstract language, for a family $\mathcal{F}$ of subspaces G covering Ω, a single-spin update Markov process consists of the following steps:

1. Select a subspace $G \in \mathcal{F}$ with the probability $P(G|C)$,
2. Select the new state C' out of G with probability $P(C'|C, G)$,

where $P(G|C) = 0$ if C is not in G. While these statements are a rather formal way of viewing the process, they in fact provide a simplifying point of view: Instead of selecting a state out of the whole phase space, we first restrict the subspace and then select a state out of this restricted space. As we will see, we can classify several important algorithms discussed in later chapters as such "two-step selection" algorithms.

We can regard the overall transition matrix of the two steps as the sum of all possible paths between the two states:

$$P(C'|C) = \sum_G P(C'|C, G)P(G|C). \tag{4.4}$$

Therefore, the detailed balance condition holds if it does so for each term in the sum; that is, it holds if

$$P(C'|C, G)P(G|C)W(C) = P(C|C', G)P(G|C')W(C'), \tag{4.5}$$

where $W(C)$ is the Boltzmann weight (equilibrium distribution).

4.4 Cluster updates

We now explore a second example of a two-step selection algorithm, so-called cluster algorithms. When first proposed, they were a marked departure from the Metropolis or heat-bath type algorithms, because they involve the flipping of clusters of spins without rejection. For many years, algorithms that did more than a single-spin update were sought to mitigate the expense often encountered in equilibrating simulations, particularly near a continuous phase transition, but simply defining a unit of local spins generally had minor, if any, benefits. In many applications, cluster algorithms (Swendsen and Wang, 1987) enjoy a much shorter computational auto-correlation time than single-spin algorithms. Sometimes they

---

Algorithm 7 Swendsen-Wang cluster algorithm for the ferromagnetic Ising model.

Input: Current Ising configuration, $K \equiv J/kT$.

    for every pair of interacting spins s_i and s_j **do**

        Place an edge connecting them with probability $\delta_{s_i,s_j}(1 - e^{-2K})$;

    end for

    for every cluster of connected spins in the graph **do**

        With probability $\frac{1}{2}$, flip all spins in the cluster ;

    end for

    return the updated configuration.

---

reduce the sign problem (Chandrasekharan et al., 1999, 2003). Another important feature is that they can be parallelized (Chapter 13), while other methods that reduce critical slowing down often do not yet share this property. While still a two-step algorithm, they gain their advantage by changing the configuration on length scales that are relevant to the physics of the problem, as opposed to updating the degrees of freedom that define the microscopic model.

4.4.1 Swendsen-Wang algorithm

The first cluster algorithm was proposed by Swendsen and Wang (1987) for the Ising model (Algorithm 7). In their algorithm, a graph whose edges connect two parallel spins characterizes the restricted phase space G.[3] A *graph G* is defined by a non null set of *vertices* and a possibly null collection of *edges*. The edges specify undirected relationships between pairs of vertices. The Ising model, for example, is defined on a graph. The vertices are the lattice sites and the edges are the bonds specifying the pairs of sites whose spins interact.

    The first step of the Swendsen-Wang algorithm is the construction of a graph. Here, the vertices of the graph are the lattice sites. The edges are a stochastically constructed subset of the bonds. The second step samples a configuration C from the ones that satisfy the restriction imposed by the graph. We do this sampling by flipping "clusters," where a cluster is a collection of connected spins in the graph.

    With the addition of graph degrees of freedom, our transition probability matrix is now of the form (4.4), thereby transferring our task of proving the detailed balance condition for the whole phase space to the task of showing the condition in the restricted phase space (4.5). To this end, we introduce a new set of variables $\{g_{ij}\}$ representing connection ($g_{ij} = 1$) and disconnection ($g_{ij} = 0$) of lattice sites (presence or absence of edges), and we let G represent the set of all g_{ij}. We relate

---

[3] In what follows, because a graph imposes a restriction on the phase space and therefore defines a subset of phase space, we use the same symbol G for subsets of phase space and for graphs defined on a lattice.

these variables to subsets of the local phase space as follows: $g_{ij} = 1$ is compatible with $(s_i, s_j) = (-1, -1)$ or $(1, 1)$, while $g_{ij} = 0$ is compatible with $(s_i, s_j) = (-1, -1)$, $(-1, 1)$, $(1, -1)$, or $(1, 1)$. To be more specific, we write the probability $P(G|C)$ in (4.5) as

$$P(G|C) = \prod_{\langle ij \rangle} P(g_{ij}|s_i, s_j), \tag{4.6}$$

$$P(1|s_i, s_j) = 1 - P(0|s_i, s_j) = \delta_{s_i, s_j}(1 - e^{-2K}). \tag{4.7}$$

We now justify these choices for the conditional probabilities.

We can view the lattice itself as a graph with an edge between each pair of neighboring lattice sites. At each site (vertex) is an Ising spin variable. Let us overlay on this graph another one that places an edge between neighboring spins only if they are both up or both down. The overlaid graph creates disconnected clusters of aligned spins, and we obtain a new spin configuration by flipping each cluster with probability $\frac{1}{2}$. In the rest of this chapter, we call this overlaid graph simply a graph, and represent it by the symbol G.

If either C or C' is incompatible with G, (4.5) is trivially satisfied since both sides of the equation vanish. If, on the other hand, both states are in G, because of the random coin-flipping nature of the second step of the algorithm, the probability of selecting the final state is always $1/2^{N_{cluster}}$, independent of the initial or final state. Therefore, in (4.5), $P(C'|C, G) = P(C|C', G)$. Thus, the condition that remains to be verified is

$$C, C' \in G \Rightarrow P(G|C)W(C) = P(G|C')W(C').$$

To show this, we use (4.6) and factorize both sides of this equation to obtain

$$\prod_{\langle ij \rangle} P(g_{ij}|s_i, s_j)w(s_i, s_j) = \prod_{\langle ij \rangle} P(g_{ij}|s_i', s_j')w(s_i', s_j'),$$

where $w(s_i, s_j) = e^{Ks_i s_j}$ (see (4.2)). With (4.7), we find by simple inspection that corresponding factors on both sides of the equation are equal, if the pairs of spins are compatible with g_{ij} (constraint imposed by G):

$$(s_i, s_j), (s_i', s_j') \in g_{ij} \Rightarrow P(g_{ij}|s_i, s_j)w(s_i, s_j) = P(g_{ij}|s_i', s_j')w(s_i', s_j'). \tag{4.8}$$

Explicitly, if $g_{ij} = 1$, then $s_i = s_j$ and $s_i' = s_j'$, so that (4.8) is satisfied. If $g_{ij} = 0$, all spin configurations are compatible with the graph and $P(0|s_i, s_j)w(s_i, s_j) = (1 - \delta_{s_i, s_j}(1 - e^{-2K}))e^{Ks_i s_j} = e^{Ks_i s_j} - \delta_{s_i, s_j}(e^K - e^{-K}) = e^{-K}$, and (4.8) is again satisfied. Thus, we have verified that the transition probability (4.6) for the Swendsen-Wang algorithm satisfies (4.5), and therefore proved that the Swendsen-Wang algorithm satisfies detailed balance. We summarize the Swendsen-Wang algorithm in Algorithm 7, and illustrate a cluster updating cycle in Fig. 4.2.

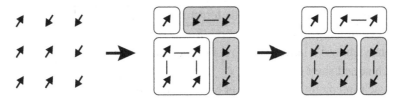

Figure 4.2 A cycle in the Swendsen-Wang cluster update of the Ising model.

To identify the clusters of connected spins, one can use for example the Hoshen-Kopelman clustering algorithm (Hoshen and Kopelman, 1976). It works as follows: For each site, one stores a cluster index and a root index. The cluster index is initially set equal to the site index. The *root index* of a site is determined iteratively by moving to the site stored in the cluster index until the cluster index becomes identical to the site number (which then defines the root index). If two sites are connected by a new bond, their root indices are computed and the cluster indices of the sites, as well as the cluster indices of the root sites, are set to the smaller root index. Once all the bonds have been inserted, the cluster index of each site is set to its root index. The sites with identical cluster indices belong to the same cluster. The algorithm is discussed in more detail in Section 13.4.

There is an even simpler and more efficient version of the Monte Carlo cluster algorithm, called Wolff algorithm (Wolff, 1989). This algorithm builds only a single cluster, starting from a randomly chosen site, and always flips it. A pseudo-code is provided in Algorithm 8. We leave it as an exercise to prove that the Wolff algorithm satisfies detailed balance.

4.4.2 Graphical representation

It is often useful to view the Swendsen-Wang cluster algorithm as a Markov process in an extended phase space where each state is expressed as $\tilde{C} = (C, G)$. We realize this extension by using a graphical decomposition of the Ising model weight originally proposed by Kasteleyn and Fortuin (1969) and Fortuin and Kasteleyn (1972),

$$e^{Ks_i s_j} = e^{-K} + \delta_{s_i,s_j}(e^K - e^{-K}). \tag{4.9}$$

The utility of this very simple formula becomes more apparent when we write the weight (4.2) as

$$w(s_i, s_j) = e^{Ks_i s_j} = e^{-K} + \delta_{s_i,s_j}(e^K - e^{-K}) = \sum_{g_{ij}=0,1} \Delta(s_i, s_j | g_{ij}) v(g_{ij})$$

with

$$v(0) = e^{-K}, \quad v(1) = e^K - e^{-K}, \quad \Delta(s_i, s_j | 0) = 1, \quad \Delta(s_i, s_j | 1) = \delta_{s_i,s_j}.$$

Algorithm 8 Wolff Algorithm for the ferromagnetic Ising model.

Input: Current Ising spin configuration, $K \equiv J/kT$.

    Randomly choose a lattice site i (root of the cluster) ;

    for each added site i **do**

        for each interacting site j not yet on the cluster **do**

            if spins at i and j are parallel **then**

                add j to the cluster with probability $1 - e^{-2K}$;

            end if

        end for

    end for

    Flip all spins in the cluster ;

    return the updated configuration.

Plugging this relation into (4.1) and using (4.2) yields

$$Z = \sum_C \sum_G \Delta(C, G) V(G), \tag{4.10}$$

with $V(G) \equiv \prod_{\langle ij \rangle} v(g_{ij})$ and $\Delta(C, G) \equiv \prod_{\langle ij \rangle} \Delta(s_i, s_j | g_{ij})$ being the generalized Kronecker delta expression that is one if and only if all spins connected by the graph are parallel to each other and zero otherwise. Hence, we have expressed the partition function as $Z = \sum_{C,G} W(C, G)$, with $W(C, G) = \Delta(C, G) V(G)$.

With these definitions, we now state the algorithm as a two-step process characterized by the transition probabilities

$$P_{\text{graph}}(C', G'|C, G) = P_{\text{graph}}(G'|C) = \frac{W(C, G)}{W(C)}$$

for the graph assignment step and by

$$P_{\text{flip}}(C', G'|C, G) = P_{\text{flip}}(C'|G) = \frac{W(C', G)}{W(G)}$$

for the cluster flipping step, with $W(C) = \sum_G W(C, G')$ and $W(G) = \sum_C W(C, G)$.

We also see that taking the partial summation over spin variables in (4.10) is easy, since the factor $\Delta(C, G)$ restricts the states to those generated by random cluster flips. If there are $N_c(G)$ connected clusters in G, then there are only $2^{N_c(G)}$ such states. All these states have the same weight. Therefore, taking the partial trace of spin variables simply produces $2^{N_c(G)}$, and we obtain

$$Z = \sum_G V(G) q^{N_c(G)}, \tag{4.11}$$

with $q = 2$ for the Ising model. Taking $q = 3, 4, \ldots$ we get the partition function of the q-state Potts model. This formula can also be used to study the q-state Potts model with non integer q (Blöte and Nightingale, 1982). Similar techniques also exist for the SU(N) quantum Heisenberg model (Beach et al., 2009).

4.4.3 Correlation functions and cluster size

The reason why the cluster algorithm can have a shorter auto-correlation time than a single spin-flip algorithm is that the cluster algorithm updates the state of the system by units whose physical size is comparable to the system's intrinsic spatial correlation length. For example, in the Swendsen-Wang algorithm, the typical range of the spatial correlations is related to the average size of the clusters by the formula

$$\langle s_i s_j \rangle = \text{Prob}(\text{``}i \text{ and } j \text{ are in the same cluster''}). \tag{4.12}$$

To prove this, we express the left-hand side as

$$\langle s_i s_j \rangle = \sum_{C,G} W(C, G) s_i s_j = \sum_G W(G) \sum_C P(C|G) s_i s_j$$

$$= \sum_G W(G) \chi (\text{``}i \text{ and } j \text{ are in the same cluster''})$$

$$= \langle \chi (\text{``}i \text{ and } j \text{ are in the same cluster''}) \rangle_{\text{MC}}$$

where $\chi(\text{``statement''}) = 1$ if the statement is true, and $\chi(\text{``statement''}) = 0$ otherwise. From the last equation (4.12) directly follows.

Equation (4.12) says that the linear size of a typical cluster is roughly proportional to the spatial correlation length. During a Monte Carlo simulation, various spin configurations appear. If the system is near or already in equilibrium, then there are many clusters of up-spins and down-spins, and in each cluster the spins are aligned. The typical size of such clusters is comparable to the spatial correlation length. When we use a single-spin update algorithm, the process leading to the appearance or disappearance of a cluster takes a long time because it usually involves a boundary propagation.

Figure 4.3 illustrates this situation for the one-dimensional Ising model in which the system relaxes through the diffusion of the kinks. This way of propagating is why the relaxation time depends on the size of the clusters and therefore on the correlation length. The autocorrelation time in fact becomes proportional to the correlation length raised to some power (usually close to 2, reflecting the diffusive nature of the domain-wall motion). In a cluster algorithm, we modify spatial structures of the size of the correlation length in just one Monte Carlo step.

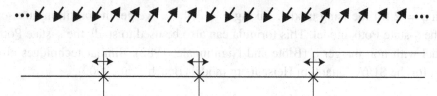

Figure 4.3 Relaxation through kink diffusion in the one-dimensional Ising model with single-spin update.

We note that the partition functions in (4.1) and (4.11) have the same value but express the Boltzmann distribution for the model as summations over two different sets of degrees of freedom. Whereas the spin configurations in (4.1) are the most obvious degrees of freedom in terms of the definition of the model, the graphs used in (4.11) are the most natural ones in terms of the physics of the model near its critical point. If we simulate away from the critical point at high temperatures, the Swendsen-Wang algorithm loses its efficiency advantage because the physics of the model changes. There, the computational cost of the cluster construction becomes a disadvantage.

4.5 Worm updates

So far we have seen two different examples of two-step Monte Carlo algorithms for many-body problems defined on lattices: the local update and the cluster update. In the case of the Ising model, because it is defined in terms of spin variables, it is natural to consider a Markov process defined in the space of spin configurations $\Omega \equiv \{C\} = \{(s_1, s_2, \ldots, s_N)\}$. However, it is in general possible, and often quite advantageous, to reformulate the original problem as one defined in terms of a new set of degrees of freedom. We now discuss such a reformulation of the problem (Prokof'ev and Svistunov, 2001).

We start from the high-temperature series expansion of the Ising model, which can be obtained by decomposing the local Boltzmann factor and using the simple identity $e^{\pm x} = (\cosh x)(1 \pm \tanh x)$,

$$Z = \sum_C \prod_{\langle ij \rangle} e^{K s_i s_j} \propto \sum_C \prod_{\langle ij \rangle} (1 + t s_i s_j),$$

where $t \equiv \tanh K$. In the last expression, we neglected the constant factor $(\cosh K)^{N_{\text{bonds}}}$, with N_{bonds} being the number of interacting spin pairs in the lattice. By expanding the parentheses, we obtain many terms. Each term corresponds to a graph G that consists of the lattice sites as vertices and bonds for the edges associated with $ts_i s_j$. (The edges for which 1 is chosen instead of $ts_i s_j$ are not included in G.) The resulting formula for the partition function is

$$Z = \sum_C \sum_G t^{|G|} \prod_i s_i^{n_i(G)}. \tag{4.13}$$

Here, the symbol G represents a graph defined on the lattice, and $|G|$ stands for the number of edges in G, while $n_i(G)$ is the number of edges in G that share the vertex i.

Let us take the summation over C before the summation over G. If $n_i(G)$ is an odd number for any given i, the summation over C kills the term because $\sum_{s_i = \pm 1} s_i^n = 0$ if n is an odd number. Therefore, we can neglect all graphs with "odd" vertices. We call G a *closed graph* if it does not have any odd vertices. With all this in place, we write the partition function, apart from an overall numerical constant, more simply as

$$Z = \sum_{G \in \Omega_0} t^{|G|}. \tag{4.14}$$

Now, the phase space Ω_0 is the set of all *closed* graphs defined on the lattice, and our task is to construct a Markov process for generating closed graphs with a frequency proportional to $t^{|G|}$. Note that both (4.11) and (4.14) express the partition function as a sum over graphs, but the graphs are quite different. Additionally, unlike the cases studied so far, the new phase space Ω_0 has a complicated structure and cannot be expressed as a simple product of many local phase spaces because of the "no-odd-vertex" constraint: We cannot propose a new state, or a closed graph, simply by choosing a pair of nearest neighbor sites at random and then removing or adding an edge connecting them, because doing so results in a state with two odd vertices and such a state does not contribute to the partition function.

What we can do is to allow in our phase space "artificial" states that do not contribute to the partition function. In the sampling, we generate states stochastically (some are closed graphs and others are artificial ones), but for the measurements, we discard all artificial states, and count only the "real" states. If we keep at most two odd vertices in the artificial states, we have a good chance of observing closed graphs with sufficient frequency. To keep this condition, when we add a new edge we make one of its ends cover an odd vertex, and when we remove an edge we remove one that has an odd end. As a result, we shift the location of the odd vertex by one lattice spacing. Algorithm 9 describes the procedure. In this algorithm, we introduce two odd vertices, which we call *head* and *tail*. We add or remove an edge by moving the head from the current site to one of its nearest neighbors.

Allowing up to two odd vertices means that we have extended the phase space from the zero odd-vertex space Ω_0 to $\Omega_0 \cup \Omega_2$, where Ω_2 is the space of graphs with

---

Algorithm 9 Worm algorithm for the Ising model.

Input: A high-temperature series configuration with or without worms.
    if there is no head and tail **then**
        Choose a site uniformly and randomly and place head and tail on it ;
    else if the head is on the same position as the tail **then**
        Remove head and tail ;
    end if
    if there are head and tail **then**
        Choose the direction of the head's motion randomly ;
        With probability p move the head in this direction and change the edge state
        ($p = 1$ if the edge exists, otherwise $p = t$) ;
    end if
    return the updated configuration.

---

two odd vertices. Since the weight of a state in Ω_2 is undefined a priori, it becomes
a parameter we can choose for our convenience. We choose a weight that not only
covers Ω_2 but also reproduces the previous definition of the weight for Ω_0:

$$W(G) = a^{\nu(G)} t^{|G|}.$$

Here a is a constant we fix later and $\nu(G)$ is the number of odd vertices in G. Now
our new task is to generate graphs G in $\tilde{\Omega} = \Omega_0 \cup \Omega_2$ with the weight $W(G)$. For a
uniform lattice, we can do this with Algorithm 9 (Prokof'ev and Svistunov, 2001).

We note that the artificial states in Ω_2 include the case where two odd vertices
happen to be on the same lattice site. Such a state is the same as a state in Ω_0 except
that the position of the imaginary pair of odd vertices is defined and contributes
a weight a^2 relative to the corresponding state in Ω_0. The first two cases in the
conditional in Algorithm 9 represent transitions between these two states, say, $G \in$
Ω_0 and $G' \in \Omega_2$. The detailed balance condition between them is

$$\frac{1}{N} \times W(G) = 1 \times W(G'), \tag{4.15}$$

from which it follows that a^2 should be $1/N$.

The description of this head creation/annihilation procedure is rather formal and
is mainly a reminder of the phase space structure we are using (in particular, that
Ω_0 is still present). In practice, we skip going back and forth between Ω_0 and Ω_2
and simplify the procedure as described in Algorithm 10.

Now we consider the third case in the conditional in Algorithm 9, where the
worm exists and the head is separated from the tail. In this case, we stochastically
move the head. We first select the trial direction at random, and then move the head

---

Algorithm 10 Worm algorithm for the Ising model on a uniform lattice (simplified).

Input: A worm configuration with head and tail.

    if the head is on the same position as the tail **then**

        Randomly select a site and move the head and tail there ;

    end if

    Choose the direction of the head's motion randomly ;

    With probability p move the head in this direction, and change the edge state ($p = 1$ if the edge exists, otherwise $p = t$) ;

    return the updated configuration.

---

in this direction with probability $p = 1$ ($p = t$) if an edge is present (no edge is present). Afterward, we change the edge state. Suppose we have no edge in the trial direction. Then, the final state of the motion, say, G', has one edge more than G, and because each edge carries the weight t, we have $W(G') = tW(G)$. The probability flow from the initial state G to the final state G' is the product of three probabilities: the probability $1/z_i$ for choosing the direction, where z_i is the coordination number of the site i at which the head is currently located, the probability t for accepting that motion, and the target probability of the state ($\propto W(G)$). We express the probability flow in the opposite direction in a similar fashion and then write the detailed balance condition as

$$\frac{1}{z_i} \times t \times W(G) = \frac{1}{z_j} \times 1 \times W(G').$$

This equality holds since $z_i = z_j$ is constant on a uniform lattice and $W(G') = tW(G)$ as we noted above. Thus, we confirmed detailed balance. In the same way, we can confirm detailed balance when an edge is in the direction of the head motion. We illustrate a cycle in the worm algorithm in Fig. 4.4.

In the above derivation, the introduction of Ω_2 might merely look like a trick for making the whole procedure work. However, it plays another important and useful role in the measurement of two-point correlation functions. As in the cluster update algorithm, the way the two-point correlation function is calculated within the present framework is the key to understanding the essential property of the worm update algorithm. In the high-temperature series expansion, the two-point correlation function is

$$\langle s_i s_j \rangle = \frac{Z_{ij}}{Z}, \quad Z_{ij} \equiv N \sum_{G \in \Omega_{ij}} W(G),$$

where Ω_{ij} is the set of graphs that has odd vertices on i and j and thus is a subset of Ω_2. Note that we have the factor N in the above expression because we have defined

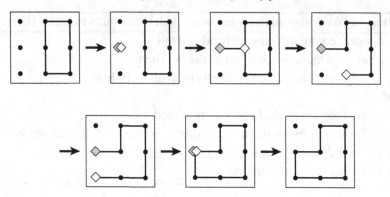

Figure 4.4 A worm update cycle for the Ising model, based on the high-temperature series representation. The head is depicted by the empty diamond and the tail by the shaded diamond.

the weight of the wormful states with this factor in (4.15). In the Markov process previously defined, the probability that the head and the tail are on i and j or vice versa occurs with a frequency proportional to $\sum_{G \in \Omega_{ij}} W(G)$, whereas the frequency of having a closed graph is proportional to $\sum_{G \in \Omega_0} W(G) = Z$. Therefore, the two-point correlation function equals the total number of the "i, j" events divided by the number of no odd-vertex events, multiplied by N. The factor N again comes from the definition of the weights of the wormful states relative to the wormless states. In a uniform system, we can further simplify the procedure: Because of translational invariance, instead of multiplying by N, we count all events where the location of the head relative to that of the tail is $r_{ij} \equiv r_i - r_j$ or is $-r_{ij}$ where r_i and r_j are the position vectors of sites i and j. Hence,

$$\langle s_i s_j \rangle = \langle \mathrm{Count}(r_{\mathrm{head}} - r_{\mathrm{tail}} = \pm r_{ij}) \rangle_{\mathrm{MC}},$$

where Count("event") is the number of times that "event" occurs during a head-creation/annihilation cycle.[4]

4.6 Closing remarks

We started with the Metropolis algorithm and focused on what is at its core as opposed to simply explaining its familiar application to the Ising model. The insight

---

[4] By now, it should be evident why the procedure is called *worm* update. Nonetheless, since it is only the current positions of the head and the tail that matter for the stochastic process that follows, and since the trajectory of the head has no significant meaning that makes it necessary to record it, the word "worm" misleads one to think about the wiggly "body" of a worm and may not be particularly appropriate. "Pac-Man" update, for example, would reflect the nature of the algorithm better. However, since the name "worm algorithm" has become standard among the people working in the field, we stick to this accepted name throughout this book.

of Metropolis et al. was the detailed balance condition. Their craftiness produced a general two-step algorithm for sampling in accord with this condition.

When we look at the "forest" and not the "trees," the cluster algorithm is a Metropolis-like two-step algorithm for a new set of degrees of freedom. These degrees of freedom are the edges of a graph and are not part of the Hamiltonian's dynamical variables. These edges link like spins on neighboring sites, thereby forming clusters of aligned spins reflecting those found in a physical system near a critical point. A nonlocal updating of spin configurations becomes possible and potentially more efficient. The graphs are introduced in such a way that they provide a path between an initial and final configuration of an overall transition probability matrix. We sample this path, that is, the graphs, in a way that satisfies detailed balance.

Finally, we discussed the worm algorithm, which is strictly speaking not a two-step Metropolis algorithm. However, it is in the same spirit as the cluster algorithm in that it also enlarges the configuration space of the model and imposes restrictions on it, by allowing loops with open ends and restricting the number of them to at most two.

The three algorithms illustrate the flexibility of the Monte Carlo approach as a tool for simulating many-body problems. Because the approach was applied to a finite-temperature problem, the algorithms needed to sample from an explicit distribution. In the construction of the algorithms, the detailed balance condition was as much an aid as it was a constraint. Although satisfying the detailed balance condition, the cluster and worm algorithms a priori are not guaranteed to be more efficient than the standard Metropolis algorithm with single-spin flips. They are in fact more efficient for only some applications, such as simulations near a second-order phase transition, for which they were designed.

In the next chapters, we move to quantum Monte Carlo algorithms. As we commented in Chapter 1, a quantum Monte Carlo algorithm is a classical Monte Carlo algorithm applied to a quantum problem recast into classical degrees of freedom. When we first learned quantum mechanics, we began by quantizing classical problems. In recasting a quantum problem into a form suitable for Monte Carlo sampling, we do not undo this step but rather "classicize" the problem. Once this is done, the algorithms discussed in this chapter reapply, or at least the basic concepts carry over to the simulation of quantum problems.

Suggested reading

J. S. Wang and R. H. Swendsen, "Cluster Monte Carlo algorithm," *Physica A* **167**, 565 (1990).

J. M. Yeomans, *Phase Transitions* (Oxford University Press, Oxford, 1992), chapters 6 and 7.

N. Prokof'ev and B. Svistunov, "Worm algorithm for problems in quantum and classical systems," in *Understanding Phase Transitions*, ed. L. D. Carr (Boca Raton, FL: Taylor and Francis, 2010).

Exercises

4.1 Single-Spin Update Algorithm. Prove that as long as $q_i(S'|S)$ satisfies the detailed balance condition itself, that is,

$$q_i(S'|S)\bar{P}(S) = q_i(S|S')\bar{P}(S'),$$

for all i with $\bar{P}(S)$ being the target distribution, the Markov matrix (4.3) satisfies the detailed balance condition regardless of the choice of p_i.

4.2 Two-Dimensional Random Walk. In a two-dimensional random walk on a square, the walker steps north, south, east, and west with equal probability. No "diagonal" moves are permitted.

1. Show that the probability of the walker returning to the starting point after $2n$ steps is

$$p_{2n} = \frac{(2n)!}{4^{2n}} \sum_{m=0}^{n} [m!\,(n-m)!]^{-2}.$$

2. Using the series expansion $(1+x)^{2n}$, establish the relations

$$\frac{(2n)!}{[m!\,(n-m)!]^2} = \binom{2n}{n}^2 \sum_{m=0}^{n} \binom{n}{m}^2$$

and hence that

$$p_{2n} = \frac{1}{4^{2n}} \binom{2n}{n}^2.$$

3. Using Sterling's large n approximation, $n! \approx \sqrt{2\pi}\, n^{n+\frac{1}{2}} e^{-n}$, establish that

$$p_{2n} \approx \frac{1}{\pi n}.$$

4.3 Show that the Wolff algorithm (Algorithm 8) satisfies detailed balance.

4.4 Worm Update for the q-State Potts Model. Generalize the worm update for the Ising model discussed in Section 4.5 to the q-state ferromagnetic Potts model,

$$H = -J \sum_{\langle ij \rangle} \delta_{\langle s_i s_j \rangle} \quad (J > 0,\ s_i = 1, 2, \ldots, q).$$

4.5 Worm Update of a Nonuniform System. In Fig. 4.4, in the direction of the head's first move there are three options, whereas in the second move there are four. Therefore, the probability flow from the first state to the second is $(1/3) \times t \times w_1$, whereas the opposite is $(1/4) \times 1 \times w_2$, with $w_1 = t^6$ and $w_2 = t^7$ being the weight of the first and the second states, respectively. How can we modify the algorithm to make the algorithm work for nonuniform cases?

5
Quantum Monte Carlo primer

In this chapter, we extend the ideas and techniques discussed in the previous chapter to introduce several basic methods for the Monte Carlo simulation of quantum many-body systems. For the sake of pedagogy, we consider only simple quantum spin-$\frac{1}{2}$ models at nonzero temperature. Some of the techniques we discuss, such as path-integral Monte Carlo methods with local updates and imaginary-time discretization, are introduced only to present key ideas and concepts in preparation for the discussion of the state-of-the-art continuous-time cluster and worm algorithms. In the remaining chapters, we discuss more advanced quantum spin algorithms and zero and finite-temperature algorithms for Fermion and Boson systems.

5.1 Classical representation

At present, all Monte Carlo simulations are performed on classical computers. With these computers we can directly manipulate only classical numbers whose products commute with each other. Quantum mechanics, however, deals with operators, most of whose products do not commute with each other. Therefore, we need to find a classical representation of our quantum system that reformulates the quantum problem in terms of classical numbers. We start by exploring a simple and obvious approach.

We are all familiar with the matrix formulation of quantum mechanics, in particular expressing the Hamiltonian operator as a Hermitian matrix. This representation is the most basic "classicization" of a quantum problem. While in general their products are noncommuting, matrices are just an ordered arrangement of classical numbers whose algebra classical computers handle easily. In particular, classical computers can diagonalize such matrices and give us all the eigenvalues and eigenvectors. This information "solves" the quantum problem.

In the quantum many-body problems of interest, the phase space is, as in the classical case, a direct product of many local phase spaces. If the size of the local space is n and we have N lattice sites, the phase space scales exponentially with the physical size of the system as n^N. The order of the Hamiltonian matrix is equal to the size of the phase space and hence exhibits an exponential scaling. Because of computer memory limitations, we can only obtain, and in fact ever expect to obtain, all the eigenvalues and eigenvectors for lattices of small size (Section 10.1). Solutions for small-sized lattices are, however, generally tainted by boundary effects and often are not indicative of the physics of the thermodynamic limit. We thus have to exploit the Monte Carlo method's ability to sample the important parts of very large phase spaces.

There are two other approaches to classicization: the high-temperature series expansion (Handscomb, 1962a) and the Feynman path-integral formulation (Suzuki, 1976a,b). The Feynman path integral converts a d-dimensional quantum problem into a $(d+1)$-dimensional classical problem. The original Hamiltonian in d dimensions acts in layers stacked along the new dimension, with a new set of interactions coupling the layers. In quantum statistical mechanics, this additional dimension is an imaginary-time axis.

There are discrete- and continuous-time versions of path-integral algorithms, and these algorithms are typically used to simulate the finite-temperature equilibrium properties of quantum many-body systems. The first generation of path-integral algorithms works with a discretized imaginary time. To eliminate discretization errors, which scale polynomially in the imaginary-time step, we have to extrapolate to the continuous imaginary-time limit. Often, we can do this at the level of the algorithm, so that the simulation results of modern algorithms are free of discretization errors.[1]

The high-temperature expansion leads to a classical representation that is formally similar to a discrete imaginary-time representation of the path-integral method. An important difference is that the high-temperature series method has an exponentially small discretization error that may be neglected for most practical purposes.

In Chapter 6, we detail both the quantum high-temperature series expansion and the path-integral formulation. In the present chapter, as part of our initial development of quantum Monte Carlo algorithms, we give a short introduction to the path-integral approach.

---

[1] In the past, imaginary-time steps were simply chosen small enough so that the statistical error masked the discretization error or else multiple Monte Carlo runs were made for different step sizes, and the results were fitted to a polynomial in the step size and then extrapolated to zero step size.

5.2 Quantum spins

We now develop our first quantum Monte Carlo algorithms. We begin with discrete-time path-integral representations of the spin-$\frac{1}{2}$ quantum Ising and XY models. By taking the continuous-time limit, we eventually replace the venerable discrete-time (world line) Monte Carlo methods with loop/cluster and worm algorithms. In doing so, we achieve an algorithmic evolution analogous to the one we presented for classical many-body problems in Chapter 4. Our objective is unveiling several basic features and techniques used in quantum Monte Carlo algorithms. While our discussion is in the context of specific simple finite-temperature examples, some of these features and techniques are also useful for zero-temperature problems.

5.2.1 Longitudinal-field Ising model

The quantum version of the classical one-dimensional Ising model (1.3) is given by the Hamiltonian

$$H = -\sum_{i=1}^{N} \left(J^z S_i^z S_{i+1}^z + H^z S_i^z \right). \tag{5.1}$$

Here, S_i^z is the z-component of the $S = \frac{1}{2}$ angular momentum operator at lattice site i, so its eigenvalue is $\frac{1}{2}$ or $-\frac{1}{2}$. If $H^z = 0$, this Hamiltonian is invariant under the transformation $S_i^z \rightarrow -S_i^z$ executed at each site i. If $H^z \neq 0$, it is invariant under this operation, provided the sign of H^z is reversed. Thus, this quantum model has the same symmetries as the classical model. As we will see below, this "quantum model" is actually equivalent to the classical Ising model.

In quantum statistical mechanics, the partition function is

$$Z = \text{Tr}\left[e^{-\beta H}\right], \tag{5.2}$$

with $\beta = 1/kT$ denoting the inverse temperature. (In the following, we set the Boltzmann constant $k = 1$.) The trace can be computed using a complete orthonormal set of basis states, which we denote as $\{|C\rangle\}$,

$$\langle C|C'\rangle = \delta(C, C'), \quad \sum_C |C\rangle\langle C| = I.$$

With such a basis, (5.2) becomes

$$Z = \sum_C \langle C| e^{-\beta H} |C\rangle.$$

If the $|C\rangle$ are also the eigenstates of H, that is, $H|C\rangle = E(C)|C\rangle$, it follows that $\langle C|e^{-\beta H}|C\rangle = e^{-\beta E(C)}$ and therefore that

$$Z = \sum_C e^{-\beta E(C)}. \tag{5.3}$$

Similarly, the thermal expectation value of some physical observable X becomes

$$\langle X \rangle = \frac{1}{Z}\text{Tr}\left[Xe^{-\beta H}\right] = \frac{1}{Z}\sum_C \sum_{C'} \langle C|X|C'\rangle\langle C'|e^{-\beta H}|C\rangle. \tag{5.4}$$

In particular, when the $\{|C\rangle\}$ are orthonormal eigenstates of the Hamiltonian, we obtain

$$\langle X \rangle = \frac{1}{Z}\sum_C X(C)e^{-\beta E(C)},$$

with $X(C) = \langle C|X|C\rangle$, an expression reminiscent of an expectation value in classical statistical mechanics.

Because knowing the eigenenergies and eigenstates of H amounts to knowing the solution to the problem, the reduction of (5.2) to (5.3) is generally possible only formally. In the case of model (5.1), however, the eigenstates are

$$|C\rangle = |s_1\rangle \otimes |s_2\rangle \otimes \cdots \otimes |s_N\rangle = |s_1 s_2 \cdots s_N\rangle, \tag{5.5}$$

where $s_i = \pm\frac{1}{2}$ and $|s_i\rangle$ are the eigenpairs of S_i^z. Then, the eigenvalues $E(C)$ are just the energies of the classical model, apart from the classical spins s_i taking the values $\pm\frac{1}{2}$ instead of ± 1,[2] and the problem of sampling the partition function (5.3) is identical to the classical Ising problem.

5.2.2 Transverse-field Ising model

We now consider a less trivial quantum-spin model, a spin-$\frac{1}{2}$ chain in a transverse magnetic field, and explicitly derive the Feynman path integral by discretizing the imaginary-time interval. The Hamiltonian of the transverse-field Ising model is

$$H = -\sum_{i=1}^N \left(J^z S_i^z S_{i+1}^z + H^x S_i^x\right), \tag{5.6}$$

which we break into the two noncommuting pieces

$$H = H_0 + H_1 \tag{5.7}$$

---

[2] We could have made the correspondence exact by using the Pauli spin operators, instead of using the angular momentum operators, in the definition of the Hamiltonian.

defined by

$$H_0 = -J^z \sum_{i=1}^{N} S_i^z S_{i+1}^z, \quad H_1 = -H^x \sum_{i=1}^{N} S_i^x. \tag{5.8}$$

The challenge in constructing a finite-temperature quantum Monte Carlo algorithm is exponentiating the Hamiltonian. Since the current model has an exact solution (Sachdev, 1999), in principle we can do this exponentiation exactly, and we do not need Monte Carlo simulations. Our intent, however, is pedagogy, so for the time being we pretend we cannot do the exponentiation.

We begin by recasting the partition function $Z = \mathrm{Tr}\, e^{-\beta H}$ into a form with classical degrees of freedom. Noting that any operator commutes with itself, we write

$$e^{-\beta H} = \underbrace{e^{-\Delta\tau H} e^{-\Delta\tau H} \cdots e^{-\Delta\tau H}}_{M \text{ factors}}, \tag{5.9}$$

where $\Delta\tau = \beta/M$, and hence

$$Z = \sum_{C_1, C_2, \ldots, C_M} \langle C_1 | e^{-\Delta\tau H} | C_M \rangle \langle C_M | e^{-\Delta\tau H} | C_{M-1} \rangle \cdots \langle C_2 | e^{-\Delta\tau H} | C_1 \rangle. \tag{5.10}$$

This equation is the Feynman path-integral expression for the quantum mechanical partition function. The matrix products of the exponential of the Hamiltonian run through an ordered sequence of basis states. Each basis state is a configuration in phase space so we can interpret the summations in (5.10) as being over an imaginary-time ordered sequence of configurations. The sequence of configurations $C_1 \to C_2 \to \cdots \to C_M \to C_1$ defines a path through phase space. Thus, the partition function is a sum over all possible (closed) paths. In sampling the configurations, the Monte Carlo algorithm is selecting the most important of these paths.

As it stands, (5.10) is a trivial result. We can easily evaluate it if we know the matrix elements of $\exp(-\Delta\tau H)$ in the chosen basis. If we know them, the path-integral formulation is a step backward because we have to compute all the matrix products, which requires more work than simply tracing the matrix elements of $\exp(-\beta H)$. The representation (5.10) is useful in cases where we are unable to express the matrix elements of $\exp(-\Delta\tau H)$ exactly. By writing $\exp(-\beta H)$ as a product of many $\exp(-\Delta\tau H)$ factors, we introduce a small parameter $\Delta\tau$ into the problem, which enables us to obtain a good approximation for each matrix element of $\exp(-\Delta\tau H)$.

The difficulty in exponentiating H stems from the fact that S_i^x and S_j^z lack on-site ($i = j$) commutivity. The two parts of H are thus not simultaneously diagonalizable. Accordingly, the states defined in (5.5) are not eigenstates of the Hamiltonian (5.6);

they are eigenstates of H_0 only. However, in this basis (or in the eigenbasis of H_1) both $\langle C'|e^{-\Delta \tau H_0}|C\rangle$ and $\langle C'|e^{-\Delta \tau H_1}|C\rangle$ are known. Unfortunately, $e^{-\Delta \tau (H_0+H_1)} \neq e^{-\Delta \tau H_0} e^{-\Delta \tau H_1}$ if $[H_0, H_1] \neq 0$. To proceed, we need to develop a strategy for handling exponentials of the type $e^{-\Delta \tau (A+B)}$ when $[A, B] \neq 0$.

Let us consider the error incurred if we make the approximation

$$e^{-\Delta \tau (A+B)} \approx e^{-\Delta \tau A} e^{-\Delta \tau B} \approx e^{-\Delta \tau B} e^{-\Delta \tau A}. \tag{5.11}$$

For operators (and matrices), $\exp(A) = I + A + \frac{1}{2!}A^2 + \frac{1}{3!}A^3 + \cdots$ is defined by its power series. Accordingly,

$$e^{-\Delta \tau (A+B)} = e^{-\Delta \tau A} e^{-\Delta \tau B} + \mathcal{O}((\Delta \tau)^2), \tag{5.12}$$

which means that the error scales as $(\Delta \tau)^2$. A higher order (symmetric) approximation is

$$e^{-\Delta \tau (A+B)} = e^{-\Delta \tau B/2} e^{-\Delta \tau A} e^{-\Delta \tau B/2} + \mathcal{O}((\Delta \tau)^3). \tag{5.13}$$

Approximations such as (5.12) and (5.13) are called *Suzuki-Trotter approximations* (Hatano and Suzuki, 2005), or just *Trotter approximations* for short.

It is natural to choose a basis in which we can easily find the matrix elements of the exponentials of H_0 and H_1. The basis (5.5) allows a trivial exponentiation of H_0. It also permits an easy calculation of the matrix elements of $\exp(-\Delta \tau H_1)$. Using the lower-order Trotter approximation (5.12), we write $e^{-\Delta \tau H} \approx e^{-\Delta \tau H_0} e^{-\Delta \tau H_1}$ and then

$$\langle C_{k+1}|e^{-\Delta \tau H}|C_k\rangle \approx e^{-\Delta \tau E_0(C_{k+1})} \langle C_{k+1}|e^{-\Delta \tau H_1}|C_k\rangle,$$

where $E_0(C_k)$ is the classical Ising energy in zero magnetic field corresponding to the state $|C_k\rangle$ (again apart from the s_i taking the values $\pm\frac{1}{2}$ instead of ± 1).

We have to evaluate the energy $E_0(C_k)$ for each time step $\tau_k = k\Delta \tau$, and we record which spin eigenvalue belongs to which time step by attaching to them an additional subscript and writing

$$E_0(C_k) = -J^z \sum_i s_{i,k} s_{i+1,k}.$$

Consequently,

$$e^{-\Delta \tau \sum_k E_0(C_k)} = e^{\Delta \tau J^z \sum_{ik} s_{i,k} s_{i+1,k}},$$

which corresponds to the Boltzmann weight for M uncoupled classical Ising-like chains.

The remaining task is to specify the matrix elements of the exponential of the transverse-field part of the Hamiltonian. First, we note that

$$\langle C | e^{-\Delta\tau H_1} | C' \rangle = \prod_{i=1}^{N} \langle s_i | e^{\Delta\tau H^x S_i^x} | s_i' \rangle. \tag{5.14}$$

Next,

$$
\begin{aligned}
e^{\Delta\tau H^x S_i^x} &= I + \Delta\tau S_i^x H^x + \tfrac{1}{2!}(\Delta\tau H^x S_i^x)^2 + \tfrac{1}{3!}(\Delta\tau H^x S_i^x)^3 + \cdots \\
&= I + (\Delta\tau H^x) S_i^x + \tfrac{1}{4\cdot 2!}(\Delta\tau H^x)^2 I + \tfrac{1}{4\cdot 3!}(\Delta\tau H^x)^3 S_i^x + \cdots,
\end{aligned}
$$

where we used the fact that $(S^x)^2 = \tfrac{1}{4}$. Summing the power series yields

$$\langle s | e^{\Delta\tau S^x H^x} | s' \rangle = \langle s | s' \rangle \cosh \tfrac{1}{2}\Delta\tau H^x + 2\langle s | S^x | s' \rangle \sinh \tfrac{1}{2}\Delta\tau H^x.$$

Noting that the only nonzero matrix elements in the second term are

$$\langle \downarrow | S^x | \uparrow \rangle = \tfrac{1}{2}, \quad \langle \uparrow | S^x | \downarrow \rangle = \tfrac{1}{2},$$

we reexpress the matrix elements of the exponential as

$$\langle s | e^{\Delta\tau S^x H^x} | s' \rangle = Q e^{K^\tau s s'}, \tag{5.15}$$

where $s, s' = \pm\tfrac{1}{2}$ and

$$Q = \sqrt{\sinh \tfrac{1}{2}\Delta\tau H^x \cosh \tfrac{1}{2}\Delta\tau H^x}, \quad K^\tau = 2\ln \coth \tfrac{1}{2}\Delta\tau H^x. \tag{5.16}$$

Ultimately,

$$Z \propto \sum_{C_1, C_2, \ldots, C_M} \exp\left(\Delta\tau J^z \sum_{i,k} s_{i,k} s_{i+1,k} + K^\tau \sum_{i,k} s_{i,k} s_{i,k+1} \right). \tag{5.17}$$

This is just the partition function for a zero-field, anisotropic, two-dimensional classical Ising-like model (apart from the factor of $\tfrac{1}{2}$ in the fields). If we set the inverse temperature $\beta_{\text{classical}}$ of this classical model to one, we have a coupling $K^x = \Delta\tau J^z$ in the x-direction, and a field-dependent exchange K^τ in the τ-direction:

$$H_{\text{classical}} = -K^x \sum_{i,k} s_{i,k} s_{i+1,k} - K^\tau \sum_{i,k} s_{i,k} s_{i,k+1}. \tag{5.18}$$

The size of the lattice is L in the spatial dimension and $M = \beta/\Delta\tau$ in the temporal dimension. To leading order in $\Delta\tau$, the coupling constants in the spatial and temporal directions are

$$K^x = \Delta\tau J^z, \quad K^\tau = -2\ln \tfrac{1}{2}\Delta\tau H^x. \tag{5.19}$$

See (5.16).

Let us take stock of what just happened. By expressing the matrix elements of $\exp(-\beta H)$ as a product of matrices representing $\exp(-\Delta\tau H)$, invoking the Trotter approximation, and using the eigenstates of H_0 as the basis, we obtained a two-dimensional lattice of Ising-like variables $s_{i,k}$. The matrices representing $\exp(-\Delta\tau H_0)$ produced the Boltzmann-like factors of independent, zero-field Ising-like chains. The spin variables in these chains interact only in the x-direction. The matrices representing $\exp(-\Delta\tau H_1)$ generated an interaction between these chains in the τ-direction. From the path-integral representation, it follows that this τ-direction is actually an imaginary-time direction. Expressing $\exp(-\beta H)$ as a product of factors $\exp(-\Delta\tau H)$ led to a discrete imaginary-time expression of the path integral. According to Feynman (Feynman and Hibbs, 1965), transforming field theory into quantum statistical mechanics is done by analytically continuing the path-integral's dynamical factor $\exp(-itH)$ by $it \to \tau$ and requiring periodicity in the imaginary-time direction. The trace expresses this periodicity. In mapping the d-dimensional quantum system onto a $(d+1)$-dimensional classical model, the Hamiltonian is not simply boosted to the extra dimension, but rather new interactions are introduced along this extra dimension to account for the imaginary-time dynamics.

In our derivation, based on the Trotter decomposition, the strategy was to separate the Hamiltonian into pieces whose exponentials we easily obtain in some basis, often the eigenbasis of one piece of the Hamiltonian. The simplest consequence is the introduction of a discretization error, which, however, as discussed below, we can completely avoid by reformulating the algorithm. A more problematic consequence (which remains even in a continuous-time formulation) is a possible *sign problem*. Because the Hamiltonian is Hermitian, its exponential is always a *positive-definite operator*, meaning that its eigenvalues are all greater than zero. This property of an operator is independent of the basis we choose to express its matrix elements. This positivity is manifest, for example, in the matrix representing the exponential of the Hamiltonian for the longitudinal-field Ising model. It is diagonal in the chosen basis, and because its diagonal elements are exponentials of some real number, they are all positive.

A positive-definite matrix differs from a *positive matrix*, that is, one having all elements greater than zero, and from a *non negative matrix*, that is, one that has no negative elements. Positivity and non negativity depend on the basis we choose to represent the operator. For Monte Carlo simulations, we want positive or non negative matrices. Quantum imaginary-time propagators, however, are only guaranteed to be positive definite. Only in a diagonal representation, in which we almost never work, are we assured that the matrix representing $\exp(-\beta H)$ is non negative. If we are unable to find a workable representation that removes the negative matrix elements, a *sign problem* may appear. This problem affects our ability to validate our results and can be so severe that the quantum Monte Carlo algorithm is unable to produce reliable estimates of the statistical error associated with computed averages. We say more about this problem in subsequent sections and chapters.

For the transverse-field model, a potential sign problem occurred, but was let pass without comment. In (5.15), the off-diagonal matrix elements are negative if H^x is negative. If so, some configurations give a negative contribution to the partition function. For the case at hand, we can avoid the sign problem by requiring, as we implicitly assumed, that $H^x > 0$. We recover the physics for $H^x < 0$ by applying a unitary transformation $(S_i^x, S_i^y, S_i^z) \rightarrow (-S_i^x, S_i^y, -S_i^z)$ to our Hamiltonian, which results in the inversion of the magnetic field $H^x \rightarrow -H^x$. This invariance is manifested by the fact that there is always an even number of spin flips caused by S_i^x along the imaginary-time direction for every spin, which allows us to simply neglect the sign associated with individual spin-flip events.

5.2.3 Continuous-time limit

The transformation of the one-dimensional quantum transverse-field Ising model to a (1+1)-dimensional classical Ising model means that we can use whatever is our favorite Monte Carlo algorithm for the classical Ising model as a quantum Monte Carlo algorithm for the transverse-field model. However, unless we take the limit $M \rightarrow \infty \, (\Delta\tau \rightarrow 0)$, (5.17) is not exact, and hence the results of our simulation have a systematic error due to the time discretization. We need to eliminate this error. A straightforward solution is to perform different Monte Carlo simulations with different values of M and extrapolate to $M \rightarrow \infty$. While this procedure works fine and was used in practice, it is no longer needed for most problems. In this section and Section 5.2.5, we extend the basic idea of the cluster algorithm discussed in Section 4.4 to two simple quantum spin models. Then, in Chapter 6, we discuss applications to various other systems.

We now explain how to take the continuous-time limit for the transverse-field Ising model (5.6). The discussion in the previous subsection showed that the discrete imaginary-time representation of this model is nothing but the two-dimensional *classical* Ising model with anisotropic interactions given by $K^x = \Delta\tau J^x$ and $K^\tau = -2\ln(\Delta\tau H^x/2)$, and with "temperature" $1/\beta_{\text{classical}} = 1$. We observe that the effective coupling constant in the time direction increases as we make $\Delta\tau$ smaller. This increase is a generic feature of the quantum-to-classical mapping that ensures that the configurations (which in this case are collections of time intervals for spin up and down) have a well-defined limit $\Delta\tau \rightarrow 0$. We can understand this by noting that the spin configurations in the discrete-time formulation are simply "low-resolution" images of the "true" continuous Feynman paths. Suppose that in the continuous limit, the average distance between spin-flips (kinks) is τ_{kink}. Then, in discrete time, we have on average one kink in every $\tau_{\text{kink}}/\Delta\tau$ time steps. To obtain such a long characteristic length (in units of $\Delta\tau$) in the temporal direction, the effective coupling in this direction must be large.

We have introduced the Swendsen-Wang algorithm as an efficient algorithm for simulating the Ising model (Section 4.4). To apply this method to the discrete-time representation of the transverse-field Ising model (5.17), we switch to Ising variables $\tilde{s}_i = \pm 1$, and couplings $\tilde{K}^x = \frac{1}{4}K^x = \frac{1}{4}\Delta\tau J^z$ and $\tilde{K}^\tau = \frac{1}{4}K^\tau = -\frac{1}{2}\ln(\frac{1}{2}\Delta\tau H^x)$. In the discrete-time cluster simulation, we then

1. Place a horizontal bond that binds two neighboring parallel spins in the space direction with probability $1 - e^{-2\tilde{K}^x} \approx \frac{1}{2}\Delta\tau J^z$.
2. Connect neighboring parallel spins in the time direction with probability $1 - e^{-2\tilde{K}^\tau} \approx 1 - \frac{1}{2}\Delta\tau H^x$.

We can replace these procedures by different ones that do not involve $\Delta\tau$ but are statistically equivalent in the limit $\Delta\tau \to 0$.

To this end, we change the "encoding system" of the configurations. In the discretized imaginary-time representation, we defined a spin on every space-time lattice point. Doing this is obviously impossible when the imaginary time is continuous, so instead of specifying the spin value on every space-time point, we specify for every lattice site only the imaginary-time locations of the kinks (the points at which the local spin value changes) and the spin values just after these kinks. Because the probability of having a kink at a given discrete-time point is proportional to $1/M$ and the number of time points is M, the total number of kinks should not strongly depend on M and should converge to some finite value in the limit $M \to \infty$. As a result, we can express a configuration C in the new Markov process as

$$C = ((\tau_1(1), \tilde{s}_1(1)), (\tau_1(2), \tilde{s}_1(2)), \cdots, (\tau_1(n_1), \tilde{s}_1(n_1)),$$
$$(\tau_2(1), \tilde{s}_2(1)), \cdots (\tau_N(n_N), \tilde{s}_N(n_N))),$$

where $\tau_i(k)$ is the imaginary-time value of the k-th kink (that is, the k-th spin-flip) on the site i, $\tilde{s}_i(k)$ is the value of the Ising spin just after the kink, and n_i is the number of kinks on the site i. (In the case where there is no kink on site i, we just store the spin state.)

The second step in the Swendsen-Wang procedure is equivalent to placing *cuts* between neighboring pairs along the time direction with probability $e^{-2\tilde{K}^\tau} = \frac{1}{2}\Delta\tau H^x$. In the limit of $\Delta\tau \to 0$, this amounts to cutting the segments of constant spins into shorter ones by inserting cuts randomly distributed on the imaginary-time axis according to the uniform probability density $\frac{1}{2}H^x$ (Fig. 5.1).

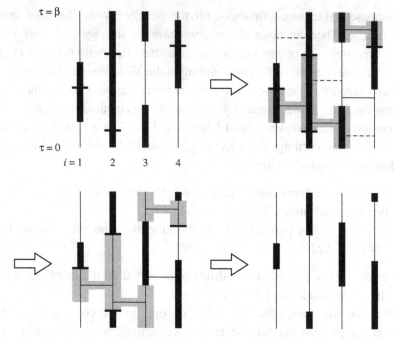

Figure 5.1 Illustration of the continuous-time cluster algorithm for the transverse-field Ising model. The vertical lines represent the imaginary-time intervals of four sites. Spin-up segments are marked by thick lines. The four steps are: (i) insertion of cuts on the imaginary-time intervals, (ii) insertion of nearest neighbor bonds (a bond, represented by a solid horizontal line, is possible only if the spin states on both sides are identical; rejected bonds are marked by dashed lines), (iii) random flipping of the clusters connected by the nearest neighbor bonds, and (iv) elimination of cuts and definition of the new segments. In the second and third panels, we highlight one of the clusters with gray shading.

In the spatial directions, on the other hand, the first step in the discrete-time Swendsen-Wang procedure has us placing *bonds* between two neighboring parallel spins with probability $1 - e^{-2\tilde{K}^x} \sim \frac{1}{2}\Delta\tau J^z$. In the continuous-time limit, these bonds bind two overlapping neighboring segments with density $\frac{1}{2}J^z$ if they have the same spin value. In practice, we propose potential bond insertion times with the uniform probability density $\frac{1}{2}J^z$ (irrespective of the spin states), and then check if the spin states on the two segments connected by these bonds are identical. Only if the spin states at the location of the bond are the same is the bond inserted (solid horizontal line in Fig. 5.1); otherwise, the proposed bond is rejected (dashed horizontal line in Fig. 5.1).

The next step in the Swendsen-Wang algorithm is to identify and randomly flip the clusters of segments bound together by the spatial bonds. To do this, we use a union-find algorithm, such as the Hoshen-Kopelman algorithm (Hoshen and

---

Algorithm 11 Continuous imaginary-time cluster algorithm for the transverse-field Ising model.

Input: A configuration consisting of spin-up and spin-down segments.

    for every site **do**

        Remove all existing cuts at which the spin state does not change ;

        Generate new cuts by selecting the cut times $\tau \in [0, \beta)$ uniformly and randomly with density $\frac{1}{2}H^x$;

    end for

    for every pair of interacting sites **do**

        Select potential bond times $\tau \in [0, \beta)$ uniformly and randomly with density $\frac{1}{2}J^z$;

        for every selected τ **do**

            if the two sites have the same spin orientation at τ **then**

                Insert a horizontal bond connecting the two segments at τ ;

            end if

        end for

    end for

    Identify clusters of connected segments ;

    for every cluster **do**

        With probability $\frac{1}{2}$, flip the spin on all the segments in the cluster ;

    end for

    return the updated configuration.

Kopelman, 1976) discussed in Section 13.4. In Fig. 5.1, one example of a cluster is indicated by the gray shading. Finally, we define new segments by removing obsolete cuts. Algorithm 11 gives the pseudocode for these steps (Rieger and Kawashima, 1999).

Since we record only the positions of the segment end points and the spin state associated with each segment, the memory requirement for this algorithm (at low temperature and for $H^x \neq 0$) is proportional to the number of cuts, which is $\mathcal{O}(\beta H^x N)$. Therefore, its memory requirement is reduced compared with the discrete-time representation by at least a factor of $M/(\beta H^x)$. Although the discrete-time simulation produces qualitatively correct results as long as this factor is greater than unity, we usually need to make it much larger than unity to reduce the systematic error and obtain quantitatively correct results.

We now discuss how to generate the uniformly and randomly distributed times in Algorithm 11. It is important to note that these times and their number n are random variables, and the times are ultimately placed in ascending order, that is, $0 < \tau_1 < \tau_2 < \cdots < \tau_n < \beta$. A stochastic process that generates discrete time-ordered events with a uniform probability density λ is called a homogenous *Poisson process* (Karlin and Taylor, 1975). Not surprisingly, then, we will need to work with the Poisson distribution (2.11),

$$P(N(\beta) = n) = \frac{(\lambda\beta)^n}{n!}e^{-\lambda\beta}, \tag{5.20}$$

where $N(\beta)$ is the number of events occurring in the interval $[0, \beta)$.

We can motivate the appropriateness of this distribution in the following way: As noted in Section 2.3.1, the average number of events occurring in the interval $[0, \beta)$ is $\lambda\beta$ and hence they occur with a uniform density λ. From the functional form, it is easy to show that the Poisson distribution has the following short-time behaviors:

1. When the interval size $\Delta\tau$ is small, the probability that only one event occurs in the time interval $[\tau, \tau + \Delta\tau)$ is $\lambda\Delta\tau$; that is, it is proportional to the length of the interval.
2. The probability that no events occur in this interval is $1 - \lambda\Delta\tau$.
3. The probability that two events occur is zero to $\mathcal{O}((\Delta\tau)^2)$.

The important observation is that the short-time behavior is consistent with our limiting probabilities. The conditions on the events with respect to the number that can occur in a time $\Delta\tau$ are the same as what we implicitly assumed. Hence, to develop our sampling algorithm for the times, instead of building up to a finite interval from short-time properties, we can simply use the Poisson distribution over the targeted finite time interval $[0, \beta)$.

The functional form of the Poisson distribution leads to many interesting properties that provide the rigorous basis for the sampling algorithm. It follows directly from the definition that $P(N(\beta) = 1) = \lambda\beta \exp(-\lambda\beta)$, the exponential distribution, and $P(N(\beta) = 0) = \exp(-\lambda\beta)$. Stated without proof (Karlin and Taylor, 1975) is the important relation that

$$P(N(\tau_1 + \tau_2) - N(\tau_2) = n) = P(N(\tau_1) = n) = \frac{(\lambda\tau_1)^n}{n!}e^{-\lambda\tau_1}, \quad \tau_1, \tau_2 > 0.$$

This relation says that we can break a Poisson process into intervals, and the distribution in each interval is a Poisson distribution. Implicit in this result is the fact that nonoverlapping intervals are statistically independent.

We now use the above properties to generate the sequence of times needed for Algorithm 11. We generate one event at a time. The probability of having one event in any finite interval of time is the exponential probability whose rate constant is λ. In Section 2.3.2, we showed how to sample from this continuous probability density by generating a uniform random number ζ and then calculating

$$\tau_1 = -\ln\zeta/\lambda$$

Algorithm 12 Sequential generation of events $\tau_k < \tau_{k+1} \in [0, \beta)$ with probability density λ.

Input: λ, β.

$\quad \tau = 0$;

$\quad k = 0$;

$\quad$**repeat**

$\qquad$ Generate a uniform random number $\zeta \in (0, 1]$;

$\qquad \tau \leftarrow \tau - \ln \zeta / \lambda$;

$\qquad k \leftarrow k + 1$;

$\qquad \tau_k \leftarrow \tau$;

$\quad$**until** $\tau \geq \beta$.

$\quad n = k - 1$;

$\quad$**return** n and $\{\tau_1, \tau_2, \ldots, \tau_n\}$.

---

to obtain the first event time τ_1. Invoking the statistical independence of Poisson time intervals, we simply repeat the process to obtain $\tau_2 - \tau_1$, the time from the current event time τ_1 to the next one. Then, by the same procedures, we obtain $\tau_3 - \tau_2$, and so on, until $\sum_i \tau_i$ becomes greater than β. This strategy is summarized in Algorithm 12.

While this algorithm is effective, an alternative exists that is generally more efficient because it eliminates the calculation of logarithms. The alternative is also a simple algorithm, but its justification is not as obvious. To develop the idea, we start by considering the case when just one event occurs in the interval $[0, \beta)$ and ask how its location is distributed. We answer this question in two steps. First, we consider a Poisson process with a rate λ that we condition on $N(\beta) = 1$, that is, on only one event occurring in $[0, \beta)$. Then, we consider the arrival time τ_1 relative to any other random time ξ uniformly distributed in the interval $[0, \beta)$. From the definition of a conditional probability and the independence of events in nonoverlapping intervals

$$
\begin{aligned}
P\left(\tau_1 \leq \xi \,|N(\beta) = 1\right) &= \frac{P(\tau_1 \leq \xi, N(\beta) = 1)}{P(N(\beta) = 1)} \\[2mm]
&= \frac{P(N(\xi) = 1, N(\beta) - N(\xi) = 0)}{P(N(\beta) = 1)} \\[2mm]
&= \frac{\lambda \xi e^{-\lambda \xi} e^{-\lambda(\beta - \xi)}}{\lambda \beta e^{-\lambda \beta}} \\[2mm]
&= \frac{\xi}{\beta}.
\end{aligned}
$$

This result says that the arrival time of a single event in the interval $[0, \beta)$ is uniformly distributed over this interval. The Poisson point is located at ξ.

If we were to generate a set of n random numbers $\{\xi_i\}$ uniformly over $[0, \beta)$, we would generate n times $\{t_{\xi_n}\}$ where each was uniformly distributed over the interval $[0, \beta)$. Of the $n!$ arrangements of these times, only one has ascending order. The set could always be placed into this order by sorting. The joint distribution (probability density) of the ordered sequence $0 < t_1 < t_2 < \cdots < t_n < \beta$ is

$$P(\tau_1 = t_1, \ldots, \tau = t_n | N(\beta) = n) = \frac{n!}{\beta^n}.$$

If we generated the needed sequence the way just implied, would this distribution be consistent with an ordered sample of n events from a Poisson distribution?

We can show that it is by again exploiting conditional probabilities and the statistical independences of a Poisson distribution:

$$P(\tau_1 = t_1, \ldots, \tau_n = t_n | N(\beta) = n) = \frac{P(t_1, t_2, \ldots, t_n, N(\beta) = n)}{P(N(\beta) = n)}$$

$$= \frac{P(t_1, t_2 - t_1, \ldots, t_n - t_{n-1}, t_{n+1} > \beta - t_n)}{P(N(\beta) = n)}$$

$$= \frac{\lambda^n e^{-\lambda\beta}}{P(N(\beta) = n)}$$

$$= \frac{n!}{\beta^n}.$$

In going from the first to the second equation, we invoked the Poisson nature of times and intervals of time. In going from the second to the third equation, we noted that in each interval the distribution is Poisson and, except of the last interval, that only one event occurs in each interval. We went from the third to fourth equation by using the definition of the Poisson distribution. Algorithm 13 expresses this result. For a given β and λ, the first part computes $N(\beta) = n$, the second part generates the unordered set of τ_i, and the last step produces the desired result by sorting the values in this set.

5.2.4 Zero-field XY model

In this section and the following three, we discuss various basic concepts using the zero-field, one-dimensional, spin-$\frac{1}{2}$ quantum XY model

$$H = -\sum_{i=1}^{N} J^{xy}\left(S_i^x S_{i+1}^x + S_i^y S_{i+1}^y\right) \tag{5.21}$$

as the example. The algorithms that we discuss for this model, the loop algorithm (Section 5.2.5) and the worm algorithm (Section 5.2.6), are representative

---

Algorithm 13 Faster generation of uniformly distributed random events $\tau_k \in [0, \beta)$ with density λ.

Input: λ, β.

  $n = 0$;

  $d = e^{-\beta\lambda}$;

  $p = d$;

  Generate a uniform random number $\zeta \in [0, 1]$;

  while $\zeta > p$ **do**

    $n \leftarrow n + 1$;

    $d \leftarrow d\beta\lambda/n$;

    $p \leftarrow p + d$;

  end while

  for $k = 1$ to n **do**

    Generate a uniform random number $\tau_k \in [0, \beta)$;

  end for

  Sort the set $\{\tau_k\}$ in ascending order ;

  return $\{\tau_k\}$.

---

of important classes of quantum Monte Carlo algorithms that have a broad range of application, as we will see in Chapter 6.

Again, we start by expressing the partition function in terms of classical degrees of freedom. While no term in this model is diagonal in the S^z basis (5.5), we continue to use this basis because of its convenience. Noting that spin operators on different sites commute, and hence all pairs of spin operators starting on odd (even) sites commute, we split H into two parts by writing

$$H = H_{\text{odd}} + H_{\text{even}},$$

where

$$H_{\text{odd}} = -\sum_{i=1}^{N/2} J^{xy}\big(S_{2i-1}^x S_{2i}^x + S_{2i-1}^y S_{2i}^y\big),$$

$$H_{\text{even}} = -\sum_{i=1}^{N/2} J^{xy}\big(S_{2i}^x S_{2i+1}^x + S_{2i}^y S_{2i+1}^y\big).$$

Making the Trotter approximation for the partition function, we find that

$$Z = \text{Tr}\Big[\big(e^{-\Delta\tau H_{\text{odd}}} e^{-\Delta\tau H_{\text{even}}}\big)^M\Big],$$

and then inserting a complete set of states between each factor, we find that the basic matrix element we need to evaluate is

$$w\begin{pmatrix} s_{i,k+1} & s_{i+1,k+1} \\ s_{i,k} & s_{i+1,k} \end{pmatrix} \equiv \langle s_{i,k+1}s_{i+1,k+1}| e^{\Delta\tau J^{xy}\left(S_i^x S_{i+1}^x + S_i^y S_{i+1}^y\right)} |s_{i,k}s_{i+1,k}\rangle.$$

Evaluating it is relatively straightforward. First, for $i \neq j$, and with $S^+ = (S_x + iS_y)$, $S^- = (S_x - iS_y)$,

$$S_i^x S_j^x + S_i^y S_j^y = \tfrac{1}{2}\left(S_i^+ S_j^- + S_i^- S_j^+\right).$$

Using this in the argument of the exponential and expressing the exponential as a power series, we then explicitly consider term by term the possible matrix elements, finding that

$$\begin{pmatrix} w\left(\substack{\uparrow\uparrow\\\uparrow\uparrow}\right) & w\left(\substack{\uparrow\downarrow\\\uparrow\uparrow}\right) & w\left(\substack{\downarrow\uparrow\\\uparrow\uparrow}\right) & w\left(\substack{\downarrow\downarrow\\\uparrow\uparrow}\right) \\ w\left(\substack{\uparrow\uparrow\\\uparrow\downarrow}\right) & w\left(\substack{\uparrow\downarrow\\\uparrow\downarrow}\right) & w\left(\substack{\downarrow\uparrow\\\uparrow\downarrow}\right) & w\left(\substack{\downarrow\downarrow\\\uparrow\downarrow}\right) \\ w\left(\substack{\uparrow\uparrow\\\downarrow\uparrow}\right) & w\left(\substack{\uparrow\downarrow\\\downarrow\uparrow}\right) & w\left(\substack{\downarrow\uparrow\\\downarrow\uparrow}\right) & w\left(\substack{\downarrow\downarrow\\\downarrow\uparrow}\right) \\ w\left(\substack{\uparrow\uparrow\\\downarrow\downarrow}\right) & w\left(\substack{\uparrow\downarrow\\\downarrow\downarrow}\right) & w\left(\substack{\downarrow\uparrow\\\downarrow\downarrow}\right) & w\left(\substack{\downarrow\downarrow\\\downarrow\downarrow}\right) \end{pmatrix}$$

$$= \begin{pmatrix} 1 & 0 & 0 & 0 \\ 0 & \cosh\tfrac{1}{2}\Delta\tau J^{xy} & \sinh\tfrac{1}{2}\Delta\tau J^{xy} & 0 \\ 0 & \sinh\tfrac{1}{2}\Delta\tau J^{xy} & \cosh\tfrac{1}{2}\Delta\tau J^{xy} & 0 \\ 0 & 0 & 0 & 1 \end{pmatrix}. \tag{5.22}$$

A number of things have happened. The odd-even site breakup of the Hamiltonian and the Trotter decomposition created a structure consisting of $2M$ layers in the imaginary-time direction. Pairs of spin operators starting at even sites act on layers $k = 1, 3, 5, \ldots$, while pairs starting at odd sites act on layers $k = 2, 4, 6, \ldots$ Two subsequent layers correspond to the propagation of the full Hamiltonian by one time-step $\Delta\tau$.

The local transition matrix depends only on the values of two adjacent spin variables on one layer and those of the same pair of sites on the next layer. These four spins are located at the positions (i, k), $(i+1, k)$, $(i+1, k+1)$, and $(i, k+1)$, which define a *plaquette*. We notice that these plaquettes naturally decompose space-time into a checkerboard structure (Fig. 5.2) and that we need to sample only the light or the dark plaquettes.

Associated with each plaquette are 16 possible spin configurations. However, (5.22) says that only six of them are allowed. The nonzero matrix elements express a *conservation law*,

$$s_{i,k} + s_{i+1,k} = s_{i,k+1} + s_{i+1,k+1}, \tag{5.23}$$

for the z-component of total angular momentum at each plaquette. Accordingly, the total z-component of angular momentum $S^z \equiv \sum_i S_i^z$ is conserved at each imaginary

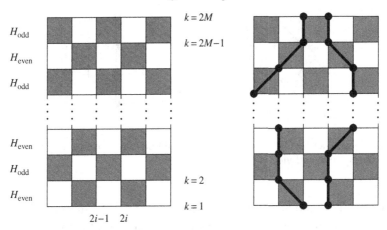

Figure 5.2 The checkerboarding of space and imaginary time in one dimension (left). World lines on the checkerboard (right).

time. In other words, the number of up-spins on a spatial plane is independent of the imaginary time. This is an example of a reduction of phase space due to symmetry. While the quantum Ising model was invariant under spin-reversal, the *XY* model is invariant under arbitrary rotations of all spins about the z-axis. This invariance conserves S^z. This conservation law demands that if we change the number of up-spins on a spatial plane during the Monte Carlo simulation, we must change it in the same way on all the other planes as well. A peculiar implication of the invariance and conservation law is that the value of S^z that results from our initialization of the simulation is the value that remains throughout, unless we introduce some "global flip" procedure.

Let us emphasize that (5.23) not only implies the conservation of the total S^z but expresses a local conservation law, involving only the four spin variables of a plaquette. This restriction of the phase space leads to a challenge in defining a sampling strategy. Going site to site and proposing local spin reversals via the Metropolis algorithm is not consistent with the conservation law. An ergodic sampling must involve multiple spins simultaneously.

If we mark the corners of the shaded plaquettes that have an up-spin with a dot and connect the dots with a line, with the rule that on the all-up plaquette the lines run up the left and the right edges (so the lines do not intersect), we obtain a collection of continuous paths, looping in the imaginary-time direction (see right panel of Fig. 5.2). Each such path is called a *world line*. Monte Carlo sampling then becomes a matter of deforming the world lines. There is a particularly convenient way of doing this while maintaining spin conservation: We visit each *unshaded* plaquette in turn, and if there is a segment running along one edge and not the

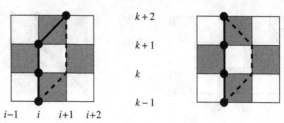

Figure 5.3 Basic world-line update. The world line may move from the solid line to the dashed one and vice versa.

other, we attempt to move it from one edge to the other. We illustrate these moves in Fig. 5.3, where we assume that the unshaded plaquette is located at

$$\begin{pmatrix} s_{i,k+1} & s_{i+1,k+1} \\ s_{i,k} & s_{i+1,k} \end{pmatrix}.$$

The move then is accepted or rejected via a Metropolis decision based on

$$P_{\text{accept}} \equiv \min[1, R_{\text{north}} R_{\text{south}} R_{\text{east}} R_{\text{west}}],$$

where

$$R_{\text{north}} = \frac{w\begin{pmatrix} s_{i,k+2} & s_{i+1,k+2} \\ -s_{i,k+1} & -s_{i+1,k+1} \end{pmatrix}}{w\begin{pmatrix} s_{i,k+2} & s_{i+1,k+2} \\ s_{i,k+1} & s_{i+1,k+1} \end{pmatrix}}, \quad R_{\text{south}} = \frac{w\begin{pmatrix} -s_{i,k} & -s_{i+1,k} \\ s_{i,k-1} & s_{i+1,k-1} \end{pmatrix}}{w\begin{pmatrix} s_{i,k} & s_{i+1,k} \\ s_{i,k-1} & s_{i+1,k-1} \end{pmatrix}},$$

$$R_{\text{east}} = \frac{w\begin{pmatrix} -s_{i+1,k+1} & s_{i+2,k+1} \\ -s_{i+1,k} & s_{i+2,k} \end{pmatrix}}{w\begin{pmatrix} s_{i+1,k+1} & s_{i+2,k+1} \\ s_{i+1,k} & s_{i+2,k} \end{pmatrix}}, \quad R_{\text{west}} = \frac{w\begin{pmatrix} s_{i-1,k+1} & -s_{i,k+1} \\ s_{i-1,k} & -s_{i,k} \end{pmatrix}}{w\begin{pmatrix} s_{i-1,k+1} & s_{i,k+1} \\ s_{i-1,k} & s_{i,k} \end{pmatrix}} \quad (5.24)$$

are the ratios of the local weights before and after the proposed change in the world-line configuration, for plaquettes to the north, south, east, and west of the unshaded plaquette. The proposal is reversing the spins on the border of the unshaded plaquette, provided the spin states on the two sides are different (Algorithm 14).

We notice from (5.22) that we have a sign problem if $J^{xy} < 0$. If our lattice is *bipartite*, that is, the sites separate into an A and B sublattice so that only B sites are nearest neighbors to A sites and vice versa, then we can remove the problem by performing a canonical transformation that reverses the spins on one sublattice (Marshall, 1955). Square lattices are bipartite, while triangular lattices

---

Algorithm 14 World-line Monte Carlo method for the one-dimensional spin-$\frac{1}{2}$ XY model with local updates. (5.22) defines the weights w in (5.24).

Input: A spin configuration $\{s_{i,k}\}$.

    for every unshaded plaquette $(s_{i,k}, s_{i+1,k}, s_{i+1,k+1}, s_{i,k+1})$ **do**

        if $s_{i,k} = s_{i,k+1}$ and $s_{i+1,k} = s_{i+1,k+1}$ and $s_{i,k} = -s_{i+1,k}$ **then**

            Compute $R_{\text{north}}, R_{\text{south}}, R_{\text{east}}, R_{\text{west}}$ by (5.24) ;

            Generate a uniform random number $\zeta \in [0, 1]$;

            if $\zeta < R_{\text{north}} R_{\text{south}} R_{\text{east}} R_{\text{west}}$ **then**

                $s_{i,k} \leftarrow -s_{i,k}$;

                $s_{i+1,k} \leftarrow -s_{i+1,k}$;

                $s_{i+1,k+1} \leftarrow -s_{i+1,k+1}$;

                $s_{i,k+1} \leftarrow -s_{i,k+1}$;

                    $\triangleright$ Need to account for boundary conditions.

            end if

        end if

    end for

    return the updated spin configuration.

---

are not. Nonbipartite lattices often frustrate local spin configurations. For frustrated systems, the sign problem generally persists.

As in the case of the discrete-time algorithm for the transverse-field Ising model, we have to remove the systematic error caused by the Trotter approximation by extrapolating to the limit $M \to \infty$. Fortunately, we can again replace the discrete-time algorithm by a continuous-time one. Since our restructuring of the partition function of the XY model did not produce one analogous to the classical model, we cannot directly apply the Swendsen-Wang algorithm and proceed from there. However, as discussed in the next subsection, we can do something analogous.

5.2.5 Simulation with loops

We now generalize the cluster scheme, which was first introduced in Section 4.4 and reformulated for the transverse-field Ising model in Section 5.2.3, to the path-integral representation of the partition function of quantum systems. Here we use the XY model as our example. The generalization to the transverse-field Ising model discussed in Section 5.2.3 was rather straight forward through the mapping of the path integral with discretized imaginary time to the $(d + 1)$-dimensional classical Ising model. The generalization to the XY model in the present section involves something more, namely, the introduction of various types of graph elements

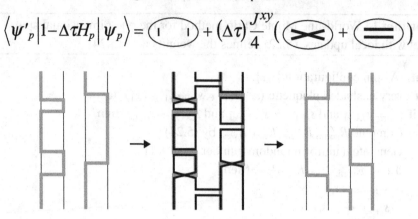

$$\left\langle \psi'_p \left| 1 - \Delta\tau H_p \right| \psi_p \right\rangle = \text{(} \text{I} \quad \text{I} \text{)} + (\Delta\tau)\frac{J^{xy}}{4}\left(\text{(}\boldsymbol{\times}\text{)} + \text{(}\boldsymbol{=}\text{)}\right)$$

Figure 5.4   Graph elements of the loop algorithm for the spin-$\frac{1}{2}$ XY model, and one Monte Carlo step.

according to the nature of the interaction terms. In the present section, we focus on "how" rather than "why" and postpone a detailed derivation and justification of the proposed cluster-type algorithm to Chapter 6.

As in the case of the transverse-field Ising model, we obtain the continuous-time version of this algorithm simply by reinterpreting stochastic assignments by distributing cuts and bonds to imaginary-time intervals with a certain density, instead of assigning them to individual plaquettes. While we focus on the one-dimensional spin-$\frac{1}{2}$ XY model, the generalization of the cluster updating scheme to higher dimensions and to other types of Hamiltonians is straightforward in many cases (Chapter 6).

The first step is to express the local imaginary-time evolution operator in terms of Kronecker deltas. We start with essentially the same matrix elements of the imaginary-time evolution operator as (5.22) but correct only up to the first order in $\Delta\tau$.[3] For convenience, we add a constant $-\frac{1}{4}J^{xy}$ to the pair Hamiltonian and then express the matrix elements as

$$\langle s'_i s'_j | 1 - \Delta\tau H_{ij} | s_i s_j \rangle$$

$$= \langle s'_i s'_j | s_i s_j \rangle + \tfrac{1}{2}\Delta\tau J^{xy} \langle s'_i s'_j | (S_i^+ S_j^- + S_i^- S_j^+) + \tfrac{1}{2} | s_i s_j \rangle$$

$$= \delta_{s'_i,s_i}\delta_{s'_j,s_j} + \tfrac{1}{4}\Delta\tau J^{xy}(\delta_{s'_i,s_j}\delta_{s'_j,s_i} + \delta_{s_i,-s_j}\delta_{s'_i,-s'_j}). \qquad (5.25)$$

There are the three terms, which we identify with the three different graph elements, encircled in Fig. 5.4. To convince ourselves that this graph assignment

---

[3] This "approximation" does not yield any systematic error in the end because we take the continuous imaginary-time limit.

is reasonable, let us examine the contributions of the different plaquette configurations. The second $\delta_{s'_i,s_j}\delta_{s'_j,s_i}$ term contributes $\frac{1}{4}\Delta\tau J^{xy}$ in the plaquette configurations

$$\begin{pmatrix} \uparrow & \uparrow \\ \uparrow & \uparrow \end{pmatrix}, \quad \begin{pmatrix} \downarrow & \downarrow \\ \downarrow & \downarrow \end{pmatrix}, \quad \begin{pmatrix} \uparrow & \downarrow \\ \downarrow & \uparrow \end{pmatrix}, \quad \begin{pmatrix} \downarrow & \uparrow \\ \uparrow & \downarrow \end{pmatrix},$$

while the third $\delta_{s_i,-s_j}\delta_{s'_i,-s'_j}$ term contributes $\frac{1}{4}\Delta\tau J^{xy}$ in the configurations

$$\begin{pmatrix} \uparrow & \downarrow \\ \uparrow & \downarrow \end{pmatrix}, \quad \begin{pmatrix} \downarrow & \uparrow \\ \downarrow & \uparrow \end{pmatrix}, \quad \begin{pmatrix} \uparrow & \downarrow \\ \downarrow & \uparrow \end{pmatrix}, \quad \begin{pmatrix} \downarrow & \uparrow \\ \uparrow & \downarrow \end{pmatrix}.$$

In all the configurations contributing to the second term, the spin variables located on the diagonals are parallel, whereas in those contributing to the third term, the spin variables in the same row are antiparallel.

Expression (5.25) leads us to the continuous-time cluster algorithm for the XY model in question. Specifically, we identify the first term with a "vertical" graph element, the second term with a "diagonal" graph element, and the third term with a "horizontal" graph element. Let us first consider the positions where the world lines have no kinks. The vertical graph element corresponds to the identity operator. As for the diagonal graph element, because its weight is $\frac{1}{4}J^{xy}\Delta\tau$ when $s_i = s_j$, we generate this graph element with density $\frac{1}{4}J^{xy}$ on the imaginary-time interval if the two spins are parallel to each other. This graph element imposes the restriction on the local spin configuration that two spins located at diagonally opposite corners must be parallel to each other, and we thus represent it by a pair of (disconnected) diagonal lines, as in Fig. 5.4. Similarly, the third term in (5.25) gives the weight $\frac{1}{4}J^{xy}\Delta\tau$ for $s_i = -s_j$. Therefore, we insert the corresponding graph element with density $\frac{1}{4}J^{xy}$ to intervals with antiparallel spins. The Kronecker delta functions representing this term indicate that we bind the antiparallel spins horizontally, again as illustrated in Fig. 5.4. At the positions where the world lines have kinks, only diagonal or horizontal elements can match. As the weight of these two are equal, we choose one of these elements with equal probability for each kink.

Because all three graph elements connect the four points on the plaquette pairwise, the resulting clusters are loops, as illustrated in the middle panel of Fig. 5.4 (hence the name *loop algorithm*). Whether a cluster algorithm becomes a loop algorithm or not depends on the nature of the Hamiltonian.

Taking the continuous-time limit $\Delta\tau \to 0$, similar to the case of the transverse-field Ising model, just becomes a matter of switching from the lattice point representation to the segment representation and placing the graph elements with appropriate density, except at the locations of the kinks. Algorithm 15 summarizes the loop algorithm in its continuous-time version.

Algorithm 15 Loop algorithm for the spin-$\frac{1}{2}$ *XY* model.

Input: A world-line configuration ;
    for all kinks **do**
        Select a diagonal or horizontal graph element with probability $\frac{1}{2}$, and assign it
        to the kink ;
    end for
    for all pairs of neighboring sites **do**
        Generate graph positions $\tau \in [0, \beta)$ with density $\frac{1}{4}J^{xy}$;
        Place diagonal (horizontal) graph elements if the spins are parallel
        (antiparallel) ;
    end for
    Identify loops ;
    for all loops **do**
        Flip the loop with probability $\frac{1}{2}$;
    end for
    return the updated world-line configuration.

5.2.6 Simulation with worms

Instead of decomposing the entire lattice into loops and computing a new world-line configuration by random flipping of these loops, we can also update the world-line configuration by creating worms and letting them move, as in the worm update for the Ising model discussed in Section 4.5. Here the essential idea is the same: Extend the configuration space by introducing points of discontinuity, generalize the weight accordingly, and construct the Markov transition matrix such that it satisfies the detailed balance with the generalized weight. In some simple cases, including the one for the $S = \frac{1}{2}$ *XY* model shown in the present section, the resulting algorithm is very close to the loop algorithm discussed in the previous subsection, with the only major difference that loops are created and flipped one by one sequentially, in contrast to the loop algorithm in which all loops are constructed simultaneously. Because of this the algorithm is also referred to as the *directed loop algorithm*.[4] In general, however, there is no direct correspondence between the worm algorithm and the loop algorithm. We look into the details of the worm and directed loop algorithms in Chapter 6. The following simple example is a "sneak preview" without a rigorous derivation of the procedure.

    The first step of the algorithm is defining the locations of potential kink positions. But instead of inserting specific graph elements, we insert "vertices." A vertex is

---

[4] The word *directed* comes from the fact that the worm head has a direction.

similar to a graph element in the loop algorithm, but the type of the graph element on a vertex is not predetermined. It is more like a scatterer at which the worm head changes its course in a nondeterministic way.

Specifically, after removing the vertices from the previous cycle, we place a vertex on every existing kink, and with density $\frac{1}{4}J^{xy}$ place additional vertices between each pair of neighboring sites. Then we put the head and tail of the worm at a randomly chosen point in space-time and randomly choose the initial direction (forward or backward in time) of the head's motion. Suppose the selected direction is "forward." In this case, the head moves forward in time until it hits the next vertex.[5] If the vertex is on a kink, then with probability $\frac{1}{2}$ we assign a graph element (diagonal or horizontal graph) to the vertex. If the vertex is not on a kink, then the graph element is determined by the world-line configuration: If the vertex is located between two parallel spins, the graph element is a diagonal graph; otherwise, it is a horizontal graph. In either case, the worm head scatters at the vertex. In the case of a diagonal graph, the scattering means that the head continues to move forward on the neighboring site, while a horizontal graph element implies that the head reverses direction and moves backward on the neighboring site. Following these rules, the worm head eventually returns to its original position and annihilates with its tail. At this point the world-line configuration has been updated by a single loop, and we place another worm head and tail at some random position and repeat the procedure. This cycle is illustrated in Fig. 5.5.

A noteworthy subtlety is that the assignment of graph elements to the vertices may change during the worm's excursion through space-time and that the worm head may scatter multiple times at the same vertex in different ways. For example, in Fig. 5.5, there is only one vertex that is actually visited by the head, and it is visited twice. At the first encounter, the head is scattered diagonally, whereas the second time, it is scattered horizontally. The graph element assigned to this vertex is not fixed but stochastically determined every time the head visits the vertex. In this sense, the algorithm is not exactly identical to a sequential execution of the loop algorithm. While we used the phrase "assign graph elements to vertices" in the earlier discussion to emphasize the connection to the loop algorithm, "scattering of the head at the vertices" is a more appropriate description of the process.

After N_{cycle} worm update cycles, we remove the vertices and start over, that is, place a new set of vertices, perform N_{cycle} worm updates for this set of vertices, and so on. A pseudocode for this algorithm is given in Algorithm 16.

In the original worm algorithm, as proposed in Prokov'ev et al. (1998), the kink positions are inserted on the fly during the worm's trajectory, using special updates

---

[5] If the worm is placed on a world line, then the head erases this world line segment; otherwise, it draws a new world line.

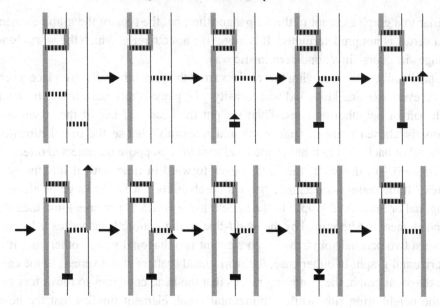

Figure 5.5　Illustration of the directed-loop algorithm with worm update for the spin-$\frac{1}{2}$ XY model.

that insert or remove kinks. Whether or not we separate the vertex generation from the worm motion usually does not make much difference in terms of efficiency. However, the parallelized implementation of the worm algorithm (Masaki et al., 2013) is based on prefixed vertex positions. On the other hand, when the system is very dilute, most of the vertices are never visited by the head, as the head motion is always accompanied by a particle creation/annihilation, which can seriously reduce the efficiency. In such cases, the vertex generation on the fly may be very advantageous because vertices in the deserted area are not even created. In Section 6.4.6, we describe a method that is essentially the same as the original worm algorithm but more similar to the algorithm depicted in Fig. 5.5.

Let us finally note that although we took the same spin-$\frac{1}{2}$ XY model as an example to make the correspondence clearer between the loop and the worm algorithm, introducing worms is not actually necessary for this model, as the loop algorithm works fine in this case. The loop update overcomes the (computational) critical slowing down from which the discrete-time world-line algorithm with local updates suffers. In addition, it is relatively easy to split the computational task of the loop update across many processors simply by splitting the whole space-time domain into small blocks.[6] For other models, however, the worm algorithm can be much

---

[6] While identifying large loops that do not fit into a single block is technically challenging, a good solution to this problem exists (Todo et al., 2012).

Algorithm 16 Directed-loop algorithm with worm update for the spin-$\frac{1}{2}$ XY model.

Input: A world-line configuration.

   Place vertices on all kinks ;

   for all pairs of neighboring sites **do**

      Generate new vertex positions $\tau \in [0, \beta)$ with density $\frac{1}{4}J^{xy}$, and place vertices

      there ;

   end for

   for $k = 1$ to N_{cycle} **do**

      Pick a point uniform-randomly in the whole space-time domain ;

      Place the head and the tail there ;

      Choose the initial direction of the head with probability $\frac{1}{2}$;

      loop

         Move the head to the next vertex ;

         if the head hits the tail **then**

            Let the head and the tail annihilate and exit the loop ;

         else if the head hits a vertex on a kink **then**

            Scatter the head diagonally or horizontally with probability $\frac{1}{2}$;

         else

            Scatter the head according to the world-line configuration ;

             ▷ There is only one way to satisfy the local conservation law.

         end if

      end loop

   end for

   return the updated world-line configuration.

---

more efficient. The performance of the loop update is generally very poor when the Hamiltonian contains mutually competing terms. A typical example is the anti-ferromagnetic Heisenberg model with a uniform external magnetic field, in which the exchange coupling favors an antiparallel alignment of neighboring spins while the external field favors the opposite. If the temperature is low, the loop algorithm experiences a *bottleneck*: The path between antiferromagnetic and ferromagnetic alignments is blocked by a high free-energy barrier, even when both give a non negligible contribution to the thermodynamic distribution. Thus, even though the notorious negative-sign problem does not exist for this example, the loop update has an equilibration problem at low temperature. A fix is possible by exploiting the aforementioned "subtlety" that the worm update does not involve "fixed" graph elements, in contrast to the loop algorithm. Namely, in the worm algorithm, all the matrix elements competing with each other can be taken into account in the scattering probability of the worms. (Hence cancellation of the frustration.) The worm algorithm is more flexible than the loop algorithm when it comes to decomposing the interactions into graph elements.

5.2.7 Ergodicity and winding numbers

Strictly speaking, the world-line method with local updates is not ergodic. It conserves the *winding numbers*. As we have discussed, the loop or worm algorithm replaces local plaquette sampling with a global sampling, a change that allows winding numbers to fluctuate and makes measurements of unequal time correlation functions and simulations in the grand canonical ensemble more accessible. While the efficiency of the algorithm may not depend strongly on the difference between the ensembles, by working in the grand canonical ensemble we obtain direct access to some interesting physical quantities.

If spatial periodic boundary conditions are used (the imaginary-time direction is always periodic), the space-time lattice is on a $(d + 1)$-dimensional torus. The winding number is the number of times the path wraps around the torus before it comes back to its starting point. The winding number in the spatial direction is closely related to the superfluid density in Bose systems and the helicity modulus in spin systems (Pollock and Ceperley, 1987). Specifically, we obtain the superfluid density ρ_s from the winding number W_μ in the μ ($= x, y, \ldots$) direction as

$$\rho_s = \frac{L_\mu^2}{\Omega} \langle W_\mu^2 \rangle, \tag{5.26}$$

where $\Omega \equiv L_\tau L_x L_y \cdots$ with $L_\tau \equiv 2t\beta$ in Boson models with hopping constant t. In the *XY* model, $L_\tau = J^{xy}\beta$ is the dimensionless length in the temporal direction. Likewise, we obtain the dimensionless compressibility $\tilde{\kappa} \equiv 2t\partial\langle n\rangle/\partial\mu$ as

$$\tilde{\kappa} = \frac{L_\tau^2}{\Omega} \langle W_\tau^2 \rangle, \tag{5.27}$$

where W_τ is the temporal winding number. These quantities are exponentially small in the normal state (ρ_s) and in the incompressible state ($\tilde{\kappa}$). Therefore, for large lattices, nonzero winding number configurations are highly improbable in such states, so the seemingly disastrous nonergodicity of the local update is actually not a serious problem. Even in superfluid or compressible states, thermodynamic quantities are usually insensitive to the difference between the canonical and the grand canonical ensembles. However, working in the grand canonical ensemble is often advantageous, as ρ_s and κ are often the relevant response functions that characterize the state of the quantum system.

5.3 Bosons and Fermions

Quantum Monte Carlo simulations are not restricted to quantum spin systems at finite temperature. Systems of Bosons and Fermions are of considerable interest, as are the ground-state properties of quantum many-body systems. Not all finite-temperature algorithms are easily formulated in continuous time via loop/cluster or worm algorithms. Even if so, the efficiency of these algorithms is not guaranteed. Most quantum Monte Carlo algorithms, whether for quantum spins, Bosons, or Fermions, are potential victims of a sign problem.

Algorithms for Fermionic systems have characteristic features that distinguish them from quantum spin and Boson algorithms. Some Boson systems, however, are simulated with algorithms that are very similar to those for quantum spins. One-dimensional systems with different quantum statistics map from one to another. In some cases this equivalence might allow algorithmic adoption as opposed to algorithmic creation. In the following subsections, we briefly address these topics.

5.3.1 Bosons

Today, the loop and worm algorithms are the standard methods for simulating quantum spin models not only in one dimension, but also in two and three. They mesh nicely with the short-range exchange interactions typical of these models. Loop and worm algorithms are similarly useful for the simulation of *hard-core Boson* models.[7] The derivations of world-line algorithms for this quantum statistics proceed in a manner analogous to that for quantum spins, with the basis states being the eigenstates of the Boson number operators.

In one dimension, a shortcut exists. Instead of deriving the transition probabilities for a Bosonic world-line algorithm anew, we use the *Matsubara-Matsuda transformation* (Matsubara and Masuda, 1956) to map a hard-core Boson model to a spin-$\frac{1}{2}$ model and then adopt the spin algorithms.

Specifically, the Matsubara-Masuda transformation

$$S_i^+ \leftrightarrow a_i^\dagger, \quad S_i^- \leftrightarrow a_i, \quad S_i^z + \tfrac{1}{2} \leftrightarrow a_i^\dagger a_i = n_i$$

maps a hard-core Boson onto a spin-$\frac{1}{2}$ and vice versa. For example, if we apply this transformation to the longitudinal-field *XXZ* model

$$H = -\sum_{i=1}^{N} \left[J^{xy} \left(S_i^x S_{i+1}^x + S_i^y S_{i+1}^y \right) + J^z S_i^z S_{i+1}^z + H^z S_i^z \right]$$

---

[7] A hard-core-Boson model is one where only a finite number of Bosons are allowed to occupy a given state.

we obtain

$$H = -t \sum_{i=1}^{N} \left(a_i^\dagger a_{i+1} + a_{i+1}^\dagger a_i \right) - \mu \sum_{i=1}^{N} n_i + V \sum_{i=1}^{N} n_i n_{i+1} + \text{const.},$$

where $t = \frac{1}{2}J^{xy}$, $V = J^z$, and $\mu = H^z - J^z$. The resulting model is called the extended hard-core *Bose-Hubbard model*. The word *extended* comes from the fact that we have a nearest neighbor coupling, whereas the word *hard-core* indicates that the on-site repulsion is infinite. In Section 6.6, we will discuss the standard Bose-Hubbard model in which we have only an on-site repulsion, which is finite.

We see that the mapping is quite direct. Not surprisingly, techniques used to eliminate potential sign problems from quantum spin algorithms are also useful here. For these two classes of quantum statistics, the sign of the matrix elements of the Hamiltonian is usually dictated by the signs of the model parameters. This distinguishes the construction of spin and Boson algorithms from the challenges of constructing them for Fermions.

5.3.2 Fermions

The derivations of world-line algorithms for Fermions proceed in a manner analogous to that for quantum spins. In exponentiating the pieces of the Hamiltonian, slight differences occur to account for the different quantum statistics. The basis states are typically the eigenstates of the Fermion number operators. It is not that algorithms cannot be constructed, but rather it is difficult to formulate ones that work well.

To better understand the source of the difficulty, we consider a special case in which the *Jordan-Wigner transformation* (Jordan and Wigner, 1928) maps one-dimensional Fermion models onto quantum spin-$\frac{1}{2}$ models. Let us apply this transformation, defined by

$$c_i = K_i S_i^+, \quad c_i^\dagger = S_i^- K_i^\dagger, \quad K_i \equiv e^{i\pi \sum_{j=1}^{i-1} S_j^+ S_j^-},$$

to the one-dimensional, noninteracting, *spinless Fermion model*

$$H = -t \sum_{i=1}^{N} \left(c_i^\dagger c_{i+1} + c_{i+1}^\dagger c_i \right).$$

The operator K_i, called the *kink operator*, counts the number of down-to-up spin flips that appear to the left of i. This automatically accounts for the Fermion sign coming from the permutation of two particles. The kink operator is unitary and self-conjugate,

$$K_i^\dagger = K_i, \quad K_i^2 = I,$$

and commutes with $S_j^{\pm}$ for $j \geq i$. Using

$$c_i^{\dagger} c_{i+1} = S_i^{+} S_{i+1}^{-}, \quad c_{i+1}^{\dagger} c_i = S_{i+1}^{+} S_i^{-}, \quad c_i^{\dagger} c_i = S_i^{+} S_i^{-}, \quad (5.28)$$

we find

$$H = -2t \sum_{i=1}^{N} \left(S_i^{x} S_{i+1}^{x} + S_i^{y} S_{i+1}^{y} \right),$$

the *XY* model with $J^{xy} = 2t$. Hence, we can simulate spinless Fermions by using the algorithms discussed for this spin model.[8] When we add Fermion spin into the problem, we have to introduce world lines for the up-spins and for the down-spins. The simulation then requires an additional Monte Carlo step to accept or reject world-line movements on the basis of the interactions between up- and down-spins (Kawashima et al., 1994).

Let us emphasize the kink operator in the mapping between the spin raising and lowering operators and the creation and destruction operators. A similar operator did not exist in the Matsubara-Matsuda transformation. We recall that our basis states are direct product states listing the lattice sites in a specific order. As long as the lattice sites are sequentially numbered, the Jordan-Wigner transformation applies to any dimension. What makes it difficult to generalize the above one-dimensional solution to higher dimensions is that it is impossible to number the sites in such a way that all interacting pairs appear next to each other in the list. (If it is possible, it means that the lattice is one-dimensional.) To be more specific, the general form of (5.28) is

$$c_i^{\dagger} c_j = e^{i\pi \sum_{k=i+1}^{j-1} S_k^{+} S_k^{-}} S_i^{+} S_j^{-},$$

with an ugly phase factor that does not vanish if $|j - i| > 1$ and makes the problem unsolvable in general. This sign problem for all practical purposes restricts the use of the world-line method to one-dimensional systems.

In short, the source of the difficulty in simulating Fermion problems is the sign arising from the permutation among Fermions that is represented by the phase factor $e^{i\pi \sum_{k=i+1}^{j-1} n_k} = \pm 1$. If we forget about this phase factor for the time being, the Fermion Hamiltonian is identical to that of hard-core Bosons or $S = \frac{1}{2}$ spin models. We can therefore apply the loop/cluster (or any other) algorithms for spins or Bosons to Fermions as well. However, to take into account the phase

---

[8] Historically, the inverse transformation

$$S_i^{+} = c_i K_i, \quad S_i^{-} = c_i^{\dagger} K_i,$$

mapping the *XY* model to the easily solvable spinless Fermion model, was used to solve the former exactly.

factor in a way that does not lead to severe cancellations between different Monte Carlo configurations is very difficult, and it is called "Fermion sign problem" or "negative-sign problem." There is no general solution to this problem to this day.

5.4 Negative-sign problem

The negative-sign problem is the biggest bottleneck in numerical simulations of quantum models. For the algorithms considered here, it originates from the negative matrix elements of the Hamiltonian. When some of the off-diagonal matrix elements of the local Hamiltonian H are negative, the weight $w(C)$ generally is negative for some of the configurations C. In such a case, we perform a Monte Carlo simulation for which the target distribution is $|w(C)|$, not $w(C)$. Then, we compute the expectation value of an observable Q using the identity

$$\langle Q \rangle = \frac{\sum_C |w(C)| \operatorname{sign}(C) Q(C)}{\sum_C |w(C)| \operatorname{sign}(C)}$$

$$= \frac{\sum_C |w(C)| \operatorname{sign}(C) Q(C)/ \sum_C |w(C)|}{\sum_C |w(C)| \operatorname{sign}(C)/ \sum_C |w(C)|}$$

$$= \frac{\langle \operatorname{sign}(C) Q(C) \rangle_{\mathrm{MC}}}{\langle \operatorname{sign}(C) \rangle_{\mathrm{MC}}}, \tag{5.29}$$

where $\operatorname{sign}(C) = \pm 1$ denotes the sign of $w(C)$ and $\langle \cdots \rangle_{\mathrm{MC}}$ is the Monte Carlo average with the weight $|w(C)|$.

The negative contribution Z_- to the partition function cancels part of the positive contribution Z_+, while the total $Z = Z_+ - Z_-$ must always be positive. In fact, in many cases, the negative contribution *almost completely* cancels the positive one. We can understand this (Loh et al., 1990; Hatano and Suzuki, 1992) by considering a fictitious Hamiltonian H', whose matrix elements are the absolute value of the corresponding matrix elements of the original Hamiltonian H: $\langle C'|H'|C \rangle \equiv |\langle C'|H|C \rangle|$. The difference in the free energy per site $\Delta f \equiv (F' - F)/N$ is of $\mathcal{O}(1)$ when the difference between the two Hamiltonians is extensive. Because $\beta F = \ln Z$ and $\beta F' = \ln Z'$, where Z' is the partition function of the fictitious system, it follows that the denominator of (5.29) can be expressed as

$$\frac{Z_+ - Z_-}{Z_+ + Z_-} = \frac{Z}{Z'} = e^{-\beta N \Delta f}.$$

It is clear from this expression that the positive and the negative parts of Z cancel almost completely at low temperature and in systems of large size, and it becomes practically impossible to calculate the expectation value $\langle Q \rangle$ from (5.29), because in both the numerator and denominator, the statistical error will be much larger than the mean.

5.5 Dynamics

Equilibrium quantum Monte Carlo results possess physically meaningful dynamical information, while classical Monte Carlo results do not. The access to *real* dynamical information is the reward for having to deal with the extra imaginary-time dimension in the simulation. To access this information, specific types of measurements must be made. By definition, the expectation value of a quantum mechanical operator X is

$$\langle X \rangle = \frac{\mathrm{Tr}\, e^{-\beta H} X}{\mathrm{Tr}\, e^{-\beta H}}.$$

The operator at imaginary time τ being $X(\tau) = e^{\tau H} X e^{-\tau H}$ we find

$$\langle X(\tau) \rangle = \frac{\mathrm{Tr}\, e^{-(\beta-\tau)H} X e^{-\tau H}}{\mathrm{Tr}\, e^{-\beta H}}.$$

As a result of the time-translation invariance of equilibrium states, $\langle X \rangle = \frac{1}{\beta} \int_0^\beta d\tau \, \langle X(\tau) \rangle$. This class of expectation values is called *equal-time expectation values*.

Dynamical probes involve the product of two (or more) operators at different times

$$\langle A(\tau) B(0) \rangle = \frac{\mathrm{Tr}\, e^{-(\beta-\tau)H} A e^{-\tau H} B}{\mathrm{Tr}\, e^{-\beta H}}.$$

These expectation values also enjoy time-translation invariance: $\langle A(\tau) B(0) \rangle = \langle A(\tau + \tau') B(\tau') \rangle$.

The importance of unequal-time measurements is their relation to spectral densities,

$$\langle A(\tau) B(0) \rangle = \int_{-\infty}^{\infty} d\omega \frac{e^{-\tau\omega} D(\omega)}{1 \pm e^{-\beta\omega}},$$

where the $\pm$ depends on whether A and B anticommute or commute. The spectral density $D(\omega)$ connects imaginary-time dynamics with real dynamical information. More specifically, the spectral density contains measurable information about the elementary excitations in the system. For example, measuring the imaginary-time dependence of the spatial Fourier transformation of $\langle S_i^z S_j^z \rangle$ and inverting the integral equation yields $D(k, \omega)$ (Kawashima et al., 1996), which is often measured in inelastic neutron scattering experiments. Peaks in this function are indicators of elementary z-component spin excitations, and the locations of the peaks as a function of k yield the excitation's dispersion relation $\omega(k)$. Inverting the integral equation is a difficult task even with data that lack the statistical errors associated with a Monte Carlo estimate. We discuss techniques for this inversion in Chapter 12. More accessible are static susceptibilities, found by integrating over imaginary time. For example, the *static longitudinal and transverse spin suscepti-bilities* are

$$\chi_{ij}^{\parallel} = \int_0^{\beta} d\tau \, \langle S_i^z(\tau) S_j^z(0) \rangle,$$

$$\chi_{ij}^{\perp} = \int_0^{\beta} d\tau \, \langle S_i^x(\tau) S_j^y(0) + S_i^y(\tau) S_j^x(0) \rangle.$$

For the measurement of diagonal correlation functions, we use the identity

$$\langle S_i^z(\tau) S_j^z(0) \rangle \equiv \langle s_i(\tau) s_j(0) \rangle_{\text{MC}},$$

where $\langle \cdots \rangle_{\text{MC}}$ denotes the simple average over the world-line configurations gen-erated by the Monte Carlo simulation. Although the measurement of off-diagonal correlation functions is not as straightforward as this, when we use the loop/cluster update or the worm update, the formula is almost as simple. For example, we can estimate the correlation function between S^x operators in the spin-$\frac{1}{2}$ XY model using

$$\langle S_i^x(\tau) S_j^x(0) \rangle \equiv \frac{1}{4} \langle \theta((i, \tau), (j, 0)) \rangle_{\text{MC}},$$

where $\theta(X, Y)$ is 1 if the space-time points X and Y are on the same loop, and 0 otherwise. By integrating over X and Y, we obtain the uniform, static transverse susceptibility,

$$\chi^{\perp} = \frac{1}{N\beta} \sum_{i,j} \int_0^{\beta} d\tau_1 d\tau_2 \langle \mathcal{T} S_i^x(\tau_2) S_i^x(\tau_1) \rangle$$

$$= \frac{1}{4} \frac{1}{N\beta} \sum_{i,j} \int_0^{\beta} d\tau_1 d\tau_2 \, \langle \theta((i, \tau_2), (j, \tau_1)) \rangle_{\text{MC}}$$

$$= \frac{1}{4} \frac{1}{N\beta} \left\langle \sum_l \Lambda_l^2 \right\rangle_{\text{MC}},$$

where $\mathcal{T}$ denotes the time-ordered product and Λ_l is the length of the loop specified by the loop index l. Similarly, we estimate the correlation function in the worm algorithm by simply measuring how frequently the head visits a specific position relative to the tail. The simple and efficient measurement of off-diagonal correlation functions is an important advantage of the loop/cluster and worm algorithms. We discuss these points in more detail in Chapter 6.

Suggested reading

M. Creutz and B. Freedman, "A statistical approach to quantum mechanics," *Ann. Phys.* **132**, 427 (1981).

R. T. Scalettar, "World-line quantum Monte Carlo," in *Quantum Monte Carlo Methods in Physics and Chemistry*, series C, volume 525, NATO Science Series, ed. M. P. Nightingale and C. J. Umrigar (Dordrecht: Kluwer Academic, 1999).

H. G. Evertz, "The loop algorithm," *Adv. Phys.* **52**, 1 (2003).

N. Prokof'ev and B. Svistunov, "Worm algorithm for problems in quantum and classical systems," in *Understanding Phase Transitions*, ed. L. D. Carr (Boca Raton, FL: Taylor and Francis, 2010).

A. W. Sandvik, "Computational studies of quantum spin systems," in *Lectures on the Physics of Strongly Correlated Systems XIV*, ed. A. Avella and F. Mancini, AIP Conference Proceedings, **1297**, 135 (2010).

Exercises

5.1 Conservation Rule. Prove that if a quantum spin Hamiltonian is invariant under the $U(1)$ rotation,

$$S_i^+ \to e^{i\phi} S_i^+, \ S_i^- \to e^{-i\phi} S_i^- \ (\text{for all } i \in V),$$

the Hamiltonian has zero matrix elements between the eigenstates of $\sum_{i \in V} S_i^z$ with different eigenvalues, where V is the area on which the Hamiltonian is defined.

5.2 The most familiar example of a Poisson process is the decay of a radioactive atom with decay rate λ. For this decay,

1. Show that the probability of the atom surviving longer than some time τ is

$$P_{\text{survive}}(\tau) = (1 - \lambda \Delta \tau)^{\tau/\Delta \tau} \to e^{-\lambda \tau} \quad \text{as } \Delta \tau \to 0.$$

2. Show that the probability of the first decay event occurring in the short-time interval $[\tau, \tau + \Delta \tau)$ is

$$P_{\text{survive}}(\tau) \lambda \Delta \tau = \lambda \Delta \tau e^{-\lambda \tau}.$$

3. Obtain the same result by using the product of two Poisson distributions: the first to calculate the probability of no events in the interval $[0, \tau)$ and the second to calculate the probability of one event in the short-time interval $[\tau, \tau + \Delta \tau)$.

5.3 Suppose we have two independent Poisson processes over a time interval τ that are characterized by (N_1, λ_1) and (N_2, λ_2) for their number of events and density of events. Let $N = N_1 + N_2$ and $\lambda = \lambda_1 + \lambda_2$.

1. Show that $P(N = 0) = e^{-\lambda \tau}$ and $P(N = 1) = \lambda \tau e^{-\lambda \tau}$.

2. Generalize the above to show that two independent Poisson processes have a joint distribution of density λ.
3. Show that conditional on N, the distribution of N_1 is a binomial distribution (2.34) of index N and parameter $p = \lambda_1/(\lambda_1 + \lambda_2)$.

5.4 Suppose x has a Poisson distribution parameterized by a density λ, and for a given x, y has a binomial distribution of index k and parameter p (Exercise 2.2). Show that the unconditional distribution of y is a Poisson distribution with density $p\lambda = E[E[y|x]]$.

5.5 Verify (5.12) and (5.13).

5.6 Show that

$$e^A C e^{-A} = C + [A, C] + \frac{1}{2!} [A, [A, C]] + \frac{1}{3!} [A [A [A, C]]] + \cdots \quad (5.30)$$

5.7 Relate the steps in Algorithm 15 with the panels in Fig. 5.4, keeping in mind some graph assignments are done probabilistically. In the middle panel of Fig. 5.4, identify the loops. There are four. Which were flipped to obtain the right figure?

5.8 What are the matrices representing $\exp(-\tau S^z)$ and $\exp(-\tau S^x)$?

5.9 To be canonical, the Matsubara-Matsuda and Jordan-Wigner transformations must conserve the operator commutation relations. Verify this for both transformations and their inverse. Using the two transformations, establish the one that transforms directly between Fermions and Bosons.

Part II

Finite temperature

6

Finite-temperature quantum spin algorithms

The simulation of quantum spin systems at finite temperatures is a very active research area not only because of the importance of the problems being simulated but also because of the availability of highly effective Monte Carlo methods, such as the the loop/cluster and worm algorithms, for many problems of interest. We previewed these algorithms in Chapters 4 and 5. In this chapter and its associated appendices, we provide a more complete and formal derivation and discussion of them. We also discuss the conditions under which simulations at low but finite temperature may be used to study ground-state properties and quantum critical phenomena. While we mainly detail the applications of these algorithms to quantum spin systems, we conclude with a discussion of their use (with minor modifications) to simulate Bosonic lattice models.

6.1 Feynman's path integral

In Section 5.2.5, we used the path-integral formalism to present a simple example of a loop/cluster algorithm for a quantum spin model. Here, we restate the same formalism in a more general and formal language to prepare for a more detailed discussion of loop/cluster updates.

We consider the Hamiltonian

$$H = \sum_{l=1}^{N_l} H_l, \tag{6.1}$$

where H_l is a *local* Hamiltonian defined on some small set of lattice points b_l. A typical local Hamiltonian is $H_l = -J S_{i_l} \cdot S_{j_l}$, where $S_{i_l} = (S_{i_l}^x, S_{i_l}^y, S_{i_l}^z)$ is a spin operator with the standard normalization $S_{i_l} \cdot S_{i_l} = S(S+1)$ in an appropriate basis ($S = 1/2, 1, 3/2, \cdots$). In this case, the local set of lattice points is a pair of nearest-neighbor sites, that is, $b_l = \{i_l, j_l\}$. However, b_l being a set of more than two sites or being just a single site is permissible.

121

In discrete imaginary time, the path integral (5.10) corresponds to the approximation

$$Z = \mathrm{Tr} e^{-\beta H} = \mathrm{Tr} e^{-\beta \sum_l H_l} \approx \mathrm{Tr} \prod_{k=1}^{M} \prod_{l=1}^{N_l} e^{-\Delta\tau H_l}, \qquad (6.2)$$

where M is some positive integer, and $\Delta\tau = \beta/M$. Although the expression on the right-hand side is approximate, its error is proportional to $(\Delta\tau)^2$, and thus vanishes in the limit $M \to \infty$. Because we eventually take this limit to obtain algorithms in continuous imaginary time, the error due to this approximation can be neglected.

We use the product states introduced in (5.5) as our complete orthonormal basis,

$$|\psi\rangle \equiv |s_1\rangle \otimes |s_2\rangle \otimes \cdots \otimes |s_N\rangle,$$

with s_i specifying the local state on the site i. By inserting the identity operator

$$1 = \sum_{\psi=\{s_1,\dots,s_N\}} |\psi\rangle\langle\psi|$$

between the $e^{-\Delta\tau H_l}$ operators in (6.2), we obtain a representation of the partition function containing only c-numbers,

$$Z \approx \sum_{C=\{\psi(k,l)\}} \prod_{k=1}^{M} \prod_{l=1}^{N_l} \langle \psi(k,l+1)| e^{-\Delta\tau H_l} |\psi(k,l)\rangle. \qquad (6.3)$$

This is a "classicized" representation in terms of the world lines

$$C \equiv \{\psi(1,1), \psi(1,2), \dots, \psi(M,N_l)\}$$

of spin configurations $\psi(k,l) \equiv \{s_1(k,l), s_2(k,l), \dots, s_N(k,l)\}$, which satisfy the periodic boundary condition

$$\psi(M+1,l) \equiv \psi(1,l)$$

in imaginary time. We also adopt the notation

$$\psi(k,N_l+1) \equiv \psi(k+1,1).$$

Factorizing (6.3) further, we write

$$\langle \psi(k,l+1)| e^{-\Delta\tau H_l} |\psi(k,l)\rangle = \left(\prod_{i\notin b_l} \delta_{s_i(k,l+1),s_i(k,l)} \right)$$

$$\times \langle \psi_{b_l}(k,l+1)| e^{-\Delta\tau H_l} |\psi_{b_l}(k,l)\rangle,$$

where $|\psi_{b_l}\rangle$ is the restriction of $|\psi\rangle$ on the set of lattice points b_l.[1] For example, when $b = \{i,j\}$, $|\psi_b\rangle = |s_i, s_j\rangle$.

---

[1] To be very formal, H_l on the left-hand side is an operator defined on the whole Hilbert space, whereas H_l on the right-hand side is an operator defined on the "local" Hilbert space that concerns b_l only. However, we

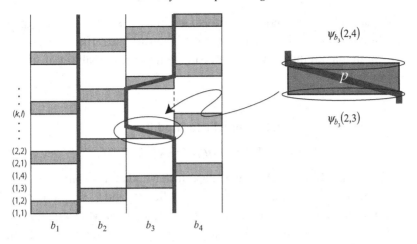

Figure 6.1 The discretized path integral. Each shaded plaquette represents the action of a single bond operator in a time-interval $\Delta\tau$. The N_l consecutive plaquettes for a given k correspond to the imaginary-time evolution operator $\exp(-\Delta\tau H)$.

We call the minimal set of the space-time points involved in a single scattering event a *plaquette*. Namely, a plaquette is a set of points that corresponds to b_l in space and (k, l) and $(k, l+1)$ in time, for example, a set of four space-time points when b_l represents a pair of sites. With $C_p \equiv \{\psi_{b_l}(k, l+1), \psi_{b_l}(k, l)\}$, where p stands for a plaquette, we can define the weight for the whole space-time configuration as

$$W(C) \equiv \prod_{p=(k,l)} w_p(C_p), \tag{6.4}$$

$$w_p(C_p) \equiv \left\langle \psi_{b_l}(k, l+1) \left| e^{-\Delta\tau H_l} \right| \psi_{b_l}(k, l) \right\rangle, \quad [p = (k, l)]. \tag{6.5}$$

Then, our path-integral expression of the partition function becomes

$$Z \approx \sum_{C:\,\Delta(C)=1} W(C), \tag{6.6}$$

with

$$\Delta(C) \equiv \prod_{k,l,i\notin b_l} \delta_{s_i(k,l+1),\,s_i(k,l)}$$

imposing the constraint on the *world-line configuration* C that the two states $\psi(k, l)$ and $\psi(k, l+1)$, successive in imaginary time, differ only on the plaquettes (Fig. 6.1).

will use the same symbol for both of them here and in what follows as the implication should be obvious from the context.

We note that (6.4), (6.5), and (6.6) correspond to a classical and local expression of the partition function. It is classical in that all the variables $s_i(k, l)$ are c-numbers, and it is local in that all the $w_p(C_p)$ factors are defined on a small set of space-time lattice points.[2]

6.2 Loop/cluster update

6.2.1 General framework

We now discuss the framework of the loop/cluster update algorithm for classical and quantum models. In Section 6.2.3 and Appendices D and E, we present detailed formulas for a few important and illustrative quantum cases. With this framework, we go beyond both the Swendsen-Wang algorithm for the Ising model (Section 4.4) and the quantum loop/cluster update for the $S = \frac{1}{2} XY$ model (Section 5.2.5).

The general framework is most easily understood by analogy with the previously discussed loop/cluster updates. First, let the space of configurations be Σ. On Σ the target weight $W(C)$ is defined and the configuration $C \in \Sigma$ should be generated by it. Now, recalling that the Swendsen-Wang algorithm had graphs allowing only parallel spins and that the loop/cluster update for the XY model had graph elements allowing only certain local world-line configurations, we introduce a set of graphs Γ such that a graph $G \in \Gamma$ is a specifier of a certain set of restrictions on the configuration C. As we did in the Swendsen-Wang algorithm, we define a new weight $W(C, G)$ on the joint phase space $\Sigma \times \Gamma$ such that

$$W(C) = \sum_{G \in \Gamma} W(C, G), \qquad (6.7)$$

and construct a Markov chain that samples $W(C, G)$. Equation (6.7) tells us that after marginalizing G, our Markov process will generate a sequence of states C with a frequency proportional to $W(C)$.

Each step of our Markov process has two parts. In the first part, we stochastically generate a graph G for the current configuration C with probability $P(G|C)$, and in the second, we generate a new configuration C' with probability $P(C'|G)$. The conditional probabilities are

$$P(G|C) \equiv \frac{W(C, G)}{W(C)}, \quad P(C|G) \equiv \frac{W(C, G)}{W(G)}, \qquad (6.8)$$

---

[2] Here, being "local" means depending only on $O(1)$ number of degrees of freedom. In this sense, long-range interactions can be treated in just the same way as short-range interactions. The problem arising from the large number of interacting pairs can also be handled efficiently (see Appendix E).

and

$$W(G) \equiv \sum_{C \in \Sigma} W(C, G).$$

When we view this combined procedure as one step, the transition probability from the current state C to the next state C' is the sum of the probabilities of all paths leading from C to C',

$$P(C'|C) = \sum_{G \in \Gamma} P(C'|G) P(G|C). \tag{6.9}$$

Using (6.9) and (6.8), we next compute the probability flow from C to C',

$$P(C'|C)W(C) = \sum_{G \in \Gamma} \frac{W(C', G)}{W(G)} \frac{W(C, G)}{W(C)} W(C) = \sum_{G \in \Gamma} \frac{W(C', G) W(C, G)}{W(G)}.$$

Given that the expression on the far right is symmetric with respect to swapping C and C', the detailed balance condition $P(C'|C)W(C) = P(C|C')W(C')$ holds. Therefore, provided the sampling is ergodic, our newly constructed Markov process converges to the correct target distribution $W(C)$. Whether ergodicity holds is something we must examine on a case-by-case basis. Here we just assume that it does.

Viewed as a single Markov step in the joint phase space $\Sigma \times \Gamma$ from (C, G) to (C', G'), the transition probability is simply

$$P\big((C', G') \,|\, (C, G)\big) = P(C'|G') P(G'|C).$$

For this transition probability, the detailed balance condition generally does not hold (Exercise 6.1). However, the Markov process still converges to the target distribution $W(C, G)$. To show this, we use the fact, just proven above, that the Markov process on Σ converges to the target distribution. Therefore, after a sufficiently large number of steps, say, n, $P^{(n)}(C) \propto W(C)$ with an arbitrarily high accuracy. The convergence of the Markov process in the joint phase space immediately follows:

$$P^{(n)}(C, G) = P(C|G)P^{(n)}(G) = P(C|G) \sum_{C'} P(G|C') P^{(n-1)}(C')$$

$$\propto P(C|G) \sum_{C'} P(G|C') W(C') = W(C, G). \tag{6.10}$$

Thus, constructing a loop/cluster update algorithm is equivalent to the problem of finding an appropriate set of graphs Γ and a function $W(C, G)$ that satisfies (6.7). We recall that a local definition of the weight makes the resulting transition procedure particularly easy to use. Therefore, we assume the factorization

$$W(C, G) = \prod_p w_p(C_p, G_p), \tag{6.11}$$

where p represents a small set of lattice sites on which the interaction terms are defined. In classical models, this set is typically a pair of nearest-neighbor spins ($p = \{i,j\}$). In quantum models, it is typically a set of four variables within the path-integral representation. Whereas C stands for the configuration of the whole system, C_p is a part of it, namely, the "local" state on a unit p. The subgraph G_p is defined similarly.

To satisfy (6.7) and (6.11), we now require that

$$w_p(C_p) = \sum_{G_p} w_p(C_p, G_p). \tag{6.12}$$

Because $\sum_G = \prod_p \sum_{G_p}$, (6.7) is satisfied.[3] Further, to update C by a "coin-tossing" process, we require for a fixed G_p that all possible C_p have the same weight, that is, we require

$$w_p(C_p, G_p) = v_p(G_p)\Delta(C_p, G_p), \tag{6.13}$$

where $\Delta(C_p, G_p) = 0, 1$ expresses the local constraints on C_p imposed by the graph element G_p.

With (6.11), (6.12), and (6.13), our transition probability (6.8) finally becomes

$$P(G|C) = \frac{W(C, G)}{W(C)} = \frac{\prod_p w_p(C_p, G_p)}{\prod_p w_p(C_p)} = \prod_p \frac{w_p(C_p, G_p)}{w_p(C_p)}$$

$$= \prod_p P_p(G_p|C_p), \tag{6.14}$$

where

$$P_p(G_p|C_p) \equiv \frac{w_p(C_p, G_p)}{w_p(C_p)} = \Delta(C_p, G_p)\frac{v_p(G_p)}{w_p(C_p)}. \tag{6.15}$$

These equations tell us how to construct a new graph: For each p we are to choose a graph G_p with relative weight $v_p(G_p)$ from the ones compatible with C_p. Then, the whole graph G is simply the union of the local graphs. For the spin configuration update process, we obtain[4]

---

[3] All the examples presented in this book, such as the Swendsen-Wang algorithm and the loop/cluster algorithms for quantum spin models, fit this definition.

[4] This equation can be derived as follows:

$$P(C|G) = \frac{W(C, G)}{W(G)} = \frac{\prod_p w_p(C_p, G_p)}{\prod_p \sum_{C_p} w_p(C_p, G_p)}$$

$$= \frac{\prod_p v_p(G_p)\Delta(C_p, G_p)}{\prod_p \sum_{C_p} v_p(G_p)\Delta(C_p, G_p)} = \frac{\prod_p \Delta(C_p, G_p)}{\prod_p \sum_{C_p} \Delta(C_p, G_p)} = \frac{\Delta(C, G)}{\sum_C \Delta(C, G)}.$$

$$P(C|G) = \frac{\Delta(C, G)}{\sum_C \Delta(C, G)},$$ (6.16)

with

$$\Delta(C, G) \equiv \prod_p \Delta(C_p, G_p).$$

Equation (6.16) means that we can update the state by a simple coin toss, that is, by choosing with equal probability a state from those compatible with the constraints imposed by G. Note that a local degree of freedom s_i may take more than two values in general. In such cases the "coin toss" means choosing among n (> 2) values with equal weight.

6.2.2 Continuous-time loop/cluster update

We now focus on quantum lattice problems, and discuss what algorithm the generic formulas for the transition probabilities (6.14) (with (6.15)) and (6.16) yield in the continuous imaginary-time limit. In doing so, our key task is specifying (6.15).

Let us start with the graphical decompositions (6.12) and (6.13). In the quantum case, the local weight is the matrix element of the local time-evolution operator (6.5). As we did for the $S = \frac{1}{2}$ XY model in Section 5.2.5, we expand the local Hamiltonian H_l as a sum of operators and represent each term in the sum by a graph element,

$$H_l = - \sum_g a_l(g) D_l(g).$$ (6.17)

Then, the local imaginary-time evolution operator becomes the operator

$$e^{-\Delta\tau H_l} \approx 1 + \Delta\tau \sum_g a_l(g) D_l(g),$$

whose *matrix elements* are

$$w_p(C_p) \approx \Delta(C_p, g = 0) + \Delta\tau \sum_g a_l(g) \Delta(C_p, g).$$ (6.18)

The symbol $\Delta(C_p, g)$ represents the matrix element of $D_l(g)$,

$$\Delta(C_p, g) \equiv \langle \psi_{b_l}(k, l+1) | D_l(g) | \psi_{b_l}(k, l) \rangle.$$

The term with $g = 0$ corresponds to the identity operator,

$$\Delta(C_p, g = 0) = \prod_{i \in b_l} \delta_{s_i(k,l+1), s_i(k,l)},$$

which is represented by the "trivial" graph of vertical segments connecting $s_i(k, l)$ to $s_i(k, l + 1)$. Similarly, the other terms with $\Delta(C_p, g \neq 0)$ are products of Kronecker delta functions that fix the relative values of the local variables bound in g.

By comparing (6.18) with the general formula (6.12) and using (6.13), we identify

$$v_p(g) = \begin{cases} 1 + \mathcal{O}(\Delta\tau) & (g = 0) \\ \Delta\tau\, a_l(g) + \mathcal{O}((\Delta\tau)^2) & (g \neq 0). \end{cases} \tag{6.19}$$

Note that we can neglect the $\mathcal{O}(\Delta\tau)$ term in the $g = 0$ case, since it adds to a constant multiplicative factor to the weight, and can neglect the $\mathcal{O}((\Delta\tau)^2)$ term in the $g \neq 0$ case, because it does not contribute to the weight at all in the $\Delta\tau \to 0$ limit.[5] Then, substituting (6.18) and (6.19) in (6.15), we obtain the graph-assignment probabilities: For C_p not a kink, that is, for the same initial and final state and a $g(\neq 0)$ compatible with C_p,

$$P_p\left(g | C_p\right) = \Delta\tau\, a_l(g) + \mathcal{O}(\Delta\tau^2). \tag{6.20}$$

If C_p is a kink and if g is compatible with C_p,[6]

$$P_p\left(g | C_p\right) = \frac{a_l(g)}{\sum_{g'} a_l(g')\Delta(C_p, g')} + \mathcal{O}(\Delta\tau). \tag{6.21}$$

The first formula (6.20) tells us what to do with the plaquettes with no kinks: To each, we must assign with probability $v_p(g) = \Delta\tau a_l(g)$ a graph element g compatible with the current state. The second formula (6.21) tells us how to update the graph element on plaquettes with kinks: We must choose with a probability proportional to $a_l(g)$ a graph element g among those compatible with the current state and replace the current graph by it.

We now extrapolate to continuous imaginary time. For operations on the kinks, we take the steps described above, that is, we switch the graph elements according to the weight $a_l(g)$. Operations on the nonkinks have a less trivial limit. We regard the k-th imaginary-time layer in the discretized formulation as the one corresponding to the imaginary-time interval $I_k \equiv \{\tau \,|\, k\Delta\tau \leq \tau < (k + 1)\Delta\tau)\}$. In other words, we replace $\psi(k, l)$ by $\psi(\tau)$ with $\tau = k\Delta\tau$. If we consider a finite imaginary-time interval of length I that does not include any kinks, we have $I/\Delta\tau$ plaquettes corresponding to each interaction term H_{b_l}. What we do to each plaquette is assign a graph element of type g with probability $v_p(g) = \Delta\tau a_l(g)$. The number of plaquettes is inversely proportional to $\Delta\tau$, whereas the probability is proportional

---

[5] Because there are only $\mathcal{O}((\Delta\tau)^{-1})$ off-diagonal graphs in the whole system, the $\mathcal{O}((\Delta\tau)^2)$ correction to the off-diagonal term makes an $\mathcal{O}(\Delta\tau)$ contribution that vanishes in the $\Delta\tau \to 0$ limit.

[6] If g is not compatible with C_p, $P_p(g|C_p) = 0$, of course.

Algorithm 17 A Markov step of the quantum loop/cluster algorithm.

Input: A world-line configuration ;
  for each graph element with no kink **do**
    Remove it ;
  end for
  for each kink **do**
    Select a graph element with a probability proportional to $a_l(g)$ from those compatible with the current local state ;
    Replace the current element by it ;
  end for
  for each interaction term l **do**
    for each type of graph element g **do**
      Select imaginary time points $\tau \in [0, \beta)$ with density $a_l(g)$;
      for each selected imaginary time τ **do**
        if the local state is compatible with g **then**
          Place a graph element of type g there ;
        end if
      end for
    end for
  end for
  Connect open ends of neighboring graph elements with vertical lines ;
  Identify loops/clusters ;
  for each loop/cluster **do**
    Flip the loop/cluster (Choose one point on the loop/cluster, select one of its possible local states with equal probability, and assign local states to all the points on the loop/cluster according to the constraint) ;
  end for
  return the updated world-line configuration.

to $\Delta\tau$. Therefore, in the $\Delta\tau \to 0$ limit, we obtain a procedure with well-defined probabilities, namely, assigning the type g graphs with density $a_l(g)$. (Everywhere else the trivial graph is to be assigned.) Algorithm 17 summarizes this procedure.

6.2.3 XXZ models

In the previous subsection, we assumed that the Hamiltonian can be decomposed into graphical terms (Eq. (6.17)). While this is not generally true, it is possible for a wide variety of Hamiltonians. An example is the $S = \frac{1}{2}$ XXZ model: $H = \sum_{\langle ij \rangle} H_{ij}$

with $\langle ij \rangle$ denoting nearest neighbors and

$$H_{ij} = -J^{xy}(S_i^x S_j^x + S_i^y S_j^y) - J^z S_i^z S_j^z. \tag{6.22}$$

The most frequently studied $S = \frac{1}{2}$ Hamiltonians, such as the XY model and the ferromagnetic and the antiferromagnetic Heisenberg models, are special cases of this model. Even the classical Ising model is a special case of it, and in fact we can show that when the quantum cluster update is applied to (6.22) in the limit $J^{xy} \to 0$, it becomes equivalent to the Swendsen-Wang algorithm for the Ising model (Exercise 6.3). Different anisotropies of the spin interaction, however, yield various types of loops and clusters (Evertz et al., 1993; Evertz and Marcu, 1994; Kawashima, 1996). (In Appendix D we detail the loop/cluster derivations for other important Hamiltonians for which loop/cluster algorithms are possible: the SU(N) model and the SU(N) J-Q model.)

When $J^{xy} < 0$, some off-diagonal matrix elements of $e^{-\Delta \tau H_{ij}}$ are negative. If the lattice includes odd cycles, as does the triangular lattice, we cannot avoid the sign problem (Section 5.4). On the other hand, if the lattice lacks odd cycles, as does a bipartite lattice, we can avoid the negative signs by applying the unitary transformation (Marshall transformation, Section 5.2.4) $(S_i^x, S_i^y, S_i^z) \to (-S_i^x, -S_i^y, S_i^z)$ to all sites in one sublattice to change the sign of J^{xy} and thus change the sign of the negative off-diagonal matrix elements. Therefore, we assume $J^{xy} \geq 0$ on bipartite lattices without loss of generality.

The matrix elements of the pair Hamiltonian (6.22) in the basis $\left\{ |\frac{1}{2}, \frac{1}{2}\rangle, |\frac{1}{2}, -\frac{1}{2}\rangle, |-\frac{1}{2}, \frac{1}{2}\rangle, |-\frac{1}{2}, -\frac{1}{2}\rangle \right\}$ are

$$\langle \psi_b' | -H_{ij} | \psi_b \rangle = \begin{pmatrix} \frac{J^z}{4} & 0 & 0 & 0 \\ 0 & -\frac{J^z}{4} & \frac{J^{xy}}{2} & 0 \\ 0 & \frac{J^{xy}}{2} & -\frac{J^z}{4} & 0 \\ 0 & 0 & 0 & \frac{J^z}{4} \end{pmatrix}.$$

We consider four different graph elements (apart from the trivial one corresponding to the identity operator): diagonal (g_d), horizontal (g_h), binding (g_b), and antibinding (g_{ab}). We illustrate these graph elements in Fig. 6.2, together with the matrix elements of the corresponding operators. With these four, the local Hamiltonian can be expressed in three different forms:

$$-H_{ij} = \begin{cases} -\frac{J^z}{4} + \frac{J^{xy}}{2}D_{ij}(g_d) + \frac{J^z - J^{xy}}{2}D_{ij}(g_b) & \text{(I)} \\[2mm] -\frac{J^{xy}}{4} + \frac{J^{xy} + J^z}{4}D_{ij}(g_d) + \frac{J^{xy} - J^z}{4}D_{ij}(g_h) & \text{(II)} \\[2mm] \frac{J^z}{4} + \frac{J^{xy}}{2}D_{ij}(g_h) + \frac{-J^{xy} - J^z}{2}D_{ij}(g_{ab}) & \text{(III).} \end{cases} \tag{6.23}$$

Symbol	Graph	$\langle s_i' s_j' \,\lvert D_{ij}(g) \rvert\, s_i s_j \rangle$
g_d	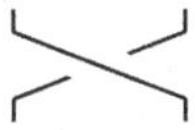	$\begin{pmatrix} 1 & 0 & 0 & 0 \\ 0 & 0 & 1 & 0 \\ 0 & 1 & 0 & 0 \\ 0 & 0 & 0 & 1 \end{pmatrix}$
g_h		$\begin{pmatrix} 0 & 0 & 0 & 0 \\ 0 & 1 & 1 & 0 \\ 0 & 1 & 1 & 0 \\ 0 & 0 & 0 & 0 \end{pmatrix}$
g_b	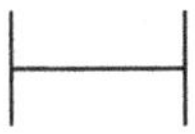	$\begin{pmatrix} 1 & 0 & 0 & 0 \\ 0 & 0 & 0 & 0 \\ 0 & 0 & 0 & 0 \\ 0 & 0 & 0 & 1 \end{pmatrix}$
g_{ab}	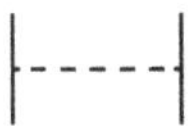	$\begin{pmatrix} 0 & 0 & 0 & 0 \\ 0 & 1 & 0 & 0 \\ 0 & 0 & 1 & 0 \\ 0 & 0 & 0 & 0 \end{pmatrix}$

Figure 6.2 The graph elements and the matrix elements of the operators $D_{ij}(g)$ corresponding to each type of graph element. The basis vectors of the two-spin Hilbert space are $\lvert \frac{1}{2}, \frac{1}{2} \rangle$, $\lvert \frac{1}{2}, -\frac{1}{2} \rangle$, $\lvert -\frac{1}{2}, \frac{1}{2} \rangle$, and $\lvert -\frac{1}{2}, -\frac{1}{2} \rangle$. To emphasize the difference in the constraints, we use dashed lines when the connected spins must be anti parallel and solid lines when they must be parallel.

The requirement that $a_l(g)$ be positive selects one of the three:

    (I)   easy-axis ferromagnetic model $(0 < J^{xy} \le J^z)$,

    (II) easy-plane model $(0 \le \lvert J^z \rvert < J^{xy})$,

    (III) easy-axis antiferromagnetic model $(0 < J^{xy} \le -J^z)$.

In case (I), the clusters consist of the diagonal graph elements g_d, the binding elements g_b, and all the straight vertical lines connecting open ends of these graph elements. A binding graph connects the four spins on a plaquette and forces them to align in the same direction. As a result, all the spins in the same cluster point in the same direction. The resulting graph is not just a collection of loops but is a collection of clusters (that is, loops bound together). The algorithm in case (III) is the same as the one in case (I) except that spins on different sublattices are bound antiparallel while those on the same sublattice are bound parallel. The clusters are simple loops as they are for the *XY* model.

6.2.4 Correlation functions

Why the loop/cluster algorithm often converges much faster than the local update algorithm is best explained by discussing the way we measure the spin-spin correlation functions in the algorithm. We estimate the value of these functions at two space-time points from the frequency with which these points appear in the same loop or cluster. This estimate means that the range of the correlations, the correlation length, corresponds to the typical size of the loops or clusters.

Let us consider a quantity Q expressible as a sum of local operators,

$$Q = \sum_l Q_l.$$

Then

$$\langle Q \rangle \equiv \frac{1}{\beta} \left[\frac{1}{Z(\eta)} \frac{d}{d\eta} Z(\eta) \right]_{\eta \to 0}, \qquad (6.24)$$

where

$$Z(\eta) \equiv \mathrm{Tr} \left[e^{-\beta(H - \eta Q)} \right]. \qquad (6.25)$$

Accordingly, with (6.5) and the simplified notations $\psi_{b_l}(k, l) \to \psi_p$, $\psi_{b_l}(k, l+1) \to \psi_p'$, $l, b_l \to b$, etc., the weight (6.4) becomes

$$W'(C) \equiv \prod_{p=(k,b)} \left\langle \psi_p' \left| 1 - \Delta\tau \left(H_b - \eta Q_b \right) \right| \psi_p \right\rangle. \qquad (6.26)$$

With the substitution of (6.26) into (6.6), and then (6.6) into (6.24), the thermal average of Q becomes equivalent to the Monte Carlo average of $Q(C)$

$$\langle Q \rangle = \langle Q(C) \rangle_{\mathrm{MC}}, \qquad (6.27)$$

where

$$Q(C) \equiv \frac{1}{\beta} \sum_{p=(k,b)} Q(C_p),$$

$$Q(C_p) \equiv \frac{\langle \psi_p' | \Delta\tau Q_b | \psi_p \rangle}{\langle \psi_p' | 1 - \Delta\tau \mathcal{H}_b | \psi_p \rangle} \approx \begin{cases} \Delta\tau \langle \psi_p | Q_b | \psi_p \rangle & (\psi_p' = \psi_p) \\ \dfrac{\langle \psi_p' | Q_b | \psi_p \rangle}{\langle \psi_p' | (-H_b) | \psi_p \rangle} & (\psi_p' \neq \psi_p). \end{cases} \qquad (6.28)$$

After taking the continuous-time limit,

$$Q(C) = \frac{1}{\beta} \int_0^\beta d\tau\, Q(\psi(\tau)) + \frac{1}{\beta} \sum_{(b,\tau):\text{kink}} Q(\psi_b(\tau^+), \psi_b(\tau^-)), \qquad (6.29)$$

with

$$Q(\psi(\tau)) \equiv \left\langle \psi(\tau) \left| Q \right| \psi(\tau) \right\rangle \tag{6.30}$$

and

$$Q(\psi_b(\tau^+), \psi_b(\tau^-)) \equiv \frac{\left\langle \psi_b(\tau^+) \left| Q_b \right| \psi_b(\tau^-) \right\rangle}{\left\langle \psi_b(\tau^+) \left| (-H_b) \right| \psi_b(\tau^-) \right\rangle}. \tag{6.31}$$

τ^+ and τ^- are the imaginary times infinitesimally after and before τ. When Q is a diagonal operator, the second term in (6.29) drops. For example, it drops when we estimate the magnetization $Q \equiv \sum_i S_i^z$ in the usual S^z basis. In general, we cannot ignore the second contribution originating from the kinks.

The fact that the loop/cluster algorithm is a Markov process in graph space suggests an alternative estimator defined in terms of graphs rather than states. We recall that in Section 4.4.3, we were able to measure the correlation functions for the susceptibility from the graphs.

To see this in the present case, we write the thermal average of Q as

$$\langle Q \rangle = \frac{\sum_C W(C) Q(C)}{\sum_C W(C)} = \frac{\sum_{C,G} W(C,G) Q(C)}{\sum_{C,G} W(C,G)}$$

$$= \frac{\sum_G W(G) Q(G)}{\sum_G W(G)} = \langle Q(G) \rangle,$$

where $Q(G)$ is the fixed-graph average of $Q(C)$ or the graphical estimator of Q:

$$Q(G) \equiv \frac{\sum_C W(C,G) Q(C)}{W(G)}. \tag{6.32}$$

These formulas and an argument similar to the one in Section 4.4.2 yield a specific form of the graphical estimator in such cases. For example, the staggered magnetic susceptibility is defined as

$$\chi_{zz}(q) \equiv \langle Q \rangle, \quad Q \equiv N^{-1} \int_0^\beta d\tau \, M_q^\dagger(\tau) M_q(0),$$

where $M_q \equiv \sum_i e^{iqr_i} S_i^z$ and $M_q(\tau) \equiv e^{\tau H} M_q e^{-\tau H}$. When we set $q = 0$, we obtain the uniform susceptibility. The ordinary estimator of this quantity is

$$Q(C) = (N\beta)^{-1} \left| \int dX \, e^{iqr_i} S^z(X) \right|^2, \tag{6.33}$$

where $X \equiv (i, \tau)$ and the integral $\int dX$ stands for $\sum_i \int d\tau$. The graphical estimator that yields exactly the same mean value is

$$Q(G) = \frac{1}{N\beta} \sum_{c \in G} |M_c|^2 \tag{6.34}$$

with

$$M_c \equiv \int_c dX \, e^{iqr_i} S^z(X),$$

where the integration is over the space-time region covered by a connected cluster c in the graph G. In the case of the easy-plane *XXZ* model, the clusters are loops, and $M_c(q = 0)$ is simply the total length of the loop. The graphical estimator often yields a more accurate estimate than the ordinary estimator.

In some cases, the difference between the two estimators is not just in the accuracy. We note that an expression such as (6.33) is valid only when the operators are diagonal. Suppose, on the contrary, that the operators involved have nonzero off-diagonal matrix elements, but the corresponding matrix elements of the Hamiltonian are zero. Such a case occurs for the *XXZ* model if we measure the transverse susceptibility

$$\chi_{xx} \equiv N^{-1} \int_0^\beta d\tau \, \langle M^x(\tau) M^x(0) \rangle,$$

where the total magnetization M^x is in the x direction. In a case like this, we cannot estimate the quantity in an "ordinary" way. With the loop algorithm, as we now show, such nondiagonal quantities are measurable (Brower et al., 1998).

With $X = (i, \tau)$ and $Y = (j, \tau')$ specifying two space-time points and $\mathcal{T}$ indicating the time-ordered product, we estimate the expectation value of

$$Q = (N\beta)^{-1} \mathcal{T} \int dX dY S^x(X) S^x(Y)$$

by expressing its thermal average as

$$\chi_{xx} \equiv \langle Q \rangle = \frac{1}{N\beta} \int dX dY \frac{1}{Z} {\sum_{C}}' W(C)$$

$$\times \langle s_i(\tau^+) | S^x(X) | s_i(\tau^-) \rangle \langle s_j(\tau^+) | S^x(Y) | s_j(\tau^-) \rangle.$$

Note that we have extended the configuration space by including two discontinuous points in the world-line configuration. The prime in the summation indicates that the summation is over all states that have discontinuities in the world lines at X and Y. Such a state is illustrated in Fig. 6.3. It is obvious that this type of configuration never appears in the loop/cluster simulation. Therefore, it is not possible to measure this quantity by a straightforward generalization of the method discussed so far.

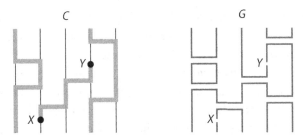

Figure 6.3 Measurement of an off-diagonal correlation function. The world-line configurations (such as "*C*") and the graphs (such as "*G*") with discontinuities at two points X and Y contribute.

Therefore, we consider the graphical estimator,

$$\chi_{xx} = \frac{1}{N\beta} \int dX dY \frac{1}{Z} \sum_{C,G}' V(G) \Delta(C,G) \frac{1}{4} \delta_{s_i(\tau^+),-s_i(\tau^-)} \delta_{s_j(\tau^+),-s_j(\tau^-)}.$$

First note that in this model there is no way to assign local spin values along a loop if only one discontinuous point lies on the loop. Hence, the summation is zero unless the two points X and Y are on the same loop in G. The result is

$$\chi_{xx} = (N\beta)^{-1} \int dX dY \frac{1}{4} \frac{\sum_G 2^{n_C(G)} V(G) \delta_{c(X),c(Y)}}{\sum_G 2^{n_C(G)} V(G)}$$

$$= \frac{1}{4N\beta} \left\langle \sum_c V_c^2 \right\rangle,$$

where $c(X)$ and $c(Y)$ specify the clusters (loops) to which the points X and Y belong. Thus we obtain a graphical estimator for χ_{xx}:

$$Q(G) \equiv \frac{1}{4N\beta} \sum_{c \in G} V_c^2.$$

The key observation that led to this result was that the summation in the numerator is over the same set of graphs as the summation in the denominator.

The final result for $q = 0$ looks identical to (6.34) even though the quantity discussed is the transverse susceptibility instead of the longitudinal one. This is not a contradiction because M_c and V_c differ by the sign of the spins. For the easy-plane *XXZ* model, the sign of the spins alternates along a loop, and M_c is defined with this alternating sign. V_c is simply the length of the loop. As a result, $|V_c| \geq |M_c|$, which in turn yields $\chi_{xx} \geq \chi_{zz}$. For the easy-axis ferromagnetic *XXZ* model, the sign of the spins does not alternate and M_c is essentially the cluster size. On the other hand, in the above model the graphical estimator of the transverse susceptibility is no longer valid. To make it valid, we replace $\delta_{c(X),c(Y)}$ by a function that is 1 only when we can split the cluster into two disjoint clusters by cutting it at the points X and Y. In other words, if X and Y are multiply connected in G, the value of the function is 0 even if X and Y are on the same cluster. As a result, χ_{xx} in general becomes smaller than the average cluster size, yielding $\chi_{xx} \leq \chi_{zz}$. Again there is no contradiction. At the isotropic point ($J^{xy} = J^z$), the two estimators are both valid and agree with each other, reflecting the equality of χ_{xx} and χ_{zz} at this point. We note that in both cases (the longitudinal susceptibility in the easy-axis case and the transverse susceptibility in the easy-plane case) the relevant susceptibility corresponds to the average cluster size. Hence, the system update occurs in "units" of clusters whose sizes are roughly the same as the correlation length. This fact is the reason why in these cases the loop/cluster update equilibrates the system very efficiently.

6.2.5 Magnetic fields

None of the quantum spin models studied so far involved an external magnetic field. Given that the frameworks developed in Sections 6.2.1 and 6.2.2 are quite general, we can easily include an external field by simply regarding the field as a one-body interaction that is part of the total interaction.

Let us add the Zeeman term $-H^z S_i^z$ to $S = \frac{1}{2}$ spin models. Then the graphical decomposition is

$$\langle s_i(\tau^+)|H^z S_i^z|s_i(\tau^-)\rangle = H^z \delta_{s_i(\tau^+),\frac{1}{2}} \delta_{s_i(\tau^-),\frac{1}{2}} - \frac{1}{2}H^z\delta_{s_i(\tau^+),s_i(\tau^-)}.$$

We can neglect the second term on the right-hand side as it only shifts the energy by a constant. The first term is a kind of a graph operator. The constraint imposed by this operator or graph element is obviously that the variable on the cluster containing the location (i, τ) is fixed to be $\frac{1}{2}$. According to the general prescription, we are to generate these "fixing" graph elements over the region where spins are aligned with the magnetic field. For the ferromagnetic *XXZ* model, for which the spins in every cluster are aligned with each other, the probability of having at least one fixed graph somewhere in a cluster of volume V is $1 - e^{-H^z V}$, assuming that the cluster is already aligned with the field. This means that we flip the cluster with the probability $\frac{1}{2}e^{-H^z V}$. The factor of $\frac{1}{2}$ comes from the fact that we flip the cluster with probability $\frac{1}{2}$ if there is no fixing element.

We can find the same flipping probability in a different way; namely, instead of taking the Zeeman weight into account in constructing the graphs, we take it into account in the flipping probability. In this approach, we neglect the Zeeman weight when constructing clusters. When we flip them, we do not simply flip them with probability $\frac{1}{2}$, but rather adjust the flipping probability so that the detailed balance condition is satisfied with respect to the Zeeman weight. When the spins in a cluster of volume V are parallel to the field, flipping the cluster against the field decreases the Zeeman weight from $e^{\frac{1}{2}H^zV}$ to $e^{-\frac{1}{2}H^zV}$. Therefore, if we follow the standard prescription of the Metropolis algorithm, the cluster flipping probability is e^{-H^zV}. Because multiplying the transition probabilities P_{ij} and P_{ji} by the same factor preserves the detailed balance condition, we can make the probability of flipping against the external field $\frac{1}{2}e^{-H^zV}$ and make it $\frac{1}{2}$ for the opposite process. This result is exactly the same as the flipping probability derived in the first approach.

While the two approaches yield the same result in the case of the ferromagnetic *XXZ* model, they do not in general yield the same result when spins are inhomogeneous within the cluster, as is the case for the antiferromagnetic *XXZ* model. For such cases, the second approach is better, although it still might not be good enough to avoid the freezing problem discussed in Section 6.4.1.

6.2.6 Large spins $\left(S > \frac{1}{2}\right)$

The loop algorithm has been discussed just for $S = \frac{1}{2}$ quantum spin models, but the framework presented in Sections 6.2.1 and 6.2.2 is more general. If we were to apply this framework to $S > \frac{1}{2}$ models, each world-line segment in the resulting algorithm would take $2S + 1$ possible values instead of just the two values it takes for the $S = \frac{1}{2}$ case. However, the graphical decomposition is generally not unique, and the efficiency of the resulting algorithm may depend on it. Unfortunately, the existence of a "good" decomposition is not guaranteed. Efficient algorithms for $S > \frac{1}{2}$ spin models that fit the general framework are known only in very few special cases. One example is the SU(N) symmetric model with a fundamental representation that is itself a special case of a $S = (N - 1)/2$ spin model. We discuss this model in Appendix D. Here, we consider an alternative strategy.

The problem can be solved by replacing each spin operator by the sum of $2S$ spins, each carrying $S = \frac{1}{2}$ (Kawashima and Gubernatis, 1994).[7] Let us "split" a spin of magnitude S into $2S$ spins each carrying $S = \frac{1}{2}$:

$$S_i^\alpha \Rightarrow \tilde{S}_i^\alpha \equiv \sum_{\mu=1}^{2S} \sigma_{i\mu}^\alpha \qquad (\alpha = x, y, z), \tag{6.35}$$

where $\sum_{\alpha=x,y,z}(\sigma_{i\mu}^\alpha)^2 = 3/4$. By this replacement, we also expand the Hilbert space so that it now has dimension 2^{2SN}, considerably larger than the original one, $(2S + 1)^N$. To obtain a formulation that exactly corresponds to the original

---

[7] See also Todo and Kato (2001) for an explicit formulation for $S \geq \frac{3}{2}$.

problem, we have to eliminate many states. A projection operator $\hat{P}$ (Kawashima and Gubernatis, 1994; Harada et al., 1998; Todo and Kato, 2001) does this job,

$$Z = \text{Tr}\left[e^{-\beta H(\{S_i\})}\right] = \text{Tr}\left[\hat{P}e^{-\beta H(\{\tilde{S}_i\})}\right].$$

This operator projects the extended Hilbert space onto the original space where all local spin states have the highest S_z value, that is, $(\tilde{S}_i)^2 = S(S+1)$. $\hat{P}$ is the product of the local projection operators $\hat{P}_i$,

$$\hat{P} = \prod_i \hat{P}_i.$$

Because the highest spin state is a symmetric state, each local projection operator is the symmetrizer

$$\hat{P}_i = \frac{1}{(2S)!}\sum_\pi D_i(\pi),$$

where the operator $D_i(\pi)$ represents a permutation π that maps $\{1, 2, \ldots, 2S\}$ onto itself. Specifically,

$$D_i(\pi)\left|s_{i,1}, s_{i,2}, \ldots, s_{i,2S}\right\rangle = \left|s_{i,\pi(1)}, s_{i,\pi(2)}, \ldots, s_{i,\pi(2S)}\right\rangle.$$

In terms of the split spins, the pair Hamiltonian (6.22) is now expressed as

$$H_{ij} = \sum_{\mu=1}^{2S}\sum_{\nu=1}^{2S} H_{i\mu,j\nu},$$

where $H_{i\mu,j\nu}$ represents the interaction between the μ-th split spin on the site i and the ν-th split spin on the site j. Each pair Hamiltonian $H_{i\mu,j\nu}$ is simply a Hamiltonian of an $S = \frac{1}{2}$ model. Therefore, we can apply the general prescription discussed previously (Section 6.2.3) directly to the new problem.[8] In the resulting method, $2S$ vertical lines represent each site i (Fig. 6.4). We must repeat the graph-assignment procedure for each of the $(2S)^2$ pairs of indices μ and ν. Apart from this repetition with respect to the split spin indices, the algorithm is identical to the $S = \frac{1}{2}$ algorithm. Specifically, we use exactly the same types of graphs and the same graph-assignment densities as in the $S = \frac{1}{2}$ algorithm.

The remaining task is to cast the projection operator into the framework of the general loop/cluster algorithm. This task is easy as the permutation operator is a type of graph operator. For example, in the case of $S = 1$, the operator $D_i(\pi)$, with π the identity permutation, corresponds to the left diagram at the top of Fig. 6.4 and consists of two vertical world-line segments, whereas if π is the permutation

---

[8] We have to interpret a pair of sites b as $((i\mu), (j\nu))$ instead of (i, j).

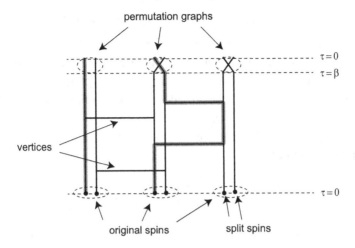

Figure 6.4 A world-line configuration for an $S = 1$ quantum spin system in the split-spin representation.

that swaps the two spins, it corresponds to the middle and the right diagram and consists of two crossing segments. Therefore, as illustrated in Fig. 6.4, inserting the projection operator amounts to picking one of the permutation diagrams compatible with the current state of split spins for each i on the boundaries $\tau = 0, \beta$ and connecting the split spins on i at $\tau = \beta$ to those at $\tau = 0$.

We now briefly mention the calculation of the Haldane gap for various S, as it illustrates the high efficiency of the algorithm. When S is integer, the one-dimensional Heisenberg antiferromagnet is equivalent to the two-dimensional nonlinear σ-model. The latter model is equivalent to the two-dimensional classical Heisenberg model, which does not have an ordered state at any finite temperature. Therefore, for the one-dimensional Heisenberg antiferromagnet with integer spins, we expect a disordered ground state. Accordingly, the excitation gap, which should be proportional to the inverse of the correlation length ξ_τ in the temporal direction, is finite, in contrast to the magnetically ordered state that is gapless due to the Goldstone modes. The gap is called the *Haldane gap* and is measured as the energy difference Δ between the ground and the first excited states. However, since the gap decreases exponentially fast as the spin magnitude increases, it becomes increasingly difficult to obtain accurate estimates of the gap for larger spins. Monte Carlo simulations (Todo and Kato, 2001) yielded $\xi = 6.0153(3), \Delta = 0.41048(6)$ for $S = 1, \xi = 49.49(1), \Delta = 0.08917(4)$ for $S = 2$, and $\xi = 637(1), \Delta = 0.01002(3)$ for $S = 3$ for the spatial spin-spin correlation length ξ, defined as $\langle S_i^z S_j^z \rangle \propto e^{-r_{ij}/\xi}$, and the excitation gap Δ.

6.3 High-temperature series expansion

In Section 6.1, we discussed the mapping of quantum problems to classical problems based on the Feynman path integral. In this section, we derive another mapping based on a high-temperature series expansion. The final results are equivalent. Introducing this second alternative provides a good starting point for discussing the worm update in the next section.

Let us start from the high-temperature series expansion of the partition function truncated at the M-th order:

$$Z_M \equiv \sum_{n=0}^{M} \frac{\beta^n}{n!} \mathrm{Tr}[(-H)^n].$$

If the dimension of the Hilbert space is finite, the series converges absolutely, and when M is increased beyond a certain finite value, it converges exponentially fast. The value of M required for a good approximation is usually of the order of $\beta \Lambda$ where Λ is the typical magnitude of the matrix elements of $-H$.

The first step in the mapping is the introduction of M "boxes," each of which can "contain" the operator $-H$. For each box, we define a variable γ_k that specifies the state of the box: $\gamma_k = 0$, if it is empty, and $\gamma_k = 1$, if it is occupied. Next, we let a state with n filled boxes represent an n-th order term in the series expansion. Since there are $M!/((M - n)!n!)$ ways of filling n boxes, we use the reciprocal of this number as the weight of each such state. Also, because the Hamiltonian is a sum of local interactions, $H = \sum_l H_l$, we partition each box into N_l "cells." The l-th cell might or might not contain H_l. Specifying the empty or filled state of the l-th cell of the k-th box by $\gamma_{kl} = 0$ or 1, we rewrite the partition function as

$$Z_M = \sideset{}{'}\sum_{\{\gamma_{kl}\}} \frac{(M - n)!}{M!} \beta^n \, \mathrm{Tr}\left[\prod_{k=1}^{M} \prod_{l=1}^{N_l} (-H_l)^{\gamma_{kl}} \right], \tag{6.36}$$

where $n \equiv \sum_{kl} \gamma_{kl}$ and the summation is taken over all filling configurations $\{\gamma_{kl}\}$. The prime indicates the restriction $\sum_l \gamma_{kl} \le 1$ for all k.

6.3.1 Stochastic series expansion

The *stochastic series-expansion* method is a Monte Carlo algorithm that defines a Markov process in the space of configurations $C \equiv \{\gamma_{kl}\}$ with the weight being the summand of (6.36) (Sandvik and Kurkijärvi, 1991). Historically, the first classical representation used in Monte Carlo simulations of a quantum system was the high-temperature series expansion for the $S = \frac{1}{2}$ isotropic Heisenberg model (Handscomb, 1962b). The method was then fully developed by Sandvik and coworkers (Sandvik and Kurkijärvi, 1991; Sandvik, 1992, 1999).

In Section 4.5, we discussed how to define a Markov process to sample a classical high-temperature series expansion. There, the expansion yielded a representation of the partition function as a sum of closed graphs. A major difference between the previous classical case and the current quantum case is that we must now deal with the noncommutivity of the quantum mechanical operators: The order of the operator products in the individual terms of the series expansion matters. To keep track of the order, we introduce an "order" space, which is naturally discrete. This additional dimension plays a role similar to the imaginary-time dimension in the path-integral formalism of Section 6.1.

There are various ways to construct a Markov sampling of the series expansion. However, since the weight is nonlocal, computing the acceptance probability of a proposed change in γ_{kl} is in general computationally expensive. To reduce this expense, we make the weight local by introducing "spin" variables. The trick is by now a familiar one: We insert the closure relation between adjacent operators, obtaining

$$Z_M = \sum_C \sum_G W(C, G), \tag{6.37}$$

where $C \equiv (\psi(1), \psi(2), \ldots, \psi(M))$, $G \equiv (\gamma_{1,1}, \gamma_{1,2}, \ldots, \gamma_{M,N_l-1}, \gamma_{M,N_l})$, and

$$W(C, G) \equiv \frac{(M - n)!}{M!} \beta^n \prod_{k=1}^{M} \prod_{l=1}^{N_l} \langle \psi(k + 1) | (-H_l)^{\gamma_{kl}} | \psi(k) \rangle.^9 \tag{6.38}$$

It is evident that the weight is local except for the constraint imposed implicitly on G, that is, $\sum_l \gamma_{kl} \leq 1$ and the factor $(M - n)!$. Fortunately, these sources of nonlocality impose only a relatively low computational cost. The Monte Carlo simulation based on the stochastic series expansion is a Markov process in the product space of C and G, with the target distribution being $W(C, G)$ defined in (6.38).

From this new expression, we now define the procedure for updating γ_{kl} while the "world line configuration" C is fixed. For each k, if there is no filled cell in the k-th box, we choose the l-th cell and fill it with probability $p_l^{(\text{fill})}$.[10] If there is a filled cell in the k-th box, we empty it with probability $p^{(\text{empty})}$. The filling and emptying probabilities satisfy the detailed balance condition

$$p_l^{(\text{fill})} \langle \psi(k + 1) | \psi(k) \rangle = p^{(\text{empty})} \frac{\beta}{M - n} \langle \psi(k + 1) | (-H_l) | \psi(k) \rangle.$$

---

[9] Note here that the notation is simplified. The symbol $-H_l$ should be interpreted as $1 \otimes 1 \otimes \cdots \otimes (-H_l) \otimes \cdots \otimes 1$; that is, it is defined as an operator acting on the whole Hilbert space, not on its local component. For spins not in the local unit b_l, it acts as an identity operator.

[10] We construct a cumulative distribution function and use it to pick l. If $\sum_l p_l^{(\text{fill})} < 1$ there is a nonzero probability of not filling any cell.

If $\psi(k+1) \neq \psi(k)$, $p^{(\text{empty})} = 0$ satisfies the equation, so we don't do anything and move to the next box. If $\psi(k+1) = \psi(k)$, on the other hand, there are nontrivial solutions for $p_l^{(\text{fill})}$ and $p^{(\text{empty})}$. The simplest may be

$$p_l^{(\text{fill})} = \frac{\beta}{M-n} \langle \psi(k) \,|\, (-H_l) \,|\, \psi(k) \rangle, \qquad p^{(\text{empty})} = 1. \qquad (6.39)$$

This choice works if M is large enough to make $\sum_l p_l^{(\text{fill})}$ less than or equal to one for any combination of $\psi(k)$ and $\psi(k+1)$, that is, if $(M-n)/\beta$ is greater than any off-diagonal matrix element of $-H$. Because the typical value of n realized in a simulation is of the order of $\beta\Lambda$, where Λ is the magnitude of a typical matrix element of $-H$, $M \gg \beta\Lambda$ is a sufficient condition for both the validity of the simple choice of the transition probability and the validity of the truncation. Since $\sum_l p_l^{(\text{fill})} \leq 1$, we might not choose any cell even if the current state has no filled cell. In this case, we simply go to the next box.

Once we have updated G, we must then update C. We may do this by a local update, that is, by changing the "world-line" configuration "locally." However, since we have the constraint that $\psi(k+1)$ and $\psi(k)$ can differ only at a filled cell, the update cannot be done for each cell or box independently. One way of updating the configuration without violating the constraints is to consider each pair of adjacent filled cells located at the same spatial position and attempt to modify the intermediate spin configuration (Fig. 6.5). We determine the probability of accepting an attempt, as usual, from the detailed balance condition. In the present case

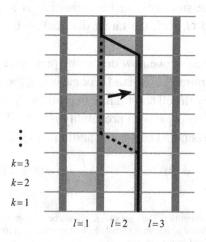

Figure 6.5 Stochastic series expansion with local update. Shaded cells correspond to $\gamma_{kl} = 1$ and white cells are those with $\gamma_{kl} = 0$.

Algorithm 18 Stochastic series expansion with local update.

Input: A world-line (cell) configuration ;

  for each box k **do**

    if $\psi(k+1) \neq \psi(k)$ **then**

      Move to the next box ;

    else

      if there is a filled cell in the box k **then**

        Empty the cell ;

      else

        With probability (6.39), fill the l-th cell ;

      end if

    end if

  end for

  for each filled cell (k, l) **do**

    Find the next filled cell (k', l) ;

    With probability (6.40), change the local states $\psi_l(k), \psi_l(k+1), \ldots, \psi_l(k')$ between the two cells ;

  end for

  return the updated configuration.

$$p_{\text{accept}} = \min(1, \mathcal{R}), \quad \mathcal{R} \equiv \prod_{(k,l)} \frac{\langle \psi'(k+1) \, |(-H_l)| \, \psi'(k) \rangle}{\langle \psi(k+1) \, |(-H_l)| \, \psi(k) \rangle}, \quad (6.40)$$

where $\psi(k)$ and $\psi'(k)$ are the states before and after the modification and the product is over all cells affected by the proposed modification. Algorithm 18 summarizes these procedures.

We can also derive loop/cluster algorithms within the present series-expansion framework. To do so, we simply expand $-H_l$ in terms of graphical operators,

$$-H_l = \sum_g a_l(g) D_l(g).$$

Correspondingly, (6.36) becomes

$$Z_M = \sum_{\{\psi(k)\} \{g_{kl}\}} {\sum}' \frac{(M-n)!}{M!} \beta^n \prod_{k=1}^{M} \left\langle \psi(k+1) \, \middle| \, \prod_{l=1}^{N_l} a_l(g_{kl}) D_{g_{kl}} \, \middle| \, \psi(k) \right\rangle,$$

with the convention that $D_l(0) = 1$ and $a_l(0) = 1$. Here, n is the total number of nontrivial graphs, that is, $n \equiv \sum_{kl}(1 - \delta_{g_{kl},0})$. The resulting loop algorithm is described in Algorithm 19.

Algorithm 19 Stochastic series expansion with loop/cluster update.

Input: A world-line (cell) configuration ;
  for each box k **do**
    if $\psi(k+1) \neq \psi(k)$ **then**
      Move to the next box ;
    else
      if there is a filled cell **then**
        Empty the cell ;
      else
        Fill the l-th cell with probability (6.39) ;
        With probability $P(g) = a_l(g) / \sum_g' a_l(g)$, select the type of graph element
        g to be placed there ;
          ▷ Here, g is chosen among the graph elements that are nontrivial and
        compatible with the current state on the cell. The summation is over all
        such graph elements.
      end if
    end if
  end for
  Identify loops/clusters ;
  for each loop or cluster **do**
    Flip the loop/cluster ;
  end for
  return the updated configuration.

6.3.2 *"Continuous-time" limit*

Let us now consider the stochastic series expansion with large M. We will see that we can replace the procedure with infinite M by a statistically equivalent one with a finite number of vertices. In the large-M limit, the discrete index k maps onto an imaginary time $\tau \equiv k\Delta\tau$ and the factor $(M-n)!/M!$ to $1/M^n$, reducing (6.37) to

$$Z_M = \sum_{\{\psi(\tau)\}} \sum_{\{(\tau_v, l_v)\}} (\Delta\tau)^n \prod_v \langle \psi(\tau_v^+) | (-H_{l_v}) | \psi(\tau_v^-) \rangle, \qquad (6.41)$$

where $\Delta\tau = \beta/M$ and $v = (\tau_v, l_v)$ specifies a *vertex* (previously called a "filled cell"). The filling probability (6.39) becomes

$$p_l^{(\text{fill})} = \Delta\tau \langle \psi(k) | (-H_l) | \psi(k) \rangle.$$

We note that the nonlocal nature of the weight due to the factor $M-n$ in (6.39) disappears because $M-n$ is replaced by M. Another source of nonlocality, that is, the constraint of having no more than one filled cell for each k also disappears, because even if we relax the constraint, the probability of having more than one filled cell is of $\mathcal{O}(\Delta \tau^2)$ and hence is negligible. Therefore, in the infinite M limit, the stochastic series-expansion becomes completely local.

The correspondence between the path-integral and the series-expansion formulations should now be obvious. To obtain the continuous-time algorithm with local update, we simply replace the vertex-assignment procedure in Algorithm 18 with the generation of space-time points, which are uniformly and randomly distributed with a density determined by the diagonal matrix element of $-H_l$, and place interaction vertices at these points. The flipping procedure does not change. We could obtain exactly the same algorithm by starting from the path-integral formulation, but we do not do this here. In the loop/cluster variant, we may again take the infinite M limit with the result being identical to Algorithm 17, which we derived from the path-integral framework.

Since we obtain identical algorithms in the continuous-time limit, the practical value of the series-expansion formulation lies in its application with finite M. Although a finite truncation error then exists, we can neglect it as long as M is much larger than $\beta \Lambda$, because the error is in this case exponentially small. This situation contrasts with that of the path-integral formulation, where the discretization error decreases only slowly (polynomially) as we increase M. Therefore, the stochastic series-expansion method with finite M is the most practical algorithm for many applications. Having a finite and fixed number of boxes is convenient, since we can use a simple array data structure, which makes the programming much simpler than in the case of the continuous-time algorithm where we must use linked-list data structures. On the other hand, the finite M algorithm is difficult to parallelize due to its nonlocal weight, its dependence on the total number of vertices n, and the constraint of having no more than one vertex in a box. These issues make it difficult to express the weight as a simple product of local factors, a fact that in turn makes it difficult to split the task into independent subtasks.

6.4 Worm update

The loop/cluster updates just discussed are very efficient when the loops or clusters created reflect the relevant physical correlations. The natural graphical decomposition of the Hamiltonian for the *XXZ* model discussed in Section 6.2.3 results in such loops or clusters. There are cases, however, in which we are not so lucky. Examples of particular importance are antiferromagnets in a uniform external field and frustrated spin systems. In these cases, the loops or clusters formed in the straightforward application of the method become much larger than the correlation length, and even diverge in size well before the critical point, resulting in the "freezing" of the simulation. For these cases, an alternative simulation approach based on the worm update can be useful. Although the algorithm is also very useful for dealing with Bosonic systems, we describe the algorithm mainly for spin models in what follows. For applications to a typical Bosonic Hamiltonian, see Section 6.6.

6.4.1 Freezing problem

Let us consider a simple example, a system of only two $S = \frac{1}{2}$ spins coupled antiferromagnetically in a uniform field,

$$H = J\vec{S}_1 \cdot \vec{S}_2 - H^z(S_1^z + S_2^z), \quad J > 0,$$

and see what happens if we apply the loop update to it at low temperature $\beta J, \beta H^z \gg 1$.

First, suppose the current state has both spins pointing up. Since the coupling represented by the horizontal graph in Fig. 6.2 is not compatible with the parallel spins, we cannot assign this graph element. Therefore, the resulting graph consists of only trivial graphs, that is, two vertical loops, each representing a spin. Next, in the loop flipping phase, we attempt to flip each spin. Because the volume of each loop is β, according to the prescription given in Section 6.2.5, the acceptance probability of flipping the loop is thus proportional to $e^{-\beta H^z}$. Since $\beta H^z \gg 1$, for all practical purposes the spins are never flipped and the state remains unchanged.

Now, suppose one of the loops is flipped, against the highly unlikely odds, and the new state has two antiparallel spins. This time, there is nothing that stops the assignment of the horizontal graph elements with density J. The probability of having no graph elements is proportional to $e^{-\beta J}$, and since $\beta J \gg 1$, for all practical purposes we always assign at least one horizontal graph element in this condition. No loop is then a vertical straight line, in contrast to the previous case. As a result, flipping loops never changes the total magnetization or places the system back into the original ferromagnetic state.

It is now clear that the transition between the $S^z = 1$ and $S^z = 0$ states never happens when the temperature is much smaller than J and H^z. The ground state of this system, however, depends on the relative values of J and H^z and changes from the singlet state to the triplet state at $H^z = J$. This observation suggests that in the vicinity of the transition point, the loop update fails to identify the system's ground state.

This example is not an exceptional case, as we encounter similar problems whenever competing interactions, such as antiferromagnetic couplings and a uniform magnetic field, coexist.

6.4.2 *Directed-loop algorithm*

We can often avoid the freezing problem by using the worm algorithm (Prokov'ev et al., 1998; Kashurnikov et al., 1999). In this section, we present a variant called the *directed loop algorithm* (Sandvik, 1999; Syljuåsen and Sandvik, 2002). While originally used with the series-expansion formulation with finite M, in the following we present it for the infinite M (that is, the continuous imaginary time) framework for conceptual simplicity.

In Section 4.5, we applied the worm update to the high-temperature series expansion of the Ising model. There, the key idea was to break the rule that every site must be shared by an even number of edges. In the worm algorithm, this rule was violated at two sites called *worm head* and *worm tail*. The state was updated by moving the head or tail. In the quantum algorithm described below, we break the rule of conservation of the total S^z at two points in space-time. In other words, world lines terminate at the worm head and tail. Initially, we create the head and tail at the same point or close to each other, and then we move them, leaving a part of a new world line behind or erasing an existing world line. When the head and tail meet, they can annihilate. If they annihilate, the resulting state is the updated spin configuration.

As we discuss below, while the head moves around, it changes its direction of motion only when it is scattered by a *vertex*. After the scattering the head moves in the direction determined in the scattering. Because of this "moment of inertia," the head in the directed-loop algorithm draws a loop more efficiently, avoiding a diffusive "back-and-forth" motion, which is an advantage compared with the usual worm algorithm.

Formally, introducing worms corresponds to adding a source term $\eta \sum_i Q_i$ to the Hamiltonian,

$$H_{\text{worm}} = H - \eta \sum_i Q_i.$$

Here, Q_i corresponds to the head or the tail. In the case of spin models, the typical choice is

$$Q_i = S_i^+ + S_i^-,$$

(in the basis set in which the S_i^z operators are diagonal).[11] In general, Q_i is a Hermitian operator that has only off-diagonal matrix elements. With the source term, the partition function in the series-expansion formulation (6.41) becomes

---

[11] For the Bose-Hubbard model, we would choose $Q_i = b_i^\dagger + b_i$ (in a basis set in which local occupation number operators are diagonal).

$$Z_M = \sum_{\{\psi(\tau)\}} \sum_{\{(\tau_v, l_v)\}} (\Delta\tau)^n \prod_v \langle\psi(\tau_v^+)|(-H_{l_v})|\psi(\tau_v^-)\rangle$$

$$\times \left(1 + \sum_{\substack{(\tau_h, i_h), \\ (\tau_t, i_t)}} (\Delta\tau)^2 \eta^2 \langle\psi(\tau_h^+)|Q_{i_h}|\psi(\tau_h^-)\rangle\langle\psi(\tau_t^+)|Q_{i_t}|\psi(\tau_t^-)\rangle\right), \quad (6.42)$$

where t and h stand for the tail and the head and (τ_t, i_t) and (τ_h, i_h) are the corresponding locations. The summation over $\{\psi(\tau)\}$ is over all configurations satisfying the condition that $\psi(\tau)$ changes only at vertices or at worms. We truncate the series at second order in η since we consider at most one worm head and tail. Including more worms in principle is possible and perhaps useful.

With a slight simplification, the above expression for the partition function becomes

$$Z = \sum_{C,G} W(C, G), \quad (6.43)$$

where $C \equiv \{\psi(\tau)\}$, $G \equiv \{(\tau_v, l_v)\}$ and

$$W(C, G) \equiv (\Delta\tau)^n \prod_v \langle\psi(\tau_v^+)|(-H_{l_v})|\psi(\tau_v^-)\rangle. \quad (6.44)$$

Here, we regard the head and the tail as special vertices and include them in the product over vertices. Correspondingly, the existence or absence of the worm and the positions of its head and tail are now in the definition of G, and the weight of the head and tail is in the definition of H_{l_v}, with l_v being the index of the site on which the head ($v = h$) or tail ($v = t$) lie. Specifically, when G has no worms, $W(C, G)$ is the weight of the stochastic series expansion discussed in Section 6.3.1. If G has a head and a tail, the weight has an extra factor, namely, the second term in the parentheses in (6.42). A "state" in the Markov process of the directed loop algorithm consists of three pieces of information: the spin configuration C, the vertex configuration G, and the direction of the head D.[12] We use the symbol Σ to represent all three, that is, $\Sigma \equiv (C, G, D)$. The directed-loop algorithm is a Markov process in Σ space.

A Markov step consists of four substeps. Starting from a configuration without worms, the first substep is the vertex assignment, the second is the creation of the worm head and tail, and the third is the worm cycle in which the head moves until it comes back to the tail. The fourth and final substep is the pair annihilation of the worm head and tail. The vertex assignment substep proceeds in exactly the same manner as the stochastic series expansion with local update discussed in

---

[12] The weight depends only on C and G, not D.

Section 6.3.2. Namely, we place the vertices over the whole system with density $\langle \psi(\tau)|(-H_I)|\psi(\tau)\rangle$. What is new is the switching between configurations with and without worms and updating the configuration with worms.

Let us consider the creation and annihilation of worms in more detail. We create a worm by first choosing a point uniformly and randomly in space time and then attempting to place the head and tail there. The attempt is accepted with a certain probability. If it is rejected, we count the current state once more as a configuration generated by the Markov process. The inverse process becomes possible when the head returns to the tail. On this occasion, we make an attempt to erase both. The attempt is accepted with a certain probability, and if it is rejected, we reverse the direction of the head and let it keep moving. The configurations before and after the creation of the worm differ only at one space-time point. Due to (6.42), the weight of the worm configuration relative to the wormless one is $(\eta \Delta \tau)^2 \langle s_i|Q_i|\sigma\rangle\langle\sigma|Q_i|s_i\rangle$, where σ is the state between the head and the tail, and s_i is the state outside of the pair. Therefore, the detailed balance condition is

$$P_{\text{create}}(s_i, \sigma)\frac{\Delta \tau}{N\beta} = P_{\text{annihilate}}(s_i, \sigma)(\eta\Delta\tau)^2\langle s_i|Q_i|\sigma\rangle\langle\sigma|Q_i|s_i\rangle.$$

$P_{\text{create}}(s_i, \sigma)$ is the probability for creating the head and the tail at the chosen position (i, τ) with the intermediate state being σ and the current local state s_i, and $P_{\text{annihilate}}(s_i, \sigma)$ is the probability of the pair annihilation when the head comes back to the tail. The factor $\Delta\tau/N\beta$ on the left is the probability that the region $\{(i, \tau')|\tau' \in \tau \le \tau' < \tau + \Delta\tau\}$ is chosen in the creation process. Because η is an artificial field, we can choose it to our advantage by making it sufficiently small so that the annihilation probability is 1. Then

$$P_{\text{create}}(s_i, \sigma) = N\beta\Delta\tau\eta^2\langle s_i|Q_i|\sigma\rangle\langle\sigma|Q_i|s_i\rangle.$$

If we choose $\eta^2 \equiv (\max_{s_i}\sum_\sigma N\beta\Delta\tau\langle s_i|Q_i|\sigma\rangle\langle\sigma|Q_i|s_i\rangle)^{-1}$, we obtain

$$P_{\text{create}}(s_i, \sigma) = \frac{\langle s_i|Q_i|\sigma\rangle\langle\sigma|Q_i|s_i\rangle}{\max_{s_i}\sum_\sigma\langle s_i|Q_i|\sigma\rangle\langle\sigma|Q_i|s_i\rangle}, \quad P_{\text{annihilate}}(s_i, \sigma) = 1. \qquad (6.45)$$

In particular, if we take $Q_i \equiv S_i^x$ for the $S = \frac{1}{2}$ XXZ model, $P_{\text{create}} = P_{\text{annihilate}} = 1$.

Later we will see that it is convenient to generalize the worm creation and annihilation process. In the above process, we have only two choices: When we have a worm, we either annihilate it or reverse the direction of the head. However, if the local degree of freedom is not a binary variable, when we attempt to create a pair, there are usually multiple choices for the state between the head and the tail (Fig. 6.6). In addition, when the attempted annihilation is rejected and the head turns around, the intermediate state left behind in general can differ from the one before the collision. In order to account for all these possibilities, we have to

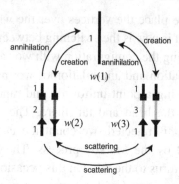

Figure 6.6   The process of creation, annihilation, and scattering of the head (triangle) at the tail (square) when there are three possible local states.

consider the detailed balance condition among a wider class of states. To make the discussion simpler, let us define

$$w_\sigma^t \equiv \begin{cases} (N\beta\eta^2\Delta\tau)^{-1} & (\sigma = s_i) \\ \langle s_i|Q_i|\sigma\rangle\langle\sigma|Q_i|s_i\rangle & (\sigma \neq s_i), \end{cases} \qquad (6.46)$$

where the case $\sigma = s_i$ represents the wormless configuration. (The superscript "t" of w^t is for the transition at the "t"ail.) Below we regard the creation/annihilation/scattering process of the head at the tail as the transition among various intermediate states σ. Our task now is to construct the transition probability $p^t(\sigma'|\sigma)$ among the states (or scattering channels) satisfying the condition[13]

$$p^t(\sigma'|\sigma)w_\sigma^t = p^t(\sigma|\sigma')w_{\sigma'}^t.$$

We solve this problem by first finding a solution $w_{\sigma'\sigma}^t$ that satisfies

$$w_{\sigma'\sigma}^t = w_{\sigma\sigma'}^t \geq 0, \quad w_\sigma^t = \sum_{\sigma'} w_{\sigma'\sigma}^t$$

and letting

$$p^t(\sigma'|\sigma) = w_{\sigma\sigma'}^t/w_\sigma^t. \qquad (6.47)$$

Then, if $\sigma \neq s_i$, $p^t(\sigma|s_i)$ is the creation probability for the intermediate state σ, and if $\sigma \neq s_i$, $p^t(s_i|\sigma)$ is the annihilation probability when the current intermediate state is σ. In addition, when neither σ nor σ' coincides with s_i, $p^t(\sigma'|\sigma)$ is the probability of turning around and switching the intermediate state to σ'. If we distinguish two states with the same configuration, except for the direction of the head motion, we note that these processes, strictly speaking, do not satisfy the detailed balance

---

[13]   Here, p^t in general depends on s_i. This dependence is not explicitly shown.

condition. However, as we discuss in Section 6.4.3, they do satisfy the time-reversal symmetry condition, and this guarantees the convergence of the Markov process to the correct target weight distribution.

Next, we consider updating the worm configurations via a stochastic motion of the head. When we create the head, we assign its initial direction such that it moves away from the tail. After traveling along the vertical line, the head eventually arrives at a vertex. There, it either does or does not change its location and direction of motion. When the vertex represents an m-body interaction, in general $2m$ or more scattering channels exist. A vertex of a two-body interaction, for example, typically has four possibilities for the location of the head after the scattering. We choose stochastically among the four possibilities, taking into account the corresponding weights.

Our problem now is satisfying the time-reversal symmetry condition among the states resulting from the scattering of the head at the vertices. Let us define

$$w_\mu \equiv \left\langle \psi_{b_l}^\mu(\tau_v^+) \left| (-H_l) \right| \psi_{b_l}^\mu(\tau_v^-) \right\rangle \left\langle \psi_{i_h}^\mu(\tau_h^+) \left| Q_i \right| \psi_{i_h}^\mu(\tau_h^-) \right\rangle, \qquad (6.48)$$

where $\psi_{b_l}^\mu(\tau_v^+)$ and $\psi_{b_l}^\mu(\tau_v^-)$ are the local states associated with the interaction unit b_l just above and below the temporal position of the vertex τ_v. The label μ specifies the scattering channel. The symbol $\psi_{i_h}^\mu(\tau_h^+)$ represents the local state on the site i_h where the head lies, with τ_h being the temporal position of the head, which is supposed to be close to the vertex. To solve our problem, we follow exactly the same discussion that led to (6.47) for the worm creation and annihilation processes; namely, if we define the mutual weight $w_{\mu\nu}$ such that

$$w_\mu = \sum_\nu w_{\nu\mu}, \text{ and } w_{\mu\nu} = w_{\nu\mu} \geq 0, \qquad (6.49)$$

then the scattering probability at vertices, p^v, defined by

$$p^v(\nu|\mu) = w_{\nu\mu}/w_\mu, \qquad (6.50)$$

automatically satisfies the time-reversal symmetry condition.

One Monte Carlo step in the directed-loop algorithm is presented in Algorithm 20. The number of cycles N_{cycle} in one step is an arbitrarily fixed number. It is usually set so that every space-time point is visited on average once.[14] For example, in the case of the $S = \frac{1}{2}$ XY model discussed in "Simulation with Worms" in Section 5.2.5, the solution of (6.49) is illustrated in Fig. 6.7. From this solution and (6.50), we obtain the scattering probabilities

$$p^v(1|4) = p^v(2|4) = \tfrac{1}{2}, \quad p^v(4|1) = p^v(4|2) = 1,$$

---

[14] A sample program that implements this procedure may be found on the Web by searching for the keyword "DSQSS."

Algorithm 20 A Markov step of the directed-loop algorithm.

Input: A world-line configuration ;
  for each vertex with no kink **do**
    Remove it ;
  end for
  for all l **do**
    Pick imaginary times uniform-randomly with density $\langle \psi(\tau)|(-H_l)|\psi(\tau)\rangle$;
    Place vertices corresponding to H_l at these times ;
  end for
  for $i = 1$ to N_{cycle} **do**
    Choose a point in space-time uniformly and randomly :
    With probability $p^{\text{t}}(\sigma|s)$ (6.47), create a head and a tail at this point with the
    intermediate state being σ ;
    if the attempt is rejected **then**
      Terminate the cycle ;
    else
      Direct the head away from the tail ;
    end if
    loop
      Let the head move until it hits a vertex or the tail ;
      if it hits a vertex **then**
        With probability $p^{\text{v}}(\nu|\mu)$, choose the scattering channel ν (6.50), with μ
        being the current local state near the vertex ;
        Let the head scatter into the chosen channel ;
      else
        With the probability $p^{\text{t}}(\sigma'|\sigma)$, choose the new intermediate state σ',
        where σ is the current intermediate state between the head and the tail ;
        if σ' coincides with the state outside of the pair **then**
          Let the head and the tail annihilate ;
          Terminate the cycle ;
        else
          Flip the head's direction and make the intermediate state σ' ;
        end if
      end if
    end loop
  end for
  return the updated configuration.

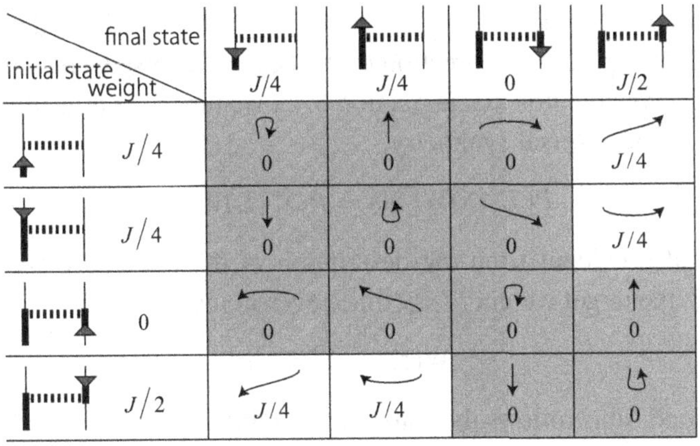

initial state \ final state / weight	J/4	J/4	0	J/2
J/4	0	0	0	J/4
J/4	0	0	0	J/4
0	0	0	0	0
J/2	J/4	J/4	0	0

Figure 6.7 The solution of the weight equation (6.49) for the head's scattering at a vertex in the $S = \frac{1}{2}$ XY model. The formula in the lower part of each cell is $w_{\mu\nu}$, defined in (6.49). The shaded entries are the prohibited scatterings. (The coupling constant J^{xy} is denoted as J for simplicity.)

and $p^v(\mu|\nu) = 0$, otherwise. This solution leads to the algorithm described in Section 5.2.5.

6.4.3 Violation of the detailed balance condition

The directed-loop algorithm does not satisfy the detailed balance condition because the head always moves in the current direction and no backward motion is allowed.[15] For example, let C and C' be two subsequent states appearing in this order, in the directed loop simulation. Note that the direction of the head is also a part of the state. The reverse process $C' \to C$ cannot happen since the head's direction in C' is such that the head moves away from its position in C. This is a clear violation of the detailed balance. Therefore, we need something extra to ensure that the process converges to the correct thermodynamic distribution.

Because the head motion violates the detailed balance condition, we need to prove the stationary condition directly (as in Section 2.6) to show the convergence of the Monte Carlo simulation to the correct stationary distribution. In the present case, the stationary condition reads

$$\sum_{\Sigma} P(\Sigma'|\Sigma)W(\Sigma) = W(\Sigma').$$

[15] It is allowed, of course, to change the direction after a scattering event and move in the new direction. What is not allowed is moving against the current direction.

Let us introduce the "time-reversal" operator " $\bar{}$ " which is defined such that $\bar{\Sigma}$ is identical to Σ except that the direction of the head is reversed. (If Σ is a wormless state, $\bar{\Sigma} = \Sigma$.) We note first that in Section 6.4.2 we tuned the transition probability such that the "time-reversal symmetry"

$$P(\bar{\Sigma}|\bar{\Sigma}')W(\bar{\Sigma}') = P(\Sigma'|\Sigma)W(\Sigma) \tag{6.51}$$

holds. As is also evident from the description of the directed-loop algorithm in Section 6.4.2, the target weight W is "time-reversal invariant,"

$$W(\bar{\Sigma}) = W(\Sigma). \tag{6.52}$$

From these conditions follows the stationary condition,

$$\sum_{\Sigma} P(\Sigma'|\Sigma)W(\Sigma) = \sum_{\Sigma} P(\bar{\Sigma}|\bar{\Sigma}')W(\bar{\Sigma}') = W(\bar{\Sigma}') = W(\Sigma').$$

Thus, the convergence is guaranteed by the theorem presented in Section 2.6.

As a matter of fact, if we ignore all intermediate states in the directed-loop algorithm, as it transitions from one wormless state to another, it satisfies detailed balance. For example, let us consider two wormless states Σ and Σ' and a sequence of states starting from Σ and ending at Σ': $\Sigma_0, \Sigma_1, \ldots, \Sigma_n$, where $\Sigma_0 = \Sigma$ and $\Sigma_n = \Sigma'$. We can deform the equilibrium probability flowing along this path by using the time-reversal symmetry conditions (6.51) and (6.52).[16] By summing over all intermediate states and the path lengths, we obtain

$$\tilde{P}(\Sigma'|\Sigma)W(\Sigma) = \tilde{P}(\Sigma|\Sigma')W(\Sigma'),$$

where $\tilde{P}$ is the "renormalized" transition probability,

$$\tilde{P}(\Sigma'|\Sigma) \equiv \sum_{n} \sum_{\substack{\Sigma_1,\ldots,\Sigma_{n-1} \\ \text{(with worm)}}} P(\Sigma'|\Sigma_{n-1}) \cdots P(\Sigma_2|\Sigma_1)P(\Sigma_1|\Sigma).$$

6.4.4 Correlation functions

The worm algorithm and the directed-loop algorithm are efficient near a critical point for similar reasons. In these algorithms, an estimator of the correlation

---

[16] In the case of $n = 3$, for example,

$$P(\Sigma'|\Sigma_2)P(\Sigma_2|\Sigma_1)P(\Sigma_1|\Sigma)W(\Sigma) = P(\Sigma'|\Sigma_2)P(\Sigma_2|\Sigma_1)P(\Sigma|\bar{\Sigma}_1)W(\Sigma_1)$$

$$= P(\Sigma'|\Sigma_2)P(\bar{\Sigma}_1|\bar{\Sigma}_2)P(\Sigma|\bar{\Sigma}_1)W(\Sigma_2) = P(\bar{\Sigma}_2|\Sigma')P(\bar{\Sigma}_1|\bar{\Sigma}_2)P(\Sigma|\bar{\Sigma}_1)W(\Sigma')$$

$$= P(\Sigma|\bar{\Sigma}_1)P(\bar{\Sigma}_1|\bar{\Sigma}_2)P(\bar{\Sigma}_2|\Sigma')W(\Sigma'),$$

where Σ and Σ' are wormless configurations and hence $\Sigma = \bar{\Sigma}$ and $\Sigma' = \bar{\Sigma}'$. From this example, it should be clear that the same is true for any n.

functions is the frequency by which the head visits a certain location. For example, to compute the correlation function

$$\Gamma(Y, X) \equiv \langle \mathcal{T} S^x(Y) S^x(X) \rangle$$

where $\mathcal{T}$ indicates the imaginary-time ordered product, we count the number of times the head passes the position $\pm(Y - X)$ relative to the tail. This estimator is quite natural in cases where the loop and directed-loop algorithms are equivalent and the trajectory of the head in the directed-loop algorithm is statistically identical to a loop in the loop algorithm. In what follows, we show that the above estimator is valid even when the directed-loop algorithm is not identical to the loop algorithm.

We start from (6.42) and (6.44) with $Q \equiv S^x$. The correlation function can be expressed as

$$(\Delta \tau \eta)^2 \Gamma(Y, X) = \frac{\sum_{\Sigma' : X, Y} W(\Sigma')}{\sum_{\Sigma' : \text{no worm}} W(\Sigma')},$$

where $\Sigma = (C, G, D)$ represents a state in the Markov chain and $W(\Sigma)$ is the weight of the worm or wormless state as defined in (6.44). The numerator is the sum over the states with the head at Y and the tail at X or vice versa. Because we encounter the state Σ with a frequency proportional to $W(\Sigma)$ in the Monte Carlo simulation, we can reexpress the correlation function as

$$\Gamma(Y, X) = \frac{1}{\eta^2} \frac{\langle \Delta_{Y,X}(\Sigma) \rangle_{\text{MC}}}{\langle \Delta_\emptyset(\Sigma) \rangle_{\text{MC}}},$$

where $\Delta_{Y,X}(\Sigma)(\Delta \tau)^2 = 1$ if and only if one discontinuity is in the interval $\Delta \tau$ centered at X and the other in the interval $\Delta \tau$ centered at Y. Now, we obtain

$$\Gamma(R) \equiv \frac{1}{N\beta} \int dX dY \Gamma(Y, X) \delta(R - (Y - X))$$

$$= \frac{1}{N\beta\eta^2} \frac{\int dX \langle \Delta_{X+R,X}(\Sigma) \rangle_{\text{MC}}}{\langle \Delta_\emptyset(\Sigma) \rangle_{\text{MC}}}$$

$$= \frac{1}{N\beta\eta^2} \langle n(R) \rangle_{\text{MC}},$$

where $n(R)$ is the average number of times the head passes, during a cycle, the point whose location relative to the tail is R or $-R$.[17]

---

[17] The denominator $\langle \Delta_\emptyset \rangle$ in the last identity is the probability of having no worm at an arbitrarily chosen Monte Carlo step. Thus, it is the inverse of the average length of the trajectory. The numerator is the probability of the relative distance between the head and the tail being R at an arbitrarily chosen Monte Carlo step. Thus, the numerator and the denominator together correspond to the probability of having R at a given step multiplied by the number of steps in one cycle, namely, the average number of times the distance becomes R during one cycle. Hence the last identity.

6.4.5 XXZ model

In this section, we explicitly describe the directed-loop update for the case of the $S = \frac{1}{2}$ XXZ model. To avoid the explicit dependence on the dimensionality or the lattice geometry, we work with the pair Hamiltonian

$$H_{ij} = -J^{xy}\left(S_i^x S_j^x + S_i^y S_j^y\right) - J^z S_i^z S_j^z - \tfrac{1}{2}h(S_i^z + S_j^z) - E_0 \qquad (6.53)$$

with the exchange coupling $J^{xy} > 0$ and the *pair* magnetic field $h > 0$. The pair magnetic field is the magnetic field per interacting spin pair, for example, $h = H^z/d$ for the d-dimensional hyper-cubic lattice. We have introduced a constant E_0 to ensure that the diagonal elements of H_{ij} are negative. The parameter space of this model divides into six regions, as shown in Fig. 6.8. In solving (6.49), we choose the constant E_0 as $J^z/4$ for region I, $(J^{xy} - h)/4 + h/2$ for regions II_1 and II_2, and $-J^z/4 + h/2$ for regions III_1, III_2, and III_3.

In the special case of the XY model without field ($J^z = 0, h = 0$), we obtain the same result as in "Simulation with Worms" in Sections 5.2.6 and 6.4.2. We construct the vertex density and the scattering probability for general S using the one for $S = \frac{1}{2}$ via the coarse-graining prescription discussed below. The results are summarized in Table 6.1. The parameters for $S = \frac{1}{2}$ are found by setting S equal to $\frac{1}{2}$ in this table. Table 6.1 lists the scattering probabilities at the vertices for three scattering directions. The directions of the arrows indicate the direction of the scattering assuming that the head enters the vertex from the lower-left corner. The symbol $p^v(\nearrow |\mu)$, for example, represents the probability for the scattering from the lower left to the upper right, when the current configuration at the vertex is μ. Similarly, the other two symbols $p^v(\downarrow |\mu)$ and $p^v(\rightarrow |\mu)$ represent the probabilities for backward scattering and horizontal scattering.

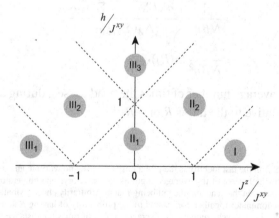

Figure 6.8 The "algorithmic phase diagram" of the XXZ model.

Table 6.1 *The vertex density ρ and the scattering probabilities $P(\Gamma|\Sigma)$ of the coarse-grained algorithm for the quantum XXZ spin model with arbitrary S. The "pair magnetic field" h is defined such that for each $S = \frac{1}{2}$ spin the sum of h over all the interacting bonds sharing it is equal to H^z, for example, $h = H^z/(2dS)$ for the d-dimensional hyper cubic lattice. Only the probabilities for nontrivial scattering are shown. (For the initial configurations (μ) not shown in the table, the nontrivial scattering probability is zero, that is, the head always passes straight through the vertex.) $\bar{l} \equiv 2S - l$, $\bar{m} \equiv 2S - m$.*

Case I $(h - J^z \leq -J^{xy})$

$\rho(l, m) = C \equiv \frac{1}{2}[lm(J^z + h) + \bar{l}\bar{m}(J^z - h)]$

μ	$p^v(\downarrow|\mu)$	$p^v(\nearrow|\mu)$	$p^v(\rightarrow|\mu)$
$\begin{pmatrix} l & m \\ l_- & m \end{pmatrix}$	$\frac{\bar{m}(-J^{xy}+J^z-h)}{2C}$	$\frac{\bar{m}J^{xy}}{2C}$	0
$\begin{pmatrix} l & m \\ l_+ & m \end{pmatrix}$	$\frac{m(-J^{xy}+J^z+h)}{2C}$	$\frac{mJ^{xy}}{2C}$	0
$\begin{pmatrix} l+1 & m \\ l_- & m+1 \end{pmatrix}$	0	$\frac{1}{\bar{l}}$	0
$\begin{pmatrix} l-1 & m \\ l_+ & m-1 \end{pmatrix}$	0	$\frac{1}{l}$	0

Case II$_1$ $(h + J^z \leq J^{xy},\ h - J^z \leq J^{xy})$

$\rho(l, m) = A \equiv \frac{1}{4}[lm(J^{xy} + J^z + 3h) + (l\bar{m} + \bar{l}m)(J^{xy} - J^z + h) + \bar{l}\bar{m}(J^{xy} + J^z - h)]$

μ	$p^v(\downarrow|\mu)$	$p^v(\nearrow|\mu)$	$p^v(\rightarrow|\mu)$
$\begin{pmatrix} l & m \\ l_- & m \end{pmatrix}$	0	$\frac{m(J^{xy}+J^z-h)}{4A}$	$\frac{m(J^{xy}-J^z-h)}{4A}$
$\begin{pmatrix} l & m \\ l_+ & m \end{pmatrix}$	0	$\frac{m(J^{xy}+J^z+h)}{4A}$	$\frac{\bar{m}(J^{xy}-J^z+h)}{4A}$
$\begin{pmatrix} l+1 & m \\ l_- & m+1 \end{pmatrix}$	0	$\frac{J^{xy}+J^z+h}{\bar{l}\cdot 2J^{xy}}$	$\frac{J^{xy}-J^z-h}{\bar{l}\cdot 2J^{xy}}$
$\begin{pmatrix} l-1 & m \\ l_+ & m-1 \end{pmatrix}$	0	$\frac{J^{xy}+J^z-h}{l\cdot 2J^{xy}}$	$\frac{J^{xy}-J^z+h}{l\cdot 2J^{xy}}$

(continued)

Table 6.1 (*cont.*)

Case II$_2$ $(J \leq h + J^z, \ -J^{xy} \leq h - J^z \leq J^{xy})$

$\rho(l,m) = A \equiv \frac{1}{4}[lm(J^{xy}+J^z+3h) + (l\overline{m}+\overline{l}m)(J^{xy}-J^z+h) + \overline{l}\,\overline{m}(J^{xy}+J^z-h)]$

μ	$p^{\mathrm{v}}(\downarrow\mid\mu)$	$p^{\mathrm{v}}(\nearrow\mid\mu)$	$p^{\mathrm{v}}(\rightarrow\mid\mu)$
$\begin{pmatrix} l & m \\ l_- & m \end{pmatrix}$	0	$\dfrac{\overline{m}(J^{xy}+J^z-h)}{4A}$	0
$\begin{pmatrix} l & m \\ l_+ & m \end{pmatrix}$	$\dfrac{m(-J^{xy}+J^z+h)}{2A}$	$\dfrac{mJ^{xy}}{2A}$	$\dfrac{\overline{m}(J^{xy}-J^z+h)}{4A}$
$\begin{pmatrix} l+1 & m \\ l_- & m+1 \end{pmatrix}$	0	$\dfrac{1}{l}$	0
$\begin{pmatrix} l-1 & m \\ l_+ & m-1 \end{pmatrix}$	0	$\dfrac{J^{xy}+J^z-h}{l\cdot 2J^{xy}}$	$\dfrac{J^{xy}-J^z+h}{l\cdot 2J^{xy}}$

Case III$_1$ $(h + J^z \leq -J^{xy})$

$\rho(l,m) = B \equiv lmh + (l\overline{m}+\overline{l}m)\dfrac{-J^z+h}{2}$

μ	$p^{\mathrm{v}}(\downarrow\mid\mu)$	$p^{\mathrm{v}}(\nearrow\mid\mu)$	$p^{\mathrm{v}}(\rightarrow\mid\mu)$
$\begin{pmatrix} l & m \\ l_- & m \end{pmatrix}$	$\dfrac{m(-J^{xy}-J^z-h)}{2B}$	0	$\dfrac{mJ^{xy}}{2B}$
$\begin{pmatrix} l & m \\ l_+ & m \end{pmatrix}$	$\dfrac{\overline{m}(-J^{xy}-J^z+h)}{2B}$	0	$\dfrac{\overline{m}J^{xy}}{2B}$
$\begin{pmatrix} l+1 & m \\ l_- & m+1 \end{pmatrix}$	0	0	$\dfrac{1}{l}$
$\begin{pmatrix} l-1 & m \\ l_+ & m-1 \end{pmatrix}$	0	0	$\dfrac{1}{l}$

Case III$_2$ $(-J^{xy} \leq h + J^z \leq J^{xy}, \ J^{xy} \leq h - J^z)$

$\rho(l,m) = B \equiv lmh + (l\overline{m}+\overline{l}m)\dfrac{-J^z+h}{2}$

μ	$p^{\mathrm{v}}(\downarrow\mid\mu)$	$p^{\mathrm{v}}(\nearrow\mid\mu)$	$p^{\mathrm{v}}(\rightarrow\mid\mu)$
$\begin{pmatrix} l & m \\ l_- & m \end{pmatrix}$	0	0	$\dfrac{m(J^{xy}-J^z-h)}{4B}$
$\begin{pmatrix} l & m \\ l_+ & m \end{pmatrix}$	$\dfrac{\overline{m}(-J^{xy}-J^z+h)}{2B}$	$\dfrac{m(J^{xy}+J^z+h)}{4B}$	$\dfrac{\overline{m}J^{xy}}{2B}$
$\begin{pmatrix} l+1 & m \\ l_- & m+1 \end{pmatrix}$	0	$\dfrac{J^{xy}+J^z+h}{\overline{l}\cdot 2J^{xy}}$	$\dfrac{J^{xy}-J^z-h}{\overline{l}\cdot 2J^{xy}}$
$\begin{pmatrix} l-1 & m \\ l_+ & m-1 \end{pmatrix}$	0	0	$\dfrac{1}{l}$

Table 6.1 (*cont.*)

Case III$_3$ ($J^{xy} \le h + J^z$, $J^{xy} \le h - J^z$)

$\rho(l,m) = B \equiv lmh + (l\overline{m} + \overline{l}m)\frac{-J^z+h}{2}$			
μ	$p^v(\downarrow \mid \mu)$	$p^v(\nearrow \mid \mu)$	$p^v(\rightarrow \mid \mu)$
$\begin{pmatrix} l & m \\ l_- & m \end{pmatrix}$	0	0	0
$\begin{pmatrix} l & m \\ l_+ & m \end{pmatrix}$	$\dfrac{m(-J^{xy}+J^z+h)+\overline{m}(-J^{xy}-J^z+h)}{2B}$	$\dfrac{mJ^{xy}}{2B}$	$\dfrac{\overline{m}J^{xy}}{2B}$
$\begin{pmatrix} l+1 & m \\ l_- & m+1 \end{pmatrix}$	0	$\dfrac{1}{l}$	0
$\begin{pmatrix} l-1 & m \\ l_+ & m-1 \end{pmatrix}$	0	0	$\dfrac{1}{l}$

An example that illustrates the utility of the directed-loop algorithm is the calculation of the magnetization of the antiferromagnetic Heisenberg chain as a function of the uniform external field. As we discussed in Section 6.4.1, a magnetic field competing with the exchange couplings causes a freezing problem in the loop algorithm, so this method cannot be used to compute a magnetization curve at low temperatures. With the directed-loop algorithm, Syljuåsen and Sandvik (2002) obtained accurate data using a reasonable amount of computer time. However, their results also showed that the autocorrelation time increases as the system size increases in the intermediate regions between successive magnetization plateaus, suggesting that the slowing down is tamed but not completely eliminated even by the worm algorithm. To date, no complete solution to this problem is known.

Now, let us consider the problem of finding the scattering probabilities for general S using the one obtained for $S = \frac{1}{2}$. In principle, we could directly solve (6.49), but the solution is not unique, and the efficiency of the resulting algorithm largely depends on which solution we choose.

Below we present the result of another approach based on the split-spin representation (Section 6.2.6). We can reformulate the model with $S > \frac{1}{2}$ in terms of $2S$ Pauli spins: $S_i \to \sum_\mu \sigma_{i\mu}$. Doing so leads to an algorithm in which the head moves in a space-time manifold of $2S$ vertical lines for each site. Now let us suppose that we visualize the simulation and make a real-time animation of it. Then, imagine we look at it on a low-resolution monitor. The $2S$ lines will blur and appear as a single thick line. In the blurred image, we cannot tell on which of the $2S$ lines the head μ is. We can tell only on which site i and at what time τ it is located. Similarly, we cannot tell on which of the $2S$ lines a particular vertex is footed, but again we

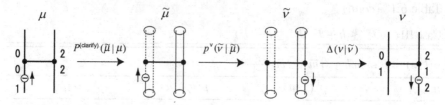

Figure 6.9 The derivation of the scattering probability of the head in the "blurred" algorithm. μ and ν are the initial and the final states in the blurred image, whereas $\tilde{\mu}$ and $\tilde{\nu}$ are the corresponding states in the split-spin image. The numbers associated with the vertical lines in the first and the last diagram represent the "brightness" of the line.

can identify the site and time. Suppose the single line in the blurred image looks brighter when in the original image we have more up-spins than down-spins on the $2S$ lines, that is, we have $2S + 1$ brightness levels (distinguishable on the monitor). The head changes the brightness level of the line by one.

Harada and Kawashima (2002) noted that such a blurred animation can be generated without first creating the sharp image and viewing it on the low-resolution monitor. Instead we could set up our simulation with a set of transition matrices defined directly in terms of the brightness. Note also that for all practical purposes what we need is only the blurred animation. In short, the split-spin representation is not necessary for describing the algorithm or writing computer codes, but as we discuss below, it is useful for the derivation of the algorithm.

It is straightforward to implement the split-spin idea by defining appropriate scattering probabilities. As an example we consider the $S = 1$ antiferromagnetic Heisenberg model. Suppose that the head has just hit a vertex in the state μ (the first diagram in Fig. 6.9). The probability of obtaining the last diagram as the final state of the scattering is

$$p^{\text{v}}(\nu|\mu) = \sum_{\tilde{\mu}\tilde{\nu}} \Delta(\nu|\tilde{\nu})p^{\text{v}}(\tilde{\nu}|\tilde{\mu})P^{(\text{clarify})}(\tilde{\mu}|\mu). \tag{6.54}$$

The probability $P^{(\text{clarify})}(\tilde{\mu}|\mu)$ that the original (sharp) image $\tilde{\mu}$ is associated with the blurred image μ is proportional to the weight of the original image,

$$P^{(\text{clarify})}(\tilde{\mu}|\mu) = \frac{\tilde{w}(\tilde{\mu})\Delta(\mu|\tilde{\mu})}{\sum_{\tilde{\mu}} \tilde{w}(\tilde{\mu})\Delta(\mu|\tilde{\mu})},$$

where $\Delta(\mu|\tilde{\mu}) = 1$ if and only if μ is the blurred image of the original one $\tilde{\mu}$. The weight $\tilde{w}(\tilde{\mu})$ is the weight in the split-spin representation,

$$\tilde{w}(\tilde{\mu}) = \sum_{\alpha\beta} \langle \sigma'_{i\alpha}\sigma'_{j\beta} | (-H_{i\alpha,j\beta}) | \sigma_{i\alpha}\sigma_{j\beta} \rangle.$$

The probability p^v is the scattering probability of the head in the split-spin representation (in the sharp image). Here we have not explicitly introduced the worm weight.[18]

In a similar fashion, we obtain the worm creation/annihilation probabilities from the blurring of the corresponding process in the split-spin representation. In the $S = \frac{1}{2}$ model, if the current local spin state at the chosen point is $+\frac{1}{2}$, we place a head corresponding to the operator S^+ above the one corresponding to S^- (switching the local state between them to $-\frac{1}{2}$). We do the opposite if the local spin state is $-\frac{1}{2}$. When coarse-grained, these placements amount to the following: When the local spin state is l, the probability of creating an S^+ head above an S^- head is $l/2s$; otherwise, we create S^- above S^+. For the worm annihilation, if S^+ is above S^- before the rendezvous and the local spin state outside of the interval is l (the local state between the two is $l - 1$), the two heads are on the same line with the probability l^{-1}, which is the annihilation probability. If, on the other hand, the S^+ is below S^- just before the rendezvous, the probability is $(2S - l)^{-1}$.

Finally, we assign the vertex density as follows: Let us consider an interval in which the local spin state is l on one of the two sites and m on the other. In the original (split-spin) image, we assign vertices with density $\langle \sigma'_{i\mu}, \sigma'_{j\nu} | (-H_{i\mu,j\nu}) | \sigma_{i\mu}, \sigma_{j\nu} \rangle$ between the two vertical lines specified by $(i\mu)$ and $(j\nu)$. Therefore, in the blurred image, we assign vertices with density

$$\rho = \sum_{\alpha\beta} \langle s_{i\alpha}, s_{j\beta} | (-H_{i\alpha,j\beta}) | s_{i\alpha}, s_{j\beta} \rangle = lm\rho_{++} + l\bar{m}\rho_{+-} + \bar{l}m\rho_{-+} + \bar{l}\bar{m}\rho_{--},$$

where $\rho_{\pm\pm}$ is the vertex density for the $S = \frac{1}{2}$ model with the local spin state $(\pm\frac{1}{2}, \pm\frac{1}{2})$.

The resulting vertex densities and the scattering probabilities are shown in Table 6.1. Note that h is now the magnetic field per Pauli-spin pair, for example, $h = H^z/(2dS)$ for the d-dimensional hypercubic lattice. The offset energies E_0 subtracted from the standard definition of the Hamiltonian are $(J^z - h)S^2 + hS$ for regions I and V, $-J^{xy}S^2 + hS$ for regions II, III, and IV, and $J^{xy}S^2 + h(S - 2S^2)$ for region VI.

6.4.6 On-the-fly vertex generation

Section 6.4.2 describes a worm update method in the framework of the series-expansion classicization in which the vertex assignment and the worm update are two separate procedures. In the present subsection, we describe an alternative way

---

[18] Alet et al. (2005) remarked that the present algorithm can be viewed as a directed-loop algorithm that is a special solution to (6.48) with worm weight $w_w \propto \langle s_{i_h}(\tau_h^+) | S_{i_h}^x | s_{i_h}(\tau_h^-) \rangle$.

of introducing worms in the world-line Monte Carlo simulation that generates inter-action vertices "on the fly." The resulting algorithm is quite similar to the original worm algorithm (Prokov'ev et al., 1998); however, the algorithm we present here is a generalization of the directed-loop algorithm and thus has a directed head motion. The "directed-ness" saves some computational time by preventing unnecessary diffusive motion of the head. Compared with the directed-loop algorithm, the on-the-fly algorithm has the same advantage as the conventional worm algorithm; that is, it needs computer memory only for storing the world lines and not for storing "empty space." This advantage is significant in the case of extremely dilute Bose gases, which are often considered in applications to cold atom systems.

The basic idea is very simple. We essentially do the same thing we did in the continuous-time directed-loop algorithm described in Section 6.4.2, but we apply the vertex generation procedure only in a restricted area near the head (Kato and Kawashima, 2009). More precisely, we consider all the interaction terms H_l involv-ing the head (or the site on which the head is currently located). For the imaginary-time interval I, delimited by the head itself and the kink ahead of it, we attempt to place a vertex for each interaction term following the rule described in Section 6.4.2; that is, we place it by a Poisson process whose rate equals the diagonal matrix element of H_l. Among such vertices, we pick the one closest to the head. As described in Section 6.4.2, we move the head to the vertex and let it scatter in exactly the same fashion as in the conventional directed-loop algorithm. Once the scattering is done, the vertices are erased and we repeat this procedure again and again until the head annihilates with the tail. It is possible that no new vertex is placed between the head's current position and the next kink. In this case, we simply move the head to the kink delimiting I and let it scatter there, again following the same rule as before.

While this procedure gives us correct results, we can improve it by skipping all the "forward-scatterings." In the procedure we just described, at each vertex there is a probability that the head scatters forward with no change in its spatial location or in its direction of motion. Therefore, when this probability is close to unity, we are forced to generate random numbers and make stochastic decisions many times with no change in the world-line configuration until finally some nonforward scattering occurs. To avoid the computational cost of this repetition, we may compute the first scattering time τ_l at which a (nonforward) scattering by the vertex characterized by H_l should take place if no other earlier scattering occurs. After generating τ_l for all interactions, we choose the smallest one, say, τ_{l^*}, let the head proceed to τ_{l^*}, and scatter at the vertex corresponding to $-H_{l^*}$. In this way, we save the computational cost of handling the forward scatterings.

An even better procedure with the same outcome is generating the first scattering time with the "total" scattering rate, that is, the sum of scattering rates of all vertices,

and then decide which is the scattering vertex. This procedure involves a Poisson
process only once for each actual nonforward scattering. This task is a little complicated because the state of spins interacting with the current spin varies in imaginary
time, which, in turn, makes the decay rate inhomogeneous in the interval I.

There are two simple solutions to this problem. One is to change the definition
of I. So far, we have taken I as the interval delimited by the next future kink. Instead,
we could define I as the interval delimited by the point at which the environment
of the head changes. In other words, we could define I as the maximal interval
with a constant "molecular field." Though I in the new definition is shorter than
the one previously defined, the new definition serves the current purpose. Now the
procedure becomes quite simple. We first generate a uniform random number ζ and
let τ_{first} be

$$\tau_{\text{first}} = -\log \zeta /a \tag{6.55}$$

with a being the total scattering amplitude. We then let the head advance by τ_{first}.
The total scattering amplitude is the sum of the nonforward scattering amplitudes
of all channels, that is,

$$a \equiv \sum_l a_l p_l^{\text{n.f.}}, \tag{6.56}$$

where the summation is taken over all interaction terms in the Hamiltoninan in
which the current site is involved, a_l is the matrix element of $-H_l$ evaluated just in
front of the head,

$$a_l \equiv \langle \psi(\tau)|(-H_l)|\psi(\tau)\rangle,$$

and $p_l^{\text{n.f.}}$ is the nonforward scattering probability at the vertex representing H_l. To
be specific, $p_l^{\text{n.f.}} \equiv 1 - p_l^{\text{v}}(\uparrow |\mu)$, with μ being the local state just in front of the
current position of the head.[19] Next, we choose the interaction term l for which the
scattering actually occurs with the probability

$$p_l \equiv \frac{a_l p_l^{\text{n.f.}}}{a}.$$

Then, we place a vertex corresponding to H_l just in front of the head and let it
scatter there. In this scattering, only nonforward scatterings should be considered
since the forward scattering is already taken into account by the fact that we skip
the interval τ_{first}. In other words, we should use a modified scattering probability,

$$p^{\text{v}'}(\nu|\mu) \equiv \frac{p^{\text{v}}(\nu|\mu)}{p_l^{\text{n.f.}}}, \tag{6.57}$$

---

[19] Here, p^{v} and μ depend on the interaction term l, though the dependence is not explicitly shown.

where $p^v(v|\mu)$ is the previously introduced scattering probability from the initial state μ to the final state v in the standard procedure (6.50) for the H_l vertex.

Another solution to the problem of a time-dependent scattering rate becomes possible by modifying the above procedure. We replace the (now generally time-dependent) scattering rate a in (6.55) by its maximum value $a_{\max}$, compute

$$\tau_{\text{first}} = -\log \zeta / a_{\max}, \tag{6.58}$$

and then compensate for the overestimate by rejecting the proposed scattering with a certain probability. To be more precise, $a_{\max}$ is any configuration-independent constant larger than the rate in (6.56). In the new prescription, instead of placing a vertex at the time $\tau + \tau_{\text{first}}$, we place only a "marker" for it. We move the head to this point and compute the total scattering rate a by (6.56). Then, with the probability $1 - a/a_{\max}$, we choose the forward scattering. Otherwise we select the scattering channel with the probability (6.57). This variant is adopted in Algorithm 21.

Because the resulting algorithm for general models looks complicated, it may be helpful to specialize for the simple case of the $S = \frac{1}{2}$ XY model. We first make $w_{\text{worm}}(\sigma)$ in (6.46) the same for all cases by adjusting the value of η, which determines the creation and annihilation probabilities of the worm, such that both are unity. The scattering rate for an interaction is always $\frac{1}{4}J^{xy}$ for each bond, making a in (6.58) $\frac{1}{4}zJ^{xy}$ where z is the coordination number. Since a does not depend on the configuration, we take $a_{\max} = a$, which avoids having to choose forward scatterings at the marker. The scattering probability of the vertex is unity if there is no kink on it, and the scattering probability at a kink is $\frac{1}{2}$ for diagonal and horizontal scatterings. The resulting algorithm is somewhat simpler than the general one, as shown in Algorithm 22.

6.5 Toward zero temperature

In Part III, we discuss algorithms specifically designed to simulate systems at zero temperature. We now briefly discuss how finite-temperature Monte Carlo methods may be used to study zero-temperature properties. A straightforward way to do this is to run simulations at temperatures that are nonzero but low enough that the behavior is approximately that of the system at zero temperature. This approach is generally successful when there is a finite gap between the ground state and the first excited state. Studying systems with vanishing excitation gap requires a more sophisticated technique. Of particular importance is the application of finite-temperature quantum Monte Carlo simulations to the study of quantum critical phenomena.

Algorithm 21 A cycle in the directed-loop algorithm with "on-the-fly" vertex generation.

Input: A world-line configuration with no worm ;

    Choose a point in space-time uniformly and randomly ;

    Choose σ with probability $p^{\mathrm{t}}(\sigma|s)$ (s is the current state) (6.47) ;

    if $\sigma = s$ **then**

        Return ;

    else

        Create a head and a tail with the intermediate state being σ ;

        Direct the head away from the tail ;

    end if

loop

    Generate τ_{first} by (6.58) ;

    Place a "marker" at $\tau + \tau_{\mathrm{first}}$;

    Move the head to the position of the first object it encounters ;

    if the next object is the marker **then**

        Compute a defined by (6.56) at the new position ;

        Choose with probability $1 - a/a_{\mathrm{max}}$ whether a forward scattering occurs ;

        if forward scattering **then**

            Return to the beginning of the loop ;

        else

            Choose the interaction term l with probability $p_l \equiv a_l p_l^{\mathrm{n.f.}}/a$;

            Place the vertex corresponding to H_l just in front of the head ;

        end if

        With probability $p^{\nu\prime}(\nu|\mu)$ (6.57), let the head scatter into channel ν ;

    else if the next object is a kink **then**

        With probability $p^{\nu}(\nu|\mu)$ (6.50), let the head scatter into channel ν ;

    else if the next object is the tail **then**

        Generate σ' with $p^{\mathrm{t}}(\sigma'|\sigma)$ (6.47), where σ is the current intermediate state between the head and the tail ;

        if σ' coincides with the state just behind the head **then**

            Let the head and the tail annihilate ;

            Terminate the loop ;

        else

            Reverse the head's direction and make the intermediate state σ' ;

        end if

    end if

end loop

return the updated world-line configuration without worm.

Algorithm 22 Algorithm 21 specialized for the $S = \frac{1}{2} XY$ model.

Input: A world-line configuration with no worm ;
  Choose a point in space-time uniformly and randomly ;
  Create a head and tail ;
  Direct the head away from the tail ;
  loop
    Generate a random number ζ ;
    $\tau_{\text{first}} \leftarrow -\log \zeta / (\frac{1}{4} z J^{xy})$;
    Place the marker at the distance τ_{first} from the current position ;
    Let the head proceed until it encounters the marker, a kink, or the tail ;
    if it encounters the marker **then**
      With equal probability, choose a site j from the nearest neighbors to the current site i ;
      Place the vertex corresponding to H_{ij} just in front of the head ;
      if the spins on i and j are parallel **then**
        Let the head scatter diagonally ;
      else
        Let the head scatter horizontally ;
      end if
    else if it encounters a kink **then**
      With probability $\frac{1}{2}$, let the head scatter diagonally or horizontally ;
    else if it encounters the tail **then**
      Let the head and the tail annihilate ;
      Terminate the loop ;
    end if
  end loop
  return the updated configuration.

6.5.1 Extrapolation to zero temperature

A system that has a finite-dimensional Hilbert space always has a finite difference ΔE between the ground state energy and the energy of the first excited states. In this case, the expectation value of an arbitrary quantity Q must have a temperature dependence that varies as

$$\langle Q \rangle \sim Q_0 + \mathcal{O}(e^{-\Delta E/kT}), \tag{6.59}$$

in the temperature region $0 < kT \ll \Delta E$. Here, Q_0 is the expectation value of Q at zero temperature. Therefore, if the gap is not too small, we can estimate this expectation value by doing a quantum Monte Carlo simulation with one of the algorithms discussed in this chapter. We simply need to reach a β much larger than $(\Delta E)^{-1}$. If this is possible, we can regard the results as zero-temperature results. Even better is to run simulations at different temperatures and extrapolate the outcome to zero temperature by using (6.59).

The magnitude of the gap ΔE, however, is usually unknown beforehand. It depends on the nature of the system and might converge to zero in the thermodynamic limit $L \rightarrow \infty$. A convergence to zero may be related to the occurrence of a spontaneous symmetry breaking. If the order parameter that characterizes the symmetry breaking commutes with the Hamiltonian, the ground state is degenerate even if the system size is finite. A simple example for this is the ferromagnetic, Ising-like XXZ model $(0 < |J^{xy}| < J^z)$ for which the ground state is either the "all-up" or "all-down" state. The order parameter is the uniform magnetization $M \equiv \sum S_i^z$ and it commutes with the Hamiltonian. Hence the degeneracy is exact. In this case, if we apply the above approach, the energy gap is the difference between the twofold degenerate ground state and the first excited state. Since this gap is $\mathcal{O}(J^z - J^{xy})$, the simulation with $kT \ll \Delta E$ is reasonably straightforward. Therefore, obtaining zero-temperature properties from a simulation at low, but finite temperature is possible.

A more nontrivial example is the easy-axis, antiferromagnetic XXZ model on a bipartite lattice. The ground state is the Néel state, and therefore it is twofold degenerate in the thermodynamic limit. The order parameter is the staggered magnetization $M_s \equiv \sum \epsilon_i S_i^z$ with $\epsilon_i = 1$ for i in one sublattice, and -1 for i in the other. This order parameter does not commute with the Hamiltonian unless $J^{xy} = 0$, which we assume is not the case. As a result the ground state need not be degenerate, and, in fact, it is nondegenerate for any finite system with an even number of spins. We write the ground state of a finite system *symbolically* as $|+\rangle \equiv (|\cdots \uparrow\downarrow\uparrow\downarrow \cdots\rangle + |\cdots \downarrow\uparrow\downarrow\uparrow \cdots\rangle)/\sqrt{2}$, where $|\cdots \uparrow\downarrow\uparrow\downarrow \cdots\rangle$ is a state with positive staggered magnetization (not a perfect, fully polarized, Néel state as the equation may erroneously suggest) and $|\cdots \downarrow\uparrow\downarrow\uparrow \cdots\rangle$ is the degenerate state obtained from the other one by inverting all spins. The first excited state is then $|-\rangle \equiv (|\cdots \uparrow\downarrow\uparrow\downarrow \cdots\rangle - |\cdots \downarrow\uparrow\downarrow\uparrow \cdots\rangle)/\sqrt{2}$, with an exponentially small excitation energy $\Delta E \propto |J^{xy}/J^z|^N$. To single out the ground state, which is unique in a finite system, the temperature must be lower than this exponentially small energy. However, we often do not need to reach such a low temperature since most quantities of interest, such as the energy and the susceptibility, do not depend on which of the two states the system is in. Therefore, the observed values do not strongly depend on the temperature as long as the temperature is below the second excitation level.

Even more subtle cases arise from the spontaneous breaking of a continuous symmetry. In this case, Goldstone's theorem (Nambu, 1960; Goldstone, 1961) predicts the existence of quasi-particles, Nambu-Goldstone Bosons, that represent long-wave length excitations. The excitation energy of a Nambu-Goldstone Boson is gapless and typically has an algebraic wave-number dependence, that is, $\epsilon_k \propto k^\mu$ where μ is a positive integer. For example, in two or more dimensions the $O(3)$ symmetry is spontaneously broken in the ferromagnetic Heisenberg model. The dispersion of the magnon excitation is $\epsilon_k \sim Ja^2k^2$, where a is the lattice constant. In a finite system with linear size L, the minimum nonzero wave number is $k_1 \equiv \pi/L$, making the first-excitation gap $\Delta E \sim J(\pi a/L)^2$. Therefore, to probe zero temperature, we need to push the simulation to a temperature low enough so that $\beta J \gg J/(\Delta E) \sim (L/a)^2$ to obtain the ground state. For the antiferromagnetic Heisenberg model, the magnon excitation is $\epsilon_k \sim Jak$. Accordingly, the condition for an effective "zero-temperature" simulation is $\beta J \gg L/a$.

6.5.2 Quantum phase transitions

Many-body problems exhibit phenomena that we cannot predict from the behavior of the individual degrees of freedom. However, when we focus on critical phenomena, we often find behavior that is independent of the details of the interactions. The ability to use simulations to probe this universality is what makes these studies so valuable. Realistic systems contain many different kinds of interactions, and their simulation looks like a formidable problem. A common strategy is to start from a simplified model, identify the critical phenomena, and then consider the effect of various correction terms.

In quantum systems, critical phenomena are just as important and often even more interesting than critical phenomena in classical systems. In the limit of low temperature, it is possible to observe critical phenomena that are specific to quantum systems. We can motivate this fact from the path-integral representation, which maps a d-dimensional quantum system to a $(d+1)$-dimensional classical system. However, we cannot study zero-temperature quantum critical phenomena using the strategy discussed in the previous subsection, since the system generally has a gapless excitation at the quantum critical point and thus lacks a finite energy scale to mark the entry into the "zero-temperature region." Instead, near the quantum critical point, the gap depends on the system size algebraically, resulting in a slow decrease of the finite-temperature error as a function of system size.

Let us consider physical properties of the one-dimensional transverse-field Ising model, for which the computational aspect was already discussed in Section 5.2.2. Equation (5.6) provides a mapping onto an anisotropic two-dimensional classical Ising model. The quantum model still has a symmetry with respect to the 180-degree rotation around the x-axis. This symmetry is called a Z_2 *symmetry*. At zero temperature and $H^x = 0$, the quantum model exhibits long-range order, that is, the

Z_2 symmetry is spontaneously broken. While the ground state at $H^x = 0$ is fully polarized in the $+z$ or $-z$ direction, as soon as we introduce a finite transverse magnetic field, this state ceases to be an eigenstate of the Hamiltonian. *Quantum fluctuations* generated by the noncommutativity of the first and the second terms in the Hamiltonian cause this change. It is not obvious whether these quantum fluctuations destroy long-range order or merely weaken it. As we discussed in Section 5.2.2, the transverse-field Ising model eventually undergoes a transition from the ordered phase to a disordered phase, as we increase the transverse field.

In Section 5.2.2, we mapped the transverse-field Ising model in one dimension to the two-dimensional classical Ising model with anisotropic couplings (5.19)

$$K^x = \Delta\tau J^z, \quad K^\tau = -2\ln\frac{\Delta\tau H^x}{2}.$$

This mapping connects the critical behavior of the two-dimensional classical anisotropic Ising model (Baxter, 1982) with that of the one-dimensional quantum model. In particular, it establishes that the one-dimensional quantum model is critical when

$$\sinh\frac{K^x}{2}\sinh\frac{K^\tau}{2} = 1. \tag{6.60}$$

By substituting the expressions for K^x and K^τ quoted above in (6.60) and taking the limit $\Delta\tau \to 0$, where the classical mapping becomes exact, we obtain the exact value of the critical field

$$H_c^x = \frac{1}{2}J^z.$$

However, this statement is true only when the thermodynamic limit is properly taken, and to do so we have to be careful, because the thermodynamic limit of the classical system does not necessarily correspond to the thermodynamic limit of the quantum system. To be more specific, if we keep β fixed, the thermodynamic limit of the original one-dimensional quantum model corresponds to the two-dimensional classical model in the limit of infinite aspect ratio in which the size L in the spatial direction is taken to infinity, while the size in the temporal direction, which is β, is kept finite. It is only in the zero-temperature limit of the quantum system that the size of the equivalent classical system becomes infinite in both directions. Therefore, in the present example, the classical mapping guarantees the existence of a critical point only for the zero-temperature phase transition.

In general, when we increase the size of the temporal direction, we observe a crossover from d-dimensional behavior to $(d + 1)$-dimensional behavior. The crossover takes place when the correlation length in the temporal direction in the $(d + 1)$-dimensional system becomes comparable to the system size in the temporal direction, that is,

$$\xi_t \propto |H^x - H_c^x|^{-zv} \sim \beta,$$

where v is the critical exponent that characterizes the divergence of the correlation length around the quantum critical point, and z is the exponent that relates the length scale to the time scale, called the *dynamical critical exponent*. In the case of the transverse-field Ising model $z = 1$. If $\xi_t \ll \beta$, the system's behavior is identical to the $(d + 1)$-dimensional system since the correlation length is as long as it would be in the infinite-size system. On the other hand, if $\xi_t \gg \beta$, the variation along the temporal direction becomes zero so that the system behaves like a d-dimensional one. This general remark also applies to the transverse-field Ising model, and it follows that this model behaves as a d-dimensional one near the quantum critical point for

$$T > T^*(H^x) \propto |H^x - H_c^x|^{zv}.$$

If $d \geq 2$, the d-dimensional classical system itself has a phase transition, and the correlation length diverges near the transition temperature. This finite-temperature phase transition belongs to the d-dimensional Ising universality class. In this case, the crossover from the $(d+1)$-dimensional behavior to the d-dimensional behavior becomes a genuine phase transition, resulting in the phase diagram schematically sketched in Fig. 6.10.

In the case of the transverse-field Ising model, it is apparent from the mapping that the imaginary-time direction is equivalent to the spatial direction apart from some scaling factor, referred to as the "light (or sound) velocity." This is why $z = 1$ for this model. However, this equivalence does not hold in general. There are cases in which the dimension of the temporal axis is not the same as that of the spatial one.

6.5.3 Finite-size scaling

Knowledge of the renormalization group is indispensable for understanding critical phenomena. Very important for the numerical study of quantum systems is that this formalism enables a finite-size scaling analysis of a zero-temperature critical point. Using finite-size scaling, we are able to characterize a zero-temperature critical point using numerical simulations at finite temperatures. There are many textbooks on this subject (e.g., Ma, 1985; Cardy, 1996; Nishimori and Ortiz, 2010) so we only briefly outline the method.

Let the parameters characterizing the thermodynamic state be $t, h, g, \ldots$, and the system size be L. In the case of the Ising model, for example, t is the deviation of

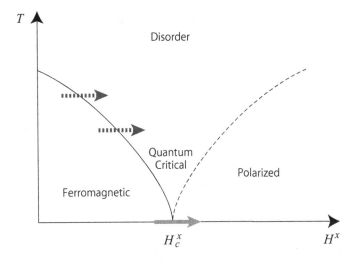

Figure 6.10 Schematic phase diagram of the transverse-field Ising model. The finite-temperature phase transition exists only in two or higher dimensions. When the phase boundary is crossed at nonzero temperature (dotted arrow), the critical phenomena belong to the d-dimensional classical Ising universality class. When it is crossed at zero temperature (solid arrow), it belongs to the $(d + 1)$-dimensional classical Ising universality class. In one dimension, there is a crossover line instead of a transition line.

the temperature from the critical temperature and h is the magnetic field. The total free energy F of the system is a function of these parameters:

$$F(t, h, g, \ldots, L).$$

Let us suppose that the system is invariant under a renormalization transformation of scale b and that the parameters transform as

$$t \to t b^{y_t}, \quad h \to h b^{y_h}, \quad g \to g b^{y_g}, \quad \ldots \quad (6.61)$$

While, in general, we must redefine the parameters to make these transformations exact, here we simply assume that (6.61) is satisfied. Then, we say that y_p ($p = t, h, g, \ldots$) is the scaling eigenvalue of the field p and that the scaling dimension of the operator Q conjugate to p is $x_Q \equiv d - y_p$. For example, the asymptotic behavior of the two-point correlation function is

$$\langle Q(x)Q(y) \rangle \propto \frac{1}{|x - y|^{2x_Q}}.$$

If $y_p > 0$, we say that the scaling field p (or the operator Q) is *relevant*. If $y_p < 0$, on the other hand, it is *irrelevant*.

When the parameters obey the scaling transformation (6.61), the free energy satisfies

$$F(t, h, g, \dots, L) = F(tb^{y_t}, hb^{y_h}, gb^{y_g}, \dots, L/b).$$

By setting $L/b = \Lambda$ and eliminating b, we get

$$F(t, h, g, \dots, L) = F(t(L/\Lambda)^{y_t}, h(L/\Lambda)^{y_h}, g(L/\Lambda)^{y_g}, \dots, \Lambda).$$

If we now regard Λ as a constant, we can eliminate it and obtain the *finite-size scaling* formula (Fisher and Barber, 1972)

$$F(t, h, g, \dots, L) = \tilde{F}(tL^{y_t}, hL^{y_h}, gL^{y_g}, \dots). \tag{6.62}$$

From this expression, the scaling form in the thermodynamic limit follows by first deforming (6.62) to

$$f(t, h, g, \dots, L) = F/L^d = t^{d/y_t}\tilde{f}(tL^{y_t}, ht^{-y_h/y_t}, gt^{-y_g/y_t}, \dots)$$

and then taking the limit $L \to \infty$ with fixed t to produce the following formula for the free-energy density:

$$f(t, h, g, \cdots) = t^{d/y_t} f(ht^{-y_h/y_t}, gt^{-y_g/y_t}, \dots). \tag{6.63}$$

The final step is justified because f must converge to some finite value independent of L. Equation (6.63) is called the *scaling form of the free energy*.

By considering a large enough L, we can neglect the dependence on the irrelevant parameters. For example, in the case of the critical point of the two-dimensional Ising model, we have only two relevant scaling fields, one associated with the temperature difference from the critical temperature $T - T_c$ and the other with the magnetic field H. Therefore, we identify t and h in (6.62) with these fields to obtain an asymptotically correct scaling form,

$$F(T, h, L) \approx F(tL^{y_t}, hL^{y_h}) \quad \text{(Ising model)}. \tag{6.64}$$

The scaling forms of other quantities follow from this equation by differentiating both sides with respect to the relevant scaling fields. For example, for the zero-field magnetic susceptibility, differentiating (6.64) twice with respect to H (or h) yields

$$\chi(T, L) \approx -L^{-d} \frac{\partial^2 F((tL^{y_t}, hL^{y_h})}{\partial h^2}\bigg|_{h=0}$$

$$= L^{-d+2y_h}\tilde{\chi}((T - T_c)L^{y_t}) \quad \text{(Ising model)}, \tag{6.65}$$

where $\tilde{\chi}(X) \equiv -(\partial^2 F(X,Y)/\partial Y^2)_{Y=0}$. From this thermodynamic relation, it follows that $\gamma = (2y_h - d)/y_t$, where γ is the critical exponent characterizing the divergence of the susceptibility at the transition point.

Equations such as (6.64) and (6.65) are of crucial importance in analyzing the results of a numerical simulation. We can use them to identify the fixed point that dominates the phase transition by fitting the computed results to these formulas. For example, suppose we carry out a Monte Carlo simulation of the Ising model for various values of T and L. As a result we will obtain many estimates of $\chi(T,L)$. If we then plot $\chi(T,L)/L^{2y_h-d}$ against $(T-T_c)L^{y_t}$, all data points, regardless of the system size, will fall onto a single curve, $Y = \tilde{\chi}(X)$, provided that L is sufficiently large. Of course, in most cases, we do not know the correct values of y_t, y_h, and T_c. Therefore, we make some guesses. To obtain the correct values, we then adjust the guessed values until all the data points collapse onto a single curve.[20]

With two modifications, we can apply the finite-size scaling discussed above to quantum systems. The first modification is switching from $F \equiv -kT \log Z$ to $\Phi \equiv -\log Z$. This is a natural step, because in the finite-size scaling the imaginary-time axis is treated just as another direction, and therefore we should remove the normalization factor $1/\beta$ to make the equivalence between space and time apparent. The second is to introduce the dynamical critical exponent z, which is the dimension of imaginary time.

In the case of the transverse-field Ising model discussed in the previous section, the imaginary-time axis is equivalent to any of the spatial axes apart from some constant, which means *Lorentz invariance*, $z = 1$. In such cases, we can regard imaginary time as an additional length scale, and when we investigate the quantum critical phenomena, we simply apply the finite-size scaling discussed above with d replaced by $d+1$. In doing so, we need to fix the aspect ratio by varying the inverse temperature β with the system size L. However, in some cases, the imaginary-time direction is not equivalent to a space direction. To extend the finite-size scaling analysis to such cases, we generalize (6.62) by adding the inverse temperature as another argument,

$$\Phi(t,\ldots,\beta,L) = \Phi(tL^{y_t},\ldots,\beta/L^z), \qquad (6.66)$$

where $t = q-q_c$ measures the distance from the quantum critical point $q = q_c$. This equation describes the zero-temperature phase transition even though it contains a finite inverse temperature. By fitting the simulation data to these forms, we obtain the dynamical scaling exponent z as well as y_t and q_c.

---

[20] Even if we choose the exact values of T_c, y_h, and y_t, the data collapse will not be perfect in practice. This imperfection is due to *corrections to scaling* that arises because we use raw variables such as T and H, not the scaling fields. One can, in principle, address this problem by including correction terms in the fitting.

6.6 Applications to Bosonic systems

In this section, we briefly discuss the application of the algorithms for quantum
spin lattices to lattice models of Bosons. Interest in studying Boson lattice models
was sparked by various low-temperature experiments (Anderson et al., 1995) on
electromagnetically trapped atoms that observed Bose-Einstein condensation. This
phenomenon, along with the phenomenon of superfluidity, had long been modeled
by systems in a continuum, and, in fact, quantum Monte Carlo simulations in the
continuum have been used to study both phenomena for quite some time. The game
changer was the experiments realizing an *optical lattice* that traps Bosonic atoms
in periodic potential wells created by standing electromagnetic waves (Greiner
et al., 2002). The result is an almost ideal realization of the Bose-Hubbard lattice
model: While the trapped Bosonic atoms condense, they remain very dilute. The
diluteness makes a δ-function description of the inter-atomic interaction plausible,
and due to the overlap of the wave functions, the atoms hop (tunnel) from one
lattice well to another. (In the optical lattices, the mean distance between the ultra-
cold atoms, however, is a thousand times larger than the mean distance between
nearest-neighbor lattice sites in a solid.)

The Bose-Hubbard model

$$ H = -t \sum_{\langle ij \rangle} (b_i^\dagger b_j + \text{h.c.}) - \frac{U}{2} \sum_i n_i(n_i - 1) - \mu \sum_i n_i, \qquad (6.67) $$

with $n \equiv b_i^\dagger b_i$, is the simplest interacting Boson system relevant to the optical
lattice systems. When considered as an approximation of a continuous-space system, it represents a system with short-range repulsive interactions where only the
s-wave scattering is relevant. However, instead of being an approximation to the
continuum, the model is more directly a representative of Bosonic atoms trapped in
an optical lattice.

It is straightforward to use the worm algorithm in Section 6.4.2 to study the
Bose-Hubbard model with different dimensionalities, lattice geometries, inhomo-
geneities, and so on. Suppose the local basis is the product basis of local Boson
occupation eigenstates $|n_i\rangle$ and that the source operator is defined as $Q_i = b_i^\dagger + b_i$.
We then have three states mutually reachable by the worm creation/annihilation
process: The state with no worm (0 state) and the two states with the worms. In the
worm states, the local occupation number between the head and the tail is larger
($+$ state) or smaller ($-$ state) than the original state. According to the general
prescription presented in Section 6.4.2, we tune the weight $w(0)$ of the no-worm
state for our convenience and make the weights of the worm states

$$ w(+) = |\langle n+1|Q_i|n\rangle| = n+1 \text{ and } w(-) = |\langle n-1|Q_i|n\rangle| = n. $$

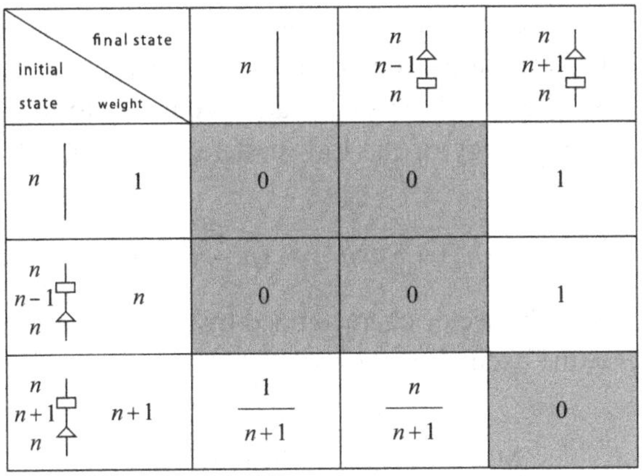

initial state / final state	weight	n	$n-1 / n$	$n+1 / n$
n	1	0	0	1
$n-1 / n$	n	0	0	1
$n+1 / n$	$n+1$	$\dfrac{1}{n+1}$	$\dfrac{n}{n+1}$	0

Figure 6.11 Creation and annihilation of a worm in a lattice Boson system in the occupation number representation. Shaded cells correspond to the prohibited transitions. The triangles and the rectangles in the diagrams represent heads and tails, respectively, and each triangle points in the direction of the head's motion (always upward in the examples shown here).

By setting $w(0) = 1$, we easily obtain

$$w(0,+) = w(+,0) = 1, \quad w(-,+) = w(+,-) = n,$$

as a solution satisfying (6.12). (The other weights vanish.) This solution yields

$$p(+|0) = 1, \quad p(0|+) = \frac{1}{n+1}, \quad p(+|-) = 1, \quad p(-|+) = \frac{n}{n+1},$$

as summarized in Fig. 6.11. In particular, creation and annihilation happen for the $+$ worm (the bottom row or last column of Fig. 6.11) with probability 1 and $1/(n+1)$. The $-$ worms are never created or annihilated by direct processes. They appear or disappear as a result of the head's motion.[21]

The algorithm enables a precise estimation of various quantities of the Bose-Hubbard model as well as other Bosonic models on lattices. For example, we might want to know the most basic property of the model, that is, its phase diagram. To find it amounts to calculating the critical chemical potential μ_c for a given hopping constant t, and we can employ the finite-size scaling method discussed in Section 6.5.3 to numerically locate critical points. To execute the scaling, we carry out Monte Carlo simulations at various values of the control parameter (μ, in the present case). However, when the corresponding term commutes with the

---

[21] Suppose a worm is created as a $+$ worm (e.g., the top-right diagram of Fig. 6.11), and the head moves upward as it first leaves the original position. After spending some time elsewhere, the head can approach the tail from the opposite direction. In such a case, the configuration near the head corresponds to the middle-left diagram of Fig. 6.11 with n replaced by $n+1$. It thus appears as a $-$ worm.

whole Hamiltonian, as the chemical potential term does in the present case, the task is simplified. In such cases, the control term corresponds to a well-defined quantum number, and we can classify all states according to its value. Therefore, for an arbitrary μ^*, the energy of the lowest state characterized by the quantum number N is

$$E_N(\mu) = E_N(\mu^*) - (\mu - \mu^*)N.$$

If the ground state is in the sector characterized by $N = N_0$ and its excited state is in the $N = N_1$ sector, the excitation gap is

$$\Delta(\mu) = \Delta(\mu^*) - (\mu - \mu^*)\Delta N$$

where $\Delta N \equiv N_1 - N_0$. Suppose μ_c is the critical value and the gap closes for some N_1. Then, by choosing $\mu^* = \mu_c$, we obtain

$$\Delta(\mu) = -(\mu - \mu_c)\Delta N.$$

This equation tells us that to estimate μ_c, we only need to estimate $\Delta(\mu)$ at some μ, namely, $\mu_c \equiv \mu + \Delta(\mu)/\Delta N$. The estimate should be independent of our choice of μ.

To use this physics, we must at least know for which quantum number N_1 the gap closes at the anticipated critical point. In the case of the Mott-insulator/superfluid phase transition in the Bose-Hubbard model, N is the total number of particles, and $N_0 = nV$ where V is the number of sites in the n-th Mott insulator phase. Suppose we increase the chemical potential starting from some point inside the n-th Mott region. As soon as we enter the superfluid phase, the system becomes compressible, and we do not need extra energy to squeeze additional particles into the system. This compressibility also means that as we increase the particle number, the gap eventually vanishes at μ_c, and we can take $N_1 = N_0 + 1$. Therefore, the corresponding gap $\epsilon_p(\mu) \equiv E_{N+1} - E_N = -(\mu - \mu_c)$ is the particle excitation energy at μ, and $\mu_c \equiv \mu + \epsilon_p(\mu)$ is the phase boundary to the superfluid phase in which particles are condensed. Likewise, $\epsilon_h(\mu) \equiv E_{N-1} - E_N = (\mu - \mu_c')$ is the hole excitation energy at μ, and $\mu_c' \equiv \mu - \epsilon_h(\mu)$ is the phase boundary to the superfluid phase in which holes are condensed. The distinction between the particle-condensed states and the hole-condensed states vanishes at large t, and the two states are actually connected to each other without a phase boundary.

Figure 6.12 presents the phase diagram of the Bose-Hubbard model on a simple cubic lattice obtained by a Monte Carlo simulation (Kato and Kawashima, 2010) with an algorithm essentially equivalent to the one discussed previously. The

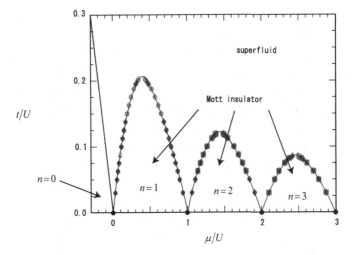

Figure 6.12 The zero-temperature phase diagram of the Bose-Hubbard model on a simple cubic lattice, based on the results of Monte Carlo simulation presented in Kato and Kawashima (2010).

excitation gap was estimated by studying the imaginary-time dependence of the correlation function with

$$\langle b(q,\tau)b^{\dagger}(q,0)\rangle \propto e^{-\epsilon_p(q,\mu)\tau}$$

for the particle excitation energy at the wave number q, and

$$\langle b^{\dagger}(q,\tau)b(q,0)\rangle \propto e^{-\epsilon_h(q,\mu)\tau}$$

for the hole excitation energy. For Fig. 6.12, three systems of linear size $L = 8, 12$, and 16 were studied at sufficiently low temperatures so that no significant temperature dependence is visible on the scale adopted in the figure. The size dependence and the statistical error is less than the line thickness.

Similar Monte Carlo techniques are being used in simulations more closely related to real optical-lattice experiments with a trapping potential, that is, a spatially inhomogeneous chemical potential. A recent example is the comparison between experiment and simulation that both trapped approximately 3×10^5 particles (Trotzky et al., 2010). The momentum distribution function was measured and computed, and they agreed with each other at various temperatures and at different potential depths. These results demonstrate the predictive power of the Bose-Hubbard model simulations for optical lattice problems.

There are a number of physical applications of the Bose-Hubbard model or its variants, other than cold atoms in an optical lattice, for which quantum Monte

Carlo simulations are uncovering the rich physics of the model. For example, pair-superfluidity is a phenomenon similar to superconductivity: Two bosonic particles of different type form a quasi-particle, and the quasi-particles condense while single particles do not. In other words, when we have two kinds of particles represented by b_1 and b_2, $\langle b_1 b_2 \rangle \neq 0$, whereas $\langle b_1 \rangle = \langle b_2 \rangle = 0$. Supercounter-flow is a similar phenomenon in which b_2 is replaced by $b_2^\dagger$; namely, particles of one kind are accompanied by holes of the other to form condensing quasi-particles. Bose systems may exhibit the pair-superfluidity or the supercounter-flow state when the repulsive force between the same kind of particles is sufficiently strong (Kuklov and Svistunov, 2003). Simulations (Kuklov et al., 2004) found these phenomena in a two-component Bose-Hubbard model on the square lattice. Supersolidity refers to a state in which a density wave (diagonal order) and a supercurrent (off-diagonal order) coexist. Although such a state has been sought for a long time since the first theoretical speculation (Andreev and Lifshitz, 1969), no conclusive evidence has been obtained so far in any real physical system (Kim and Chan, 2012). However, simulations have found that the hard-core ($U = \infty$) Bose-Hubbard model with nearest neighbor repulsion on the triangular lattice clearly exhibits the coexistence of the two orders (Boninsegni and Prokof'ev, 2005). A simple picture of this phenomenon is that the majority of the particles sustains the crystalline component, while the excess particles act as interstitials that carry the super-current. This simple two-component picture, however, does not seem to hold for the Bose-Hubbard model with nearest-neighbor repulsion on the cubic lattice. A Monte Carlo simulation shows super-solidity even at half-filling (Ohgoe et al., 2012) where no interstitials or excess particles exist except those created by fluctuations.

Suggested reading

M. P. A. Fisher, P. B. Weichman, G. Grinstein, and D. S. Fisher, "Boson localization and the superfluid-insulator transition," *Physical Review B* **40**, 546–70 (1989).

D. M. Ceperley, "Path integrals in the theory of condensed helium," *Review of Modern Physics* **67**, 279 (1995).

J. Cardy, *Scaling and Renormalization in Statistical Physics* (Cambridge University Press, 1996).

N. Kawashima and K. Harada, "Recent developments of worldline Monte Carlo methods," *J. Phys. Soc. Japan* **73**, 1379 (2004).

A. Leggett, *Quantum Liquids: Bose Condensation and Cooper Pairing in Condensed-Matter Systems* (Oxford University Press, 2006).

A. W. Sandvik, "Computational studies of quantum spin systems," in *Lectures on the Physics of Strongly Correlated Systems XIV*, ed. A. Avella and F. Mancini, AIP Conference Proceedings, **1297**, 135 (2010).

Exercises

6.1 Detailed Balance in the General Loop/Cluster Update. Show an example of a loop/cluster update in which the detailed balance condition does not hold when viewed as a Markov process in the joint space $\Sigma \times \Gamma$.

6.2 Ising Triangle. Consider three Ising spins coupled antiferromagnetically to each other and apply the Swendsen-Wang algorithm to it. If the temperature is much lower than the coupling constant, how soon will the system relax to the equilibrium distribution?

6.3 Loop/Cluster Update in the Ising Limit. Consider the $S = \frac{1}{2}$ Ising-like *XXZ* model ($J^{xy} \gg J^z > 0$) at zero field. Obtain an explicit formula for the density of graph elements. Show that in the limit $J^{xy}/J^z \to 0$ the algorithm reduces to the conventional Swendsen-Wang algorithm.

6.4 Griffiths Inequality. Using Fortuin and Kasteleyn's graphical representation of the partition function of the ferromagnetic Ising model and the graphical observable of the two-point correlation function, prove that the correlation function $\langle S_i S_j \rangle$ always increases when we increase the coupling constant J^z_{kl}, regardless of the choice of i, j, k, and l. Compose a similar proposition for quantum spin models and prove it.

6.5 Spinons in the One-Dimensional Ising Model. Compute the expectation value of the energy for the one-dimensional Ising model of length L ($\gg 1$) with periodic boundary conditions. Note that the energy gap is J^z as the first excited state has two domain boundaries and each of them costs $Jz/2$. (Here, we use the $S_i = \pm\frac{1}{2}$ convention.) Confirm (6.59) when $0 < T \ll J^z/\log L$. Also confirm that (6.59) holds for $\Delta E = J^z/2$ instead of J^z for $J^z/\log L \ll T \ll J^z$. How can this apparent paradox be solved?

6.6 A Single-Site Problem. Suppose we want to apply the algorithm described by Fig. 6.11 to a single-site problem. Let the effect of the chemical potential be taken into account in the form of the probability for reaching the next scatterer. In the present case, the next scatterer is the tail. Therefore, assume that the probability of reaching the tail, increasing the particle number by one, is $e^{-\beta\mu}$. (The move that decreases the particle number is always accepted.) By direct inspection, show that this procedure produces the correct particle number distribution: $p_n \propto e^{-\mu n}$.

7

Determinant method

This chapter introduces a finite-temperature algorithm for the simulation of interacting electrons on a lattice. Because this algorithm was developed by Blankenbecler, Scalapino, and Sugar (1981; Scalapino and Sugar, 1981), it is sometimes called the *BSS algorithm*. The method uses a Hubbard-Stratonovich transformation to convert the interacting electron problem into a noninteracting one coupled to an imaginary-time-dependent auxiliary field. For this reason, it is also called the *auxiliary-field method*. We use here yet another name, the *determinant method*, which is fitting because the transformation to a problem of noninteracting electrons generates determinants as the statistical weights. The finite-temperature determinant algorithm is a general-purpose electron algorithm that enables computations of a wide variety of local observables and correlation functions. For a discussion of a zero-temperature determinant method, refer to Appendix I.

7.1 Theoretical framework

Feynman and Hibbs (1965) formulated quantum mechanics in terms of integrals over all paths in configuration space. In real time, each path contributes a phase to the integral that is determined by the classical action along the path. Two paths can interfere constructively or destructively. In the classical limit, only the stationary-phase path is important. Being characterized by many interfering paths, real-time quantum dynamics more than challenges importance sampling. Statistical mechanics, on the other hand, involves path integrals in imaginary time. Contributions to the integrals vary exponentially in magnitude but not in phase. Thus, the path integral is dominated by paths of large magnitude. The tasks of a quantum Monte Carlo method are identifying these important paths and sampling them efficiently.

In this chapter, we address the classicization of many-electron problems at finite temperatures via a Feynman path integral. The result is a method often called the *determinant method* as the weights of the paths can be expressed as determinants, hardly classical-looking weights, but ones quite suggestive of the antisymmetry of Fermion states. Sampling these weights efficiently and in a stable manner requires special techniques. We begin with a brief overview to motivate the general form of the classical representation and the weights we need to sample.

7.1.1 Hubbard-Stratonovich transformations

As noted in Chapter 5, most quantum Monte Carlo algorithms treat the Boltzmann operator $\exp(-\beta H)$ by writing its matrix elements as an imaginary-time path integral

$$\langle \psi_L | e^{-\beta H} | \psi_R \rangle = \sum_{\psi_1, \psi_2, \ldots, \psi_{N-1}} \langle \psi_L | e^{-\Delta \tau H} | \psi_{N-1} \rangle \cdots$$
$$\langle \psi_2 | e^{-\Delta \tau H} | \psi_1 \rangle \langle \psi_1 | e^{-\Delta \tau H} | \psi_R \rangle, \quad (7.1)$$

where $\Delta \tau = \beta / N_\tau$. For finite-temperature statistical mechanics, the partition function

$$Z = \text{Tr} \left[e^{-\beta H} \right] = \sum_\psi \langle \psi | e^{-\beta H} | \psi \rangle \qquad (7.2)$$

thus becomes the integral over all paths that are periodic in imaginary time.

In Chapter 5, we also learned that evaluating the matrix elements of the exponentials of the Hamiltonian is aided by decomposing the Hamiltonian into its individual terms, $H = H_1 + H_2 + \cdots + H_n$, and using the Trotter approximation (5.12) so that

$$e^{-\Delta \tau H} = e^{-\Delta \tau H_1} e^{-\Delta \tau H_2} \cdots e^{-\Delta \tau H_n} + \mathcal{O}((\Delta \tau)^2). \qquad (7.3)$$

The matrix elements of these exponential factors are then expressed in some convenient basis where the exponential operators are either known or easily approximated. We now add the extra requirement that the decomposition is the first step toward transforming the problem into one where the only matrix elements needed are those of a noninteracting problem. To achieve this, we single out the terms in the decomposition that describe direct interactions between the electrons.

As an illustrative example, we consider an impurity Hamiltonian where there is a Coulomb interaction acting between a pair of up and down electrons that simultaneously occupy the impurity orbital.[1] We write

---

[1] For a more detailed discussion of the impurity models typically studied in condensed matter physics, see Section 8.1.

$$H = H_1 + U \left(n_\uparrow - \tfrac{1}{2} \right) \left(n_\downarrow - \tfrac{1}{2} \right), \qquad (7.4)$$

where H_1 is a Hamiltonian for some noninteracting problem, that is, it is a Hamiltonian quadratic in the electron creation and destruction operators. More deeply, this statement also implies that $H_1 = \sum_\sigma H_1^\sigma$ and that the different H_1^σ operators commute. Via the Trotter approximation (5.12),

$$e^{-\Delta\tau H} \approx e^{-\Delta\tau H_1} e^{-\Delta\tau U \left(n_\uparrow - \tfrac{1}{2} \right) \left(n_\downarrow - \tfrac{1}{2} \right)}, \qquad (7.5)$$

where n_σ is the electron number operator for electron spin $\sigma = \uparrow$ or $\downarrow$.

Having separated the interaction term from the rest, we now transform its exponential into an integral over a set of auxiliary fields. If the Coulomb interaction U is repulsive, we write the argument of the exponential (7.5) as

$$\left(n_\uparrow - \tfrac{1}{2} \right) \left(n_\downarrow - \tfrac{1}{2} \right) = -\tfrac{1}{2} \left(n_\uparrow - n_\downarrow \right)^2 + \tfrac{1}{4}.$$

Then, we make the following *Hubbard-Stratonovich transformation* (Fetter and Walecka, 1971; Negele and Orland, 1988; Fulde, 1991; Enz, 1992)

$$e^{-\Delta\tau U \left(n_\uparrow - \tfrac{1}{2} \right) \left(n_\downarrow - \tfrac{1}{2} \right)} = e^{-\tfrac{1}{4}\Delta\tau U} e^{\tfrac{1}{2}\Delta\tau U (n_\uparrow - n_\downarrow)^2}$$

$$= e^{-\tfrac{1}{4}\Delta\tau U} \frac{1}{\sqrt{2\pi}} \int_{-\infty}^{\infty} dx\, e^{-\tfrac{1}{2}x^2 + x\sqrt{\Delta\tau U}(n_\uparrow - n_\downarrow)}, \qquad (7.6)$$

which introduces the auxiliary field x into the analysis. The Hubbard-Stratonovich transformation is merely the operator generalization of a shifted Gaussian integral[2]

$$\frac{1}{\sqrt{2\pi}} \int_{-\infty}^{+\infty} dx\, e^{-\tfrac{1}{2}(x \pm y)^2} = 1 \quad \Rightarrow \quad e^{\tfrac{1}{2}y^2} = \frac{1}{\sqrt{2\pi}} \int_{-\infty}^{+\infty} dx\, e^{-\tfrac{1}{2}x^2 \mp xy}. \qquad (7.7)$$

The result of the transformation is a Gaussian-weighted integral over the auxiliary-field variable, where the integrand is the exponential of an operator proportional to the auxiliary field and quadratic in the electron creation and destruction operators $(n_\sigma = c_\sigma^\dagger c_\sigma)$. Because it is quadratic, it does not explicitly express an interaction among the electrons. In brief, the transformation has replaced the repulsive Coulomb interaction in the impurity orbital by a scalar (Bosonic) field coupled to the net spin $n_\uparrow - n_\downarrow$ of the impurity orbital. After the transformation, the effect of a positive value of U is the creation of a fluctuating spin moment at this orbital. We chose the form of the transformation such that the auxiliary field is always real. If we had not done so, we would have produced a complex-valued integrand.

---

[2] Since the rest of the integral is an even function in x, the choice of the sign in the argument of the one exponential is arbitrary. We choose the plus sign.

This result would be problematic, because we want to interpret the integrand as a probability density.

If U is attractive, we instead write

$$\left(n_{i\uparrow} - \tfrac{1}{2}\right)\left(n_{i\downarrow} - \tfrac{1}{2}\right) = \tfrac{1}{2}\left(n_{i\uparrow} + n_{i\downarrow} - 1\right)^2 - \tfrac{1}{4},$$

leading to

$$e^{-\Delta\tau U\left(n_\uparrow - \frac{1}{2}\right)\left(n_\downarrow - \frac{1}{2}\right)} = e^{+\frac{1}{4}\Delta\tau U} \frac{1}{\sqrt{2\pi}} \int_{-\infty}^{\infty} dx\, e^{-\frac{1}{2}x^2 + x\sqrt{\Delta\tau|U|}(n_\uparrow + n_\downarrow - 1)}. \tag{7.8}$$

Here, the auxiliary field couples to a charge degree of freedom. After such a transformation, the effect of a negative value of U is to create charge fluctuations in the correlated orbital.

There is an alternative form of the transformation due to Hirsch (1983), which is especially useful for lattice models that have the Coulomb interaction given as a sum of products of number operators. Hirsch's transformation introduces a *discrete auxiliary field* that in practice can be sampled more efficiently than the continuous field. His transformation is

$$e^{-\Delta\tau U\left(n_\uparrow - \frac{1}{2}\right)\left(n_\downarrow - \frac{1}{2}\right)} = \begin{cases} \frac{1}{2}e^{-\frac{1}{4}\Delta\tau U} \sum_{x=\pm 1} e^{\alpha x(n_\uparrow - n_\downarrow)}, & U > 0, \\ \frac{1}{2}e^{+\frac{1}{4}\Delta\tau U} \sum_{x=\pm 1} e^{\alpha x(n_\uparrow + n_\downarrow - 1)}, & U < 0, \end{cases} \tag{7.9}$$

with $\cosh\alpha = \exp(\Delta\tau|U|/2)$. Again, for positive U, the auxiliary field couples to a spin degree of freedom, while for negative U, it couples to the charge. Additional Hubbard-Stratonovich transformations are discussed in Appendix G.

Whether continuous or discrete auxiliary fields are introduced, the partition function becomes a nested sum

$$Z = \sum_C \sum_\psi w_C[\psi], \tag{7.10}$$

with $C = \{x_1, x_2, \ldots, x_{N_\tau}\}$ denoting the set of all possible auxiliary-field configurations as a function of imaginary time and

$$w_C[\psi] \equiv \langle\psi| e^{-\beta H} |\psi\rangle, \tag{7.11}$$

with $e^{-\beta H}$ being discretized by the Trotter approximation (7.3). The transformation breaks the interaction for each time-slice into a sum of noninteracting terms, one for each spin, and thus generates an effective noninteracting Hamiltonian. As each spin piece of this Hamiltonian commutes, its exponential factors into products of exponentials, one set of products for each spin.

The states $|\psi\rangle$ in the trace are generally an occupation number basis. In an occupation number basis, each basis state is a direct product

$$\left|n_{1\uparrow}, n_{2\uparrow}, \ldots, n_{N\uparrow}\right\rangle \left|n_{1\downarrow}, n_{2\downarrow}, \ldots, n_{N\downarrow}\right\rangle$$

of a spin-up and a spin-down state, where $n_{i\sigma} = 0$ or 1, $\sum_i n_{i\sigma} = N_\sigma$, and

$$\left|n_{1\sigma}, n_{2\sigma}, \ldots, n_{N\sigma}\right\rangle = c_{i_{N\sigma}}^\dagger \cdots c_{i_2}^\dagger c_{i_1}^\dagger |0\rangle.$$

For notational compactness, we write

$$\left|\psi_{N_\sigma}\right\rangle = |n_{1\sigma}, n_{2\sigma}, \ldots, n_{N\sigma}\rangle$$

to symbolize a basis state with N_σ electrons and use

$$\left|\psi_{N_{\text{elec}}}\right\rangle = \left|n_{1\uparrow}, n_{2\uparrow}, \ldots, n_{N\uparrow}\right\rangle \left|n_{1\downarrow}, n_{2\downarrow}, \ldots, n_{N\downarrow}\right\rangle = \left|\psi_{N\uparrow}\right\rangle \left|\psi_{N\downarrow}\right\rangle$$

to symbolize direct product states with $N_{\text{elec}} = \sum_\sigma N_\sigma$. We see that in this basis the spin states also factorize. Ultimately,

$$Z = \sum_C \sum_{N_\downarrow + N_\uparrow = N_{\text{elec}}} \prod_\sigma \sum_{\psi_{N_\sigma}} w_C^\sigma [\psi_{N_\sigma}]. \tag{7.12}$$

It is important to note that the spin factors in the product of the weights share the auxiliary-field configuration. This sharing is what reconstructs the interactions. Although the impurity problem discussed here is particularly simple, the determinant method expresses more general and complex Hamiltonians in the same form.

By placing a term that is quadratic in the electron creation and destruction operators in the argument of the exponential, the Hubbard-Stratonovich transformation enables us to trace out the electronic degrees of freedom for a fixed configuration of auxiliary-field variables. In other words, it transforms the given problem into one we know how to solve. If we were explicitly to trace out the electron degrees of freedom, we would find

$$Z = \sum_C \prod_\sigma w_C^\sigma. \tag{7.13}$$

with the new weights, denoted as w_C^σ with no argument, expressible as determinants. It is these weights we need to specify.

While a stepping stone, this result is not what we need. The starting partition function (7.2) is in the canonical ensemble that fixes the number of electrons.

It is more convenient to use the grand canonical ensemble for electrons at finite temperature,

$$Z = \text{Tr}\left[e^{-\beta(H-\mu N_{\text{elec}})}\right] = \sum_{N_{\text{elec}}} \sum_{\psi_{N_{\text{elec}}}} \langle \psi_{N_{\text{elec}}} | e^{-\beta(H-\mu N_{\text{elec}})} | \psi_{N_{\text{elec}}} \rangle, \tag{7.14}$$

where we have to sum over all possible electron numbers. Clearly, we can repeat the analysis just performed for each electron number, finding for each a determinantal weight. As we will also see, the sum of these determinants is another determinant, a determinant representation of the Fermi-Dirac partition function.

In the next subsection, we derive this grand canonical determinantal weight. Central to this development is the imaginary-time single-electron propagator. With it, we are able to construct the imaginary-time many-electron propagator, and with this propagator, we in turn can construct imaginary-time single-particle Green's functions. This construction is important: After the next subsection, we begin our presentation of the determinant method where we adopt a Metropolis sampling of auxiliary-field configurations. This sampling requires computing the ratio of two determinants to decide whether to accept or reject proposed changes in the auxiliary fields. To do the sampling efficiently, we need to use special methods that require the imaginary-time, single-particle Green's function. Having this Green's function is extra handy as it also facilitates computing many-body correlation functions. One strength of the determinant method is the ease with which a large variety of such functions can be computed.

7.1.2 Determinantal weights

From the path-integral perspective, the matrix elements of the Boltzmann factor in (7.1) are formed from a sequence of matrix multiplications where each matrix represents the imaginary-time evolution from one time step to another. The Trotter approximation (7.3) and the Hubbard-Stratonovich transformation (7.6), (7.8), or (7.9) reduce the complexity of this propagation to that of the propagation of free electron states. Central to the numerical framework of the determinant method are matrices expressing the free propagation of multi-electron states from imaginary time τ_1 to τ_2 that are built upon a matrix B whose elements $B_{ij}(\tau_2, \tau_1)$ propagate a single electron from state j at time τ_1 to a state i at time τ_2. For a given spin σ and auxiliary-field configuration, this single-particle propagator is defined as

$$B_{ij}(\tau_2, \tau_1) = \left\langle 0 \left| c_i \left[\mathcal{T} \exp\left(-\int_{\tau_1}^{\tau_2} H(\tau) d\tau\right) \right] c_j^\dagger \right| 0 \right\rangle, \tag{7.15}$$

where $\mathcal{T}$ is the time-ordering operator, $H(\tau)$ is the auxiliary-field-dependent noninteracting Hamiltonian ($\int d\tau H(\tau)$ is the Euclidean action), and $|0\rangle$ is the vacuum state. This propagator represents the solution to the equations of motion for the electron creation and destruction operators[3]

$$c_i(\tau_2) = \sum_j B_{ij}(\tau_2, \tau_1) c_j(\tau_1),$$

$$c_i^\dagger(\tau_2) = \sum_j c_j^\dagger(\tau_1) B_{ji}^{-1}(\tau_2, \tau_1). \tag{7.16}$$

We note that $c_i(\tau)$ and $c_i^\dagger(\tau)$ are not Hermitian conjugates because we are working in imaginary time. This means the propagation is not unitary. As a consequence, orbitals that initially were orthonormal lose this property.

The description of an N_σ-electron system requires a multi-electron propagator

$$B(\tau_2, \tau_1) \equiv \left\langle 0 \left| c_{i_1} c_{i_2} \cdots c_{i_{N_\sigma}} \left[\mathcal{T} \exp\left(-\int_{\tau_1}^{\tau_2} H(\tau)\, d\tau \right) \right] c_{j_{N_\sigma}}^\dagger \cdots c_{j_2}^\dagger c_{j_1}^\dagger \right| 0 \right\rangle$$

$$= \det \begin{pmatrix} B_{i_1 j_1}(\tau_2, \tau_1) & B_{i_1 j_2}(\tau_2, \tau_1) & \cdots & B_{i_1 j_{N_\sigma}}(\tau_2, \tau_1) \\ B_{i_2 j_1}(\tau_2, \tau_1) & B_{i_2 j_2}(\tau_2, \tau_1) & \cdots & B_{i_2 j_{N_\sigma}}(\tau_2, \tau_1) \\ \vdots & \vdots & \ddots & \vdots \\ B_{i_{N_\sigma} j_1}(\tau_2, \tau_1) & B_{i_{N_\sigma} j_2}(\tau_2, \tau_1) & \cdots & B_{i_{N_\sigma} j_{N_\sigma}}(\tau_2, \tau_1) \end{pmatrix}, \tag{7.17}$$

which is the determinant of a matrix whose elements are single-particle propagators. We derive this identity in Appendix H. For the moment, we dwell on its physical meaning. Because the evolving electrons are independent, we should expect the many-body propagator to be a product of single-particle propagators. We inject N_σ identical particles at time τ_1 into the orbitals $j_1, j_2, \ldots, j_{N_\sigma}$, and we do not know which specific particles arrive at the orbitals $i_1, i_2, \ldots, i_{N_\sigma}$ at time τ_2. The determinant is the sum over all possible arrivals with the appropriate minus signs accounting for an even or odd number of exchanges of identical electrons.

Now we return to the expression for the many-electron partition function in the grand canonical ensemble. The Hubbard-Stratonovich transformation (7.9) adds a summation over all configurations C of auxiliary fields and together with the Trotter approximation (7.3) with $\Delta\tau = \beta/N_\tau$ generates a spin factorization

$$Z = \sum_C \prod_\sigma \sum_{N_\sigma} \sum_{\psi_{N_\sigma}} \left\langle \psi_{N_\sigma} \left| e^{-\Delta\tau c^\dagger M_{N_\tau}^\sigma c} \cdots e^{-\Delta\tau c^\dagger M_2^\sigma c} e^{-\Delta\tau c^\dagger M_1^\sigma c} \right| \psi_{N_\sigma} \right\rangle, \tag{7.18}$$

---

[3] The matrix B represents the exponential $\exp(-\Delta\tau H)$ where H is some noninteracting Hamiltonian. The matrix B^{-1} represents $\exp(\Delta\tau H)$. In Section 7.2.1, we discuss ways to construct B. Rather than performing a numerical inversion of B, we use the same strategy with $+\Delta\tau$ as we do with $-\Delta\tau$ to construct B^{-1}.

where M_i^σ is the matrix that defines the auxiliary-field-dependent noninteraction Hamiltonian for each spin (Appendix H). For the inner summations, repeated use of (7.17) yields

$$\sum_{N_\sigma} \sum_{\psi_{N_\sigma}} \langle \psi_{N_\sigma} | e^{-\Delta\tau c^\dagger M_\tau^\sigma c} \cdots e^{-\Delta\tau c^\dagger M_2^\sigma c} e^{-\Delta\tau c^\dagger M_1^\sigma c} | \psi_{N_\sigma} \rangle =$$

$$\det(1) + \det\left(B_{11}^\sigma\right) + \det\left(B_{22}^\sigma\right) + \cdots + \det\left(B_{NN}^\sigma\right)$$

$$+ \det\begin{pmatrix} B_{11}^\sigma & B_{12}^\sigma \\ B_{21}^\sigma & B_{22}^\sigma \end{pmatrix} + \det\begin{pmatrix} B_{11}^\sigma & B_{13}^\sigma \\ B_{31}^\sigma & B_{33}^\sigma \end{pmatrix}$$

$$+ \cdots + \det\begin{pmatrix} B_{N-1,N-1}^\sigma & B_{N-1,N}^\sigma \\ B_{N,N-1}^\sigma & B_{N,N}^\sigma \end{pmatrix} + \cdots$$

$$+ \det\begin{pmatrix} B_{11}^\sigma & B_{12}^\sigma & \cdots & B_{1N}^\sigma \\ B_{21}^\sigma & B_{22}^\sigma & \cdots & B_{2N}^\sigma \\ \vdots & \vdots & \ddots & \vdots \\ B_{N1}^\sigma & B_{N2}^\sigma & \cdots & B_{NN}^\sigma \end{pmatrix}.$$

In the above, $B_{ij}^\sigma = B_{ij}^\sigma(\beta, 0)$ (Appendix H). As is easily verified by direct expansion, the right-hand side equals

$$\det\left[I + B^\sigma(\beta,0)\right] = \det\begin{pmatrix} 1 + B_{11}^\sigma & B_{12}^\sigma & \cdots & B_{1N}^\sigma \\ B_{21}^\sigma & 1 + B_{22}^\sigma & \cdots & B_{2N}^\sigma \\ \vdots & \vdots & \ddots & \vdots \\ B_{N1}^\sigma & B_{N2}^\sigma & \cdots & 1 + B_{NN}^\sigma \end{pmatrix},$$

so the partition function (7.18) simplifies to

$$Z = \sum_C \prod_\sigma w_C^\sigma, \tag{7.19}$$

with

$$w_C^\sigma = \det\left[I + B^\sigma(\beta,0)\right]. \tag{7.20}$$

As promised, we have shown that the weights are determinants. We note that

$$B^\sigma(\beta,0) \equiv B^\sigma(\beta, \beta - \Delta\tau) \cdots B^\sigma(2\Delta\tau, \Delta\tau) B^\sigma(\Delta\tau, 0) \tag{7.21}$$

is a string of matrix multiplications (Appendix H).

7.1.3 Single-particle Green's function

The standard many-body physics definition of the matrix elements of the single-particle equal-time Green's function (7.22) is[4]

$$G_{ij}^{\sigma}(\tau, \tau) \equiv \left\langle c_{i\sigma}(\tau)c_{j\sigma}^{\dagger}(\tau) \right\rangle = \delta_{ji} - \left\langle c_{j\sigma}^{\dagger}(\tau)c_{i\sigma}(\tau) \right\rangle, \qquad (7.22)$$

where the expectation values are over the grand canonical ensemble. To connect this definition with (7.17), we start by coupling the action to the operator $c_{j\sigma}^{\dagger}c_{i\sigma}$ for an instant of imaginary time

$$\mathcal{T} \exp\left(-\int_0^{\beta} H(\tau)d\tau\right) \rightarrow$$

$$\mathcal{T} \exp\left(-\int_{\tau}^{\beta} H(\tau)d\tau\right) \exp(hc_{j\sigma}^{\dagger}c_{i\sigma})\mathcal{T} \exp\left(-\int_0^{\tau} H(\tau)d\tau\right).$$

With the insertion of this interaction, (7.15) and (7.20) yield a modified weight,

$$w_C' = \det\left[I + B^{\sigma}(\beta, \tau)e^{hQ}B^{\sigma}(\tau, 0)\right] \det\left[I + B^{-\sigma}(\beta, \tau)B^{-\sigma}(\tau, 0)\right],$$

where Q is a null matrix except for $Q_{ji} = 1$. From this expression it follows that

$$\left\langle c_{j\sigma}^{\dagger}(\tau)c_{i\sigma}(\tau) \right\rangle = \frac{\partial}{\partial h} \ln w_C'\bigg|_{h=0}.$$

We now use the identity

$$\ln \det A = \operatorname{Tr} \ln A, \qquad (7.23)$$

and the cyclic property of a trace, $\operatorname{Tr}ABC = \operatorname{Tr}BCA = \operatorname{Tr}CAB$, to find

$$\frac{\partial}{\partial h} \ln w_C'\bigg|_{h=0} = \operatorname{Tr}\frac{\partial}{\partial h} \ln\left[I + B^{\sigma}(\beta, \tau)e^{hQ}B^{\sigma}(\tau, 0)\right]\big|_{h=0}$$

$$= \operatorname{Tr}\left[B^{\sigma}(\tau, 0)\left(I + B^{\sigma}(\beta, \tau)B^{\sigma}(\tau, 0)\right)^{-1} B^{\sigma}(\beta, \tau)Q\right]$$

$$= \left[B^{\sigma}(\tau, 0)\left(I + B^{\sigma}(\beta, \tau)B^{\sigma}(\tau, 0)\right)^{-1} B^{\sigma}(\beta, \tau)\right]_{ji}$$

$$= \left[\left(I + B^{\sigma}(\beta, \tau)^{-1}B^{\sigma}(\tau, 0)^{-1}\right)^{-1}\right]_{ji}$$

$$= \delta_{ji} - \left[\left(I + B^{\sigma}(\tau, 0)B^{\sigma}(\beta, \tau)\right)^{-1}\right]_{ji}.$$

---

[4] Note the transposition of indices and the choice of sign.

Thus,

$$G^\sigma(\tau, \tau) = I - [I + B^\sigma(\tau, 0)B^\sigma(\beta, \tau)]^{-1}. \tag{7.24}$$

This Green's function plays a central role in the determinant method, as it is used for two different key purposes. The first use is for sampling of the auxiliary fields. The variable $x_i(\tau)$ for orbital i and imaginary time τ couples only to $n_\uparrow(\tau)$ and $n_\downarrow(\tau)$, information embedded in $G_{ii}^\sigma(\tau, \tau) = 1 - \langle n_{i\sigma}(\tau) \rangle$. The second is for measuring expectation values of operators. The noninteracting character of the simulation enables Wick's theorem (Negele and Orland, 1988) to reduce such expectation values to sums of products of Green's functions.

7.2 Finite temperature algorithm

The methods in this chapter are most conveniently applied to interacting electron models such as

$$H = -\sum_{ij\sigma} T_{ij}^\sigma c_{i\sigma}^\dagger c_{j\sigma} + \sum_{ij} V_{ij} n_i n_j \tag{7.25}$$

that have the Coulomb interaction expressed as sums of number operator products (density-density interaction).[5] In (7.25), the spin-dependent symmetric matrix elements $T_{ij}^\sigma = T_{ji}^\sigma$ are hopping integrals, and the $c_{i\sigma}^\dagger$ and $c_{i\sigma}$ are the creation and destruction operators for an electron of spin $\sigma = \uparrow$ and $\downarrow$ on site i. The Coulomb interaction is parameterized by the spin-independent symmetric matrix $V_{ij} = V_{ji}$, and the on-site charge density is $n_i = n_{i\uparrow} + n_{i\downarrow}$. This Hamiltonian is quite generic and includes such prominent models as the Hubbard and extended Hubbard models, the single-impurity and lattice Anderson models, and the Falicov-Kimball model.

As a working example, we adopt the *repulsive Hubbard model*,

$$\begin{aligned} H &= H_1 + H_2 + H_3 \\ &= -\sum_{ij\sigma} T_{ij} c_{i\sigma}^\dagger c_{j\sigma} + U \sum_i \left(n_{i\uparrow} - \tfrac{1}{2}\right)\left(n_{i\downarrow} - \tfrac{1}{2}\right) - \mu \sum_i n_i, \end{aligned} \tag{7.26}$$

where $U > 0$ is the strength of the on-site Coulomb interaction, μ is the chemical potential, and the hopping matrix is spin-independent. While the hopping integrals in the Hubbard model are customarily restricted to nearest neighbors with a value of t, there is no a priori need to make this assumption. In what follows, we use N to designate the number of lattice sites, work in an occupation number basis, and assume periodic boundary conditions.

---

[5] The use of more general Hubbard-Stratonovich transformations (Appendix G) extends the applicability of the methods to other classes of Hamiltonians, although they cannot in general prevent a sign problem.

7.2.1 Matrix representation

Using the Trotter approximation (7.3), we first break the time evolution operator for a single time step into two factors

$$e^{-\Delta \tau H} \approx e^{-\Delta \tau (H_2+H_3)} e^{-\Delta \tau H_1}. \tag{7.27}$$

Both pieces of H in the first factor on the right are diagonal in the occupation number representation and thus commute. Consequently, this factor can be written as

$$e^{-\Delta \tau (H_2+H_3)} = \prod_i e^{-\Delta \tau U \left(n_{i\uparrow} - \frac{1}{2} \right) \left(n_{i\downarrow} - \frac{1}{2} \right)} \prod_i e^{-\Delta \tau \mu n_{i\uparrow}} e^{-\Delta \tau \mu n_{i\downarrow}}. \tag{7.28}$$

To the first product on the right, we now apply the discrete Hubbard-Stratonovich transformation (7.9) at each lattice site i. This transformation adds a summation over the auxiliary fields, but places a quadratic number operator into the arguments of the exponentials and allows us to factor the exponential for each site into commuting up- and down-spin terms.

For a given configuration of the auxiliary-field variables, we have

$$B^\sigma(\tau', \tau) = B^\sigma(\tau', \tau' - \Delta \tau) \cdots B^\sigma(\tau + 2\Delta \tau, \tau + \Delta \tau) B^\sigma(\tau + \Delta \tau, \tau), \tag{7.29}$$

where

$$B^\sigma(\tau, \tau - \Delta \tau) = e^{\Delta \tau \mu} A^\sigma(\tau) \exp(\Delta \tau T) \tag{7.30}$$

and

$$A^\sigma(\tau) = \begin{pmatrix} e^{\sigma \alpha x_1(\tau)} & & & & 0 \\ & e^{\sigma \alpha x_2(\tau)} & & & \\ & & \ddots & & \\ 0 & & & & e^{\sigma \alpha x_{N_\tau}(\tau)} \end{pmatrix}. \tag{7.31}$$

The exponent arguments are $\sigma \alpha x = \pm \alpha x$, depending on whether $\sigma = \uparrow$ or $\downarrow$. The parameter α is found from $\cosh(\alpha) = \exp(\frac{1}{2} \Delta \tau U)$. The matrix A^σ (7.31) is diagonal because the auxiliary fields couple only to local electron degrees of freedom. In the imaginary-time propagation, the effect of $\exp(\Delta \tau T)$ is diffusive, $A^\sigma(\tau)$ acts like an external scattering potential, and $\exp(\Delta \tau \mu)$ is an attenuation factor. We recall that the Trotter approximation (7.27) makes B^σ an asymmetric matrix.

We can compute $\exp(\Delta \tau T)$ by finding the eigenvalues λ_i and eigenvectors v_i of T. The eigenvectors form the columns of the similarity transformation matrix $V = (v_1, v_2, \ldots, v_N)$ that diagonalizes T. With this transformation,

$$\exp\left(\Delta\tau T\right) = V \begin{pmatrix} e^{\Delta\tau\lambda_1} & & & 0 \\ & e^{\Delta\tau\lambda_2} & & \\ & & \ddots & \\ 0 & & & e^{\Delta\tau\lambda_N} \end{pmatrix} V^T. \tag{7.32}$$

Alternately, we can approximate $\exp(\Delta\tau T)$ via a checkerboard decomposition (Section 5.2.4),

$$\exp\left(\Delta\tau T\right) = \exp\left(\Delta\tau \sum_{ij} T^{(ij)}\right) = \prod_{ij} \exp\left(\Delta\tau T_{ij}^{(ij)}\right),$$

where the $T^{(ij)}$ are sparse matrices with only $T_{ij}^{(ij)} = T_{ji}^{(ij)} = T_{ij}$ nonzero,

$$T^{(ij)} = \begin{pmatrix} 0 & \cdots & 0 & \cdots & 0 & \cdots & 0 \\ \vdots & \ddots & \vdots & & \vdots & & \vdots \\ 0 & \cdots & 0 & \cdots & T_{ij} & \cdots & 0 \\ \vdots & & \vdots & \ddots & \vdots & & \vdots \\ 0 & \cdots & T_{ij} & \cdots & 0 & \cdots & 0 \\ \vdots & & \vdots & & \vdots & \ddots & \vdots \\ 0 & \cdots & 0 & \cdots & 0 & \cdots & 0 \end{pmatrix},$$

and thus

$$\exp\left(\Delta\tau T\right) = \prod_{ij} \begin{pmatrix} 1 & \cdots & 0 & & \cdots & 0 & & \cdots & 0 \\ \vdots & \ddots & \vdots & & & \vdots & & & \vdots \\ 0 & \cdots & \cosh\left(\Delta\tau T_{ij}\right) & \cdots & \sinh\left(\Delta\tau T_{ij}\right) & \cdots & 0 \\ \vdots & & \vdots & & & \vdots & & & \vdots \\ 0 & \cdots & \sinh\left(\Delta\tau T_{ij}\right) & \cdots & \cosh\left(\Delta\tau T_{ij}\right) & \cdots & 0 \\ \vdots & & \vdots & & & \vdots & \ddots & \vdots \\ 0 & \cdots & 0 & & \cdots & 0 & & \cdots & 1 \end{pmatrix}.$$

Each $\exp(\Delta\tau T^{(ij)})$ is a sparse matrix with only the $ii, ij, ji,$ and jj elements differing from those of the identity matrix. This construction of an approximate exponential of the hopping matrix is particularly useful if T itself is sparse, for example, when the hopping is only between nearest neighbors. In such cases, the sparsity reduces the N^3 operations in a matrix-matrix multiplication to $N_{bonds}N^2$ where N_{bonds} is the number of sites j connected to site i. In most Hubbard-like models, $N_{bonds} \ll N$. The gain in the reduction of computation time per step, however, must be balanced by the need to use smaller $\Delta\tau$ to maintain the accuracy of the Trotter approximation and hence the need to take more imaginary-time steps.

7.2.2 Metropolis sampling

Having identified the statistical weight (7.20) of the auxiliary fields, we now want to sample them. Here, we adopt the Metropolis algorithm[6] (Section 2.5.1), and at each imaginary time, we go from lattice site to lattice site proposing a flip of the Ising-like discrete auxiliary-field variable x. We accept or reject the proposal depending on whether the ratio $\mathcal{R} = w_{C'}/w_C$ is greater than some random number ζ selected from the uniform distribution. The proposal affects only the time evolution through the matrix A defined in (7.31). For a flip from x to x' at time τ at site i,

$$
A^\sigma(\tau) \to
\begin{pmatrix}
e^{\sigma \alpha x_1(\tau)} & & & & 0 \\
 & \ddots & & & \\
 & & e^{\sigma \alpha x_i'(\tau)} & & \\
 & & & \ddots & \\
0 & & & & e^{\sigma \alpha x_N(\tau)}
\end{pmatrix}
\equiv \left[I + \Delta^\sigma(i, \tau) \right] A^\sigma(\tau),
$$

where the matrix $\Delta^\sigma(i, \tau)$ is null except for the $\Delta^\sigma_{ii}(i, \tau)$ element, which is

$$
\Delta^\sigma_{ii}(i, \tau) = \exp[\sigma \alpha (x_i'(\tau) - x_i(\tau))] - 1.
$$

The accompanying change in the propagator is

$$
B^\sigma(\beta, 0) = B^\sigma(\beta, \tau) B^\sigma(\tau, 0) \to B^\sigma(\beta, \tau) \left[I + \Delta^\sigma(i, \tau) \right] B^\sigma(\tau, 0)
$$

so that

$$
\mathcal{R}^\sigma \equiv \frac{w_{C'}^\sigma}{w_C^\sigma} = \frac{\det \left[I + B^\sigma(\beta, \tau) \left[I + \Delta^\sigma(i, \tau) \right] B^\sigma(\tau, 0) \right]}{\det \left[I + B^\sigma(\beta, \tau) B^\sigma(\tau, 0) \right]}. \tag{7.33}
$$

This expression perhaps seems a bit daunting to evaluate, but with the use of two standard determinant relations, the expression simplifies considerably. One relation is

$$
\det A / \det B = \det B^{-1} A = \det A B^{-1}, \tag{7.34}
$$

and the other is

$$
\det(I + AB) = \det(I + BA). \tag{7.35}
$$

The algebra is straightforward. First, we write (7.33) as

$$
\mathcal{R}^\sigma = \frac{\det \left[I + B^\sigma(\beta, 0) + B^\sigma(\beta, \tau) \Delta^\sigma(i, \tau) B^\sigma(\tau, 0) \right]}{\det \left[I + B^\sigma(\beta, 0) \right]}.
$$

---

[6] The heat-bath algorithm works just as well. The resulting modifications to the discussion are minor.

Then, we use the relation (7.34) to rewrite this equation as

$$\mathcal{R}^\sigma = \det\left[I + (I + B^\sigma(\beta,0))^{-1}B^\sigma(\beta,\tau)\Delta^\sigma(i,\tau)B^\sigma(\tau,0)\right].$$

Next, we use (7.35) to write

$$\mathcal{R}^\sigma = \det\left[I + \Delta^\sigma(i,\tau)B^\sigma(\tau,0)(I + B^\sigma(\beta,0))^{-1}B^\sigma(\beta,\tau)\right].$$

Finally, using

$$A(I + BA)^{-1}B = (I + (AB)^{-1})^{-1} = I - (I + AB)^{-1}$$

with $A \equiv B^\sigma(\tau,0)$ and $B \equiv B^\sigma(\beta,\tau)$, we obtain

$$\mathcal{R}^\sigma = \det\left[I + \Delta^\sigma(i,\tau)\left(I - G^\sigma(\tau,\tau)\right)\right], \tag{7.36}$$

where

$$G^\sigma(\tau,\tau) = \left[I + B^\sigma(\tau,0)B^\sigma(\beta,\tau)\right]^{-1}. \tag{7.37}$$

$G^\sigma(\tau,\tau)$ is the equal-time single-particle Green's function (7.24). Because of the sparseness of Δ^σ,

$$\mathcal{R} = \mathcal{R}^\uparrow\mathcal{R}^\downarrow = \prod_\sigma \left[1 + \Delta_{ii}^\sigma(i,\tau)\left(1 - G_{ii}^\sigma(\tau,\tau)\right)\right], \tag{7.38}$$

a scalar, as opposed to a matrix, expression that is exceedingly simple to calculate, provided $G_{ii}^\sigma(\tau,\tau)$ is known. Computing G^σ from (7.29) and (7.37) for every accepted flip of $x_i(\tau)$ requires a significant computational effort. Matrix algebra, however, provides a technique to reduce this effort by a factor of N. We now proceed to derive this technique.

Under a proposed flip in the sign of $x_i(\tau)$,

$$\begin{aligned}
G^\sigma(\tau,\tau) = (I + B^\sigma)^{-1} &\rightarrow \left[I + (I + \Delta^\sigma(\tau))B^\sigma\right]^{-1} \\
&= \left[(I + B^\sigma) + \Delta^\sigma(\tau)B^\sigma(I + B^\sigma)^{-1}(I + B^\sigma)\right]^{-1} \\
&= (I + B^\sigma)^{-1}\left[I + \Delta^\sigma(\tau)B^\sigma(I + B^\sigma)^{-1}\right]^{-1} \\
&= G^\sigma\left[I + \Delta^\sigma(\tau)(I - G^\sigma)\right]^{-1},
\end{aligned} \tag{7.39}$$

where we wrote B^σ as shorthand for $B^\sigma(\tau,0)B^\sigma(\beta,\tau)$. The equations of motion (7.16) for the electron operators yield

$$G^\sigma(\tau',\tau) = B^\sigma(\tau',\tau)G^\sigma(\tau,\tau) = G^\sigma(\tau',\tau')B^\sigma(\tau',\tau), \tag{7.40}$$

which are a pair of relations consistent with the propagator character of B^σ. Similarly, for $\tau' > \tau$, we advance the equal-time Green's function to a later time via

$$G^\sigma(\tau', \tau) = B^\sigma(\tau', \tau) G^\sigma(\tau, \tau) B^\sigma(\tau', \tau)^{-1}. \tag{7.41}$$

We still must show how to update $G^\sigma(\tau, \tau)$ efficiently if the Metropolis proposal is accepted.

Analytically,[7]

$$\left[I + \Delta^\sigma(\tau)\,(I - G^\sigma) \right]^{-1} = I - \frac{1}{\mathcal{R}^\sigma} \Delta^\sigma(\tau)\,(I - G^\sigma), \tag{7.42}$$

where $\mathcal{R}^\sigma$ ($\sigma = \uparrow, \downarrow$) is the ratio of the weights before and after the flip. In a compact notation,

$$\mathcal{R}^\sigma = 1 + \Delta_{ii}^\sigma \left(1 - G_{ii}^\sigma \right). \tag{7.43}$$

Hence, under the proposed flip,

$$G^\sigma \rightarrow G^\sigma - \frac{1}{\mathcal{R}^\sigma} G^\sigma\, \Delta^\sigma(\tau)\,(I - G^\sigma). \tag{7.44}$$

In terms of matrix elements,

$$G_{jk}^\sigma \rightarrow G_{jk}^\sigma - \frac{1}{\mathcal{R}^\sigma} G_{ji}^\sigma\, \Delta_{ii}^\sigma(\tau)\,(I - G^\sigma)_{ik}. \tag{7.45}$$

The significance of (7.44) becomes clearer: If we know G_{ij}^σ and accept the flip, we are able to compute the Green's function for the new configuration of auxiliary fields with a number of arithmetic operations of the order of the number of elements of G^σ, which is N^2. This number is a factor of N less than the $\mathcal{O}(N^3)$ operations associated with a matrix inversion. The gain is significant particularly because the updating of G^σ sits in the inner loop of the algorithm.

7.2.3 The algorithm

We now discuss the algorithm and several issues related to it. The overall Monte Carlo algorithm structure is the same as that discussed in Section 3.9. In the initialization, we not only set the auxiliary-field variables x_i to values of ± 1 for each site i at each time τ, but we also build the matrices $B^\sigma(\tau, \tau - \Delta\tau)$ for all times and the two

---

[7] We can prove (7.42) as follows:

$$(I + \Delta(I - G))(I - \mathcal{R}^{-1}\Delta(I - G)) = I + \frac{\mathcal{R} - 1}{\mathcal{R}}\Delta(I - G) - \frac{1}{\mathcal{R}}\Delta(I - G)\Delta(I - G)$$

$$= I + \frac{\mathcal{R} - 1}{\mathcal{R}}\Delta(I - G) - \frac{1}{\mathcal{R}}\Delta_{ii}(1 - G_{ii})\Delta(I - G) = I,$$

where we have used the sparseness of Δ and (7.43).

spin components of the Green's function matrix for the initial time. We can build this Green's function using (7.37), by performing all the matrix multiplications and then computing the required matrix inversion. The steps seem quite direct, but as we discuss below, they have a hidden complexity.

Algorithm 23 details a Monte Carlo sweep. To facilitate the discussion of the sweep procedure, we use the more convenient notations $G_i \equiv G(\tau_i, \tau_i)$ and $B_i^\sigma \equiv B^\sigma(\tau_i, \tau_{i-1})$ for the i-th imaginary-time step $\tau_i = i\Delta\tau$. With these notations,

---

Algorithm 23 Finite-temperature determinant Monte Carlo algorithm.

Input: The $N_\tau \times N$ auxiliary field variables $x_{i,j}$, the $N \times N$ matrices B_i^σ, and the $N \times N$ matrices G_1^σ.

for $i = 1$ to N_τ **do**
  if mod(i, n_{stab})=0 **then**
    for $\sigma = \uparrow$ to $\downarrow$ **do**
      Make G_i^σ from (7.46) ;
    end for
  end if
  for $j = 1$ to N **do**
    Compute $\mathcal{R}^\sigma(x_{ij})$ from (7.43) ;
    Draw ζ uniform randomly in $[0, 1]$;
    if $\mathcal{R}^\sigma > \zeta$ **then**
      $x_{ij} \leftarrow -x_{ij}$;
      for $\sigma = \uparrow$ to $\downarrow$ **do**
        Update G_i^σ via (7.45) ;
      end for
    end if
  end for
  if $i < N_\tau$ **then**
    for $\sigma = \uparrow$ to $\downarrow$ **do**
      $G_{i+1}^\sigma \leftarrow B_{i+1}^\sigma G_i^\sigma [B_{i+1}^\sigma]^{-1}$;
    end for
  else
    for $\sigma = \uparrow$ to $\downarrow$ **do**
      $G_1^\sigma \leftarrow B_1^\sigma G_{N_\tau}^\sigma [B_1^\sigma]^{-1}$;
    end for
  end if
end for
return G_i^σ for $i = 1, 2, \ldots, N_\tau$.

$$G_i^\sigma = \left[I + B_i^\sigma \cdots B_1^\sigma B_{N_\tau}^\sigma \cdots B_{i+1}^\sigma \right]^{-1}. \tag{7.46}$$

To update the auxiliary fields, we use the Metropolis algorithm and visit each lattice site at each imaginary time. If the proposed change is accepted, we update the Green's function using (7.44). After visiting all lattice sites, we advance the equal-time Green's function to the next time using (7.41) and then revisit the lattice sites. One sweep is finished after we complete the sampling for all times. We note that (7.37) implies $G^\sigma(\beta, \beta) = G^\sigma(0, 0)$ and that (7.41) rolls $G^\sigma(\beta, \beta)$ over to $G^\sigma(\Delta\tau, \Delta\tau)$, as it must because of the periodic boundary conditions in imaginary time. When a proposed change is accepted, it changes the matrix A_i^σ in the definition (7.30) of the matrix B_i^σ. In advancing G^σ to the next time, we use the B^σ with the new auxiliary-field configuration.

As a Fermion algorithm, the finite-temperature determinant method generally has a sign problem (Sections 5.4 and 11.1). A sign problem arises when w_C^σ is negative while $w_C^{-\sigma}$ is positive. From (7.43), we see that for a given x a w_C^σ is negative when a $1 - G_{ii}^\sigma$ is. Since $1 - G_{ii}^\sigma = n_{i\sigma}$, this situation is clearly unphysical. Gubernatis and Zhang (1994) investigated whether the source of this situation was numerical imprecision in computing the Green's function. They concluded that while the lack of precision can cause a sign problem, imprecision was not the dominant source of the problem. They noted that in some cases the topology of the Hubbard-Stratonvich fields is directly connected to the topology of the Fermion fields and suggested as the source of the sign problem topological defects (zero modes) that develop in the Hubbard-Stratonovich configurations. We also comment that in some cases, for example, low electron fillings, the sign problem is not debilitating.

Algorithm 23 works well if we can perform all arithmetic operations with infinite numerical precision. With finite-precision arithmetic, the repeated updating of the Green's function and the stepping forward in time build up numerical errors. To combat this loss of accuracy, as indicated in Algorithm 23, we use (7.46) to rebuild the Green's functions periodically from the current values of the auxiliary fields. One strategy is to determine the period n_{stab} empirically: Fortunately when numerical accuracy breaks down, it does so rather suddenly. Consequently, we can try several values of n_{stab} until the breakdown occurs and then pick a value below this number.

The most serious loss of accuracy occurs in the inversion of the matrix in (7.46). As the temperature is lowered, this matrix becomes very ill-conditioned, and inverting it with a standard matrix-inversion routine produces a result that may have lost all numerical accuracy. In Section 7.4, we describe techniques, called *matrix product stabilization methods*, that enable this inversion. In brief, these methods accumulate the products of the B-matrices in a factored form in such a way as to reduce the loss of information when the identity I and the product of the B's are added. The factored form is easy to invert, but the need to use these methods requires more coding than is implied in Algorithm 23. The stabilization methods add only a small cost to the total computation. *In the algorithm, where we say "Make G_i^σ from (7.46)" we assume that this step is done with the matrix product stabilization methods.* The description of these techniques is technical, yet essential. Because of their technical nature, we postpone their discussion until later in the chapter (Section 7.4). Using them is also important for accurate estimates of observables, especially those that require unequal-time Green's functions. We now discuss the basics of observable estimation.

7.2.4 Measurements

In analytic calculations of noninteracting models, Wick's theorem (Fetter and Walecka, 1971; Negele and Orland, 1988; Fulde, 1991; Enz, 1992) converts expectation values of products of operators into sums of products of all possible pair-wise contractions of the creation and destruction operators. The expectation values of these contractions are elements of the (noninteracting) Green's function. Thus, Wick's theorem expresses results of measurements in terms of the central quantity in the simulation.

As an example of the use of Wick's theorem in the determinant method, we consider the *anti ferromagnetic structure factor*,

$$S^{zz}(\pi,\pi) = \frac{1}{N}\sum_{ij}(-1)^{i-j}\langle(n_{i\uparrow}-n_{i\downarrow})(n_{j\uparrow}-n_{j\downarrow})\rangle,$$

where N is the number of lattice sites, $n_{i\sigma}$ is the number of electrons on site i with spin σ, and $(-1)^{i-j}$ is $+1$ if the sites i and j are on the same sublattice and -1 if they are not. The Monte Carlo simulation provides an estimate of $S^{zz}(\pi.\pi)$ by averaging over M independent samples of the auxiliary-field variables so that

$$S^{zz}(\pi,\pi) = \frac{1}{M}\sum_{\{C:C_1,C_2,...,C_M\}}\frac{1}{N}\sum_{ij}(-1)^{i-j}$$
$$\times\left[\langle n_{i\uparrow}n_{j\uparrow}\rangle_C - \langle n_{i\downarrow}n_{j\downarrow}\rangle_C - \langle n_{i\downarrow}n_{j\uparrow}\rangle_C + \langle n_{i\uparrow}n_{j\downarrow}\rangle_C\right],$$

where the quantum mechanical expectation values are for a specific configuration C of the auxiliary-field variables. Because the Hubbard-Stratonovich transformation factorizes the problem in the electron spin, contractions between different spin pairs, such as $\langle c_\sigma^\dagger c_{-\sigma}\rangle_C$, are zero. Hence,

$$\langle n_{i,\sigma}n_{j,-\sigma}\rangle_C = \langle n_{i,\sigma}\rangle_C\langle n_{j,-\sigma}\rangle_C = (1 - G_{ii}^\sigma)(1 - G_{jj}^{-\sigma}).$$

Like-spin terms produce more nonzero contractions

$$\langle n_{i\sigma} n_{j\sigma} \rangle_C = \langle c_{i\sigma}^\dagger c_{i\sigma} c_{j\sigma}^\dagger c_{j\sigma} \rangle_C$$
$$= \langle c_{i\sigma}^\dagger c_{i\sigma} \rangle_C \langle c_{j\sigma}^\dagger c_{j\sigma} \rangle_C + \langle c_{i\sigma}^\dagger c_{j\sigma} \rangle_C \langle c_{i\sigma} c_{j\sigma}^\dagger \rangle_C$$
$$= \left(1 - G_{ii}^\sigma \right) \left(1 - G_{jj}^\sigma \right) + \left(\delta_{ij} - G_{ji}^\sigma \right) G_{ij}^\sigma.$$

We note that the $i = j$ case reduces to $(1 - G_{ii}^\sigma)$, as is necessary because $\langle n_{i\sigma} n_{i\sigma} \rangle_C = \langle n_{i\sigma} \rangle_C$.

When measurements are made, we may exploit the symmetries of the Hamiltonian to gather at once many possible estimates of a quantity of interest. Doing this is especially practical for the current algorithms because the measurement calculations typically execute in considerably less computer time than a Monte Carlo sweep. To estimate the kinetic energy for a translationally invariant system, for example, we may average the measurements of $\langle c_{i,\sigma}^\dagger(\tau) c_{i+\delta,\sigma}(\tau) \rangle_C$ over all lattice positions i, symmetry-related sites $i + \delta$, electron spins σ, and imaginary times τ. While individual measurements do not obey the Hamiltonian symmetries, averages over many configurations will. Averaging over as many equivalent measurements as possible reduces the variance of the final estimate.

Features in the simulation might break a symmetry of the Hamiltonian artificially. For example, in the Hubbard model, the *staggered magnetization*

$$M^\alpha (\pi, \pi) = \sum_i (-1)^i \begin{pmatrix} c_{i\uparrow}^\dagger & c_{i\downarrow}^\dagger \end{pmatrix} \sigma^\alpha \begin{pmatrix} c_{i\uparrow} \\ c_{i\downarrow} \end{pmatrix}, \qquad \alpha = x, y, z,$$

where σ^α is a Pauli spin matrix, is rotationally invariant. Because it couples the auxiliary field to the z-component of the local spin $n_{i\uparrow} - n_{i\downarrow}$, the Hubbard-Stratonovich transformation breaks this symmetry, making the longitudinal estimator

$$\frac{1}{N} \langle M^z(\pi, \pi)^2 \rangle$$

much noisier than the transverse one (Hirsch, 1987)

$$\frac{1}{N} \langle M^x(\pi, \pi)^2 + M^y(\pi, \pi)^2 \rangle.$$

Using the latter is often more advisable.

7.3 Hirsch-Fye algorithm

In this section, we revisit the updating equation for the Green's function from a general point of view. The more general viewpoint does not lead to a more efficient algorithm for lattice models, but it does lead to an interesting variant of the updating called the *Hirsch-Fye algorithm*. This algorithm, which is particularly useful for the simulation of impurity models, provides a different perspective on the updating just presented for nonimpurity problems.

Updating the Green's function $G^\sigma(x_1, x_2, \ldots, x_i, \ldots, x_N)$ for a set of values of Hubbard-Stratonovich fields at a given imaginary-time τ via (7.45) generates the Green's function $G^\sigma(x_1, x_2, \ldots, x_i', \ldots, x_N)$ with just one field changed at the same imaginary time. The procedure exactly solves the problem of changing the field at one site at a given time. We could start with the noninteracting Green's function and use the same update equation with $\Delta_{11}^\sigma(i, \tau) = \exp[\sigma \alpha x_1(\tau)] - 1$ to produce $G^\sigma(x_1)$ for an interaction activated at one site and then use this Green's function to produce $G^\sigma(x_1, x_2)$ for the Green's function with the interaction activated at two sites, and so on, until the interaction is active at all sites. The final Green's function would inherit the imaginary time associated with the Hubbard-Stratonovich fields. To update the Hubbard-Stratonovich fields for all sites requires $\mathcal{O}(N^3)$ floating-point operations, that is, $\mathcal{O}(N^2)$ operations for N sites. To do so for all times requires $\mathcal{O}(N_\tau N^3)$ operations. In Section 7.2.2, we also discussed procedures that scale as $\mathcal{O}(N^3)$ for moving the equal-time Green's function from one imaginary time to another and for generating unequal-time Green's functions from equal-time ones.

In the general case, the path integral for the partition function for discrete Hubbard-Stratonovich fields has the form (Blankenbecler et al., 1981; Negele and Orland, 1988)

$$Z = \sum_C \det X(C) \tag{7.47}$$

where $X(C) = X^\sigma(C) X^{-\sigma}(C)$ is the matrix representation of the noninteracting electron dynamics and its coupling to the auxiliary field. It is of order $N_\tau \times N$ but remarkably sparse:

$$X^\sigma = \begin{pmatrix} I & 0 & \cdots & \cdots & 0 & B_{N_\tau}^\sigma \\ -B_1^\sigma & I & \ddots & \ddots & 0 & 0 \\ 0 & -B_2^\sigma & I & \ddots & 0 & 0 \\ \vdots & \ddots & \ddots & \ddots & \ddots & \vdots \\ \vdots & \ddots & \ddots & \ddots & I & 0 \\ 0 & 0 & 0 & \cdots & -B_{N_\tau-1}^\sigma & I \end{pmatrix}, \tag{7.48}$$

that is, it is an $N_\tau \times N_\tau$ block matrix where each block is an $N \times N$ matrix. The Green's function $\bar{G}^\sigma$ for all sites and times is the inverse of $X^\sigma(x)$ (Negele and Orland, 1988). While these statements are not obvious, with a little algebra they can be made palatable. For example, starting with (7.47), choosing conveniently various permutations of the block rows and columns, and using the following formulas for the *block inversion* of a matrix

$$
\begin{pmatrix} A & B \\ C & D \end{pmatrix}^{-1} = \begin{pmatrix} A^{-1} + A^{-1}BS^{-1}CA^{-1} & -A^{-1}BS^{-1} \\ -S^{-1}CA^{-1} & S^{-1} \end{pmatrix}
$$

$$
= \begin{pmatrix} T^{-1} & -T^{-1}BD^{-1} \\ -D^{-1}CT^{-1} & D^{-1} + D^{-1}CT^{-1}BD^{-1} \end{pmatrix}, \qquad (7.49)
$$

where

$$
S = D - CA^{-1}B, \quad T = A - BD^{-1}C,
$$

we can generate all the equal-time (7.24) and unequal-time (see (7.56) below) expressions for the Green's function. Hence, taking advantage of the structure and sparseness of $X^\sigma(x)$ reduces the $(N_\tau N)^3$ scaling of the cost of inverting (7.48) to the more efficient $N_\tau N^3$ scaling of the algorithm previously derived.

Related to (7.49), let us also note the useful relation

$$
\det \begin{pmatrix} A & B \\ C & D \end{pmatrix} = \det A \, \det \left(D - CA^{-1}B \right) = \det D \, \det \left(A - BD^{-1}C \right). \qquad (7.50)
$$

Using this relation, it is easy to show the equivalence of $\det X(C)$ with (7.19), (7.20), and (7.21).

In the bigger space of dimension $N_\tau \times N$, what is the updating procedure for the total Green's function? To derive it, we start with the definition of the M^σ-matrix and use the definition (7.30) of the B-matrices to define

$$
B_i^\sigma \equiv e^{-\Delta\tau V_i^\sigma} e^{\Delta\tau T},
$$

where we absorbed the chemical potential factor into the hopping matrix T, and where from (7.31) we have the $N \times N$ diagonal matrix

$$
e^{-\Delta\tau V_i^\sigma} = \begin{pmatrix} e^{\sigma\alpha x_1(\tau_i)} & & & 0 \\ & e^{\sigma\alpha x_2(\tau_i)} & & \\ & & \ddots & \\ 0 & & & e^{\sigma\alpha x_N(\tau_i)} \end{pmatrix}.
$$

This definition separates the auxiliary-field-dependent terms from the terms with no such dependence. Next, we define the $N_\tau \times N_\tau$ block diagonal matrix

$$
e^{-V^\sigma} = \begin{pmatrix} e^{-\Delta\tau V_1^\sigma} & & & \\ & e^{-\Delta\tau V_2^\sigma} & & \\ & & \ddots & \\ & & & e^{-\Delta\tau V_{N_\tau}^\sigma} \end{pmatrix}
$$

and then introduce still another definition, the matrix $\hat{G}^\sigma \equiv \bar{G}^\sigma e^{V^\sigma}$, that is,

$$
\hat{G}^\sigma = \begin{pmatrix}
e^{-\Delta\tau V_1^\sigma} & & & & e^{\Delta\tau T} \\
-e^{\Delta\tau T} & e^{-\Delta\tau V_2^\sigma} & & & \\
& -e^{\Delta\tau T} & \ddots & & \\
& & \ddots & \ddots & \\
& & & -e^{\Delta\tau T} & e^{-\Delta\tau V_{N_\tau}^\sigma}
\end{pmatrix}^{-1}.
$$

From this last expression and the operator identity

$$
1/(A+B) = 1/A - (1/A)\,B\,[1/(A+B)],
$$

it is straightforward to show that if $\exp(-V^\sigma)$ is updated to $\exp(-V'^\sigma)$, the updated Green's function $\hat{G}'^\sigma$ satisfies the Dyson equation

$$
\hat{G}'^\sigma = \hat{G}^\sigma + \hat{G}^\sigma \left(e^{-V'^\sigma} - e^{-V^\sigma} \right) \hat{G}'^\sigma.
$$

Finally, the substitution of the definition of $\hat{G}^\sigma$ into this equation delivers

$$
\bar{G}'^\sigma = \bar{G}^\sigma + \bar{G}^\sigma \left(e^{-(V'^\sigma - V^\sigma)} - I \right)\left(I - \bar{G}'^\sigma \right) \tag{7.51}
$$

for the Green's function $\bar{G}^\sigma$ of interest. This equation generalizes (7.39). Instead of a procedure for updating all sites at a given time, the new Dyson equation updates all sites at all times, a $\mathcal{O}((N_\tau N)^3)$ operation.

The generality of (7.51) allows us to define a procedure to update a given site for all times. This capability is especially useful for an impurity problem, that is, for a problem where one site is physically different from the rest; see, for example, (7.4). From (7.51), we can easily convince ourselves that if the interaction is localized at a particular site, say, 0, then the general update equation reduces to an $N_\tau \times N_\tau$ matrix equation just for that site's Green's function for all times:

$$
\bar{G}'^\sigma_{00} = \bar{G}^\sigma_{00} + \bar{G}^\sigma_{00}\left[e^{-(V'^\sigma - V^\sigma)} - I \right]_{00}\left(I - \bar{G}'^\sigma_{00} \right).
$$

Following similar algebra as previously presented reduces this updating equation to one with the same form as (7.45). The updates here are done one τ value at a time. As the matrix is $N_\tau \times N_\tau$, the computation time needed to update all times scales as N_τ^3.

This impurity updating procedure is the *Hirsch-Fye algorithm*. It was first developed for impurity problems (Hirsch and Fye, 1986) and then adapted for lattice problems. The algorithm easily generalizes to a cluster of N_c impurity sites leading to an algorithm scaling as $(N_c N_\tau)^3$. At a certain cluster size, the "all sites at a given time" procedure becomes more efficient. Before the development of the continuous-time impurity solvers (Chapter 8), the Hirsch-Fye algorithm was widely used to solve single-site and cluster impurity problems.

There are several important points about the Hirsch-Fye algorithm: Although only the impurity Green's function is needed to execute the Monte Carlo updating, once this Green's function is known, the general update equation tells us how to use the impurity Green's function and the Green's function in the absence of the impurity to compute the Green's function with spatial matrix elements between the impurity and the other sites in the lattice (Gubernatis et al., 1987). This capability is the spatial analog of finding the unequal-time Green's function for all sites for one imaginary time.

At the beginning of the section, we noted that for a lattice of interacting sites, it was possible to activate the interactions one by one if the noninteracting Green's function is known. In the impurity problem something similar, but better, is possible. To create the noninteracting impurity Green's function, we need the Green's function for the problem without interactions on the impurity site. Once we have the impurity Green's function, we can activate the Hubbard-Stratonovich fields. Then, we do not need the other (spatially nonlocal) Green's function. For the Green's function without the impurity, we can use the one for an infinite lattice, if available, or calculate it for a very large lattice. In short, we are able to simulate the impurity problem without finite-size effects. In the next chapter, we discuss methods to eliminate the error caused by discretizing imaginary time.

7.4 Matrix product stabilization

The matrix methods described in this section are an essential part of the determinant method, but they have little to do with many-electron physics per se. They are principally used to calculate the equal-time Green's function. In this calculation, the product of the B-matrices is accumulated and added to the identity in a way that adequate precision remains after the matrix sum is inverted.[8] The problem is that the matrix needing inversion is very ill-conditioned.

The *condition number* κ of a matrix M is $\kappa = \|M\| \|M^{-1}\|$, where $\|M\|$ is some matrix norm (Golub and Loan, 1989; Meyer, 2000). The largest absolute value of the eigenvalues of the matrix is one such norm. The condition number measures how close a matrix is to being singular, that is, to having one or more zero eigenvalues. If $\kappa = 10^d$, then when a matrix is inverted, for example, using Gaussian elimination with partial pivoting, d represents roughly the number of

---

[8] Because the Hirsch-Fye algorithm does not accumulate the product of these matrices, it does not need these methods.

decimal places lost (Golub and Loan, 1989). We can get a feel of the difficulties in the current applications by considering the $U = 0$ case and restricting the hopping to nearest neighbors. For a square lattice, the eigenvalues of the nearest-neighbor hopping matrix T are well known to be $\lambda(k_x, k_y) = -2t[\cos(k_x) + \cos(k_y)]$ $(0 \leq k_{x,y} \leq 2\pi)$, so that $-4t \leq \lambda(k_x, k_y) \leq 4t$. With $\mu = 0$,

$$B(\beta, 0) = [I + e^{\beta T}]^{-1} \tag{7.52}$$

is a matrix whose condition number is roughly $\exp(8\beta t)$. When the temperature is about $t/5$, a small fraction of the electron's band width of $8t$, this condition number implies a loss of 17 decimal places of precision, a number just beyond what is representable in double-precision arithmetic.

The basic difficulty in thermodynamic studies of many-electron systems at low temperatures is that the Boltzmann distribution $e^{-\beta(H_1 - \mu N)}$ gives the lowest-energy states, relative to the chemical potential, exponentially large weights. The Pauli exclusion principle, however, prevents these states from becoming macroscopically occupied and forces them to be singly occupied up to some "Fermi energy." The matrix elements of a diagonalized $B(\beta, 0)$ are just 1 minus the Fermi-Dirac distribution,

$$[1 + \exp(-\beta(\lambda - \mu))]^{-1} = 1 - [1 + \exp(\beta(\lambda - \mu))]^{-1},$$

for each eigenvalue λ. Properly adding 1 to $\exp[-\beta(\lambda - \mu)]$ is the first step toward ensuring a proper Fermi-Dirac distribution of the electrons. The procedure for inverting $I + B(\beta, 0)$, in effect, is about properly adding the identity I to the Boltzmann distribution $\exp[-\beta(H - \mu N)]$ so that the inversion of the matrix sum preserves the Fermi-Dirac character of the electrons. The single particle propagators B^{σ} for any configuration of the auxiliary fields attenuate the "high-energy" scales while amplifying the "low-energy" scales. Accordingly, the extraction of information about "intermediate" energies near the chemical potential is a very difficult numerical task, but one that must be done accurately as it is these energy states that control the finite-temperature physics. Without the procedures to be described, the Fermionic character of the simulation gets lost.

The numerical limitation in inverting a matrix embodying very diverse numerical scales is quite different from that for inverting a single number. For a single number, double-precision arithmetic typically allows the scale to range between 10^{-307} and 10^{+309}. The strategy devised (Loh et al., 1989; Loh and Gubernatis, 1992; Loh et al., 2005) for inverting our ill-conditioned matrix is stratifying the matrix multiplications with a representation of the B matrices that places these numerical scales into diagonal matrices. We can accurately invert them as doing so just requires inverting the individual numbers. The procedures also accumulate the matrix multiplications in a way that avoids mixing the scales. Only a final set of matrix multiplications mixes the scales and generates a loss of accuracy. The remaining accuracy is enough that the measurement error is at least as small as the statistical error.

The procedure employed decomposes the matrices into the form UDV where the diagonal matrix D contains the diverse scales and U and V are reasonably well conditioned and readily invertible. The following illustrates the character of the factored form and the problem we are trying to avoid:

$$
UDV = \begin{pmatrix} x & x & x & x \\ x & x & x & x \\ x & x & x & x \\ x & x & x & x \end{pmatrix} \begin{pmatrix} X & & & \\ & x & & \\ & & x & \\ & & & x \end{pmatrix} \begin{pmatrix} x & x & x & x \\ x & x & x & x \\ x & x & x & x \\ x & x & x & x \end{pmatrix}
$$

$$
= \begin{pmatrix} X & X & X & X \\ X & X & X & X \\ X & X & X & X \\ X & X & X & X \end{pmatrix},
$$

where we are using the size of the character x to represent different numerical scales. The matrices U and V have unit scales. The three matrices in the first line of the above equation represent what we want to achieve in the factorization. The second line illustrates what happens upon multiplication – the largest numerical scale dominates and small-scale information is lost.

On the other hand, if a matrix M has its numerical scales stratified into columns, we can factorize it in a stable way, preserving its small-scale information:

$$
U^{-1}MV^{-1} = \begin{pmatrix} x & x & x & x \\ x & x & x & x \\ x & x & x & x \\ x & x & x & x \end{pmatrix} \begin{pmatrix} X & x & x & x \\ X & x & x & x \\ X & x & x & x \\ X & x & x & x \end{pmatrix} \begin{pmatrix} x & x & x & x \\ x & x & x & x \\ x & x & x & x \\ x & x & x & x \end{pmatrix}
$$

$$
= \begin{pmatrix} X & & & \\ & x & & \\ & & x & \\ & & & x \end{pmatrix} = D.
$$

Multiplication on the left of M by U^{-1} only combines the elements in a given column and thus causes no loss of information. The multiplication on the right by V^{-1} combines columns of different scales but does not overwrite any small-scale information *as long as large-scale columns are first scaled down* before they are added into the columns of smaller scale. Using $MV^{-1} = (V^{-T}M^{T})^{T}$ reduces the mixing.

To compute the product (7.29) of many matrices in a stable manner, we separate the many scales throughout the calculation. As an illustration, we first imagine we have $B(\tau, 0) = UDV$. If we want this propagator at $\tau + \tau_0$, where τ_0 is the size of a few imaginary time steps,[9] we can extend the propagator with standard matrix multiplication methods. We write

$$B(\tau + \tau_0, 0) = B(\tau + \tau_0, \tau)UDV = [B(\tau + \tau_0, \tau)UD] V$$

$$= \left[B(\tau + \tau_0, \tau)U \begin{pmatrix} X & & & \\ & x & & \\ & & x & \\ & & & x \end{pmatrix} \right] V = \begin{pmatrix} X & x & x & x \\ X & x & x & x \\ X & x & x & x \\ X & x & x & x \end{pmatrix} V$$

$$= (U'D'V') V = U'D' (V'V), \tag{7.53}$$

thus obtaining the UDV factorization of $B^\sigma(\tau + \tau_0, 0)$. What we did was to decompose the stratified matrix $B^\sigma(\tau + \tau_0)UD$ into $U'D'V'$. The V matrices must be sufficiently well conditioned so that we can multiply many of them together in a stable way. Repeating this procedure as often as necessary enables the computation of a long string of products of the B matrices (7.21) by multiplying from the left.

After the matrix products are assembled in factored form, we compute the inverse $[I + UDV]^{-1}$ by the following steps:

$$G = [I + UDV]^{-1} = V^{-1} \left[U^{-1}V^{-1} + D \right]^{-1} U^{-1}$$

$$= V^{-1} \left[U'D'V' \right]^{-1} U^{-1} = (V'V)^{-1} (D')^{-1} (UU')^{-1}.$$

We note that after forming the sum $U^{-1}V^{-1} + D$, we decompose it into $U'D'V'$ and perform the inverses of the factors separately.

An alternative to the above is performing the factorization of the two partial products

$$B^\sigma(\tau, 0) = U_1 D_1 V_1,$$

$$B^\sigma(\beta, \tau) = V_2 D_2 U_2,$$

where in the second equation we reverse the UDV factors because we intend to build up the partial product by multiplications from the left. We now have

$$G^\sigma(\tau, \tau) = [I + U_1 D_1 V_1 V_2 D_2 U_2]^{-1}$$

$$= U_2^{-1} \left[U_1^{-1} U_2^{-1} + D_1 V_1 V_2 D_2 \right]^{-1} U_1^{-1}. \tag{7.54}$$

---

[9] We do not apply the factorization at each time step but rather with a period τ_0 in imaginary time.

Schematically, the piece to be inverted is

$$
\begin{pmatrix}
x & x & x & x \\
x & x & x & x \\
x & x & x & x \\
x & x & x & x
\end{pmatrix}
+
\begin{pmatrix}
X & & & \\
& x & & \\
& & x & \\
& & & x
\end{pmatrix}
\begin{pmatrix}
x & x & x & x \\
x & x & x & x \\
x & x & x & x \\
x & x & x & x
\end{pmatrix}
\begin{pmatrix}
X & & & \\
& x & & \\
& & x & \\
& & & x
\end{pmatrix}
$$

$$
=
\begin{pmatrix}
x & x & x & x \\
x & x & x & x \\
x & x & x & x \\
x & x & x & x
\end{pmatrix}
+
\begin{pmatrix}
XX & Xx & Xx & Xx \\
xX & xx & xx & xx \\
xX & xX & xx & xx \\
xX & xX & xx & xx
\end{pmatrix}.
$$

Because we kept the diagonal matrices on the outside, the small-scale cut-off occurs only in the last step when we add together elements of different scales. We invert the resulting matrix by first doing another UDV factorization.

How do we create the UDV factorization? A number of standard numerical matrix methods work, including the modified Gram-Schmidt, QR, and the singular value decomposition methods (Golub and Loan, 1989; Meyer, 2000). A UDV^T factorization is the standard result of the singular value decomposition method. The standard modified Gram-Schmidt and QR algorithms are easily rewritten to produce this result. The modified Gram-Schmidt method, though slightly less stable numerically, is noticeably faster and usually more than adequate. How do we invert the various matrices? Diagonal matrices are trivial to invert. In fact, so are U and V. The modified Gram-Schmidt method produces a column orthonormal U and an upper-unit triangular V. The first we invert by transposition, and the second by a back-substitution solution of the linear system of equations $VV^{-1} = I$ for the unknown V^{-1}, which is also a triangular matrix.

The Gram-Schmidt orthogonalization method is the following: Given the set of vectors $a_1, a_2, \ldots, a_M$, which may be the columns of a matrix, find a set of vectors $v_1, v_2, \ldots, v_M$ that are orthonormal and span the same space. The idea is quite simple: Take a_1 and divide it by its length $\|a_1\| = \sqrt{a_1^T a_1}$ to get v_1. Then subtract the component of a_2 in the direction of v_1. The remainder is orthogonal to v_1. Next, divide the remainder by its length to get v_2. Repeat the process by subtracting the component of a_3 in the directions of v_1 and v_2. Dividing the remainder by its length gives v_3. By continuing the process, we obtain the desired set of orthonormal vectors. Unfortunately, this process is numerically unstable if any of the a_i are nearly linearly dependent. Fortunately, the fix is simple: The computation is reformulated so that the a_i are never used once the v_i are obtained. The algorithm is given in Algorithm 24. In a more computer-language-like fashion, Algorithm 25 describes the orthonormalization and the matrix factorization in pseudocode.

Algorithm 24 Modified Gram-Schmidt method.

Input: M vectors $a_1, a_2, \ldots, a_M$.
  for $i = 1$ to M **do**
    $a_i \leftarrow a_i / \|a_i\|$;
    for $j = i + 1$ to M **do**
      $a_j \leftarrow a_j - (a_i^T a_j) a_i$;
    end for
  end for
  return M orthonormal vectors $a_1, a_2, \ldots, a_M$.

Algorithm 25 Modified Gram-Schmidt method: *UDV* factorization.

Input: $N \times N$ array U.
  for $k = 1$ to N **do**
    $d(i) = 0$;
    for $i = 1$ to N **do**
      $d(k) \leftarrow d(k) + U(i,k)U(i,k)$;
    end for
    $d(k) \leftarrow \sqrt{d(k)}$;
    for $i = 1$ to N **do**
      $U(i,k) \leftarrow U(i,k)/d(k)$;
    end for
    $V(k,k) = 1$;
    for $j = k + 1$ to N **do**
      $V(k,j) = 0$;
      for $i = 1$ to N **do**
        $V(k,j) \leftarrow V(k,j) + U(i,k)U(i,j)$;
      end for
      for $i = 1$ to N **do**
        $U(i,j) \leftarrow U(i,j) - V(k,j)U(i,k)$;
      end for
      $V(k,j) \leftarrow V(k,j)/d(k)$;
    end for
  end for
  return the $N \times N$ orthonormal matrix U, the N diagonal elements d of a $N \times N$ diagonal matrix D, and the $N \times N$ unit upper triangular matrix V.

While the procedure for updating the equal-time Green's function enables convenient Monte Carlo sampling, the goal of any sampling remains the estimation of various physical observables. The determinant method permits the computation of these quantities as a function of imaginary time. For many such quantities, knowing just equal-time Green's functions suffices. For others, such as *susceptibilities*,

$$\chi(i\omega_n) = \int_0^\beta \langle A^\dagger(\tau)A(0)\rangle e^{i\omega_n \tau} d\tau. \tag{7.55}$$

Wick's theorem, generates contractions of the electron operator pairs at different imaginary times. Thus, for the measurement part of the simulation, we often need the matrix elements of unequal-time Green's functions.

The unequal-time Green's functions are

$$G^\sigma(\tau',\tau) = B^\sigma(\tau',\tau)\left[I + B^\sigma(\tau,0)B^\sigma(\beta,\tau)\right]^{-1}. \tag{7.56}$$

For small changes $\tau' \to \tau''$, multiplication by the single-particle propagator produces the appropriate Green's functions

$$G^\sigma(\tau'',\tau) = B^\sigma(\tau'',\tau')G^\sigma(\tau',\tau).$$

As with the equal-time Green's function, this procedure propagates with adequate accuracy for some time τ_0. To extend this range, we write

$$G^\sigma(\tau,0) = B^\sigma(\tau,0)\left[I + B^\sigma(\beta,0)\right]^{-1} = \left[B^\sigma(\tau,0)^{-1} + B^\sigma(\beta,\tau)\right]^{-1}$$

and accumulate the two propagators in factored form,

$$G^\sigma(\tau,0) = \left[V_1^{-1}D_1^{-1}U_1^{-1} + V_2 D_2 U_2\right]^{-1}, \tag{7.57}$$

where $U_1 D_1 V_1$ is the factorization of $B^\sigma(\tau,0)$. Then, we isolate the most ill-conditioned diagonal matrix, for example, D_2, and write

$$G^\sigma(\tau,0) = U_1\left[D_1^{-1} + V_1 V_2 D_2 U_2 U_1\right]^{-1} V_1.$$

The matrix inverse of the sum is done by Gaussian elimination with partial pivoting.

A more stable procedure, related to (7.54), is to reexpress (7.57) as

$$G^\sigma(\tau,0) = U_2^{-1}\left[D_1^{-1}U_1^{-1}V_2^{-1} + V_1 V_2 D_2\right]^{-1} V_1,$$

then factor each diagonal matrix D as $D^{\max}D^{\min} = D^{\min}D^{\max}$ and rewrite (7.4) as

$$G^\sigma(\tau,0) = U_2^{-1}\left(D_2^{\max}\right)^{-1}\left[\left(D_1^{\max}\right)^{-1}U_1^{-1}V_2^{-1}\left(D_2^{\max}\right)^{-1}\right.$$
$$\left. +\left(D_1^{\min}\right)^{-1}V_1 V_2\left(D_2^{\min}\right)^{-1}\right]^{-1}\left(D_1^{\min}\right)^{-1} V_1.$$

$D^{\max}$ is D with its nonzero elements less than 1 replaced by 1, and $D^{\min}$ is D with elements greater than 1 replaced by 1. The matrix inverse of the sum is done by Gaussian elimination with partial pivoting.

7.5 Comments

The birth of supercomputers gave life to the determinant method. In the mid-1980s, it pushed the fastest and biggest computers to their limits. While processor speeds, the size of internal and external memory, and the access rate of these memories have all improved dramatically, the range of applications of these quantum Monte Carlo methods has remained somewhat static because for many interesting applications, its true computation time scales exponentially with the problem size due to the sign problem (Loh et al., 1990). Today, for systems lacking a sign problem, or ones where it is minor, the execution cost of this method is relatively cheap.

While for finite-temperature Fermion systems a good way to reduce the scaling of the sign problem is still lacking, there are ways to reduce the computational cost when there is no sign problem. Reducing the number of matrix-matrix multiplications and matrix inversions is the most direct route. In Section 7.2.1, we saw how using a sparse matrix (checkerboard) representation for the exponential of the hopping matrix can reduce the computational scaling from N^3 to N^2. Even the use of a matrix stabilization factorization producing triangular as opposed to dense matrices helps. Sometimes we can also achieve significant gains by computing, storing, and craftily using strings of partial matrix products rather than repeatedly recomputing the entire string. We illustrate one such approach for the zero-temperature determinant method described in Appendix I.

A single spin-flip algorithm, such as the Metropolis or heat bath algorithm, causes the determinant method to share the same broken ergodicity problems as the Monte Carlo simulation of a classical spin model in a strong external magnetic field. The Hubbard-Stratonovich transformation (7.9) in effect couples a local fluctuating quantum spin to an imaginary-time-dependent magnetic field that depends on the magnitude of U. As such, when U becomes large, sampling ergodically becomes a problem. Furthermore, the Trotter approximation restricts the use of this method to the weak and intermediate coupling regimes. Global spin moves (Scalettar et al., 1991) promote ergodicity and increase the magnitude of U, which can be simulated. In the past, to keep the cost of the simulation manageable, $\Delta\tau$'s of $0.25t$ or larger were used where t is the magnitude of the nearest-neighbor hopping. The recently developed continuous-time version of the finite-temperature determinant method (Iazzi and Troyer, 2014) is discussed in Appendix M.[10] These algorithms do not use a Trotter decomposition, but instead sample configurations from a weak-coupling perturbation expansion in continuous imaginary time, using techniques very similar to those that we introduce for impurity models in the following chapter.

The advantages of global or multiple-site updating methods of auxiliary and Bosonic fields have led to other algorithms, most notably the hybrid quantum Monte

---

[10] For the zero-temperature implementation, see Wang et al. (2015).

Carlo method (Scalettar et al., 1987). It updates at once all sites at all times. This method employs a continuous Hubbard-Stratonovich transformation, and it is conceivable that this method could be increasingly used for simulations of complex electron lattice models requiring multiple on-site or longer-ranged Coulomb interactions, cases when the discrete auxiliary fields become numerous. In these situations, the use of the continuous Hubbard-Stratonovich transformation is potentially more advantageous (Appendix G).

Briefly, the hybrid method uses continuous Hubbard-Stratonovich fields and reformulates (7.47) as

$$Z = \int \mathcal{D}x \, \mathcal{D}\phi \, e^{-[S(x) + \phi^T O^{-1}\phi]}, \tag{7.58}$$

where $\exp[-S(x)]$ is the weight of the fields and the matrix $O = X(x)^T X(x)$. Next, fictitious momentum fields p, canonical to the auxiliary fields x, are introduced and a Gaussian integral over the p is added to the path integral without changing the physical content of the partition function,[11]

$$Z = \int \mathcal{D}p \, \mathcal{D}x \, \mathcal{D}\phi \, e^{-H_{\text{eff}}(p,x,\phi)}.$$

Here

$$H_{\text{eff}}(p, x, \phi) = \sum_i \frac{p_i^2}{2m} + V_{\text{eff}}(x, \phi), \quad V_{\text{eff}}(x, \phi) = S(x) + \phi^T O^{-1}(x)\phi.$$

The simulation strategy is to sample all p and x fields, keeping the ϕ fields fixed, and then to sample all ϕ fields, keeping the p and x fields fixed. When the ϕ fields are fixed, the hybrid Monte Carlo method (Liu, 2001) is used. In this method, molecular dynamics evolves the system of particles described by H_{eff} in a fictitious time to generate a proposed change of all the p and x fields. Normally, molecular dynamics (Allen and Tildesley, 1987) conserves energy so the Metropolis algorithm would always accept the change. The intent here, however, is not to conserve energy well but rather to use a rough molecular dynamics simulation with a time step a little too large. The step size is chosen by adjusting it and the pseudo-mass m to cause an energy nonconservation so the Metropolis acceptance rate is around 80 to 90%. We recall that the Metropolis algorithm does not specify how the proposal is to be made. However it is made, the Metropolis algorithm ensures we are sampling the auxiliary fields with a probability $\exp(-V_{\text{eff}})$. Sampling the ϕ fields from $\exp(-\phi^T O^{-1}\phi)$ with $O = X^T X$ is done by sampling fields R^σ from a multidimensional Gaussian distribution $\exp(-\frac{1}{2}R^{\sigma T} R^\sigma)$ to find the sample of ϕ^σ via $\phi^\sigma = X^\sigma R^\sigma$ (Appendix G.1).

---

[11] Recall the factorization of the Gaussian integration of momentum of the partition function in classical statistical mechanics.

An intermediate step of the molecular dynamics is the computation of V_{eff}. It requires computing

$$Y^\sigma = [O^\sigma]^{-1}\phi^\sigma = [X^\sigma]^{-1}R^\sigma,$$

which is done by solving the large sparse linear system of equations $X^\sigma Y^\sigma = R^\sigma$. We note that the elements of $[X^\sigma]^{-1}$ are the imaginary-time elements of the Green's function. For example,

$$[X_i^\sigma]^{-1} = G_i^\sigma = [I + B_i^\sigma \cdots B_1^\sigma B_{N_\tau}^\sigma \cdots B_{i+1}^\sigma]^{-1}.$$

Accordingly, both solving the linear system defined by X^σ and computing the inverse of $I + B_i^\sigma \cdots B_{N_\tau}^\sigma B_1^\sigma \cdots B_{i-1}^\sigma$ involve matrices that are very ill-conditioned. For this reason, preconditioned, not the standard, conjugate-gradient methods (Golub and Loan, 1989; Meyer, 2000) are used to solve the system of equations. Giving more details will take us too far from our central focus.

In closing, we comment that embedded in our discussion of the hybrid quantum Monte Carlo method is a basic strategy for simulating systems where the electrons are coupled to Bosonic fields such as phonons or magnons: We fix the Bosonic fields and then use an appropriate Monte Carlo method to sample the scalar variables representing the electron degrees of freedom. Then, we fix these scalar variables and use an appropriate simulation method to sample the Bosonic degrees of freedom. The two sampling methods need not be the same. Combining Bosonic fields with the scalar auxiliary field is also possible with the determinant method.

Suggested reading

E. Y. Loh, Jr. and J. E. Gubernatis, "Stable simulations of models of interacting electrons in condensed-matter physics," in *Electronic Phase Transitions*, ed. W. Hanke and Yu. V. Kopaev (Amsterdam: North-Holland, 1992), chapter 4.

A. Muramatsu, "Quantum Monte Carlo for lattice Fermions," in *Quantum Monte Carlo Methods in Physics and Chemistry*, ed. M. P. Nightingale and C. J. Umrigar, NATO Science Series C, vol. 525 (Dordrecht: Kluwer Academics, 1999), chapter 13.

E. Y. Loh, Jr., J. E. Gubernatis, R. T. Scalettar, S. R. White, D. J. Scalapino, and R. L. Sugar, "Numerical stability and the sign problem in the determinant Monte Carlo method," *Int. J. Mod. Phys. C* **16**, 1319 (2005).

Exercises

7.1 For the Hamiltonian

$$H = U(n_\uparrow - \tfrac{1}{2})(n_\downarrow - \tfrac{1}{2}), \tag{7.59}$$

where $U > 0$, analytically evaluate the grand canonical partition function

$$Z = \text{Tr } e^{-\beta(H-\mu N)} \tag{7.60}$$

1. By summing over all the basis states $|00\rangle$, $|01\rangle$, $|10\rangle$, and $|11\rangle$.
2. By making the continuous Hubbard-Stratonovich transformation (7.6), tracing out the electron degrees of freedom, and performing the remaining Gaussian integral over the auxiliary field.
3. By making the discrete Hubbard-Stratonovich transformation (7.9), tracing out the electron degrees of freedom, and performing the remaining sum over the auxiliary field.

7.2   What unequal-time Green's functions are needed to calculate the susceptibility (7.55) when $A = n_\uparrow - n_\downarrow$?

7.3   On a bipartite lattice, with no magnetic field, the *particle-hole transformation* is

$$c_{i\uparrow} \to d_{i\uparrow},$$

$$c_{i\downarrow}^\dagger \to \begin{cases} +d_{i\downarrow}^\dagger, & i \in A, \\ -d_{i\downarrow}^\dagger, & i \in B. \end{cases}$$

A *bipartite lattice* is one composed of two sublattices (A and B) where the nearest-neighbor sites of one lattice belong to the other.

1. Show that under this transformation, the Hubbard Hamiltonian transforms as

$$H(t, U, N) \to H(t, -U, N) + UN_\uparrow,$$

where $N = N_\uparrow + N_\downarrow$ is the total number of electrons. This relation allows us to make statements about the repulsive Hubbard model by studying the attractive model and vice versa. Note that the N_σ are conserved.
2. Similarly, show that under this transformation the total charge Q and the z-component S^z of the total spin transform as $Q \to S^z + 1$ and $S^z \to Q - 1$. These relations connect the behavior of particles and spins within a repulsive or attractive model and between these models.
3. Show that under the particle-hole transformation, the Hubbard Hamiltonian also transforms as $H(N) = H(2L - N) + U(N - L)$.
4. What does the last relation say about the symmetry of the ground state and the excited states for the square lattice at half-filling? Are they the same? For a square lattice, half-filling corresponds to one electron per site and the number of up and down electrons being equal.

7.4   Verify the validity of the discrete Hubbard-Stratonovich transformations (7.9).

7.5 Prove the three determinant relations (7.34), (7.35), and (7.23). Note the first two require the matrices to be square; the last one does not.

7.6 The operator product $n_\uparrow n_\downarrow$ may be written as the sum of two squares $\frac{1}{4}[(n_\uparrow + n_\downarrow)^2 - (n_\uparrow - n_\downarrow)^2]$. What discrete Hubbard-Stratonvich transformation follows from this expression for the interaction?

7.7 Show that the diagonal and off-diagonal elements of the inverse of M^σ (7.48) are (7.37) and (7.40).

7.8 For the noninteracting problem, (7.32) represents a convenient and obvious choice of a UDV factorization of $\exp(\Delta\tau T)$. For

$$T = \begin{pmatrix} 1 & -1 \\ -1 & 1 \end{pmatrix},$$

compute analytically its eigenvalues and eigenvectors plus

$$[I + \exp(\beta T)]^{-1}.$$

Then compare this exact result with its numerical computation via the Trotter breakup

$$[I + \exp(\Delta\tau T)\exp(\Delta\tau T)\cdots\exp(\Delta\tau T)]^{-1}$$

with and without the procedure defined by (7.53) for different values of β and $\Delta\tau$. From the point of view of matrix analysis and linear algebra, what are the differences between this computation and the one found in a quantum Monte Carlo simulation?

7.9 Repeat the above using the modified Gram-Schmidt factorization instead of the eigenvalue diagonalization.

8

Continuous-time impurity solvers

This chapter, like the previous one, describes quantum Monte Carlo methods for the simulation of a system of interacting electrons at nonzero temperature. Here, we target a particular class of problems, characterized by a small number of correlated sites or orbitals, that we call impurity problems. While we could apply the methods of the previous chapter to such impurity models, the algorithms discussed here are more advantageous not only because they work in continuous imaginary time and hence lack the Trotter approximation error inherent to the previous methods, but also because they manipulate determinants of matrices of smaller size, which can be handled more efficiently. At the same time, the continuous-time approach is applicable to broader classes of impurity problems.

Driving the development of these continuous-time impurity solvers was the desire to perform more efficient simulations of correlated lattice models within the framework of dynamical mean-field theory. This formalism maps the lattice problem onto an impurity problem, whose parameters are determined by a self-consistency condition. In some cases, the self-consistent solution gives a good approximation of the properties of the original lattice problem. When we combine this dynamical mean-field approximation with the local-density approximation for electronic structure calculations, we obtain a powerful scheme for electronic structure calculations of strongly correlated materials. In this chapter we discuss the most important classes of impurity models, sketch the basics of the dynamical mean-field approximation, and detail the different variants of continuous-time impurity solvers.

8.1 Quantum impurity models

A quantum impurity model describes an atom or molecule embedded in some host with which it exchanges electrons, spin, and energy. This exchange allows the impurity to make transitions between different quantum states, and in the presence of interactions within or between the impurity orbitals, these transitions lead to nontrivial dynamical properties. Quantum impurity models play a prominent role, for example, in the theoretical description of the magnetic and electric properties of dilute metal alloys and in theoretical studies of quantum dots and molecular conductors. These models also appear as an auxiliary problem whose solution yields the *dynamical mean-field* description of correlated lattice models.

The iconic impurity models are the single orbital Anderson impurity model

$$H = \sum_{k\sigma} \epsilon_k n_{k\sigma} + \sum_{k\sigma} \left(V_{k\sigma} c_{k\sigma}^\dagger d_\sigma + V_{k\sigma}^* d_\sigma^\dagger c_{k\sigma} \right) + \epsilon_d \sum_\sigma n_{d\sigma} + U n_{d\uparrow} n_{d\downarrow} \qquad (8.1)$$

and the Kondo impurity model

$$H = \sum_{k\sigma} \epsilon_k n_{k\sigma} + \sum_{kk'} J_{kk'} \vec{S}_{kk'} \cdot \vec{s}. \qquad (8.2)$$

In the original context, ϵ_k is the energy dispersion of a conduction electron of Bloch momentum k and spin σ (creation operator $c_{k\sigma}^\dagger$, $n_{k\sigma} = c_{k\sigma}^\dagger c_{k\sigma}$). In the Anderson impurity model the $V_{k\sigma}$ describe the hybridization of an electron in an impurity state (creation operator $d_\sigma^\dagger$, $n_{d\sigma} = d_\sigma^\dagger d_\sigma$) with those in the conduction band. The impurity level energy is ϵ_d, and a repulsive Coulomb energy U acts if two electrons occupy this level simultaneously. In the Kondo impurity model, the impurity is a localized quantum spin $\vec{s}$ coupled to the spin of the conduction band,

$$S_{kk'}^\alpha = \frac{1}{2} \left(\begin{array}{cc} c_{k\uparrow}^\dagger & c_{k\downarrow}^\dagger \end{array} \right) \sigma^\alpha \left(\begin{array}{c} c_{k'\uparrow} \\ c_{k'\downarrow} \end{array} \right).$$

Here, σ^α with $\alpha = x, y$, and z are the Pauli spin matrices, and the coupling is via a wave-number-dependent exchange interaction $J_{kk'}$. These two models were the subjects of decades of numerical and analytical attacks that resulted in an understanding of the Kondo effect. The properties of these models and their generalizations are still of considerable interest.

A canonical transformation (Section 9.2.2), called the Schrieffer-Wolff transformation, shows that the physics of the Anderson impurity model and Kondo impurity model are equivalent when the hybridization is weak, the impurity level is below the conduction band, and the Coulomb energy is the dominant energy scale. These conditions combine to restrict the electron occupation of the impurity to one electron. The Kondo exchange interaction is antiferromagnetic and roughly equals the momentum average of the square of the band energy divided by the Coulomb energy.

While these iconic models allow us to address general theoretical issues, they are somewhat unrealistic. Impurities (e.g., transition metal atoms) typically have multiple orbitals. While generalizations to such situations are easily made, and simulations of such models with the determinant method of Chapter 7 have occurred (e.g., Bonča and Gubernatis, 1993a, 1994), the desire to treat more complex models generated a need for more efficient and more flexible impurity solvers.

The Hamiltonian of a general impurity model has the form

$$H = H_{\text{loc}} + H_{\text{bath}} + H_{\text{mix}}, \tag{8.3}$$

where H_{loc} describes the impurity, characterized by a small number of degrees of freedom (typically spin and orbital degrees of freedom denoted by a, b, $\ldots$), and H_{bath} describes an infinite reservoir of free electrons, labeled by a continuum of quantum numbers p, and a discrete set of quantum numbers v (typically spin). Finally, H_{mix} describes the exchange of electrons between the impurity and the bath in terms of hybridization amplitudes V_{pv}^a. Explicitly, the three terms are

$$H_{\text{loc}} = \sum_{ab} \epsilon^{ab} d_a^\dagger d_b + \frac{1}{2} \sum_{abcd} U^{abcd} d_a^\dagger d_b^\dagger d_c d_d, \tag{8.4}$$

$$H_{\text{bath}} = \sum_{pv} \varepsilon_p c_{pv}^\dagger c_{pv}, \tag{8.5}$$

$$H_{\text{mix}} = \sum_{pav} \left[V_{pv}^a d_a^\dagger c_{pv} + (V_{pv}^a)^* c_{pv}^\dagger d_a \right]. \tag{8.6}$$

Note that in this general context, the bath energies ε_p can be arbitrary; that is, they are not necessarily related to any lattice dispersion.

8.1.1 Chain representation

For notational brevity, we concentrate on the general version of the single-orbital Anderson model. In this case, the Hilbert space of the local problem,

$$H_{\text{loc}} = H_\mu + H_U, \tag{8.7}$$

$$H_\mu = -\mu(n_\uparrow + n_\downarrow), \tag{8.8}$$

$$H_U = U n_\uparrow n_\downarrow, \tag{8.9}$$

has dimension four. The discrete quantum number labeling the impurity states is the spin σ, $n_\sigma = d_\sigma^\dagger d_\sigma$ is the density operator for impurity electrons, and the chemical potential is $\mu = -\epsilon_d$. The bath and mixing terms are

$$H_{\text{bath}} = \sum_{p\sigma} \varepsilon_p c_{p\sigma}^\dagger c_{p\sigma}, \tag{8.10}$$

$$H_{\text{mix}} = \sum_{p\sigma} \left[V_{p\sigma} d_\sigma^\dagger c_{p\sigma} + V_{p\sigma}^* c_{p\sigma}^\dagger d_\sigma \right]. \tag{8.11}$$

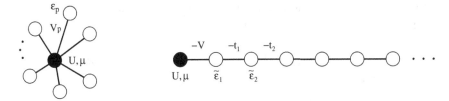

Figure 8.1 Left: Schematic representation of an Anderson impurity model. Spin up and down electrons on the impurity (black dot) interact with an on-site energy U and hop to a continuum of noninteracting bath levels with energy ε_p. The amplitudes for these transitions are given by the hybridization parameters V_p. Right: Chain representation of the Anderson impurity model. The hopping parameters V and t_i can be made positive by using a suitable gauge transformation (see Exercise 8.1).

An illustration of the Anderson impurity model is shown in the left-hand panel of Fig. 8.1.

This model can be mapped onto a semi-infinite chain whose first site is the impurity (right-hand panel of Fig. 8.1).[1] The mapping corresponds to a transformation of the operators $\{d, c_{p_1}, c_{p_2}, \ldots\}$ to new operators $\{d, c_1, c_2, \ldots\}$ such that $H_\mu + H_{\text{bath}} + H_{\text{mix}}$ becomes tridiagonal (Appendix J):

$$
\begin{pmatrix}
-\mu & V_{p_1} & V_{p_2} & V_{p_3} & \cdots \\
V_{p_1}^* & \varepsilon_{p_1} & & & \\
V_{p_2}^* & & \varepsilon_{p_2} & & \\
V_{p_3}^* & & & \varepsilon_{p_3} & \\
\vdots & & & & \ddots
\end{pmatrix}
\rightarrow
\begin{pmatrix}
-\mu & -V & & & \\
-V & \tilde{\varepsilon}_1 & -t_1 & & \\
& -t_1 & \tilde{\varepsilon}_2 & -t_2 & \\
& & -t_2 & \tilde{\varepsilon}_3 & \ddots \\
& & & \ddots & \ddots
\end{pmatrix}.
$$

In the chain representation, the impurity orbital is unchanged, and the hopping amplitude from the impurity to the first site of the chain is $V = (\sum_p |V_p|^2)^{1/2}$. We chose the phase factors in this transformation such that all hopping parameters are positive ($V \geq 0$, $t_i \geq 0$, $i = 1, 2, \ldots$). In Section 8.7, we use this fact to prove that continuous-time quantum Monte Carlo simulations of the Anderson impurity model do not have a sign problem.

8.1.2 Action formulation

As we did for the determinant methods, we formulate our impurity solvers in terms of imaginary-time Green's functions (or hybridization functions). Because of time-

---

[1] This representation was used by Wilson in his famous numerical renormalization group studies of the Anderson and Kondo impurity models (Wilson, 1975).

translation invariance, we can deconvolute the Dyson equation in imaginary time by using the Matsubara transformation

$$G(i\omega_n) = \int_0^\beta d\tau\, e^{i\omega_n \tau} G(\tau), \quad G(\tau) = \frac{1}{\beta} \sum_n e^{-i\omega_n \tau} G(i\omega_n),$$

where the Matsubara frequencies are $\omega_n = (2n+1)\pi/\beta$ and $\beta = 1/T$ is the inverse temperature.

It is also useful to express the partition function and the imaginary-time Green's function in terms of the imaginary-time action. In Appendix K, we integrate out the bath degrees of freedom in the path-integral formalism and express the partition function of the Anderson impurity model as

$$Z = \mathrm{Tr}_d\big[\mathcal{T}e^{-S}\big],$$

with the action $S = S_{\mathrm{mix}} + S_{\mathrm{loc}}$ given by

$$S_{\mathrm{mix}} = \sum_\sigma \int_0^\beta d\tau\, d\tau'\, d_\sigma^\dagger(\tau') \Delta^\sigma(\tau' - \tau) d_\sigma(\tau), \tag{8.12}$$

$$S_{\mathrm{loc}} = \int_0^\beta d\tau\Big[-\mu(n_\uparrow(\tau) + n_\downarrow(\tau)) + U n_\uparrow(\tau) n_\downarrow(\tau)\Big]. \tag{8.13}$$

$\mathcal{T}$ is the time-ordering operator. The Green's function becomes[2]

$$G(\tau) = -\langle \mathcal{T} d(\tau) d^\dagger(0)\rangle_S = -\frac{1}{Z}\mathrm{Tr}_d\big[\mathcal{T}e^{-S} d(\tau) d^\dagger(0)\big].$$

The hybridization function $\Delta^\sigma(\tau' - \tau)$ in (8.12) represents the amplitude for hopping from the impurity into the bath at time τ and back onto the impurity at time τ'. It is a function of the bath energies and hybridization amplitudes and is most conveniently expressed in Matsubara frequency space:

$$\Delta^\sigma(i\omega_n) = \sum_p \frac{|V_{p\sigma}|^2}{i\omega_n - \varepsilon_p}. \tag{8.14}$$

It is also useful to introduce the Green's function of the noninteracting impurity,[3] $\mathcal{G}_0$, which is related to the hybridization function by

$$[\mathcal{G}_0^\sigma]^{-1}(i\omega_n) = i\omega_n + \mu - \Delta^\sigma(i\omega_n). \tag{8.15}$$

This function appears in the expression of the partition function in terms of Grassmann variables. We will use both $\mathcal{G}_0$ and Δ in the subsequent sections.

---

[2] This definition uses a sign convention different from the one used in the previous chapter.

[3] This function is sometimes called the "bath Green's function" (Georges et al., 1996). Since this name can be confusing, we avoid it in this book and instead use the term "Weiss Green's function."

8.2 Dynamical mean-field theory

Quantum impurity models are a key ingredient of the dynamical mean-field theory (DMFT), which provides an approximate description of correlated lattice models (Georges et al., 1996). The success of DMFT created a demand for more powerful and flexible impurity solvers and triggered the development of the continuous-time impurity solvers. We now briefly introduce the DMFT approximation, which maps an interacting lattice model, such as the Hubbard model, onto an effective impurity problem subject to a self-consistency condition for the bath.

8.2.1 Single-site effective model

To appreciate the basic strategy, we first recall the static mean-field approximation of the classical Ising model, illustrated in the left-hand panels of Fig. 8.2. There, we focus on one particular spin s_0 of the lattice and follow the venerable Weiss self-consistent molecular field prescription by replacing the remaining degrees of freedom by an effective external magnetic field $h_{\text{eff}} = zmJ$, where z is the coordination number and m is the magnetization per site. The lattice system with interaction J between nearest-neighbor spins,

$$H^{\text{Ising}} = -J \sum_{\langle ij \rangle} s_i s_j,$$

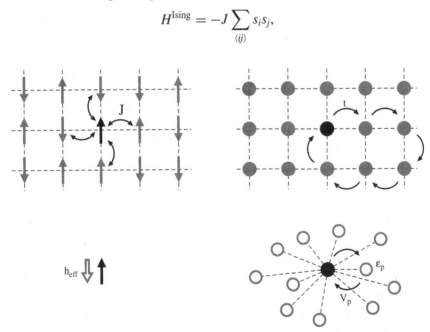

Figure 8.2 Left panels: Mapping of the classical Ising model to an effective single-site model (spin in an external magnetic field). Right panels: Mapping of the Hubbard model to an effective single-site model (one correlated site in an uncorrelated bath).

thus maps to the single-site effective model

$$H_{\text{eff}}^{\text{Ising}} = -h_{\text{eff}} s_0.$$

For this model, the magnetization is

$$m_{\text{eff}} = \tanh(\beta h_{\text{eff}}).$$

If we identify the magnetization m of the lattice problem with the magnetization m_{eff} of the single-site effective model, we obtain the self-consistency condition

$$m \equiv m_{\text{eff}} = \tanh(\beta z J m), \tag{8.16}$$

which implicitly determines the *mean field* h_{eff}. One can find the self-consistent solution by iteration.

We now turn to the Hubbard model and apply the same strategy. The model is

$$H_{\text{Hubbard}} = -t \sum_{\langle ij \rangle \sigma} (d_{i\sigma}^{\dagger} d_{j\sigma} + d_{j\sigma}^{\dagger} d_{i\sigma}) + U \sum_i n_{i\uparrow} n_{i\downarrow} - \mu \sum_{i\sigma} n_{i\sigma}$$

and describes electrons hopping between nearest neighbor sites of some lattice with amplitude t. Two electrons on the same site interact with energy U. We added a chemical potential term because we will work in the grand canonical ensemble. The noninteracting dispersion is obtained as the Fourier transform of the hopping matrix. For example, in the case of a one-dimensional lattice with lattice spacing a, $\epsilon_k = -2t \cos(ka)$.

Inspired by the Weiss molecular-field strategy, we focus on one particular site of the lattice (black dot in the right hand panels of Fig. 8.2) and replace the remaining degrees of freedom of the model by a bath of noninteracting levels and a hybridization term that connects the interacting site to the bath. The effective single-site problem thus becomes an Anderson impurity model,[4]

$$H_{\text{imp}} = \sum_{p\sigma} \epsilon_p c_{p\sigma}^{\dagger} c_{p\sigma} + \sum_{p\sigma} (V_{p\sigma} d_{\sigma}^{\dagger} c_{p\sigma} + V_{p\sigma}^* c_{p\sigma}^{\dagger} d_{\sigma}) + U n_{\uparrow} n_{\downarrow} - \mu (n_{\uparrow} + n_{\downarrow}).$$

Here, the $d^{\dagger}$ create electrons on the impurity (black dot), $n_{\sigma} = d_{\sigma}^{\dagger} d_{\sigma}$, and the $c_p^{\dagger}$ create electrons in bath states (empty dots) labeled by a quantum number p. In this effective single-site model, hoppings from the impurity into the bath and back (bottom right panel of Fig. 8.2) represent processes in the original model where an electron hops from the black site into the lattice and returns to it after some excursion through the lattice (top right panel of Fig. 8.2). The hybridization parameters V_p give the amplitudes for such transitions.

---

[4] In the DMFT context, the bath energy levels ϵ_p of the impurity model are not directly related to the dispersion of the lattice model. We will denote the latter by ϵ_k. It is important to keep this distinction in notation in mind.

Our task now is to optimize the parameters ε_p and V_p such that the bath of the Anderson impurity model mimics the lattice environment as closely as possible. We do not yet know ε_p and V_p. They are the analogs of h_{eff} and thus are the parameters we need to adjust self-consistently. If we work with the impurity action, the bath properties are encoded in $\Delta(\tau)$ or $\mathcal{G}_0(\tau)$, and these functions thus play the role of the mean field. It is a *dynamical* mean field, because the hybridization function or Weiss Green's function depends on (imaginary) time or frequency.

The self-consistent solution is constructed in such a way that the impurity Green's function $G_{\text{imp}}(i\omega_n)$ reproduces the *local* lattice Green's function $G_{\text{loc}}(i\omega_n) \equiv G_{i,i}(i\omega_n)$. In other words, if $G(k, i\omega_n)$ is the momentum-dependent lattice Green's function of the Hubbard model, we seek bath parameters and hybridizations such that[5]

$$\int (dk) G(k, i\omega_n) \equiv G_{\text{imp}}(i\omega_n). \tag{8.17}$$

8.2.2 DMFT approximation

We obtain the solution of (8.17) iteratively. However, in contrast to the Ising case (8.16), it is not immediately clear how we can use the self-consistency condition (8.17) to update the dynamical mean field. To define a practical procedure, we have to relate the left-hand side of (8.17) to impurity model quantities. This step involves, as the essential approximation of the DMFT method, a significant simplification of the momentum-dependence of the lattice self-energy.

The self-energy describes the effect of interactions on the propagation of electrons. In the noninteracting model, the lattice Green's function is $G_0(k, i\omega_n) = [i\omega_n + \mu - \epsilon_k]^{-1}$, with ϵ_k being the Fourier transform of the hopping matrix. The Green's function of the interacting model is $G(k, i\omega_n) = [i\omega_n + \mu - \epsilon_k - \Sigma(k, i\omega_n)]^{-1}$ with $\Sigma(k, i\omega_n)$ the lattice self-energy. Therefore

$$\Sigma(k, i\omega_n) = G_0^{-1}(k, i\omega_n) - G^{-1}(k, i\omega_n).$$

Similarly, we obtain the impurity self-energy

$$\Sigma_{\text{imp}}(i\omega_n) = \mathcal{G}_0^{-1}(i\omega_n) - G_{\text{imp}}^{-1}(i\omega_n),$$

with $\mathcal{G}_0^{-1}$ defined in (8.15). The DMFT approximation is the identification of the lattice self-energy with the momentum-independent impurity self-energy,

$$\Sigma(k, i\omega_n) \approx \Sigma_{\text{imp}}(i\omega_n).$$

This approximation allows us to rewrite the self-consistency equation (8.17) as

---

[5] $\int (dk)$ denotes a normalized integral over the Brillouin zone.

$$\int (dk)[i\omega_n + \mu - \epsilon_k - \Sigma_{\text{imp}}(i\omega_n)]^{-1} \equiv G_{\text{imp}}(i\omega_n). \qquad (8.18)$$

Since both $G_{\text{imp}}(i\omega_n)$ and $\Sigma_{\text{imp}}(i\omega_n)$ are determined by the impurity model parameters ε_p and V_p (or the function $\Delta(\tau)$ or $\mathcal{G}_0(\tau)$), (8.18) defines a self-consistency condition for these parameters (or functions).

8.2.3 DMFT self-consistency loop

We now formulate the self-consistency loop for the Weiss Green's function $\mathcal{G}_0(i\omega_n)$. Starting from an arbitrary initial $\mathcal{G}_0(i\omega_n)$ (e.g., the local Green's function of the noninteracting lattice model), we iterate the following steps until convergence (Algorithm 26):

1. Solve the impurity problem, that is, compute the impurity Green's function $G_{\text{imp}}(i\omega_n)$ for the given $\mathcal{G}_0(i\omega_n)$.
2. Extract the self-energy of the impurity model:
   $\Sigma_{\text{imp}}(i\omega_n) = \mathcal{G}_0^{-1}(i\omega_n) - G_{\text{imp}}^{-1}(i\omega_n)$.
3. Identify the lattice self-energy with the impurity self-energy,
   $\Sigma(k, i\omega_n) = \Sigma_{\text{imp}}(i\omega_n)$ (DMFT approximation), and compute the local lattice
   Green's function $G_{\text{loc}}(i\omega_n) = \int (dk)[i\omega_n + \mu - \epsilon_k - \Sigma_{\text{imp}}(i\omega_n)]^{-1}$.
4. Apply the DMFT self-consistency condition, $G_{\text{loc}}(i\omega_n) = G_{\text{imp}}(i\omega_n)$, and use it
   to define a new Weiss Green's function $\mathcal{G}_0^{-1}(i\omega_n) = G_{\text{loc}}^{-1}(i\omega_n) + \Sigma_{\text{imp}}(i\omega_n)$.

The computationally expensive step is the solution of the impurity problem (step 1). When the loop converges, the bath contains information about the topology of the lattice (through the density of states) and about the phase (metal, Mott insulator, antiferromagnetic insulator, etc.). The impurity, which exchanges

---

Algorithm 26 DMFT self-consistency loop. (Using Eq. (8.15), the loop can also be formulated in terms of the hybridization function Δ, instead of the Weiss Green's function $\mathcal{G}_0$.)

Input: Some initial guess for the Weiss Green's function $\mathcal{G}_0(i\omega_n)$, noninteracting dispersion ϵ_k, chemical potential μ, inverse temperature β.

repeat

    Compute $G_{\text{imp}}(i\omega_n)$ for given $\mathcal{G}_0(i\omega_n)$;   ▷ Solve impurity problem

    $\Sigma_{\text{imp}}(i\omega_n) = \mathcal{G}_0^{-1}(i\omega_n) - G_{\text{imp}}^{-1}(i\omega_n)$;

    $G_{\text{loc}}(i\omega_n) = \int (dk)[i\omega_n + \mu - \epsilon_k - \Sigma_{\text{imp}}(i\omega_n)]^{-1}$;

    $\mathcal{G}_0^{-1}(i\omega_n) \leftarrow G_{\text{loc}}^{-1}(i\omega_n) + \Sigma_{\text{imp}}(i\omega_n)$;

until G_{loc} is converged.

return G_{loc}.

---

electrons with the bath, thus behaves, at least to some extent, as if it were a site of the lattice.

Obviously, a single-site impurity model does not capture all the physics. In particular, the DMFT approximation neglects all spatial fluctuations. These fluctuations are important, for example, in low-dimensional systems. The DMFT formalism is believed to provide a qualitatively correct description of three-dimensional unfrustrated lattice models. It becomes exact in the limit of infinite dimension (Metzner and Vollhardt, 1989) or infinite coordination number (where spatial fluctuations are negligible), in the noninteracting limit ($U = 0$ implies $\Sigma = 0$), and in the atomic limit ($t = 0$ implies $\Delta = 0$).

8.2.4 Simulation of strongly correlated materials

The DMFT formalism describes band-like behavior (renormalized quasi-particle bands) *and* atomic-like behavior (Hubbard bands). It captures the competition between electron localization and delocalization, which plays a crucial role in the physics of strongly correlated materials. One possible approach to simulate real compounds is to combine the DMFT formalism with electronic structure calculations in the local density approximation (LDA). The resulting formalism is called "LDA+DMFT" (Kotliar et al., 2006).

The idea is to use the Kohn-Sham eigenvalues $\epsilon_{n,k}^{KS}$, obtained from an LDA calculation, in the self-consistency equation (8.18). To describe the strong correlation effects experienced by the d-electrons (e.g., in a transition metal oxide) or the f-electrons (e.g., in an actinide), we add a frequency-dependent, but local self-energy to the corresponding orbitals. This self-energy is obtained from a self-consistently defined impurity problem.

An important issue in practical implementations of the LDA+DMFT scheme is the choice of orbitals. The LDA calculation yields the extended Kohn-Sham wave functions $\psi_{n,k}$. The impurity model, on the other hand, describes the interactions between electrons that occupy localized, atomic-like orbitals. One common choice of orbitals is a Wannier basis, for example, the *maximally localized Wannier orbitals* (Marzari and Vanderbilt, 1997) that minimize the expectation value of the operator r^2. In this basis $\{\phi_l\}$, we identify the subset $\{\phi_\alpha\}$ of "strongly correlated" orbitals with d- or f-character and then define an interaction term

$$H_U = \frac{1}{2} \sum_{i\alpha\beta\gamma\delta} U^{\alpha\beta\gamma\delta} d_{i\alpha}^\dagger d_{i\beta}^\dagger d_{i\gamma} d_{i\delta}.$$

The interaction parameters $U^{\alpha\beta\gamma\delta}$ are either treated as adjustable parameters or extracted from the LDA band structure using techniques called "constrained LDA" or "constrained RPA."[6] Let us emphasize that the interaction parameters depend on the choice of orbitals. More localized orbitals result in larger interaction terms. In a Wannier-orbital basis, this also means that the number of (strongly and weakly correlated) bands kept in the simulation affects the interaction parameters, because a larger number of bands implies more localized and atomic-like orbitals.

The Kohn-Sham Hamiltonian

$$H^{KS} = \sum_{klm} (h_k^{KS})_{lm} d_{kl}^\dagger d_{km},$$

which is diagonal in the band basis, becomes a nontrivial matrix in the localized orbital basis $\{\phi_l\}$. Explicitly, $h_k^{KS} = O_k \epsilon_k^{KS} O_k^{-1}$, where O_k is the matrix describing the basis transformation from $\{\psi_{n,k}\}$ to $\{\phi_l\}$. By combining H^{KS}, the local interaction term H_U, and a "double counting term"

$$H_{DC} = -\sum_{i\alpha} e_{DC}^\alpha d_{i\alpha}^\dagger d_{i\alpha},$$

we obtain the model to be solved within the DMFT approximation:

$$H = H^{KS} + H_U + H_{DC}. \tag{8.19}$$

We need the double counting correction for the following reason: While density functional theory (on which the LDA approximation is based) does not take into account all electronic correlation effects, it does capture some of them. If we then explicitly describe the local interactions in the strongly correlated orbitals via the H_U-term, some interaction contributions appear twice. The double counting term is supposed to compensate for this by shifting the correlated orbitals by e_{DC}.

Currently, there is no clean and consistent solution to the double counting problem. The exchange-and-correlation potential, which describes the correlation effects within density functional theory, is (within the LDA) a *nonlinear* functional

---

[6] In the *constrained LDA* technique (Anisimov and Gunnarsson, 1991), we calculate the interaction as the derivative of the energy with respect to the charge. This approach requires different LDA calculations, in which the charges in the orbitals are fixed at different values. In the *constrained RPA* technique (Aryasetiawan et al., 2004), on the other hand, we compute polarization functions using the random phase approximation (RPA), that is, by summing bubble diagrams made of LDA propagators. The total polarization P is split into the polarization P_d associated with transitions between "strongly correlated" bands, and the remaining polarization $P_r = P - P_d$. We then define a *partially screened* interaction W_r, which has the property that if it is screened further by the polarization P_d, it gives the fully screened interaction $W = v/(1 - Pv)$ (v here is the bare Coulomb interaction): $W = W_r/(1 - P_d W_r)$. One finds that $W_r = v/(1 - P_r v)$. The interaction parameters $U^{\alpha\beta\gamma\delta}$ are then defined as the matrix elements of W_r in the localized basis $\{\phi_\alpha\}$. We note that the polarizations are *frequency dependent*, and thus the interaction parameters $U^{\alpha\beta\gamma\delta}$ are also frequency dependent. This result makes physical sense: At high frequencies (above the plasmon frequency in a metal) screening is not effective, and the interaction is essentially the bare Coulomb interaction v (typically about 20 eV in transition metal compounds), while slow charge fluctuations are screened by the surrounding electrons, resulting in a static interaction of typically only a few eV.

of the *total* density, so we cannot determine the contribution of the d- or f-electrons to the exchange-and-correlation energy. In practice, we invoke double counting terms such as $e_{DC}^{\alpha} = \langle U \rangle (\langle n_{\text{corr}} \rangle - \frac{1}{2})$,[7] with $\langle U \rangle$ being the average of the interaction parameters and $\langle n_{\text{corr}} \rangle$ being the average occupancy of the correlated orbitals. Such an orbital-independent shift preserves the crystal-field splittings in the LDA band structure.

Finally, we add a chemical potential term and adjust the chemical potential μ to obtain the correct total number of electrons in the correlated and uncorrelated orbitals.

To compute the (local) self-energy within the DMFT approximation, we solve an impurity model with interaction terms identical to those in H_U. This solution yields the matrices $[G_{\text{imp}}]_{\alpha\beta}$ and $[\Sigma_{\text{imp}}]_{\alpha\beta}$, which are defined in the subspace of correlated orbitals. To write the self-consistency condition, we define the matrix Σ in the full space of localized orbitals $\{\phi_l\}$:

$$\Sigma(i\omega_n) = \left(\begin{array}{c|c} \Sigma_{\text{imp}}(i\omega_n) & 0 \\ \hline 0 & 0 \end{array} \right), \qquad (8.20)$$

where the first diagonal block corresponds to the strongly correlated orbitals (five, in the case of a full d-shell) and the second diagonal block to the weakly correlated ones. Similarly, the double counting term is a (diagonal) matrix acting on the full space:

$$E_{DC} = \left(\begin{array}{c|c} -e_{DC} & 0 \\ \hline 0 & 0 \end{array} \right). \qquad (8.21)$$

The self-consistency condition, which fixes the dynamical mean field $\mathcal{G}_0$ or Δ, becomes

$$\left[\int (dk) \left[(i\omega_n + \mu)I - h_k^{\text{KS}} - \Sigma(i\omega_n) - E_{DC} \right]^{-1} \right]_{\alpha\beta} \equiv [G_{\text{imp}}(i\omega_n)]_{\alpha\beta}. \qquad (8.22)$$

Note that while the self-consistency condition involves only the correlated block of the local lattice Green's function, the weakly correlated orbitals enter the calculation through the matrix inversion. We summarize the computational scheme for the LDA+DMFT calculation in Algorithm 27.

---

[7] This double counting prescription follows from a mean-field estimate for the interaction energy: $E_{\text{corr}} \approx \frac{1}{2}\langle U \rangle \langle n_{\text{corr}} \rangle (\langle n_{\text{corr}} \rangle - 1)$. For a more detailed discussion of the double counting issue, see Anisimov et al. (1991); Aichhorn et al. (2011); Haule (2015).

Algorithm 27 LDA+DMFT self-consistency loop.

Input: Some ($N_{corr} \times N_{corr}$) Weiss Green's function matrix $\mathcal{G}_0(i\omega_n)$, Kohn-Sham Hamiltonian h_k^{KS} in a localized basis ($N_{tot} \times N_{tot}$), total number of electrons n_{tot}^{target}, some initial guess for the chemical potential μ, inverse temperature β.

repeat

   repeat

      Compute $G_{imp}(i\omega_n)$ for given $\mathcal{G}_0(i\omega_n)$;   ▷ Solve impurity problem

      $\Sigma_{imp}(i\omega_n) = \mathcal{G}_0^{-1}(i\omega_n) - G_{imp}^{-1}(i\omega_n)$;

      Define the $N_{tot} \times N_{tot}$ matrices Σ (8.20) and E_{DC} (8.21) ;

      $G_{loc}(i\omega_n) = \int (dk)[(i\omega_n + \mu)I - h_k^{KS} - \Sigma(i\omega_n) - E_{DC}]^{-1}$;

      Extract $N_{corr} \times N_{corr}$ correlated block $G_{loc}^{corr}(i\omega_n)$ from $G_{loc}(i\omega_n)$;

      $\mathcal{G}_0^{-1}(i\omega_n) \leftarrow (G_{loc}^{corr}(i\omega_n))^{-1} + \Sigma_{imp}(i\omega_n)$;

   until G_{loc} is converged.

   Compute $n_{tot} = - \sum_{i=1}^{N_{tot}} [G_{loc}(\beta_-)]_{ii}$;

   if $n_{tot} < n_{tot}^{target}$ **then**

      Increase μ ;

   else

      Decrease μ ;

   end if

until n_{tot} is converged.

return G_{loc}.

8.2.5 Cluster extensions

To capture the effect of short-range spatial fluctuations, cluster extensions of dynamical mean field theory were developed (Maier et al., 2005). In these extensions, a cluster of several sites, instead of a single site, is embedded in a self-consistently determined bath. This embedding allows us to describe the short-range spatial correlations on the cluster explicitly, while treating the longer range correlations on a mean-field level. This approach is the analog of the Bethe-Peierls cluster molecular-field approximation (Bethe, 1935), which extended the Weiss molecular field theory for classical spin models from a single site to a cluster of sites to capture short-range spin correlations.

In a cluster DMFT, we divide the lattice into a superlattice of clusters containing n_c sites and apply the DMFT procedure to the superlattice (Lichtenstein and Katsnelson, 2000). The cluster Green's functions and self-energies are now matrices of size $n_c \times n_c$, while ϵ_k becomes a matrix $\hat{t}(k)$, defined as the Fourier transform of the hopping matrix on the superlattice. The momenta k are those of the reduced Brillouin zone of the superlattice.

Figure 8.3 Decomposition of the one-dimensional Hubbard chain into two-site clusters (dashed boxes).

To be more specific, let us consider a one-dimensional Hubbard chain, with lattice spacing a, that we decompose into two-site clusters (Fig. 8.3). The hopping matrix then has the form

$$\begin{pmatrix} \ddots & -t & & & \\ -t & 0 & -t & & \\ & -t & 0 & -t & \\ & & -t & 0 & -t \\ & & & -t & 0 & -t \\ & & & & -t & \ddots \end{pmatrix},$$

which after Fourier transformation on the superlattice with spacing $2a$ becomes

$$-\hat{t}(k) = e^{ik0}\begin{pmatrix} 0 & t \\ t & 0 \end{pmatrix} + e^{ik(2a)}\begin{pmatrix} 0 & 0 \\ t & 0 \end{pmatrix} + e^{ik(-2a)}\begin{pmatrix} 0 & t \\ 0 & 0 \end{pmatrix}$$

$$= \begin{pmatrix} 0 & t(1 + e^{-i2ka}) \\ t(1 + e^{i2ka}) & 0 \end{pmatrix}. \tag{8.23}$$

The self-consistency condition, which fixes the 2×2 matrix of the dynamical mean field $\mathcal{G}_0$ or Δ, is now

$$\int_{\text{reduced BZ}} (dk)[(i\omega_n + \mu)\hat{I} - \hat{t}(k) - \hat{\Sigma}_{\text{imp}}(i\omega_n)]^{-1} = \hat{G}_{\text{imp}},$$

with the reduced Brillouin zone $-\pi/2a \leq k < \pi/2a$.

As is evident from the hopping matrix (8.23), the cluster DMFT formalism breaks translational invariance within the cluster. There is an alternative cluster DMFT, called the *dynamical cluster approximation* (DCA), which enforces this symmetry (Hettler et al., 1998). The two-site DCA corresponds to the hopping matrix

$$-\hat{t}_{\text{DCA}}(k) = \begin{pmatrix} 0 & 2t\cos(ka) \\ 2t\cos(ka) & 0 \end{pmatrix}. \tag{8.24}$$

In the "momentum basis," $\{d_{K=0} = \frac{1}{\sqrt{2}}(d_1 + d_2), d_{K=\frac{\pi}{a}} = \frac{1}{\sqrt{2}}(d_1 - d_2)\}$, the Green's functions and self-energies are diagonal matrices, and the self-consistency condition for each "momentum sector" K is

$$\int_{\text{sector } K} (dk)[i\omega_n + \mu - \epsilon_k - \Sigma_{\text{imp}}^K(i\omega_n)]^{-1} = G_{\text{imp}}^K, \tag{8.25}$$

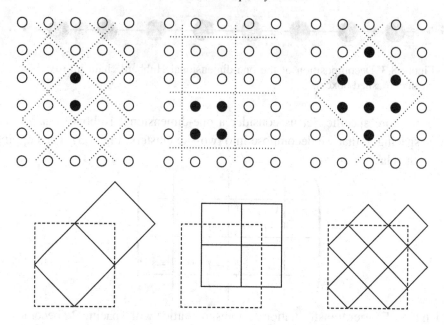

Figure 8.4 Two-, four-, and eight-site DCA approximation for the two-dimensional Hubbard model. The top panels show the (periodized) real-space clusters, and the bottom panels illustrate the corresponding decompositions of the first Brillouin zone (dashed square) into equal-sized sectors.

with $\epsilon_k = -2t\cos(ka)$. In our two-site example, sector $K = 0$ corresponds to $-\pi/2a \leq k < \pi/2a$, while sector $K = \pi/a$ corresponds to $\pi/2a \leq k < 3\pi/2a$. The first Brillouin zone thus decomposes into two sectors of equal size, centered at the reciprocal lattice vectors of the periodized two-site cluster. Figure 8.4 illustrates how the DCA concept extends to the two-dimensional square lattice. The top panels show the (periodized) real-space clusters with two, four, and eight sites and the bottom panels the corresponding tiling of the first Brillouin zone.[8]

The periodized real-space clusters of the DCA impurity model involve renormalized hoppings, which can be obtained as the Fourier transform of the patch-averaged dispersion (see Exercise 8.8).

8.3 General strategy

Quantum impurity models are (0+1)-dimensional quantum field theories and as such are computationally much more tractable than interacting lattice models. Our objective is computing the impurity Green's function[9]

---

[8] In DCA, the matrix $[\mathcal{G}_0]_{ij}$ connects all pairs of sites, and it is translation invariant within the cluster. The equivalent pairs of sites can be identified by considering the superlattice of clusters, indicated by the dotted lines in Fig. 8.4.

[9] The sign convention here differs from the one used in the previous chapter.

$$G(\tau) = -\langle \mathcal{T} d(\tau) d^\dagger(0) \rangle = -\frac{1}{Z} \text{Tr}\left[e^{-(\beta-\tau)H} d e^{-\tau H} d^\dagger \right], \tag{8.26}$$

where $Z = \text{Tr}[e^{-\beta H}]$ is the impurity model partition function, β is the inverse temperature, $\mathcal{T}$ is the (imaginary) time-ordering operator, and $\text{Tr} = \text{Tr}_d \text{Tr}_c$ is the trace over the impurity and bath states. In the last expression we assumed that $0 \le \tau < \beta$.

Continuous-time Monte Carlo algorithms expand the partition function into a series of "diagrams" and stochastically sample these diagrams (Gull et al., 2011). Following the general procedure outlined in Chapter 2, we represent the partition function as a sum (or more precisely as an integral) over configurations C with weight w_C,

$$Z = \sum_C w_C, \tag{8.27}$$

and implement a random walk $C_1 \to C_2 \to C_3 \to \cdots$ in configuration space in such a way that ergodicity and detailed balance are satisfied. Using sign-weighted averages (Section 5.4), we estimate the impurity Green's function from a finite number M of measurements as

$$G = \sum_C \frac{w_C G_C}{Z} = \frac{\sum_C |w_C| \text{sign}_C G_C}{\sum_C |w_C| \text{sign}_C} \approx \frac{\sum_{i=1}^M \text{sign}_{C_i} G_{C_i}}{\sum_{i=1}^M \text{sign}_{C_i}} \equiv \frac{\langle \text{sign} \cdot G \rangle_{\text{MC}}}{\langle \text{sign} \rangle_{\text{MC}}}. \tag{8.28}$$

Our first step toward detailing the continuous-time solvers is to express the partition function as an imaginary-time-ordered exponential in an *interaction representation*. To do this, we split the Hamiltonian into two parts, $H = H_1 + H_2$, and define the imaginary-time-dependent operators in the interaction representation as $O(\tau) = e^{\tau H_1} O e^{-\tau H_1}$. In this representation, the partition function becomes $Z = \text{Tr}\left[e^{-\beta H_1} \mathcal{T} e^{-\int_0^\beta d\tau H_2(\tau)} \right].$[10]

Next, we expand the time-ordered exponential into a power series,

$$Z = \sum_{n=0}^\infty \int_0^\beta d\tau_1 \cdots \int_{\tau_{n-1}}^\beta d\tau_n \text{Tr}\left[e^{-(\beta-\tau_n)H_1} (-H_2) \cdots e^{-(\tau_2-\tau_1)H_1} (-H_2) e^{-\tau_1 H_1} \right].$$
$$\tag{8.29}$$

We now have a representation of the partition function of the form (8.27), namely, as an infinite sum over the weights of certain configurations. The configurations are collections of time points on the imaginary-time interval: $C = \{\tau_1, \ldots, \tau_n\}$, $n = 0, 1, \ldots$, where we assume the imaginary-time ordering $\tau_i < \tau_{i+1}$ and the restriction $\tau_i \in [0, \beta)$. In contrast to the sampling we discussed in

---

[10] We can understand this formula by defining the operator $A(\beta) = e^{\beta H_1} e^{-\beta H}$ and writing the partition function as $Z = \text{Tr}[e^{-\beta H_1} A(\beta)]$. The operator $A(\beta)$ satisfies $dA/d\beta = e^{\beta H_1}(H_1 - H)e^{-\beta H} = -H_2(\beta)A(\beta)$, the solution of which is $A(\beta) = \mathcal{T} \exp\left[-\int_0^\beta d\tau H_2(\tau) \right]$.

Chapter 7, where each configuration had a fixed number of local variables, here the number of time points in a configuration varies, reflecting the sampling of different orders in the power series. The expression for the Monte Carlo weights is

$$w_C = \text{Tr}\Big[e^{-(\beta-\tau_n)H_1}(-H_2)\cdots e^{-(\tau_2-\tau_1)H_1}(-H_2)e^{-\tau_1 H_1}\Big](d\tau)^n. \tag{8.30}$$

There are two complementary continuous-time Monte Carlo techniques: (i) the *weak-coupling approach*, which scales favorably with system size (that is, the number of correlated sites or orbitals in the impurity model) and allows the efficient simulation of relatively large impurity clusters with simple interactions, and (ii) the *strong-coupling approach*, which can handle impurity models with strong interactions among multiple orbitals. For simplicity, we continue to focus on the single-orbital Anderson impurity model defined in (8.8)–(8.11). In this case, the weak-coupling continuous-time Monte Carlo approach expands Z in powers of the interaction U in an interaction representation where the imaginary-time evolution is determined by the *quadratic* part $H_\mu + H_{\text{bath}} + H_{\text{mix}}$ of the Hamiltonian. The complementary strong-coupling approach expands Z in powers of the impurity-bath hybridization term H_{mix} in an interaction representation where the imaginary-time evolution is determined by the *local* part $H_\mu + H_U + H_{\text{bath}}$ of the Hamiltonian. The details of how we sample the weights (8.30) and how we measure observables depend on which continuous-time approach we are using.

8.4 Weak-coupling approach

The weak-coupling continuous-time impurity solver (Rubtsov et al., 2005) expands the partition function in powers of $H_2 = H_U$.[11] Equation (8.30) then gives the weight of a configuration of n *interaction vertices*. Since $H_1 = H - H_2 = H_\mu + H_{\text{bath}} + H_{\text{mix}}$ is quadratic, we use Wick's theorem to evaluate the trace. The result is a product of two determinants of $n \times n$ matrices (one for each electron spin). The elements of these matrices are the Weiss Green's functions $\mathcal{G}_0^\sigma$ for the time intervals defined by the vertex positions:

$$\frac{w_C}{Z_0} = (-U d\tau)^n \frac{1}{Z_0}\text{Tr}\Big[e^{-(\beta-\tau_n)H_1} n_\uparrow n_\downarrow \cdots e^{-(\tau_2-\tau_1)H_1} n_\uparrow n_\downarrow e^{-\tau_1 H_1}\Big]$$

$$= (-U d\tau)^n \prod_\sigma \det M_\sigma^{-1},$$

where

$$[M_\sigma^{-1}]_{ij} = \mathcal{G}_0^\sigma(\tau_i - \tau_j),$$

---

[11] A related algorithm, based on an expansion in powers of $H_U - K/\beta$ (with K some nonzero constant), is the *continuous-time auxiliary field method* discussed in Appendix L.

$\mathcal{G}_0^\sigma(\tau) = -\text{Tr}[e^{-\beta H_1}\mathcal{T}d(\tau)d^\dagger(0)]/Z_0$, and $Z_0 = \text{Tr}[e^{-\beta H_1}]$ is the partition function of the noninteracting model.[12] For the diagonal elements, we adopt the convention $[M_\sigma^{-1}]_{ii} = \mathcal{G}_0^\sigma(0^-)$.

At this point, we notice a potential sign problem. In the paramagnetic phase, where $\mathcal{G}_0^\uparrow = \mathcal{G}_0^\downarrow$, the product of determinants is positive, which means that for a repulsive interaction $(U > 0)$ odd perturbation orders yield negative weights. Except in the particle-hole symmetric case, where odd perturbation orders vanish, these odd order configurations ostensibly cause a sign problem. Fortunately, we can solve this sign problem by shifting the chemical potentials for up- and down-spins in an appropriate way. To do so, we rewrite the interaction term as (Assaad and Lang, 2007)

$$H_U = \frac{U}{2}\sum_s \prod_\sigma (n_\sigma - \alpha_\sigma(s)) + \frac{U}{2}(n_\uparrow + n_\downarrow) + U\left[\left(\frac{1}{2} + \delta\right)^2 - \frac{1}{4}\right], \quad (8.31)$$

with

$$\alpha_\sigma(s) = \frac{1}{2} + \sigma s\left(\frac{1}{2} + \delta\right). \quad (8.32)$$

Here, δ is some constant and $s = \pm 1$ is an auxiliary Ising variable. This construction is not a Hubbard-Stratonovich transformation, but simply a shift in the zero of energy. The constant $U[(\frac{1}{2} + \delta)^2 - \frac{1}{4}]$ in (8.31) is irrelevant and will be ignored in the following. We absorb the contribution $\frac{1}{2}U(n_\uparrow + n_\downarrow)$ into the noninteracting Green's function by shifting the chemical potential as $\mu \to \mu - \frac{1}{2}U$. Explicitly, we redefine the Weiss Green's function as[13]

$$[\mathcal{G}_0^\sigma]^{-1} = i\omega_n + \mu - \Delta^\sigma \to [\tilde{\mathcal{G}}_0^\sigma]^{-1} = i\omega_n + \mu - \frac{1}{2}U - \Delta^\sigma.$$

The introduction of an Ising variable s_i at each vertex position τ_i enlarges the configuration space exponentially. A configuration C now corresponds to a collection of auxiliary spin variables defined on the imaginary-time interval: $C = \{(\tau_1, s_1), (\tau_2, s_2), \ldots, (\tau_n, s_n)\}$. The probability of these configurations is

$$w_C = \tilde{Z}_0(-Ud\tau/2)^n \prod_\sigma \det \tilde{M}_\sigma^{-1}, \quad (8.33)$$

where

$$[\tilde{M}_\sigma^{-1}]_{ij} = \tilde{\mathcal{G}}_0^\sigma(\tau_i - \tau_j) - \alpha_\sigma(s_i)\delta_{ij}. \quad (8.34)$$

---

[12] We note that in the DMFT framework discussed in Section 8.2.3, the function $\mathcal{G}_0^\sigma$ is determined directly by the self-consistency loop, without reference to a Hamiltonian. For the purpose of the present discussion, we may, however, assume that we know H_{bath} and H_{mix} terms whose parameters yield $\mathcal{G}_0^\sigma$ through (8.14) and (8.15).

[13] In a DMFT calculation, this means that the shifted chemical potential is used within the self-consistency loop.

In Section 8.7 we show that for $\delta \geq 0$ all Monte Carlo configurations for the Anderson impurity model have positive weights.

8.4.1 Sampling

For ergodicity it is sufficient that the sampling inserts the auxiliary spins with random orientation at random times and removes randomly chosen spins. By adopting the Metropolis-Hastings algorithm (Section 2.5.2) to generate the random walk in configuration space, we have an acceptance matrix

$$A(C' \leftarrow C) = \min[1, \mathcal{R}(C' \leftarrow C)],$$

where

$$\mathcal{R}(C' \leftarrow C) = \frac{w(C')T(C \leftarrow C')}{w(C)T(C' \leftarrow C)}$$

and $T(C' \leftarrow C)$ denotes the proposal probability for the move from C to C'. We use (8.33) to compute the ratio of the weights. To complete the description of the sampling we need to specify proposal probabilities for the insertion and removal of an auxiliary spin. There is some flexibility in choosing them. We illustrate reasonable choices in Fig. 8.5. For the insertion, we pick a random time in $[0, \beta)$ and a random orientation for the new spin, while for the removal, we simply pick a random spin. The corresponding proposal probabilities are

$$T(n+1 \leftarrow n) = \tfrac{1}{2}(d\tau/\beta), \quad T(n \leftarrow n+1) = 1/(n+1). \tag{8.35}$$

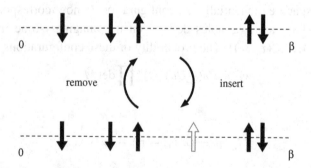

Figure 8.5 Local update in the weak-coupling method. The dashed line represents the imaginary-time interval $[0, \beta)$. We increase the perturbation order by adding an auxiliary spin with random orientation at a random time and decrease it by removing a randomly chosen auxiliary spin.

The first step is choosing with equal probability whether we insert or remove. If we insert, then we are going from a configuration with n spins to a configuration with $n + 1$ spins, and from (8.33) and the above choices for T, the acceptance matrix becomes $A(n + 1 \leftarrow n) = \min[1, \mathcal{R}_{\text{insert}}(n + 1 \leftarrow n)]$ with

$$\mathcal{R}_{\text{insert}}(n + 1 \leftarrow n) = \frac{-\beta U}{n + 1} \prod_\sigma \frac{\det[\tilde{M}_\sigma^{(n+1)}]^{-1}}{\det[\tilde{M}_\sigma^{(n)}]^{-1}}. \tag{8.36}$$

The acceptance probability for the removal follows from

$$\mathcal{R}_{\text{remove}}(n \leftarrow n + 1) = 1/\mathcal{R}_{\text{insert}}(n + 1 \leftarrow n). \tag{8.37}$$

8.4.2 Determinant ratios and fast matrix updates

From (8.36), we see that each update requires the calculation of a ratio of two determinants. We encountered a similar problem in the determinant method: Computing the determinant of a matrix of size $n \times n$ is an $O(n^3)$ operation. However, each insertion or removal of a vertex (or spin) merely changes one row and one column of the matrix M_σ^{-1} (or $\tilde{M}_\sigma^{-1}$).[14] As in the case of the determinant method, we use a result from linear algebra to evaluate this ratio in a time $O(n^2)$ for insertion and $O(1)$ for removal. The trick here differs only slightly from the one used in the determinant method.

To explain it, we first note that the objects we store and manipulate, besides the lists of the times $\{\tau_i\}$ (or times and spins $\{(\tau_i, s_i)\}$), are the matrices $M_\sigma = [\mathcal{G}_0^\sigma]^{-1}$, not $M_\sigma^{-1} = [\mathcal{G}_0^\sigma]$. Inserting a vertex (or auxiliary spin) adds a new row and column to M_σ^{-1}. We imagine inserting this row and column on the border of the given matrix and write the resulting matrix in a block matrix form (omitting the σ index for simplicity):

$$[M^{(n+1)}]^{-1} = \begin{pmatrix} [M^{(n)}]^{-1} & Q \\ R & S \end{pmatrix}.$$

We furthermore define the analogous blocks of the M matrix as

$$M^{(n+1)} = \begin{pmatrix} \tilde{P} & \tilde{Q} \\ \tilde{R} & \tilde{S} \end{pmatrix}. \tag{8.38}$$

Here Q, R, and S are $n \times 1$, $1 \times n$, and 1×1 matrices that contain the functions $\mathcal{G}_0$ evaluated at time intervals determined by the position of the new vertex (spin). They can be easily computed. We want to find $\tilde{P}$, $\tilde{Q}$, $\tilde{R}$, $\tilde{S}$, and the ratio of determinants.

---

[14] In the following, we write the formulas without the tildes, that is, for the sampling of interaction vertices. For the algorithm with auxiliary spins, it suffices to replace $M \rightarrow \tilde{M}$ and $\mathcal{G}_0 \rightarrow \tilde{\mathcal{G}}_0$.

To do this we use the expression (7.49) for the block inversion of a matrix and (7.50) for the determinant of a block matrix. With the help of these formulas we can show that the determinant ratio needed for the acceptance probability is

$$\frac{\det[M^{(n+1)}]^{-1}}{\det[M^{(n)}]^{-1}} = \det(S - RM^{(n)}Q) = S - RM^{(n)}Q. \tag{8.39}$$

Because we store $M^{(n)}$, computing the acceptance probability of an insertion move is just an $\mathcal{O}(n^2)$ operation. If the move is accepted, we compute the new matrix $M^{(n+1)}$ from $M^{(n)}$, Q, R, and S, also in a time $\mathcal{O}(n^2)$:

$$\tilde{S} = (S - [R][M^{(n)}Q])^{-1}, \tag{8.40}$$

$$\tilde{Q} = -[M^{(n)}Q]\,\tilde{S}, \tag{8.41}$$

$$\tilde{R} = -\tilde{S}[RM^{(n)}], \tag{8.42}$$

$$\tilde{P} = M^{(n)} + [M^{(n)}Q]\,\tilde{S}[RM^{(n)}]. \tag{8.43}$$

In the case of removing a spin we imagine removing a bordering row and column. It follows from (8.39) and (8.40) that

$$\frac{\det[M^{(n)}]^{-1}}{\det[M^{(n+1)}]^{-1}} = \det \tilde{S} = \tilde{S}. \tag{8.44}$$

$\tilde{S}$ is just a 1×1 matrix, so its determinant is trivial to compute. The above formulas also imply that the elements of the reduced matrix are

$$M^{(n)} = \tilde{P} - [\tilde{Q}][\tilde{R}]/\tilde{S}. \tag{8.45}$$

The calculation of the removal probability is thus $\mathcal{O}(1)$, while the calculation of the new $M^{(n)}$ matrix is $\mathcal{O}(n^2)$.

Algorithm 28 describes the elementary updates in the weak-coupling approach, namely, the insertion and removal of a spin, and shows how the fast matrix update formulas are used within these procedures.

8.4.3 Measurement of the Green's function

To compute the contribution of a configuration C to the Green's function, $G_C^\sigma(\tau)$, we insert in the right-hand side of (8.30) a creation operator $d^\dagger$ at time 0 and an annihilation operator d at time τ and divide by w_C. Wick's theorem and (8.39) then lead to the expression (Rubtsov et al., 2005)

$$G_C^\sigma(\tau) = \mathcal{G}_0^\sigma(\tau) - \sum_k \mathcal{G}_0^\sigma(\tau - \tau_k) \sum_l [M_\sigma]_{kl} \mathcal{G}_0^\sigma(\tau_l). \tag{8.46}$$

Our estimate for the impurity Green's function for a given imaginary time then follows from (8.28). To avoid unnecessary and time-consuming summations during

---

Algorithm 28 Vertex (spin) insertion/removal in the weak-coupling continuous-time impurity solver.

Input: Time-ordered spin configuration $C = \{(\tau_1, s_1), \ldots, (\tau_n, s_n)\}$, inverse Weiss Green's function matrix M, Weiss Green's function $\mathcal{G}_0$, interaction U, inverse temperature β.

Draw a uniform random number $\zeta \in [0, 1]$;

if $(\zeta < 0.5)$ **then**

    ▷ Try to insert a spin

  Randomly choose spin orientation s ;

  Randomly choose spin position $\tau \in [0, \beta)$;

  Compute acceptance probability $A(n + 1 \leftarrow n)$ using (8.36) and (8.39) ;

  Draw a uniform random number $\zeta' \in [0, 1]$;

  if $(\zeta' < A(n + 1 \leftarrow n))$ **then**

    Insert (τ, s) into C ;

    Update M using $\mathcal{G}_0$ and (8.40)–(8.43) ;

  end if

else if $(n > 0)$ **then**

    ▷ Try to remove a spin

  Randomly choose one of the spins in C ;

  Compute acceptance probability $A(n - 1 \leftarrow n)$ using (8.37) and (8.44) ;

  Draw a uniform random number $\zeta' \subset [0, 1]$;

  if $(\zeta' < A(n - 1 \leftarrow n))$ **then**

    Remove the chosen spin from C ;

    Update M using (8.45).

  end if

end if

return the updated C and M.

---

the Monte Carlo simulation (evaluation of (8.46) for many τ-values), we accumulate the quantity (Gull et al., 2008)

$$S_\sigma(\tilde{\tau}) \equiv \sum_k \delta(\tilde{\tau} - \tau_k) \sum_l [M_\sigma]_{kl} \mathcal{G}_0^\sigma(\tau_l),$$

by binning the time points $\tilde{\tau}$ on a fine grid. After the simulation is finished, we compute the Green's function as[15]

$$G^\sigma(\tau) = \mathcal{G}_0^\sigma(\tau) - \int_0^\beta d\tilde{\tau} \mathcal{G}_0^\sigma(\tau - \tilde{\tau})\langle S_\sigma(\tilde{\tau})\rangle_{\text{MC}}. \tag{8.47}$$

---

[15] Comparison of this equation with the impurity Dyson equation $G = \mathcal{G}_0 + \mathcal{G}_0 \star \Sigma \star G$ (where the $\star$ symbol denotes a convolution in imaginary time) shows that his procedure amounts to measuring $-\Sigma \star G$.

It is also possible to measure the Matsubara components of the Green's function directly. Using the imaginary-time translational invariance of the Green's functions, we obtain

$$G_C^\sigma(i\omega_n) = \mathcal{G}_0^\sigma(i\omega_n) - \mathcal{G}_0^\sigma(i\omega_n) \sum_{kl} \frac{1}{\beta} e^{i\omega_n(\tau_k-\tau_l)} [M_\sigma]_{kl} \mathcal{G}_0^\sigma(i\omega_n),$$

so that

$$G^\sigma(i\omega_n) = \mathcal{G}_0^\sigma(i\omega_n) - \frac{1}{\beta} (\mathcal{G}_0^\sigma(i\omega_n))^2 \left\langle \sum_{kl} e^{i\omega_n(\tau_k-\tau_l)} [M_\sigma]_{kl} \right\rangle_{MC}. \tag{8.48}$$

We note that because the Weiss Green's function has the high-frequency behavior $\mathcal{G}_0(i\omega_n) \sim 1/i\omega_n$, the measured impurity Green's function automatically inherits the correct high-frequency tail.

8.4.4 Multi-orbital and cluster impurity problems

The generalization of the weak-coupling method to impurity clusters is straightforward. All we have to do is to add a site index to the interaction vertices (or auxiliary Ising spin variables) and sample the vertices (auxiliary spins) on a family of n_{sites} imaginary-time intervals. In principle, we could even use the weak-coupling solver to simulate lattice models by making the cluster size the size of the lattice.[16] However, as discussed in Section 7.3, the $\mathcal{O}(n_{\text{sites}}^3 \beta^3)$ scaling of the computational effort is not competitive with the $\mathcal{O}(n_{\text{sites}}^3 \beta)$ scaling of the BSS determinant method.[17]

General four-Fermion terms as in (8.4) are, at least in principle, also easily dealt with. We simply expand the partition function in powers of the interactions U^{abcd}. The trace over the impurity and bath degrees of freedom again yields a determinant of a matrix whose order equals the total perturbation order. In general there is a sign problem. To reduce the sign problem, it is advantageous to introduce auxiliary fields α and replace

$$\frac{1}{2} \sum_{abcd} U^{abcd} d_a^\dagger d_b^\dagger d_c d_d \rightarrow -\frac{1}{2} \sum_{abcd} U^{abcd} (d_a^\dagger d_c - \alpha_{ac})(d_b^\dagger d_d - \alpha_{bd}),$$

with an appropriate shift in the quadratic part of the Hamiltonian. However, in general, we cannot completely eliminate the sign problem by a suitable choice of α parameters. Furthermore, because the number of interaction terms grows like $\mathcal{O}(n_{\text{orbitals}}^4)$, the computational cost rapidly escalates. In practice, the strong-coupling approach we discuss next is a more suitable approach for single-site multi-orbital problems with general interactions.

---

[16] Rombouts et al. (1999b) proposed such an algorithm in a pioneering work that introduced the idea of a determinant-based continuous-time Monte Carlo sampling for Fermion systems.

[17] In Appendix M we discuss a more suitable lattice Monte Carlo method that combines the weak-coupling continuous-time approach with elements of the determinant algorithm of Chapter 7, and thereby recovers the linear-in-β scaling.

8.5 Strong-coupling approach

While the Monte Carlo weights in the weak-coupling method are expressed in terms of the Weiss Green's function $\mathcal{G}_0$, the strong coupling method, which is in many ways complementary to the weak-coupling approach, naturally involves the hybridization function Δ. It follows from (8.15) that the Weiss Green's function and hybridization function contain the same information, and the DMFT procedure sketched in Algorithm 26 could be written just as well as a self-consistency loop fixing the hybridization function Δ.

The strong-coupling approach (Werner et al., 2006) is based on an expansion of the partition function in powers of the impurity-bath hybridization term.[18] Here, we decompose the Hamiltonian as $H_2 = H_{\text{mix}}$ and $H_1 = H - H_2 = H_\mu + H_U + H_{\text{bath}}$. Because $H_2 \equiv H_2^{d^\dagger} + H_2^d = \sum_{p\sigma} V_{p\sigma} d_\sigma^\dagger c_{p\sigma} + \sum_{p\sigma} V_{p\sigma}^* c_{p\sigma}^\dagger d_\sigma$ has two terms, corresponding to electrons hopping from the bath to the impurity and from the impurity back to the bath, only even perturbation orders contribute to (8.29). Furthermore, at perturbation order $2n$, only the $(2n)!/(n!)^2$ terms corresponding to n creation operators $d^\dagger$ and n annihilation operators d contribute. We therefore write the partition function as a sum over configurations $\{\tau_1, \ldots, \tau_n; \tau_1', \ldots, \tau_n'\}$ that are collections of imaginary-time points corresponding to these n annihilation and n creation operators:

$$Z = \sum_{n=0}^{\infty} \int_0^\beta d\tau_1 \cdots \int_{\tau_{n-1}}^\beta d\tau_n \int_0^\beta d\tau_1' \cdots \int_{\tau_{n-1}'}^\beta d\tau_n'$$
$$\times \text{Tr}\left[e^{-\beta H_1} \mathcal{T} H_2^d(\tau_n) H_2^{d^\dagger}(\tau_n') \cdots H_2^d(\tau_1) H_2^{d^\dagger}(\tau_1') \right]. \tag{8.49}$$

Because the imaginary-time evolution operator $e^{-\tau H_1}$ does not rotate the spin in the case of the Anderson impurity model, the configurations must contain an equal number of creation and annihilation operators for each spin. Taking this additional constraint into account and using the explicit expressions for H_2^d and $H_2^{d^\dagger}$, we find

$$Z = Z_{\text{bath}} \sum_{\{n_\sigma\}} \prod_\sigma \int_0^\beta d\tau_1^\sigma \cdots \int_{\tau_{n_\sigma-1}^\sigma}^\beta d\tau_{n_\sigma}^\sigma \int_0^\beta d\tau_1'^\sigma \cdots \int_{\tau_{n_\sigma-1}'^\sigma}^\beta d\tau_{n_\sigma}'^\sigma$$
$$\times \text{Tr}_d\left[e^{-\beta H_{\text{loc}}} \mathcal{T} \prod_\sigma d_\sigma(\tau_{n_\sigma}^\sigma) d_\sigma^\dagger(\tau_{n_\sigma}'^\sigma) \ldots d_\sigma(\tau_1^\sigma) d_\sigma^\dagger(\tau_1'^\sigma) \right]$$
$$\times \frac{1}{Z_{\text{bath}}} \text{Tr}_c\left[e^{-\beta H_{\text{bath}}} \mathcal{T} \prod_\sigma \sum_{p_1 \ldots p_{n_\sigma}} \sum_{p_1' \ldots p_{n_\sigma}'} V_{p_1\sigma}^* V_{p_1'\sigma} \cdots V_{p_{n_\sigma}\sigma}^* V_{p_{n_\sigma}'\sigma} \right.$$
$$\left. c_{p_{n_\sigma}\sigma}^\dagger(\tau_{n_\sigma}^\sigma) c_{p_{n_\sigma}'\sigma}(\tau_{n_\sigma}'^\sigma) \ldots c_{p_1\sigma}^\dagger(\tau_1^\sigma) c_{p_1'\sigma}(\tau_1'^\sigma) \right],$$

---

[18] The technically correct term is therefore *hybridization expansion* approach, rather than strong-coupling approach, but we will use both terms interchangeably.

where to separate the d and c operators we used the fact that H_1 does not mix the impurity and the bath. The local Hamiltonian H_{loc} is defined in (8.7) and $Z_{\text{bath}} = \text{Tr}_c[e^{-\beta H_{\text{bath}}}]$.

Introducing the β-antiperiodic hybridization function (8.14), which in the time domain reads

$$\Delta^\sigma(\tau) = \sum_p \frac{|V_{p\sigma}|^2}{e^{\varepsilon_p\beta} + 1} \begin{cases} -e^{-\varepsilon_p(\tau-\beta)} & 0 < \tau < \beta \\ e^{-\varepsilon_p\tau} & -\beta < \tau < 0 \end{cases},$$

we reexpress the trace over the bath states as

$$\frac{1}{Z_{\text{bath}}} \text{Tr}_c\left[e^{-\beta H_{\text{bath}}} \mathcal{T} \prod_\sigma \sum_{p_1\cdots p_{n_\sigma}} \sum_{p'_1\cdots p'_{n_\sigma}} V^*_{p_1\sigma} V_{p'_1\sigma} \cdots V^*_{p_{n_\sigma}\sigma} V_{p'_{n_\sigma}\sigma} \right.$$

$$\left. c^\dagger_{p_{n_\sigma}\sigma}(\tau^\sigma_{n_\sigma}) c_{p'_{n_\sigma}\sigma}(\tau'^\sigma_{n_\sigma}) \cdots c^\dagger_{p_1\sigma}(\tau^\sigma_1) c_{p'_1\sigma}(\tau'^\sigma_1) \right] = \prod_\sigma \det M_\sigma^{-1},$$

where M_σ^{-1} is the $(n_\sigma \times n_\sigma)$ matrix with elements

$$[M_\sigma^{-1}]_{ij} = \Delta^\sigma(\tau'^\sigma_i - \tau^\sigma_j).$$

In the hybridization expansion approach, the configuration space consists of all sequences $C = \{\tau^\uparrow_1, \ldots, \tau^\uparrow_{n_\uparrow}; \tau'^\uparrow_1, \ldots, \tau'^\uparrow_{n_\uparrow} | \tau^\downarrow_1, \ldots, \tau^\downarrow_{n_\downarrow}; \tau'^\downarrow_1, \ldots, \tau'^\downarrow_{n_\downarrow}\}$ of $n_\uparrow$ creation and annihilation operators for spin up ($n_\uparrow = 0, 1, \ldots$) and $n_\downarrow$ creation and annihilation operators for spin down ($n_\downarrow = 0, 1, \ldots$). The weight of this configuration is

$$w_C = Z_{\text{bath}} \text{Tr}_d\left[e^{-\beta H_{\text{loc}}} \mathcal{T} \prod_\sigma d_\sigma(\tau^\sigma_{n_\sigma}) d^\dagger_\sigma(\tau'^\sigma_{n_\sigma}) \cdots d_\sigma(\tau^\sigma_1) d^\dagger_\sigma(\tau'^\sigma_1) \right]$$

$$\times \prod_\sigma \det M_\sigma^{-1}(d\tau)^{2n_\sigma}. \tag{8.50}$$

The trace factor represents the contribution of the impurity, which fluctuates between different quantum states as electrons hop in and out. The determinants sum up all bath evolutions that are compatible with the given sequence of transitions. We discuss the importance of this summation in Section 8.7.

To evaluate the trace factor, we may, for example, use the eigenbasis of H_{loc}. In this basis, the imaginary-time evolution operator $e^{-\tau H_{\text{loc}}}$ is diagonal, while the operators d_σ and $d^\dagger_\sigma$ produce transitions between eigenstates with amplitude ± 1. Because the time evolution does not flip the electron spin, the creation and annihilation operators for a given spin alternate. This observation allows us to separate the operators for spin up from those for spin down and to depict the time evolution by

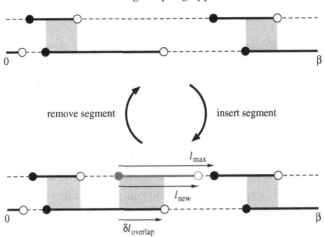

Figure 8.6 Local update in the segment picture. The two segment configurations correspond to spin-up and-down electrons. Each segment depicts a time interval in which an electron of the corresponding spin resides on the impurity. The segment end points are the locations of the operators $d^\dagger$ (full circles) and d (empty circles). We increase the perturbation order by adding a segment or antisegment of random length for random spin and decrease it by removing a randomly chosen segment or antisegment.

a *collection of segments* with each segment representing an imaginary-time interval in which an electron of spin up or down resides on the impurity (Fig. 8.6). We call an unoccupied time interval between two segments an "antisegment."

At each time, the eigenstate of the impurity follows immediately from the segment representation, and the trace factor becomes

$$\mathrm{Tr}_d\left[e^{-\beta H_{\mathrm{loc}}} \mathcal{T} \prod_\sigma d_\sigma(\tau_{n_\sigma}^\sigma) d_\sigma^\dagger(\tau_{n_\sigma}^{'\sigma}) \cdots d_\sigma(\tau_1^\sigma) d_\sigma^\dagger(\tau_1^{'\sigma})\right] =$$

$$\mathcal{S} \exp\left[\mu(l_\uparrow + l_\downarrow) - U l_{\mathrm{overlap}}\right],$$

with $\mathcal{S}$ being a permutation sign, l_σ the total "length" of the segments for spin σ, and l_{overlap} the total length of the overlap between up and down segments. The lower panel of Fig. 8.6 shows a configuration with three segments for spin up and two segments for spin down. The time intervals where segments overlap, indicated by gray rectangles, correspond to a doubly occupied impurity and cost a repulsion energy U.

In the segment formalism, it is natural to decompose a configuration with zero segments (for a given spin σ) into the "full line" and "empty line" contributions corresponding to the occupied and empty σ states.

8.5.1 Sampling

For ergodicity, it is sufficient to insert and remove pairs of creation and annihilation operators (segments or antisegments) for spin up and down. One possible strategy for inserting a segment is the following: We select a random time in $[0, \beta)$ for the creation operator. If it falls on an existing segment, the impurity is already occupied and the move is rejected. If it falls on an empty space, we compute l_{max}, the length from this selected time to the next segment (in the direction of increasing τ).[19] Then we choose the position of the new annihilation operator randomly in this interval of length l_{max} (Fig. 8.6). If in the inverse procedure we propose to remove a randomly chosen segment for this spin, then the proposal probabilities for the insertion and removal are

$$T(n_\sigma + 1 \leftarrow n_\sigma) = \frac{d\tau}{\beta} \frac{d\tau}{l_{max}}, \quad T(n_\sigma \leftarrow n_\sigma + 1) = \frac{1}{n_\sigma + 1}.$$

The acceptance probability for the insertion of a segment becomes $A(n_\sigma + 1 \leftarrow n_\sigma) = \min[1, \mathcal{R}_{insert}(n_\sigma + 1 \leftarrow n_\sigma)]$, with

$$\mathcal{R}_{insert}(n_\sigma + 1 \leftarrow n_\sigma) = \frac{\beta l_{max}}{n_\sigma + 1} e^{\mu l_{new} - U \delta l_{overlap}} \frac{\det \left[M_\sigma^{(n_\sigma + 1)} \right]^{-1}}{\det \left[M_\sigma^{(n_\sigma)} \right]^{-1}}, \tag{8.51}$$

while the acceptance probability for a removal is obtained from

$$\mathcal{R}_{remove}(n_\sigma \leftarrow n_\sigma + 1) = 1 / \mathcal{R}_{insert}(n_\sigma + 1 \leftarrow n_\sigma). \tag{8.52}$$

Here, l_{new} is the length of the new segment, and $\delta l_{overlap}$ is the change in the overlap. Again, we compute the ratio of determinants using the fast update formulas discussed in Section 8.4.2. Algorithm 29 describes the steps needed for the insertion or removal of a segment. Analogous procedures are used for the insertion or removal of antisegments.[20]

8.5.2 Measurement of the Green's function

The strategy is to create configurations that contribute to the Green's function measurement by decoupling the bath from a given pair of creation and annihilation operators in a configuration C. We start by expressing the expectation value for the Green's function as

$$G(\tau) = -\frac{1}{Z} \sum_C w_C^{d(\tau)d^\dagger(0)} = -\frac{1}{Z} \sum_C w_C^{(\tau,0)} \frac{w_C^{d(\tau)d^\dagger(0)}}{w_C^{(\tau,0)}},$$

[19] If there are no segments for the given spin, $l_{max} = \beta$.
[20] At order $n_\sigma = 1$, the removal of the antisegment leads to a full line.

Algorithm 29 Insertion/removal of a segment in the strong-coupling continuous-time impurity solver.

Input: Configuration $C = \{(\tau_1'^\uparrow, \tau_1^\uparrow), \ldots, (\tau_{n_\uparrow}'^\uparrow, \tau_{n_\uparrow}^\uparrow); (\tau_1'^\downarrow, \tau_1^\downarrow), \ldots, (\tau_{n_\downarrow}'^\downarrow, \tau_{n_\downarrow}^\downarrow)\}$, which at order $n_\sigma = 0$ encodes full or empty line, inverse hybridization function matrices M_σ, hybridization functions Δ_σ, interaction U, chemical potential μ, inverse temperature β.

Randomly choose the spin σ ;
Draw a uniform random number $\zeta \in [0, 1]$;
if $(\zeta < 0.5)$ **then**
    ▷ Try to insert a segment of spin σ
  Randomly choose the start time $\tau_\sigma' \in [0, \beta)$ for the new segment ;
  if τ_σ' does not lie on a spin-σ segment or full line of C **then**
    Compute length l_{max} to the next spin-σ creation operator ;   ▷ Must consider periodic boundary conditions
    Randomly choose the end point τ_σ of the new segment in the interval of length l_{max} starting at τ_σ' ;
    Compute the length l_{new} of the new segment $(\tau_\sigma', \tau_\sigma)$;
    Compute the overlap $\delta l_{overlap}$ of the new segment $(\tau_\sigma', \tau_\sigma)$ with the segments of opposite spin ;
    Compute $A(n_\sigma + 1 \leftarrow n_\sigma)$ using (8.51) and (8.39) ;
    Draw a uniform random number $\zeta' \in [0, 1]$;
    if $(\zeta' < A(n_\sigma + 1 \leftarrow n_\sigma))$ **then**
      Insert the segment $(\tau_\sigma', \tau_\sigma)$ into C ;
      Update M_σ using Δ^σ and (8.40)–(8.43) ;
    end if
  end if
else if $(n_\sigma > 0)$ **then**
    ▷ Try to remove a segment of spin σ
  Randomly choose one of the segments of spin σ in C ;
  Compute the length of this segment ;
  Compute the overlap of this segment with segments of opposite spin ;
  Compute $A(n_\sigma - 1 \leftarrow n_\sigma)$ using (8.52) and (8.44) ;
  Draw a uniform random number $\zeta' \in [0, 1]$;
  if $(\zeta' < A(n_\sigma - 1 \leftarrow n_\sigma))$ **then**
    Remove the chosen segment from C ;
    Update M_σ using (8.45).
  end if
end if
return the updated C and M.

---

where $w_C^{d(\tau)d^\dagger(0)}$ denotes the weight of the configuration C with an additional operator $d^\dagger(0)$ and $d(\tau)$ in the trace factor, and $w_C^{(\tau,0)}$ denotes the complete weight corresponding to the enlarged operator sequence (including enlarged hybridization determinants). Because the trace factors of both weights are identical, up to a permutation sign $(-1)^{i+j}$,

$$\frac{w_C^{d(\tau)d^\dagger(0)}}{w_C^{(\tau,0)}} = \frac{(-1)^{i+j}\det\left[M_C\right]^{-1}}{\det\left[M_C^{(\tau,0)}\right]^{-1}} = \left[M_C^{(\tau,0)}\right]_{ji},$$

with i and j denoting the row and column corresponding to the additional operators $d^\dagger$ and d in the enlarged $[M_C^{(\tau,0)}]^{-1}$. Hence, the measurement formula for the Green's function becomes[21]

$$G(\tau) = -\frac{1}{Z}\sum_C w_C^{(\tau,0)}[M_C^{(\tau,0)}]_{ji} = -\frac{1}{Z}\sum_{\tilde{C}} w_{\tilde{C}}\tilde{n}^2\delta(\tau_{\tilde{n}} - \tau)\delta(\tau'_{\tilde{n}} - 0)[M_{\tilde{C}}]_{\tilde{n}\tilde{n}}$$

$$= -\frac{1}{Z}\sum_{\tilde{C}} w_{\tilde{C}}\tilde{n}^2\frac{1}{\beta}\delta(\tau, \tau_{\tilde{n}} - \tau'_{\tilde{n}})[M_{\tilde{C}}]_{\tilde{n}\tilde{n}},$$

with $\delta(\tau, \tau') = \delta(\tau - \tau')$ for $\tau' > 0$, and $\delta(\tau, \tau') = -\delta(\tau - \tau' - \beta)$ for $\tau' < 0$. In the first step, we went from a sum over configurations C with n creation and annihilation operators in addition to $d(\tau)$ and $d^\dagger(0)$ to a sum over configurations $\tilde{C}$ with $\tilde{n} = n + 1$ operator pairs, while in the last step, we used the translational invariance and the β-antiperiodicity of the Green's function. We finally replace the factor $\tilde{n}^2$ (which comes from the $1/(n!)^2$ factor in the Monte Carlo weights without time ordering) by a sum over all pairs i, j of creation and annihilation operators, to obtain the measurement formula $G(\tau) = -\frac{1}{Z}\sum_{\tilde{C}} w_{\tilde{C}}\sum_{ij}\frac{1}{\beta}\delta(\tau, \tau_j - \tau'_i)[M_{\tilde{C}}]_{ji}$, or

---

[21] For the purpose of this derivation, it is convenient to use configurations C and $\tilde{C}$ without time ordering, that is, we write the Green's function as

$$G(\tau) = -\frac{Z_{\text{bath}}}{Z}\sum_n \frac{1}{n!^2}\int_0^\beta d\tau_1\cdots d\tau_n \int_0^\beta d\tau'_1\cdots d\tau'_n$$

$$\times \text{Tr}_d\left[e^{-\beta H_{\text{loc}}}\mathcal{T}d(\tau)d^\dagger(0)d(\tau_n)d^\dagger(\tau'_n)\cdots d(\tau_1)d^\dagger(\tau'_1)\right]\det\left[M^{(\tau,0)}\right]^{-1}\left[M^{(\tau,0)}\right]_{n+1,n+1}$$

$$= -\frac{Z_{\text{bath}}}{Z}\sum_n \frac{(n+1)^2}{(n+1)!^2}\int_0^\beta d\tau_1\cdots d\tau_{n+1}\int_0^\beta d\tau'_1\cdots d\tau'_{n+1}\delta(\tau_{n+1} - \tau)\delta(\tau'_{n+1} - 0)$$

$$\times \text{Tr}_d\left[e^{-\beta H_{\text{loc}}}\mathcal{T}d(\tau_{n+1})d^\dagger(\tau'_{n+1})d(\tau_n)d^\dagger(\tau'_n)\cdots d(\tau_1)d^\dagger(\tau'_1)\right]\det\left[M^{(\tau,0)}\right]^{-1}\left[M^{(\tau,0)}\right]_{n+1,n+1}$$

$$= -\frac{Z_{\text{bath}}}{Z}\sum_{\tilde{n}}\frac{\tilde{n}^2}{(\tilde{n}!)^2}\int_0^\beta d\tau_1\cdots d\tau_{\tilde{n}}\int_0^\beta d\tau'_1\cdots d\tau'_{\tilde{n}}\delta(\tau_{\tilde{n}} - \tau)\delta(\tau'_{\tilde{n}} - 0)$$

$$\times \text{Tr}_d\left[e^{-\beta H_{\text{loc}}}\mathcal{T}d(\tau_{\tilde{n}})d^\dagger(\tau'_{\tilde{n}})\cdots d(\tau_1)d^\dagger(\tau'_1)\right]\det\left[M^{(\tilde{n})}\right]^{-1}\left[M^{(\tilde{n})}\right]_{\tilde{n}\tilde{n}}.$$

$$G(\tau) = \left\langle -\sum_{ij} \frac{1}{\beta} \delta(\tau, \tau_i - \tau'_j) M_{ij} \right\rangle_{MC}. \tag{8.53}$$

Fourier transformation of (8.53) yields the measurement formula

$$G(i\omega_n) = \left\langle -\sum_{ij} \frac{1}{\beta} e^{i\omega_n(\tau_i - \tau'_j)} M_{ij} \right\rangle_{MC} \tag{8.54}$$

for the Fourier coefficients of the Green's function. Note that in contrast to the weak-coupling approach, where we measure the Green's function as a $\mathcal{O}(1/(i\omega_n)^2)$ correction to the Weiss Green's function, (8.54) does not automatically yield the correct high-frequency tail.

An elegant way to suppress the noise in $G(i\omega_n)$ at large ω_n and to obtain a compact representation of the Green's function is to measure the expansion coefficients in a basis of orthogonal polynomials (Boehnke et al., 2011). A suitable choice are the *Legendre polynomials* $P_l(x)$ defined on $x \in [-1, 1]$ through the recursion relation

$$P_0(x) = 1,$$
$$P_1(x) = x,$$
$$(l+1)P_{l+1}(x) = (2l+1)xP_l(x) - lP_{l-1}(x).$$

The P_l furthermore satisfy $\int_{-1}^{1} dx P_k(x) P_l(x) = \frac{2}{2l+1} \delta_{kl}$. Defining $x(\tau) = 2\tau/\beta - 1$, we may thus express the Green's function on the interval $\tau \in [0, \beta]$ as

$$G(\tau) = \sum_{l \geq 0} \frac{\sqrt{2l+1}}{\beta} P_l(x(\tau)) G_l, \tag{8.55}$$

$$G_l = \sqrt{2l+1} \int_0^\beta d\tau P_l(x(\tau)) G(\tau). \tag{8.56}$$

The advantage of the Legendre representation over the Matsubara representation is a much faster decay of the expansion coefficients with increasing order. The Matsubara Fourier transform requires anti periodization of the Green's function with discontinuities at $\tau = m\beta$, which leads to slowly decaying Matsubara coefficients ($G(i\omega_n) \sim 1/i\omega_n$ for large ω_n). On the other hand, the Legendre basis represents the smooth function $G(\tau)$ on the interval $[0, \beta]$. In practice, 30–50 Legendre coefficients are enough to reproduce the Green's function with high precision, and neglecting the higher orders acts as a convenient noise filter.

From (8.53) and (8.56) it follows that

$$G_l = \left\langle -\sum_{ij} \frac{\sqrt{2l+1}}{\beta} \tilde{P}_l(\tau_i - \tau'_j) M_{ij} \right\rangle_{MC}, \tag{8.57}$$

with $\tilde{P}_l(\tau) = P_l(x(\tau))$ for $\tau > 0$ and $\tilde{P}_l(\tau) = -P_l(x(\tau + \beta))$ for $\tau < 0$.

The Matsubara coefficients are obtained from the Legendre coefficients as $G(i\omega_n) = \sum_{l \geq 0} T_{nl} G_l$, with the unitary transformation T_{nl} given by $T_{nl} = (-1)^n i^{l+1} \sqrt{2l+1} j_l(\frac{1}{2}\beta\omega_n)$ involving the spherical Bessel functions $j_l(z)$. In the limit $n \to \infty$, T_{nl} decays $\sim 1/(i\omega_n)$ for n even and $\sim 1/(i\omega_n)^2$ for n odd.

8.5.3 Generalization – Matrix formalism

It is obvious from the derivation in Section 8.5 that the hybridization expansion formalism is applicable to general classes of impurity models (Werner and Millis, 2006). Because we compute the trace factor in the weight (8.50) exactly, H_{loc} can contain arbitrary local interactions (e.g., spin-exchange terms in multi-orbital models), degrees of freedom (e.g., spins in Kondo-lattice models), or constraints (e.g., "no double occupancy" in the *t-J* model).

For multi-orbital impurity models with H_{loc} diagonal in the occupation number basis, such as models with density-density interactions, the segment formalism illustrated in Fig. 8.6 is still applicable. We now have a collection of segments for each *flavor* α (orbital, spin, etc.), and we still compute the trace factor from the length of the segments (the chemical potential contribution) and the overlaps between segments of different flavor (the interaction contribution). This allows a very efficient simulation of models with five, seven, and in principle even more orbitals, despite the fact that the corresponding Hilbert spaces ($4^5 = 1024$ for five orbitals, $4^7 = 16,384$ for seven orbitals) are quite large.

If H_{loc} is not diagonal in the occupation number basis defined by the $d_\alpha^\dagger$, the calculation of

$$\mathrm{Tr}_d \left[e^{-\beta H_{\mathrm{loc}}} \mathcal{T} \prod_\alpha d_\alpha(\tau_{n_\alpha}^\alpha) d_\alpha^\dagger(\tau_{n_\alpha}'^\alpha) \cdots d_\alpha(\tau_1^\alpha) d_\alpha^\dagger(\tau_1'^\alpha) \right] \tag{8.58}$$

becomes rather involved and for a model with a large Hilbert space also computationally expensive. An obvious idea is to evaluate the trace in the eigenbasis where the imaginary-time evolution operators $e^{-H_{\mathrm{loc}}\tau}$ become diagonal. On the other hand, the operators d_α and $d_\alpha^\dagger$, which are simple and sparse in the occupation number basis, become complicated matrices in the eigenbasis. The evaluation of the trace factor in the eigenbasis thus involves the multiplication of matrices whose size scales as the dimension of the Hilbert space of the local problem. Because the dimension of this Hilbert space grows *exponentially* with the number of flavors, the calculation of the trace factor becomes the computational bottleneck of the simulation, and the matrix formalism is therefore restricted to a relatively small number of flavors.

It is important to identify and use *conserved quantities* (Haule, 2007). Typically, these are particle number for spin up and spin down and momentum. If we group the eigenstates of H_{loc} according to these quantum numbers, the operator matrices acquire a sparse block structure. For example, the operator $d^{\dagger}_{\uparrow,q}$ connects states corresponding to the quantum numbers $m = \{n_{\uparrow}, n_{\downarrow}, k, \ldots\}$ to those with $m' = \{n_{\uparrow} + 1, n_{\downarrow}, k + q, \ldots\}$ (if they exist). Checking the compatibility of the operator sequence with the different starting blocks allows us to identify the blocks that contribute to the trace without performing any expensive matrix-matrix multiplications.

Let us take as a simple example a two-orbital model with conserved quantum numbers $n_{\uparrow}$ and $n_{\downarrow}$. The operator sequence $d^{\dagger}_{\uparrow}(\tau_4)d^{\dagger}_{\uparrow}(\tau_3)d_{\uparrow}(\tau_2)d_{\uparrow}(\tau_1)$ (with $\tau_1 < \tau_2 < \tau_3 < \tau_4$) is compatible with the starting blocks $\{n_{\uparrow} = 2; n_{\downarrow} = 0, 1, 2\}$, since the quantum numbers evolve as

$$\{n_{\uparrow} = 2; n_{\downarrow}\} \underset{d_{\uparrow}}{\rightarrow} \{n_{\uparrow} = 1; n_{\downarrow}\} \underset{d_{\uparrow}}{\rightarrow} \{n_{\uparrow} = 0; n_{\downarrow}\} \underset{d^{\dagger}_{\uparrow}}{\rightarrow} \{n_{\uparrow} = 1; n_{\downarrow}\} \underset{d^{\dagger}_{\uparrow}}{\rightarrow} \{n_{\uparrow} = 2; n_{\downarrow}\},$$

whereas the blocks $\{n_{\uparrow} = 0, 1; n_{\downarrow} = 0, 1, 2\}$ do not contribute to the weight, since, for example,

$$\{n_{\uparrow} = 1; n_{\downarrow}\} \underset{d_{\uparrow}}{\rightarrow} \{n_{\uparrow} = 0; n_{\downarrow}\} \underset{d_{\uparrow}}{\rightarrow} \emptyset.$$

Having identified the contributing blocks, the trace calculation reduces to a block matrix multiplication of the form

$$\sum_{\substack{\text{contributing} \\ m}} \text{Tr}_m\left[\cdots [O]_{m''m'}[e^{-(\tau'-\tau)H_{\text{loc}}}]_{m'}[O]_{m'm}[e^{-\tau H_{\text{loc}}}]_m\right], \tag{8.59}$$

where O is either a creation or an annihilation operator, m denotes the index of the matrix block, and the sum runs over those starting sectors that are compatible with the operator sequence.

Using the block structure imposed by the conserved quantum numbers, we can simulate three-orbital models or four-site clusters efficiently. However, because the matrix blocks are dense and the largest blocks grow exponentially with system size, the simulation of five-orbital models already becomes quite expensive, and the simulation of seven-orbital models with five, six, or seven electrons is doable only if we truncate the size of the blocks.

In fact, one should distinguish two types of truncations:

1. Restriction of the trace $\sum_{\text{contributing } m} \text{Tr}_m[\ldots]$ to those quantum number sectors or states that give the dominant contribution
2. Reduction of the size of the operator blocks $[O]_{m'm''}$ by eliminating high-energy states.

Truncations of type 1 have little effect at low enough temperature, because they restrict the possible states only at a single point on the imaginary-time interval. Truncations of the type 2 are more problematic and possibly lead to systematic errors that are difficult to estimate and control when the system size is large.

Accumulating a histogram of the states or quantum number sectors visited during the sampling can be very instructive. For example, in the study of correlated materials with multiple partially filled orbitals, interesting issues are the typical valence or the dominant spin state, and the importance of fluctuations to other charge and spin states. Dynamical mean-field theory allows us to address these issues by adopting a real-space representation of the solid as a collection of atoms and treating the local fluctuations on a given site through the effective impurity model construction. The strong-coupling solver, which treats the local part of the impurity problem exactly, is ideally suited for such an analysis.

As an illustration, we show in Fig. 8.7 a histogram accumulated in a DMFT simulation of a half-filled two-orbital model (Werner and Millis, 2007). The energies of the two orbitals are shifted relative to each other by a "crystal field splitting" δ. For large δ, it is energetically favorable to put the two electrons into the lower orbital and hence into a low-spin ($S = 0$) configuration. For small δ, the Hund's

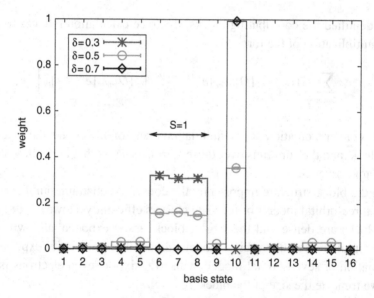

Figure 8.7 Histogram of local quantum states in a DMFT simulation of a half-filled two-orbital model with strong interactions and Hund's coupling, and with a crystal field splitting δ between the orbitals. The band widths are 4, and the temperature is $\frac{1}{50}$. The histogram shows the average relative contribution of each of the 16 eigenstates of H_{loc} to the weight of the Monte Carlo configurations. (Adapted from Werner and Millis (2007).)

interaction favors a high-spin ($S = 1$) state with equal occupation of both orbitals. At large interaction, we may thus anticipate a transition from a high-spin to a low-spin insulating phase, as the crystal field splitting δ increases, with possibly a metallic phase appearing in between. What is shown in the figure is the weight of the $4^2 = 16$ atomic states (eigenstates of H_{loc}) or more precisely the average contribution of each of these states to the trace factor (8.58). We see that for the smallest crystal field splitting, three states ($|6\rangle$, $|7\rangle$, $|8\rangle$) dominate; these are the three degenerate, half-filled $S = 1$ states. The eigenstates $|10\rangle$ and $|11\rangle$ are linear combinations of the two states with two electrons in one orbital and none in the other, with state $|10\rangle$ having the lower energy. This half-filled, low-energy $S = 0$ state becomes the dominant state at large δ. Hence, by looking at the histogram of atomic states, we can immediately identify the high-spin and low-spin character of the two phases at small and large δ. Because they are insulating, fluctuations to other charge states are suppressed.

The histogram for the intermediate value of δ exhibits a different character. Here, the half-filled spin singlet and spin triplet states all contribute significantly to the trace, and charge fluctuations to states with one (e.g., $|4\rangle$, $|5\rangle$) or three (e.g., $|14\rangle$, $|15\rangle$) electrons are not negligible. This histogram corresponds to a metallic solution that appears close to the level-crossing of the half-filled high-spin and low-spin atomic states.

As this simple example shows, by looking at the state histogram and by identifying the dominant states and the degeneracies that appear in these histograms, we gain valuable insights into the nature of phases and phase transitions in the DMFT solution of complicated (multi-orbital) lattice models.

8.5.4 Generalization – Krylov formalism

An alternative strategy (Läuchli and Werner, 2009) to evaluate the trace factor (8.58) is to:

1. Adopt the occupation number basis in which we can easily apply the d_α and $d_\alpha^\dagger$ operator matrices to any state and in which we can exploit the sparse nature of H_{loc} during the imaginary-time evolutions
2. Approximate the trace by a sum over the lowest energy states, that is, by a truncation of type (1) described in the previous subsection.

Instead of evaluating the matrix corresponding to the product of operators, we propagate each retained state in the trace through the sequence of time-evolution, creation, and annihilation operators. This computation involves only matrix-vector multiplications of the type $d_\alpha|v\rangle$, $d_\alpha^\dagger|v\rangle$, and $H_{\mathrm{loc}}|v\rangle$ with *sparse* operators d_α, $d_\alpha^\dagger$, and H_{loc} and is thus possible for systems for which the multiplication of dense matrix blocks becomes prohibitively expensive. Furthermore, the approach does not require any truncation of type (2), so all excited states remain accessible at intermediate τ. While the sparsity of H_{loc} depends on the number of interaction terms, this number grows at most proportionally to the number of orbitals squared. In contrast, the dimension of the matrix grows exponentially with the number of orbitals.

The expensive step is the calculation of the time evolution from one operator to the next. We evaluate the matrix exponentials applied to a vector, $\exp(-\tau H_{\text{loc}})|v\rangle$, by iteratively constructing the *Krylov space*

$$\mathcal{K}_p(|v\rangle) = \text{span}\{|v\rangle, H_{\text{loc}}|v\rangle, H_{\text{loc}}^2|v\rangle, \ldots, H_{\text{loc}}^p|v\rangle\}$$

and by approximating the full matrix exponential by the matrix exponential of the Hamiltonian projected onto $\mathcal{K}_p(|v\rangle)$. The iteration number p is determined by tracking the convergence of $\exp(-\tau H_{\text{loc}})|v\rangle$ and stopping the calculation if the difference between iteration p and $p+1$ drops below some cutoff value. The number of iterations depends on the time interval τ, but typically, convergence occurs for very small iteration numbers $p \ll N_{\text{dim}}$, with N_{dim} the dimension of the Hilbert space.

In the limit where the dimension of the local Hilbert space N_{dim} is large, the Krylov approach is more efficient than an implementation based on a matrix representation of the operators d_α, $d_\alpha^\dagger$, and an evaluation of the trace of the matrix product. If the Monte Carlo configuration has n creation and n annihilation operators and we perform the trace over $N_{\text{tr}} \leq N_{\text{dim}}$ states, the Krylov calculation of the trace scales as

$$\mathcal{O}(N_{\text{tr}}N_{\text{dim}}2n(1 + \langle p \rangle)),$$

where the first term comes from the application of the creation and annihilation operators and the second term, proportional to the average dimension $\langle p \rangle$ of the Krylov space, from the application of the time-evolution operators. If we retain all the states in the trace calculation, $N_{\text{tr}} = N_{\text{dim}}$, and the trace calculation scales as N_{dim}^2. If we restrict the trace to a small number of low-energy states, then N_{tr} is $\mathcal{O}(1)$ and the trace computation becomes *linear* in N_{dim}. This scaling should be compared with a computational effort of $\mathcal{O}(2nN_{\text{dim}}^3)$ for the evaluation of the trace based on matrix multiplications (without truncation of the matrix blocks).[22]

While in theory the Krylov space approach is the method of choice due to its superior N_{dim} scaling, in practice, the precise numbers of N_{tr}, $\langle p \rangle$, and N_{dim} determine which one of the two approaches performs better for a given problem. Experience shows that for five-orbital problems the Krylov approach becomes superior to the matrix method.

---

[22] In the truncated trace approach, it is important to measure the various local observables at $\tau = \frac{1}{2}\beta$ where they are least affected by the truncation at $\tau = 0$ and $\tau = \beta$. Also, it is important not to destroy the multiplet structure when truncating the trace.

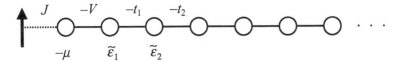

Figure 8.8 Chain representation of the quantum impurity model in the Kondo limit. The impurity states are restricted to the singly occupied states $|\uparrow\rangle$ and $|\downarrow\rangle$. This spin-$\frac{1}{2}$ degree of freedom couples via the exchange integral J to the spin $\vec{S} = \frac{1}{2}\psi_c^\dagger \vec{\sigma}\psi_c$ on lattice site 0, represented in the figure by the first site of the chain.

8.6 Infinite-U limit: Kondo model

In the limit of very strong interaction, we cannot efficiently simulate the half-filled Anderson impurity model using the weak- or strong-coupling continuous-time Monte Carlo solvers. The weak-coupling approach is unsuitable because the perturbation order becomes very large, and the strong-coupling approach may encounter a problem with the sampling efficiency, because hybridization events correspond to transitions into doubly occupied or empty states with very high energy. While transitions into states with occupancy different from unity are acceptable, as long as the excursion is very short-lived, and the strong-coupling algorithm in principle handles arbitrarily short segments or anti segments, it is more appropriate and more efficient to project out the charge fluctuations and consider a low-energy effective model in which the singly-occupied impurity (represented by a spin $\frac{1}{2}$) exchanges spin with the bath (Fig. 8.8). This projection from the Anderson impurity model leads us to the second iconic impurity model, the Kondo impurity model defined in (8.2). To solve this model, we have to compute the Green's function of the bath at site 0 (the location of the impurity). The Green's function of the impurity then follows from the T-matrix of the bath (Costi, 2000). In the following sections, we discuss two complementary continuous-time solvers for the Kondo impurity model: (i) the weak-coupling approach based on an expansion in powers of the exchange interaction J and (ii) the strong-coupling approach based on an expansion in powers of the hybridization V between the site 0 and the rest of the bath.

8.6.1 Weak-coupling approach

In the weak-coupling simulation (Otsuki et al., 2007), we Fermionize the local spin $\vec{s}$ by introducing creation operators $d_\sigma^\dagger$ and writing

$$\vec{s} = \frac{1}{2}\psi_d^\dagger \vec{\sigma}\psi_d,$$

with $\vec{\sigma} = (\sigma^x, \sigma^y, \sigma^z)$ and $\psi_d^\dagger = (d_\uparrow^\dagger, d_\downarrow^\dagger)$. We then express the Hamiltonian (8.2) with a wave-number-independent coupling J as

$$H = \sum_{k\sigma} \epsilon_k c_{k\sigma}^\dagger c_{k\sigma} + \frac{1}{2}J\left[s^z(c_\uparrow^\dagger c_\uparrow - c_\downarrow^\dagger c_\downarrow) + s^+ c_\downarrow^\dagger c_\uparrow + s^- c_\uparrow^\dagger c_\downarrow\right], \qquad (8.60)$$

where $s^+ = d_\uparrow^\dagger d_\downarrow$, $s^- = d_\downarrow^\dagger d_\uparrow$, $s^z = \frac{1}{2}(d_\uparrow^\dagger d_\uparrow - d_\downarrow^\dagger d_\downarrow)$, and $c_\sigma^\dagger = \frac{1}{\sqrt{N}}\sum_k c_{k\sigma}^\dagger$ is the creation operator for conduction electrons at site 0 (the first site in the chain representation). Using the constraint $d_\uparrow^\dagger d_\uparrow + d_\downarrow^\dagger d_\downarrow = 1$, we rewrite the Hamiltonian again, to obtain

$$H = \sum_{k\sigma} \epsilon_k c_{k\sigma}^\dagger c_{k\sigma} - \frac{1}{4}J\sum_\sigma c_\sigma^\dagger c_\sigma + \frac{1}{2}J\sum_{\sigma\sigma'} d_\sigma^\dagger d_{\sigma'} c_{\sigma'}^\dagger c_\sigma. \qquad (8.61)$$

For the weak-coupling simulation, we split this Hamiltonian into the exactly solvable part $H_1 = \sum_{k\sigma} \epsilon_k c_{k\sigma}^\dagger c_{k\sigma} - \frac{1}{4}J\sum_\sigma c_\sigma^\dagger c_\sigma$ and the remainder $H_2 = \frac{J}{2}\sum_{\sigma\sigma'} d_\sigma^\dagger d_{\sigma'} c_{\sigma'}^\dagger c_\sigma$ and then expand the partition function in powers of H_2. The trace over the Fermionic degrees of freedom yields the weight of the Monte Carlo configuration with perturbation order n and a given sequence of "diagonal" ($c_\sigma^\dagger c_\sigma$) and "off-diagonal" ($c_\sigma^\dagger c_{\bar\sigma}$) operators:

$$w_C = (-Jd\tau/2)^n \text{Tr}_d\left[\mathcal{T} d_{\sigma_1}^\dagger(\tau_1)d_{\sigma_1'}(\tau_1)\cdots d_{\sigma_n}^\dagger(\tau_n)d_{\sigma_n'}(\tau_n)\right]$$

$$\times \prod_\sigma Z_c \frac{1}{Z_c}\text{Tr}_c\left[e^{-\beta H_1}\mathcal{T} c_\sigma^\dagger(\tau_1')c_\sigma(\tau_1'')\cdots c_\sigma^\dagger(\tau_{n_\sigma}')c_\sigma(\tau_{n_\sigma}'')\right]\mathcal{S}.$$

Here, the times $0 < \tau_1 < \cdots < \tau_n < \beta$ mark the H_2-operator positions, $0 < \tau_1' < \cdots < \tau_{n_\sigma}' < \beta$ the locations of the bath creation operators, and $0 < \tau_1'' < \cdots < \tau_{n_\sigma}'' < \beta$ the locations of the bath annihilation operators ($\sum_\sigma n_\sigma = n$). $\mathcal{S}$ is a permutation sign associated with the separation of the spin-up and -down operators and the grouping of creation and annihilation operators into pairs. Z_c is the partition function for H_1, an irrelevant constant factor that we introduced in order to evaluate the expectation values over the bath states.

The trace over the d-states imposes a constraint on the type of operators we can insert. We can insert either "diagonal operators" $c_\sigma^\dagger c_\sigma$ with σ identical to the spin of the d-Fermion or pairs of spin-flip operators $c_\uparrow^\dagger c_\downarrow$ and $c_\downarrow^\dagger c_\uparrow$. The expectation values $\prod_\sigma \frac{1}{Z_c}\text{Tr}_c[\ldots]$ yield a product of determinants of two matrices, $\det \tilde{M}_\uparrow^{-1} \det \tilde{M}_\uparrow^{-1}$. Due to the scattering term in (8.61), the elements of these matrices are noninteracting Green's functions $\tilde{\mathcal{G}}_0^\sigma$, which are related to the Weiss Green's functions $\mathcal{G}_0^\sigma$ by

$$\tilde{\mathcal{G}}_0^\sigma(i\omega_n) = \frac{\mathcal{G}_0^\sigma(i\omega_n)}{1 + \frac{1}{4}J\mathcal{G}_0^\sigma(i\omega_n)}.$$

Specifically, $[\tilde{M}_\sigma^{-1}]_{ij} = \tilde{\mathcal{G}}_0^\sigma(\tau_i'' - \tau_j')$.

Figure 8.9 shows a possible sequence of operators. The upper two timelines represent the imaginary-time evolution of the spin, and the lower two timelines

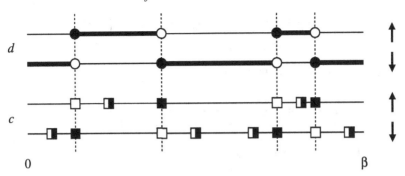

Figure 8.9 Monte Carlo configuration corresponding to four off-diagonal and six diagonal operators. The upper two lines represent the time evolution of the d-states with $\sigma = \uparrow, \downarrow$ (black segments indicate the orientation of the spin s). Full circles represent creation operators and empty circles annihilation operators. A sequence of site 0 creation and annihilation operators for spin up and down is shown on the lower two lines (full and empty squares). The insertion of a diagonal operator $c_\sigma^\dagger(\tau)c_\sigma(\tau)$ is possible only if the spin s is in state σ.

show a sequence of bath creation and annihilation operators. Creation and annihilation operators are represented by full and empty circles or squares. Half-full squares correspond to the diagonal operators $c_\sigma^\dagger(\tau)c_\sigma(\tau)$ whose σ must be identical to the spin represented by the d-line at time τ.

The Monte Carlo sampling proceeds via the insertion and removal of pairs of spin flips (the vertical dashed lines in Fig. 8.9) and via the insertion and removal of diagonal operators (half-filled squares). The spin-flip updates are analogous to the segment configuration updates discussed in Section 8.5.1. We pick a random time for the first spin flip and define an interval $l_{\max}$ up to the next operator, which is either a diagonal operator or a spin-flip event. The second spin flip is then chosen at a random point on this interval. The removal of a pair of adjacent spin flips is possible only if there is no diagonal operator in between. We insert and remove the diagonal operators individually, but an insertion is possible only if the spin is compatible with the state of the d-segment.

The measurement of the bath Green's function at site 0 works just as we described it in the weak-coupling part of Section 8.4.3. This measurement amounts to the accumulation of the T-matrix (see Eq. (8.48))

$$\tilde{t}_\sigma(i\omega_n) = \left\langle -\frac{1}{\beta} \sum_{kl} e^{i\omega_n(\tau_k - \tau_l)} [\tilde{M}_\sigma]_{kl} \right\rangle_{\text{MC}}, \qquad (8.62)$$

where the tilde reminds us that this T-matrix is with respect to $\tilde{\mathcal{G}}_0^\sigma$. It is related to the bath Green's function G^σ and the true T-matrix t^σ by

$$G^\sigma(i\omega_n) = \tilde{\mathcal{G}}_0^\sigma(i\omega_n) + \tilde{\mathcal{G}}_0^\sigma(i\omega_n)\tilde{t}^\sigma(i\omega_n)\tilde{\mathcal{G}}_0^\sigma(i\omega_n)$$
$$= \mathcal{G}_0^\sigma(i\omega_n) + \mathcal{G}_0^\sigma(i\omega_n)t^\sigma(i\omega_n)\mathcal{G}_0^\sigma(i\omega_n).$$

Hence, the T-matrix of the bath becomes

$$t^\sigma(i\omega_n) = \frac{-\frac{1}{4}J}{1 + \frac{1}{4}J\mathcal{G}_0^\sigma(i\omega_n)} + \frac{\tilde{t}^\sigma(i\omega_n)}{\left[1 + \frac{1}{4}J\mathcal{G}_0^\sigma(i\omega_n)\right]^2}, \tag{8.63}$$

and this T-matrix directly yields the impurity Green's function.

8.6.2 Strong-coupling approach

We can also simulate the Kondo impurity model (8.2) efficiently by using the strong-coupling method discussed in Section 8.5. In this approach, we treat the local part of the Hamiltonian,

$$H_{\text{loc}} = -\mu \sum_\sigma c_\sigma^\dagger c_\sigma + J\vec{s} \cdot \left(\frac{1}{2}\psi_c^\dagger \vec{\sigma} \psi_c\right), \tag{8.64}$$

exactly and expand the partition function in powers of the hoppings between the bath orbital 0 (to which the spin s couples) and the rest of the bath (Werner and Millis, 2006). In the chain representation (Fig. 8.8), this hopping amplitude is the parameter V. We integrate out the noninteracting bath sites to obtain an effective action where the eight-dimensional local problem (8.64) is coupled to hybridization functions $\Delta^\sigma(\tau)$ (see Fig. 8.10).

We next diagonalize H_{loc} in a basis labeled by the total number of electrons, the total spin, and the z-component of the total spin. If the particle number is 0 or 2, then the spin state is that of the local moment s. If the number is 1, the spin state is a singlet S or a triplet T_{m_z} with $m_z = 1$, 0, or -1. Accordingly we label the eigenstates as shown in Table 8.1, where the first entry is the number of electrons and the second entry is the spin state. The singlet state S is $(|\uparrow, \downarrow\rangle - |\downarrow, \uparrow\rangle)/\sqrt{2}$ with the first entry being the conduction electron and the second entry the local moment

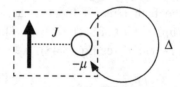

Figure 8.10 Effective action for the Kondo impurity model. The eight-dimensional local problem (dashed box) describes the local spin s coupled to the spin S of the conduction electrons on site 0. The hopping of the conduction electrons from site 0 into the rest of the lattice and back is represented by the hybridization function Δ.

Table 8.1 *Eigenstates and eigenenergies for the local part of the Kondo impurity model (8.64). The first entry labels the number of electrons, and the second entry, the spin state: either impurity spin* $\uparrow, \downarrow$ *if the number of electrons is 0 or 2 or the total spin S (singlet) or* T_{m_z} *(triplet with* $m_z = 1, 0, -1$*) if* $n = 1$.

Eigenstates	Energy
$\lvert 1 \rangle = \lvert 0, \uparrow \rangle$	0
$\lvert 2 \rangle = \lvert 0, \downarrow \rangle$	0
$\lvert 3 \rangle = \lvert 1, S \rangle$	$-\frac{3}{4}J - \mu$
$\lvert 4 \rangle = \lvert 1, T_1 \rangle$	$\frac{1}{4}J - \mu$
$\lvert 5 \rangle = \lvert 1, T_0 \rangle$	$\frac{1}{4}J - \mu$
$\lvert 6 \rangle = \lvert 1, T_{-1} \rangle$	$\frac{1}{4}J - \mu$
$\lvert 7 \rangle = \lvert 2, \uparrow \rangle$	-2μ
$\lvert 8 \rangle = \lvert 2, \downarrow \rangle$	-2μ

spin direction. In this basis, the imaginary-time evolution operator is diagonal, that is, $e^{-H_{\text{loc}}\tau}\lvert n \rangle = e^{-E_n \tau}\lvert n \rangle$ with the eigenenergies E_n listed in Table 8.1.

The creation operators for electron spin up and down become the sparse block matrices

$$
c_\uparrow^\dagger =
\left(
\begin{array}{cc|cccc|cc}
0 & 0 & 0 & 0 & 0 & 0 & 0 & 0 \\
0 & 0 & 0 & 0 & 0 & 0 & 0 & 0 \\
\hline
0 & \frac{1}{\sqrt{2}} & 0 & 0 & 0 & 0 & 0 & 0 \\
1 & 0 & 0 & 0 & 0 & 0 & 0 & 0 \\
0 & \frac{1}{\sqrt{2}} & 0 & 0 & 0 & 0 & 0 & 0 \\
0 & 0 & 0 & 0 & 0 & 0 & 0 & 0 \\
\hline
0 & 0 & \frac{-1}{\sqrt{2}} & 0 & \frac{1}{\sqrt{2}} & 0 & 0 & 0 \\
0 & 0 & 0 & 0 & 0 & 1 & 0 & 0 \\
\end{array}
\right),
\tag{8.65}
$$

$$
c_\downarrow^\dagger =
\left(
\begin{array}{cc|cccc|cc}
0 & 0 & 0 & 0 & 0 & 0 & 0 & 0 \\
0 & 0 & 0 & 0 & 0 & 0 & 0 & 0 \\
\hline
\frac{-1}{\sqrt{2}} & 0 & 0 & 0 & 0 & 0 & 0 & 0 \\
0 & 0 & 0 & 0 & 0 & 0 & 0 & 0 \\
\frac{1}{\sqrt{2}} & 0 & 0 & 0 & 0 & 0 & 0 & 0 \\
0 & 1 & 0 & 0 & 0 & 0 & 0 & 0 \\
\hline
0 & 0 & 0 & -1 & 0 & 0 & 0 & 0 \\
0 & 0 & \frac{-1}{\sqrt{2}} & 0 & \frac{-1}{\sqrt{2}} & 0 & 0 & 0 \\
\end{array}
\right).
\tag{8.66}
$$

The horizontal and vertical lines define the block structure corresponding to the conservation of the total number of electrons. With these sparse operator matrices, the sampling then proceeds as described in Section 8.5.3, and the measurement procedure for the bath Green's function at site 0 is identical to that described in Section 8.5.2. We finally extract the T-matrix from $G^\sigma(i\omega_n) = \mathcal{G}_0^\sigma(i\omega_n) + \mathcal{G}_0^\sigma(i\omega_n)t^\sigma(i\omega_n)\mathcal{G}_0^\sigma(i\omega_n)$, with $\mathcal{G}_0^\sigma(i\omega_n)$ defined in (8.15), and obtain the impurity Green's function.

While the weak-coupling approach is efficient in the regime of small J, the strong-coupling approach easily captures the singlet-formation occurring at larger J.

8.7 Determinant structure and sign problem

8.7.1 Combination of diagrams into a determinant

The determinants appearing both in the weak- and strong-coupling impurity solvers are a consequence of Wick's theorem for the noninteracting part of the Hamiltonian (weak-coupling approach) or the noninteracting bath (strong-coupling approach). In a manner similar to what we saw in Section 7.1.2, a determinant of an $n \times n$ matrix corresponds to a collection of $n!$ diagrams for the partition function. In this collection, both connected and disconnected diagrams appear. What the $n!$ diagrams have in common are the positions on the imaginary-time interval of the n interaction vertices (weak-coupling approach) or the positions of the n creation and n annihilation operators (strong-coupling approach). Figure 8.11 illustrates all second-order

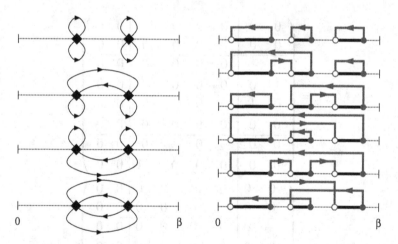

Figure 8.11 Left panel: weak-coupling diagrams summable into a determinant. The diamonds represent interaction vertices and the lines Weiss Green's functions for spin-up and -down electrons. Right panel: strong-coupling diagrams summable into a determinant. The empty circles represent creation operators and the full circles annihilation operators. Pairs of creation and annihilation operators are connected by hybridization lines.

(weak-coupling) and third-order (strong-coupling) contributions corresponding to some fixed operator positions. We note that the Fermionic nature of the operators leads to individual diagrams with anticommutivity signs. The determinants allow us to sum up $n!$ diagrams with proper signs and, at least in simple models such as the Anderson impurity model, to completely absorb the cancellation effects between positive and negative weight contributions.

To underscore the crucial role of the determinants, we consider a simple model of a noninteracting (spin-less) impurity coupled to one bath site with energy $\varepsilon = 0$ and hybridization V. The action reads

$$S = \int_0^\beta d\tau \int_0^\beta d\tau' \, d^\dagger(\tau') \Delta(\tau' - \tau) d(\tau),$$

with the hybridization function (8.14) given by $\Delta(i\omega_n) = |V|^2/i\omega_n$, which in imaginary time becomes

$$\Delta(\tau) = \begin{cases} -\frac{1}{2}|V|^2 & 0 < \tau < \beta, \\ \frac{1}{2}|V|^2 & -\beta < \tau < 0. \end{cases}$$

In this simple model, the absolute values of the diagram weights depend only on the perturbation order n and not on the operator positions, so the integral over the operator positions merely produces a factor $2\beta^{2n}/(2n)!$. The combined weight of the $n!$ topologically distinct diagrams corresponding to a set of n creation and n annihilation operators is the determinant

$$\det \begin{pmatrix} \frac{1}{2}|V|^2 & \frac{1}{2}|V|^2 & \frac{1}{2}|V|^2 & \cdots \\ -\frac{1}{2}|V|^2 & \frac{1}{2}|V|^2 & \frac{1}{2}|V|^2 & \cdots \\ -\frac{1}{2}|V|^2 & -\frac{1}{2}|V|^2 & \frac{1}{2}|V|^2 & \cdots \\ \vdots & \vdots & \vdots & \ddots \end{pmatrix} = \frac{1}{2}|V|^{2n}, \tag{8.67}$$

when we take into account the signs from the time ordering. This means that in the Monte Carlo sampling based on determinants, there is no sign problem and configurations with order n are generated with a probability

$$p_{\text{determinants}}(n) \sim (\beta|V|)^{2n}/(2n)!.$$

On the other hand, the probability distribution obtained in a sampling of individual diagrams (based on the absolute value of their weights) is

$$p_{\text{diagrams}}(n) \sim (n!/2^n)(\beta|V|)^{2n}/(2n)!.$$

Figure 8.12 compares these two distribution functions for inverse temperatures $\beta|V| = 2$ and $\beta|V| = 20$. At the lower temperature, a simulation based on a sampling of individual diagrams would spend most of the time generating configurations of order ~ 50. However, as the distribution for the determinants shows,

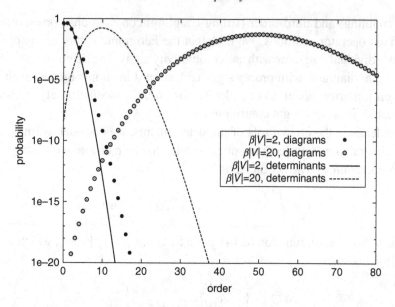

Figure 8.12 Perturbation order distribution (per spin) for the noninteracting Anderson impurity model with one bath site (energy $\varepsilon = 0$, hybridization V, inverse temperature β). The circles show the distribution obtained in a sampling of individual strong-coupling diagrams, while the lines show the distribution in the sampling of determinants.

these configurations contribute negligibly to the partition function (and hence to physical observables) due to sign cancellations. The perturbation orders that are actually relevant are much lower, as the distribution function for the determinants peaks at order ~ 10. There are thus two related reasons why the summation of diagrams into determinants is so important, especially at low temperature:

1. The determinants sum an enormous number of individual diagrams. The determinant of a 100×100 matrix is easily computed numerically and corresponds to the combined weight of $100! = 10^{158}$ diagrams! In practice we can treat average perturbation orders up to ~ 1000.
2. The determinant-based sampling generates configurations with lower perturbation orders than a sampling of individual diagrams. This gain in efficiency is the result of sign cancellations between high-order diagrams.

8.7.2 *Absence of a sign problem*

To prove the absence of a sign problem in the weak- and strong-coupling continuous-time Monte Carlo simulation of the Anderson impurity model, we use the chain representation of the impurity model introduced in Section 8.1.1. We first discuss the weak-coupling approach (Yoo et al., 2005), assuming that the interaction term is rewritten according to (8.31) and (8.32). In the chain representation the quadratic part of the Hamiltonian is

$$\tilde{H}_0 = \sum_\sigma \sum_{j=0}^{\infty} \left[\tilde{\varepsilon}_j c_{j\sigma}^\dagger c_{j\sigma} - t_j (c_{j+1,\sigma}^\dagger c_{j\sigma} + c_{j\sigma}^\dagger c_{j+1,\sigma}) \right].$$

Here, c_j is the annihilation operator for site j of the chain, $c_0 \equiv d$, $t_0 \equiv V$, and $\tilde{\varepsilon}_0 = -\mu + \frac{1}{2}U$ (see Fig. 8.1). We choose the chain basis such that all hopping parameters t_j are nonnegative. By adding an appropriate term $\Lambda(N_\uparrow + N_\downarrow)$, with $\Lambda \geq 0$ and $N_\sigma = \sum_j c_{j\sigma}^\dagger c_{j\sigma}$, we further ensure that all diagonal elements of $\tilde{H}_0 - \Lambda(N_\uparrow + N_\downarrow)$ are negative or zero. Hence, all elements of the matrix $\exp[-(\tilde{H}_0 - \Lambda(N_\uparrow + N_\downarrow))]$ are positive or zero in the chain basis. Since $\tilde{H}_0$ conserves N_σ,

$$e^{-\tau \tilde{H}_0} = e^{-\tau(\tilde{H}_0 - \Lambda(N_\uparrow + N_\downarrow))} e^{-\tau \Lambda(N_\uparrow + N_\downarrow)}$$

is a product of two matrices with elements ≥ 0, and therefore the time-evolution operator has no negative elements in the chain basis.

The weight of a weak-coupling Monte Carlo configuration is

$$w_C = \text{Tr} \left[e^{-(\beta - \tau_n)\tilde{H}_0} A(s_n) e^{-(\tau_n - \tau_{n-1})\tilde{H}_0} A(s_{n-1}) \cdots \right] (d\tau)^n,$$

where

$$A(s) = (-U/2) \left[n_\uparrow - \tfrac{1}{2} - s(\tfrac{1}{2} + \delta) \right] \left[n_\downarrow - \tfrac{1}{2} + s(\tfrac{1}{2} + \delta) \right].$$

What we still need to show is that the matrix $A(s)$ has only nonnegative elements when $\delta \geq 0$ and $U \geq 0$. We do this by considering the two values of the auxiliary spin variable s and by factorizing the interaction term into a product of two diagonal operators:

$$s = 1: \underbrace{(-U/2)}_{\leq 0} \underbrace{(n_\uparrow - 1 - \delta)}_{\leq 0} \underbrace{(n_\downarrow + \delta)}_{\geq 0},$$

$$s = -1: \underbrace{(-U/2)}_{\leq 0} \underbrace{(n_\uparrow + \delta)}_{\geq 0} \underbrace{(n_\downarrow - 1 - \delta)}_{\leq 0}.$$

Hence, in the chain basis, neither the imaginary-time evolution operators $e^{-\tau \tilde{H}_0}$ nor the "interaction vertices" $A(s)$ have negative elements. The weight is the trace of a product of matrices with nonnegative elements, and therefore must be nonnegative. We recall that in the case of attractive U, no auxiliary-field decoupling is required, and the weak-coupling weights are evidently positive.

The lack of a sign problem proof for the strong-coupling formalism is also based on the chain basis (Kaul, 2007). Here, the weight of a Monte Carlo configuration has the form

$$w_C = \text{Tr}\left[e^{-(\beta-\tau_n)(H_{\text{loc}}+H_{\text{bath}})}(-H_{\text{mix}}^{d^\dagger}) \right.$$

$$\left. \cdots e^{-(\tau_2-\tau_1)(H_{\text{loc}}+H_{\text{bath}})}(-H_{\text{mix}}^{d})e^{-\tau_1(H_{\text{loc}}+H_{\text{bath}})} \right](d\tau)^{2n}, \qquad (8.68)$$

with $-H_{\text{mix}}^{d^\dagger} = Vc_0^\dagger c_1$, $-H_{\text{mix}}^{d} = Vc_1^\dagger c_0$ ($c_0 \equiv d$). In the chain basis, the hybridization operators do not produce any negative signs ($V \geq 0$).[23] In the imaginary-time evolution operators, H_{loc} is diagonal, while H_{bath} has off-diagonal elements $-t_i \leq 0$ ($i = 1, 2, \ldots$). Writing

$$e^{-\tau(H_{\text{loc}}+H_{\text{bath}})} = \lim_{N\to\infty}\left(I - \frac{\tau}{N}[H_{\text{loc}} + H_{\text{bath}}]\right)^N,$$

we see that inside the parentheses on the right-hand side the diagonal terms (dominated by 1) are positive, and the off-diagonal terms (originating from $-\frac{\tau}{N}H_{\text{bath}}$) are nonnegative. Hence, the imaginary-time evolution operator has no negative elements. The Monte Carlo weights are therefore also the trace of a product of matrices with nonnegative elements.

8.8 Scaling of the algorithms

In the weak- and strong-coupling algorithms, the average expansion orders have a simple physical interpretation. In a DMFT calculation, they yield highly accurate measurements for the potential and kinetic energy.

Let us first consider the weak-coupling algorithm, where after the introduction of auxiliary fields (Equations (8.31) and (8.32)) and the shifting of the chemical potential $H = H_1 + H_2$, with $H_1 = H_\mu + \frac{1}{2}U(n_\uparrow + n_\downarrow) + H_{\text{bath}}$ and $H_2 = Un_\uparrow n_\downarrow - \frac{1}{2}U(n_\uparrow + n_\downarrow)$.[24] It follows from (8.29) that

$$\langle -H_2 \rangle = \frac{1}{\beta}\int_0^\beta d\tau \langle -H_2(\tau)\rangle$$

$$= \frac{1}{\beta}\frac{1}{Z}\sum_{n=0}^{\infty}\frac{n+1}{(n+1)!}\int_0^\beta d\tau \int_0^\beta d\tau_1 \cdots \int_0^\beta d\tau_n$$

$$\times \text{Tr}\left[e^{-\beta H_1}\mathcal{T}(-H_2(\tau))(-H_2(\tau_n))\cdots(-H_2(\tau_1)) \right]$$

$$= \frac{1}{\beta}\frac{1}{Z}\sum_C n_C w_C = \frac{1}{\beta}\langle n\rangle, \qquad (8.69)$$

[23] Here, we don't actually need this sign convention, as these factors come in complex conjugate pairs.
[24] For simplicity, we have chosen $\delta = 0$.

and therefore the average perturbation order $\langle n \rangle$ is related to the potential energy by

$$\langle n \rangle_{\text{weak-coupling}} = -\beta U \langle n_\uparrow n_\downarrow \rangle + \tfrac{1}{2} \beta U \langle n_\uparrow + n_\downarrow \rangle = -\beta E_{\text{pot}} + \tfrac{1}{2} \beta U \langle n_\uparrow + n_\downarrow \rangle. \quad (8.70)$$

We also learn from this formula that the average perturbation order is roughly proportional to the inverse temperature β and the interaction strength U.

In the strong-coupling case, the average perturbation order is proportional to the kinetic energy. In single-site DMFT, we can express the kinetic energy

$$E_{\text{kin}} = \sum_{k\sigma} \epsilon_k G_{k\sigma}(0^-)$$

in terms of the local Green's function and hybridization function:[25]

$$E_{\text{kin}} = \sum_\sigma \int_0^\beta d\tau \, G_\sigma(\tau) \Delta^\sigma(-\tau).$$

Substituting the strong-coupling measurement formula (8.53) for G into this expression, we find

$$E_{\text{kin}} = \sum_\sigma \int_0^\beta d\tau \left\langle -\sum_{ij} \frac{1}{\beta} \delta(\tau, \tau_i - \tau_j')[M_\sigma]_{ij} \right\rangle_{\text{MC}} \Delta^\sigma(-\tau)$$

$$= -\sum_\sigma \left\langle \frac{1}{\beta} \sum_{ij} [M_\sigma]_{ij} \Delta^\sigma(\tau_j' - \tau_i) \right\rangle_{\text{MC}}.$$

---

[25] The first step in the derivation of this formula is to switch to the Fourier representation:

$$E_{\text{kin}} = \sum_{k\sigma} \epsilon_k G_{k\sigma}(0^-) = \sum_{k\sigma} \epsilon_k \frac{1}{\beta} \sum_n e^{-i\omega_n 0^-} G_{k\sigma}(i\omega_n)$$

$$= \sum_{k\sigma} \epsilon_k \frac{1}{\beta} \sum_n e^{i\omega_n 0^+} \frac{1}{i\omega_n + \mu - \epsilon_k - \Sigma_\sigma(i\omega_n)}.$$

Introducing the density of states $\mathcal{D}(\varepsilon)$, we can then write

$$E_{\text{kin}} = \sum_\sigma \frac{1}{\beta} \sum_n e^{i\omega_n 0^+} \int d\epsilon \frac{\epsilon}{i\omega_n + \mu - \epsilon - \Sigma_\sigma(i\omega_n)} \mathcal{D}(\epsilon)$$

$$= \sum_\sigma \frac{1}{\beta} \sum_n e^{i\omega_n 0^+} \int d\epsilon \frac{-[i\omega_n + \mu - \epsilon - \Sigma_\sigma(i\omega_n)] + [i\omega_n + \mu - \Sigma_\sigma(i\omega_n)]}{i\omega_n + \mu - \epsilon - \Sigma_\sigma(i\omega_n)} \mathcal{D}(\epsilon)$$

$$= \sum_\sigma \frac{1}{\beta} \sum_n e^{i\omega_n 0^+} (-1 + [i\omega_n + \mu - \Sigma_\sigma(i\omega_n)] G_{\text{loc}}^\sigma(i\omega_n)),$$

with G_{loc} the local lattice Green's function, which after convergence of the DMFT calculation is identical to the impurity Green's function G. The latter is related to the hybridization function by $G = [i\omega_n + \mu - \Sigma - \Delta]^{-1}$. Hence, we obtain

$$E_{\text{kin}} = \sum_\sigma \frac{1}{\beta} \sum_n e^{i\omega_n 0^+} G_\sigma(i\omega_n) \Delta^\sigma(i\omega_n) = \sum_\sigma \int d\tau G_\sigma(\tau) \Delta^\sigma(-\tau).$$

Now we use that $[M_\sigma]_{ij} = (-1)^{i+j} \det M_\sigma^{-1}[j,i] / \det M_\sigma^{-1}$, where $\det M_\sigma^{-1}[j,i]$ denotes the hybridization matrix with row j and column i removed. Hence, the sum

$$\sum_j (-1)^{i+j} \det M_\sigma^{-1}[j,i] \Delta^\sigma(\tau_j' - \tau_i) = \det M_\sigma^{-1}$$

appearing in the nominator is nothing but the expansion of the determinant of the hybridization matrix along column i. The expression for the kinetic energy thus simplifies to

$$E_{\text{kin}} = -\sum_\sigma \left\langle \frac{1}{\beta} \sum_i \frac{\det M_\sigma^{-1}}{\det M_\sigma^{-1}} \right\rangle_{\text{MC}} = -\frac{1}{\beta} \sum_\sigma \langle n_\sigma \rangle,$$

and the average total perturbation order $\langle n \rangle$ of the Monte Carlo configuration is related to the kinetic energy by

$$\langle n \rangle_{\text{strong-coupling}} = -\beta E_{\text{kin}}.$$

While the average expansion order in both the weak- and strong-coupling methods scales as β, the scaling of the expansion order with the interaction strength is very different. In the weak-coupling approach it grows roughly proportional to U, while in the strong-coupling approach, it decreases with increasing U (Fig. 8.13). In the case of the Anderson impurity model, this behavior leads to a significant computational speed-up for the strong-coupling approach in the intermediate- and

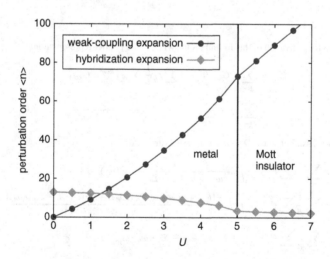

Figure 8.13 Average perturbation order for the weak-coupling and strong-coupling (hybridization expansion) algorithms. These results correspond to the DMFT solution of the one-band Hubbard model with semicircular density of states of bandwidth 4 and temperature $T = 1/30$. The bath is therefore different for each data point. (Figure adapted from Gull et al. (2007).)

Table 8.2 *Scaling of the different impurity solvers with inverse temperature β and system size L. In the case of the segment algorithm, we assume that the calculation of the determinant ratios dominates the overlap calculations. In the matrix or Krylov case, we assume that the trace calculation dominates the calculation of the determinant ratios.*

Solver	Scaling		Use
Weak-coupling	β^3	L^3	Impurity clusters with density-density interaction
Hybridization expansion (segment formalism)	β^3	L	Single-site multi-orbital models with density-density interaction
Hybridization expansion (matrix/Krylov formalism)	β	$\exp(L)$	Single-site multi-orbital models with general U_{ijkl}

large-U regime. Since local updates are $\mathcal{O}(n^2)$, a full sweep (update of all vertices in a configuration) is order $\mathcal{O}(n^3)$.

For impurity clusters, or models with complicated interaction terms, which require the matrix or Krylov formalisms discussed in Sections 8.5.3 and 8.5.4, the strong-coupling method scales exponentially with system size, and we can apply it only to relatively small systems. Here, the weak-coupling approach, if applicable, can be the method of choice. Table 8.2 gives a summary of the different scalings (assuming a diagonal hybridization) and indicates which solver is appropriate for which type of problem. The weak-coupling solvers are mainly used in cluster DMFT or DCA calculations of the Hubbard model, where the polynomial scaling allows us to treat clusters of up to 100 sites (Fuchs et al., 2011), at least in parameter regimes where there is no serious sign problem. The strong-coupling approach, on the other hand, is useful in particular for the study of (single-site) multi-orbital problems with complicated local interactions. Such problems typically have to be solved in single-site DMFT studies of strongly correlated materials or in realistic simulations of transition metal impurities (Surer et al., 2012).

Suggested reading

A. Georges, G. Kotliar, W. Krauth, and M. J. Rozenberg, "Dynamical mean-field theory of strongly correlated Fermion systems and the limit of infinite dimensions," *Rev. Mod. Phys.* **68**, 13 (1996).

A. Georges, "Strongly Correlated Electron Materials: Dynamical Mean-Field Theory and Electronic Structure," in *Lectures on the Physics of Highly Correlated Electron Systems VIII*, ed. A. Avella and F. Mancini, AIP Conference Proceedings, **715**, (2004).

T. Maier, M. Jarrell, T. Pruschke, and M. H. Hettler, "Quantum cluster theories," *Rev. Mod. Phys.* **77**, 1027 (2005).

G. Kotliar, S. Y. Savrasov, K. Haule, V. S. Oudovenko, O. Parcollet, and C. A. Marianetti, "Electronic structure calculations with dynamical mean-field theory," *Rev. Mod. Phys.* **78**, 865 (2006).

E. Gull, A. J. Millis, A. L. Lichtenstein, A. N. Rubtsov, M. Troyer, and P. Werner, "Continuous-time Monte Carlo methods for quantum impurity models," *Rev. Mod. Phys.* **83**, 349 (2011).

Exercises

8.1 Consider the semi-infinite spinless Fermion chain with Hamiltonian

$$H = \sum_{j=0}^{\infty} \left[\varepsilon_j c_j^\dagger c_j - t_j^* c_j^\dagger c_{j+1} - t_j c_{j+1}^\dagger c_j \right] \tag{8.71}$$

in a site representation. For $j \geq 1$, we define the gauge transformed operators $\tilde{c}_j$ by $c_j = \exp[i \sum_{k<j} \varphi_j] \tilde{c}_j$, where φ_j is the phase of t_j ($t_j = |t_j| e^{i\varphi_j}$). Show that in terms of the transformed operators, the Hamiltonian becomes ($c_0 \equiv \tilde{c}_0$)

$$H = \sum_{j=0}^{\infty} \left[\varepsilon_j \tilde{c}_j^\dagger \tilde{c}_j - |t_j| \tilde{c}_j^\dagger \tilde{c}_{j+1} - |t_j| \tilde{c}_{j+1}^\dagger \tilde{c}_j \right], \tag{8.72}$$

which means that all hopping terms are nonnegative.

8.2 Using the fast-update formula (8.39) for determinant ratios, derive the measurement formula (8.46) for the Green's function in the weak-coupling algorithm.

8.3 Show that the Matsubara transform of the β-antiperiodic constant function

$$\Delta(\tau) = \begin{cases} -\frac{1}{2}|V|^2 & 0 < \tau < \beta \\ \frac{1}{2}|V|^2 & -\beta < \tau < 0 \end{cases} \tag{8.73}$$

is $\Delta(i\omega_n) = |V|^2/i\omega_n$. Prove (8.67) by induction.

8.4 Show that the function

$$\Delta(\tau) = \sum_p \frac{|V_p|^2}{e^{\varepsilon_p \beta} + 1} \begin{cases} -e^{-\varepsilon_p(\tau-\beta)} & 0 < \tau < \beta \\ e^{-\varepsilon_p \tau} & -\beta < \tau < 0 \end{cases} \tag{8.74}$$

is β-antiperiodic and compute its Matsubara transform; that is, derive the formula $\Delta(i\omega_n) = \sum_p |V_p|^2/(i\omega_n - \varepsilon_p)$.

8.5 By taking into account the Fermionic anticommutivity signs, verify for perturbation orders 1, 2, and 3 that the weights of all the weak-coupling diagrams corresponding to a fixed position of the interaction vertices can be summed up into a determinant.

8.6 Verify that the weights of the six strong-coupling diagrams shown in the right-hand panel of Fig. 8.11 sum to a determinant. For a time-independent hybridization function of the form (8.73), show that the weights of the first five diagrams are positive, while the weight of the last diagram is negative. Connect this observation to the number of crossing hybridization lines.

8.7 For the local Hamiltonian H_{loc} of the Anderson impurity model and Kondo impurity model, write down the eigenstates and group them according to conserved quantum numbers.

8.8 In DCA, the cluster hybridization function Δ_K is related to the bath Green's function $\mathcal{G}_{0,K}$ by

$$i\omega_n + \mu - \bar{\epsilon}_K - \Delta_K(i\omega_n) = \mathcal{G}_{0,K}^{-1}(i\omega_n), \qquad (8.75)$$

where $\bar{\epsilon}_K$ is the average of the dispersion ϵ_k over momentum patch K. Prove this relation by calculating the high-frequency expansion of $\mathcal{G}_{0,K}^{-1}(i\omega_n) = (G_{\text{imp}}^K)^{-1}(i\omega_n) + \Sigma^K(i\omega_n)$, using (8.25), and by imposing $\Delta_K(i\omega_n) \propto \frac{1}{i\omega_n} + O(\frac{1}{(i\omega_n)^2})$. The latter condition ensures that the hybridization function has no instantaneous hopping-type contribution.

Part III

Zero temperature

9

Variational Monte Carlo

One of the reasons for the popularity of the variational Monte Carlo method is its versatility. It can be used to calculate the ground state properties of Fermion, Boson, and quantum spin systems without suffering from a sign problem. Even computing a few excited states is sometimes possible. While we can use this method as an independent quantum Monte Carlo method, we frequently resort to variational Monte Carlo calculations to generate an optimized trial state that becomes the starting point for other ground state methods. We discuss these other methods in Chapters 10 and 11. Here, we detail the variational Monte Carlo method, emphasizing its foundations and surveying the types of trial states often used. We also discuss the techniques used for the optimization of the trial state.

9.1 Variational Monte Carlo

9.1.1 The variational principle

In its simplest form, the variational Monte Carlo method uses Monte Carlo sampling to estimate the integrals implied in the Rayleigh-Ritz variational bound for the ground state energy,

$$E_0 \leq E_T = \frac{\langle \psi_T | H | \psi_T \rangle}{\langle \psi_T | \psi_T \rangle}, \tag{9.1}$$

and uses a deterministic or empirical method to adjust the trial-state's parameters to bring E_T as close to E_0 as possible. We first establish the variational principle.

Let us suppose that we are given a trial state $|\psi_T\rangle$ that is a good estimate of some eigenstate of a Hamiltonian H, an eigenstate that may or may not be the ground state. Because this trial state is in general not an eigenstate, we cannot use the standard eigenvalue-eigenstate relation $H|\psi_T\rangle = E_T|\psi_T\rangle$ to associate an energy E_T with $|\psi_T\rangle$. We may however define the residual state

$$|\phi\rangle = (H - E_T) |\psi_T\rangle,$$

whose norm provides a measure of how far $|\psi_T\rangle$ is from a true eigenstate with energy E_T. Clearly, this norm is zero only when E_T and $|\psi_T\rangle$ are some eigenpair of H. If $|\psi_T\rangle$ is not an eigenstate, we instead seek a value of E_T that minimizes the norm

$$\langle\phi|\phi\rangle = \langle(H - E_T)\psi_T|(H - E_T)\psi_T\rangle. \tag{9.2}$$

From (9.2) we have

$$\langle\phi|\phi\rangle = \langle H\psi_T|H\psi_T\rangle - E_T \langle H\psi_T|\psi_T\rangle - E_T \langle\psi_T|H\psi_T\rangle + E_T^2,$$

which after defining $\lambda = \langle\psi_T|H\psi_T\rangle/\langle\psi_T|\psi_T\rangle$ becomes

$$\langle\phi|\phi\rangle = \langle H\psi_T|H\psi_T\rangle + \langle\psi_T|\psi_T\rangle\left[(E_T - \lambda)^2 - \lambda^2\right]. \tag{9.3}$$

Because both $\langle H\psi_T|H\psi_T\rangle$ and $\langle\psi_T|\psi_T\rangle$ are positive, we minimize (9.3) by choosing E_T so that the expression in the brackets is as negative as possible. The choice

$$E_T = \lambda = \frac{\langle\psi_T|H\psi_T\rangle}{\langle\psi_T|\psi_T\rangle}$$

does this, and since H is Hermitian, we can finally write

$$E_T = \frac{\langle\psi_T|H|\psi_T\rangle}{\langle\psi_T|\psi_T\rangle}. \tag{9.4}$$

The ratio on the right-hand side is called *Rayleigh's quotient*. Its numeric value is E_T.

A standard result from linear algebra is that the minimum and maximum eigenvalues of H bound Rayleigh's quotient. To show this, we first let $E_0, E_1, \ldots, E_N$ be the eigenvalues of H in ascending order. Expressing $|\psi_T\rangle$ as a linear combination of the corresponding orthonormal eigenvectors $|\psi_i\rangle$, we have

$$E_T = \frac{\sum_{ij} a_i^* a_j \langle\psi_i |H| \psi_j\rangle}{\sum_{ij} a_i^* a_j \langle\psi_i | \psi_i\rangle} = \frac{\sum_i |a_i|^2 E_i}{\sum_i |a_i|^2}.$$

Now,

$$E_0 = \frac{\sum_i |a_i|^2 E_0}{\sum_i |a_i|^2}, \quad E_N = \frac{\sum_i |a_i|^2 E_N}{\sum_i |a_i|^2},$$

so that

$$E_0 - E_T = \frac{\sum_i |a_i|^2 (E_0 - E_i)}{\sum_i |a_i|^2} \leq 0,$$

$$E_N - E_T = \frac{\sum_i |a_i|^2 (E_N - E_i)}{\sum_i |a_i|^2} \geq 0,$$

or

$$E_0 \leq E_T \leq E_N.$$

This is a pretty powerful result, as the Rayleigh quotient for any admissible trial function must satisfy these bounds.

There are several facts that follow from the Rayleigh quotient that deserve mention. The first is that if the trial state is accurate to $\mathcal{O}(\varepsilon)$, where ε is small, then the variational energy E_T is accurate to $\mathcal{O}(\varepsilon^2)$; that is, it is accurate to a higher order. Thus, making $|\psi_T\rangle$ as close to $|\psi_0\rangle$ as possible makes E_T even closer to E_0. However, on the flip side of the coin, a value of E_T close to E_0 does not necessarily mean that $|\psi_T\rangle$ represents the physics of $|\psi_0\rangle$ as accurately as we might hope.

The second fact, or more accurately a set of related facts, has to do with the convergence to an optimal energy as a function of the parameters of the trial wave function. If the set of parameters for the trial wave function were $p = (p_1, p_2, \ldots, p_{N_{\text{opt}}})$, then we would hope that increasing the number of parameters yields a sequence $E_0 \leq \cdots \leq E_T(p_1, p_2, p_3) \leq E_T(p_1, p_2) \leq E_T(p_1)$. For example, if the ground state has a single peak in a limited region of phase space, then we would expect a trial state parameterized by the location and width of a Gaussian would yield a lower energy than a trial state parameterized by just the location or the width of a Gaussian. With just one parameter, we would need a good guess of the other to do as well as using two parameters.

One case where we can prove the systematic convergence of the trial state with increasing number of parameters is the case where we express the trial state as a linear combination of n basis states. Any basis spanning the Hilbert space of the given Hamiltonian will do. If the dimension of this space is N, but we construct a trial state $|\psi_T\rangle = \sum_{i=1}^{n} p_i |\phi_i\rangle$, where $\{|\phi_i\rangle\}_{i=1}^{N}$ is some basis in the Hilbert space, which is not necessarily orthogonal, then for $n < N$ we can show that the set of parameters $\{p_i\}_{i=1}^{n}$ that minimizes the Rayleigh quotient satisfies the generalized eigenvalue problem

$$Hp = \lambda Sp,$$

where we define the Hamiltonian matrix H and overlap matrix S by

$$H_{ij} = \langle \phi_i | H | \phi_j \rangle \text{ and } S_{ij} = \langle \phi_i | \phi_j \rangle.$$

The overlap matrix is positive-definite.

If the basis is orthonormal, then the overlap matrix is the identity matrix, and we recover the standard eigenvalue equation. We could have obtained this same equation by minimizing the quadratic form $Q = \sum_{ij} H_{ij} p_i p_j$ subject to the constraint $\sum_i p_i^2 = 1$. Let $\lambda_1 \leq \lambda_2 \leq \cdots \leq \lambda_n$ be the eigenvalues of Q. Next, suppose we reduce Q by j linear, homogeneous constraints to a quadratic form $\bar{Q}$ whose eigenvalues are $\bar{\lambda}_1 \leq \bar{\lambda}_2 \leq \cdots \leq \bar{\lambda}_{n-j}$. Then we have (Courant and Hilbert, 1965)

$$\lambda_1 \leq \bar{\lambda}_1 \leq \lambda_2 \leq \bar{\lambda}_2 \cdots \leq \bar{\lambda}_{n-j} \leq \lambda_{n+1-j}. \tag{9.5}$$

In particular, if we take $j = 1$ and $p_n = 0$, then the quadratic form Q goes to that of its first principal minor, and thus we have the result that the i-th eigenvalue of the first principal minor is at least equal to the i-th eigenvalue of the original quadratic form and at most equal to the original $(i + 1)$-th eigenvalue. We obtain a similar interlacing of eigenvalues if we repeat this process and reduce the $(n - 1)$-minor to the $(n - 2)$-minor. The importance of these interlacing results is that within the Rayleigh variational framework, we need to use a basis at least of order n to be able to bound the n-th exact eigenvalue. If we increase the basis from n to $n + 1$, we have that

$$\lambda_i \leq \lambda_i(n + 1) \leq \lambda_i(n) \leq \lambda_{i+1}$$

for a general adjacent pair of exact eigenvalues. If we keep increasing the number of basis functions, as $n \to N$, we converge to the exact result.

In Section 9.2, we discuss various common options for the choice of many-body trial states. In some cases the trial state is a single state embodying all the parameters. In others, it is a linear combination of basis states. In still others, it could be a mix of the two other cases. In Section 9.3, we present a numerical technique to execute the minimization for the mixed case. Next we discuss the basics of using the Monte Carlo method in the variational method.

9.1.2 Monte Carlo sampling

We now discuss how to evaluate (9.4) by a Monte Carlo method. With a complete orthonormal basis $\{|C\rangle\}$ satisfying $\sum_C |C\rangle\langle C| = I$ and $\langle C|C'\rangle = \delta_{CC'}$, which we call the *configuration basis*, (9.4) becomes

$$E_T = \frac{\sum_C \langle \psi_T|C\rangle \, \langle C|H|\psi_T\rangle}{\sum_C |\langle C|\psi_T\rangle|^2}.$$

We rewrite this as

$$E_T = \sum_C E(C) p(C)$$

with

$$E(C) = \frac{\langle C|H|\psi_T\rangle}{\langle C|\psi_T\rangle} \quad \text{and} \quad p(C) = \frac{|\langle C|\psi_T\rangle|^2}{\sum_C |\langle C|\psi_T\rangle|^2}.$$

$E(C)$ is called the *configuration energy*. Sampling configurations from $p(C)$ allows us to invoke the standard Monte Carlo estimate for the variational energy:

$$E_T \approx \bar{E}_T \equiv \frac{1}{M} \sum_{i=1}^{M} E(C_i). \tag{9.6}$$

We can use the Metropolis-Hastings (2.28) or a similar algorithm to sample the configurations C_i from $p(C)$. These algorithms require a user-defined proposal probability $T(C'|C)$ for changing the state $|C\rangle$ to $|C'\rangle$. We recall that the Metropolis-Hastings algorithm then *fixes* the acceptance probability $A(C'|C)$ for the proposed change to be

$$A(C' \leftarrow C) = \min\left\{1, \frac{p(C')T(C|C')}{p(C)T(C'|C)}\right\} = \min\left\{1, \frac{|\langle C'|\psi_T\rangle|^2 T(C|C')}{|\langle C|\psi_T\rangle|^2 T(C'|C)}\right\}.$$

With the same samples, we may estimate other observables by

$$\langle \mathcal{O} \rangle \approx \frac{1}{M} \sum_{i=1}^{M} \mathcal{O}(C_i), \tag{9.7a}$$

with

$$\mathcal{O}(C) = \frac{\langle C|\mathcal{O}|\psi_T\rangle}{\langle C|\psi_T\rangle}. \tag{9.7b}$$

In general, these other estimates are not variational bounds.

While the energy variational functional (9.1) is a natural starting point for optimizing a trial state, other choices are possible and at times lead to more convenient and efficient algorithms. The most common alternative uses the definition of the variance of the energy associated with H and $|\psi_T\rangle$ (Umrigar et al., 1988),

$$\sigma_E^2 \equiv \frac{\langle \phi|\phi \rangle}{\langle \psi_T|\psi_T \rangle} = \frac{\langle \psi_T|(H - E_T)^2|\psi_T \rangle}{\langle \psi_T|\psi_T \rangle}. \tag{9.8}$$

This functional is actually the same as the one we used to derive the energy variational functional (9.2). It makes explicit the fact that as $|\psi_T\rangle$ converges to the ground state, the energy variance, associated with the Rayleigh quotient, converges to zero. This fact is called the *zero variance principle*.[1] The zero variance principle also has a Monte Carlo implication: As $|\psi_T\rangle$ converges to the ground state, the configurational energy $E(C)$ converges to the ground state energy E_0 and hence becomes independent of the configurations. If it is independent of the configurations, it has zero variance.

---

[1] We note that the zero variance property is true for any eigenstate and not just the ground state.

Zero is a lower bound of (9.8), and its having a known lower bound is a feature distinguishing a variational calculation based on the variance functional from one based on the energy functional (9.1). While a lower bound for the energy functional exists, it is unknown.

We can easily show that

$$\sigma_E^2 = \sum_C [E(C) - E_T]^2 p(C),$$

with

$$E_T = \sum_C E(C)p(C) \quad \text{and} \quad p(C) = \frac{|\langle C|\psi_T\rangle|^2}{\sum_C |\langle C|\psi_T\rangle|^2}.$$

This implies that the variance computation requires the same energy estimator and sampling as the calculation based on (9.4), not a surprising result because the Rayleigh quotient is the energy estimator that minimizes the energy variance. Formally, as $|\psi_T\rangle \rightarrow |\psi_0\rangle$, the two functionals yield the same E_T and $|\psi_T\rangle$, that is, they both yield E_0 and $|\psi_0\rangle$. In practice, the computational efficiency of the two can vary widely.

Ideally, the variational Monte Carlo method would use a readily available $|\psi_T\rangle$, which is as good an estimate of $|\psi_0\rangle$ as possible. After running the simulation, the job is done. Typically, a good guess has a set of adjustable parameters. If the number of parameters is small, the minimum is most easily found graphically. If the number of parameters is large, a combination of Monte Carlo and deterministic methods is necessary. In this combination, the energy is estimated simultaneously with the optimization of the wave function; that is, we do not first optimize the wave function and then estimate the energy.

In the next section we discuss the different classes of trial states that have proven useful in variational Monte Carlo simulations of lattice models. Then, in Section 9.3, we discuss a procedure to optimize the trial state. As the parameterization of the trial states becomes more complex, this procedure becomes more useful.

9.2 Trial states

The use of trial states gives the variational Monte Carlo method, and any variational calculation for that matter, a "what you get out is what you put in" character. This character manifests itself in two ways. First, if we know the symmetry of the ground state and construct a trial state matching it, then we can bring the variational energy as close to the ground state energy as possible by increasing the parameterization of the trial state. How close we get depends on how much effort we put into crafting the trial state. For quantum lattice problems, we typically know only the symmetries of the Hamiltonian and are interested in whether one of these symmetries breaks. The variational calculation, however, does not break symmetries.

A vexing problem with a variational calculation is that while different *Ansätze* for the ground state may be orthogonal, their variational energies may be nearly identical. Which one is the correct state? Unfortunately, even if the energies differ significantly, we cannot conclude that the state with lower energy necessarily represents the ground state. Choosing an appropriate ansatz for the ground state requires additional information. In general, we craft the trial state to embrace whatever we exactly know about the ground state and then experiment. This exact information may be just that the ground state is real and square integrable.

Researchers working on electronic structure calculations for atoms, molecules, and solids in the continuum, confident that its ground state is a spin singlet, have finely honed the variational Monte Carlo method by developing powerful procedures to optimize trial states with a large number of parameters. While a variational Monte Carlo simulation on a lattice shares many of the challenges of one in the continuum, an additional complexity arises by the need to compute well more than the ground state energy. Probing novel phases of matter is one need; identifying possible quantum phase transitions is another. Addressing both these needs requires computing correlation functions very accurately. Additionally, besides models for electronic properties, lattice models for quantum spins and Bosons have to be considered. Accordingly, the lattice simulations require more classes of trial states. In the electronic continuum, up to recently, there has been the Slater-Jastrow trial state, but as we will see, this situation is evolving. We start our discussion of trial states by discussing this Slater-Jastrow class as a point for comparison and contrast, and then we discuss several classes of trial states useful for lattice models. Several of these functions have a form similar to the Slater-Jastrow construction, but others evoke quite different concepts.

9.2.1 Slater-Jastrow states

Variational Monte Carlo calculations of the electronic structure of atoms, molecules, and solids use a Slater-Jastrow trial state. In a first quantized representation this state has the form

$$\psi_T(R) = e^{J(R)} D(R),$$

where R represents the spatial coordinates $(r_1, r_2, \ldots, r_N)$ of N electrons, $J(R)$ is the Jastrow function, and $D(R)$ is a Slater determinant. A common Jastrow function has the form

$$J(R) = \sum_{i=1}^{N} \sum_{i<j}^{N} \frac{a_1 r_{ij} + a_2 r_{ij}^2 + \cdots}{1 + b_1 r_{ij} + b_2 r_{ij}^2 + \cdots}, \tag{9.9}$$

where $r_{ij} = |r_i - r_j|$, although more elaborate ones are currently being used. The *Slater determinant* is the product of a spin-up and spin-down determinant,

$$D(R) = \det\left[\psi_n^\uparrow(r_{i\uparrow})\right] \det\left[\psi_m^\downarrow(r_{j\downarrow})\right],$$

where $\psi_n^\uparrow$ and $\psi_m^\downarrow$ are single-particle orbitals for up- and down-spin electrons. The $\psi_n^\sigma(r_{i\sigma})$ in the arguments of the determinants signify a $N_\sigma \times N_\sigma$ matrix built from these orbitals. More generally,

$$D(R) = \sum_k d_k \det\left[\psi_{k,n}^\uparrow(r_{i\uparrow})\right] \det\left[\psi_{k,m}^\downarrow(r_{j\downarrow})\right]. \tag{9.10}$$

In a given application, we fix the Slater-Jastrow parameters so the trial state satisfies translational and rotational symmetries. An essential symmetry is the coordinate antisymmetrization the Pauli exclusion principle requires with respect to exchanges of electrons with the same spin. The Slater determinants clearly respect this, but for the complete trial state to do so, the Jastrow factor must be symmetric (Bosonic) with respect to such changes.

The determinants have the form of a single-electron approximation to the many-electron problem. We try to construct them by using the best possible single-particle orbitals for the physical system at hand. In principle, using relatively arbitrary functional forms is possible. In practice, we use the solutions of some mean-field theory. Mean-field theories, such as Hartree-Fock, fail to capture electron correlations adequately. The purpose of the Jastrow factor is to remedy this deficiency. For example, one important use of the Jastrow factor is capturing the correlation energy in the cusp of the wave function caused by the interaction of the electrons with the nuclei.

Typically the parameters of an assumed functional form, such as (9.9), determine the Jastrow function, while the Slater determinants contain the parameters modeling the single-particle orbitals. Both E_T and ψ_T inherit the nonlinear dependence on the parameters of the Jastrow factor and Slater determinant. The only ostensible linear dependence is on the d_k coefficients if we use a multiple state determinant ansatz (9.10).

9.2.2 Gutzwiller projected states

Hubbard model

For electronic lattice problems, the Gutzwiller trial state has a form analogous to the Slater-Jastrow trial state. Gutzwiller proposed this state in conjunction with his proposal for a multi-electron model for ferromagnetism that we now call the *Hubbard model* (Gutzwiller, 1965),

$$H = \sum_{k\sigma} \varepsilon_k n_{k\sigma} + U \sum_i n_{i\uparrow} n_{i\downarrow}. \qquad (9.11)$$

The Gutzwiller trial state has the form

$$|\psi_T\rangle = P_G |\psi_0\rangle, \qquad (9.12)$$

where the projector P_G is defined as

$$P_G = \exp\left[-\eta \sum_i n_{i\uparrow} n_{i\downarrow}\right] = \prod_i \left[1 - (1-g)\, n_{i\uparrow} n_{i\downarrow}\right]. \qquad (9.13)$$

Here, $g = \exp(-\eta)$ and $|\psi_0\rangle$ is some estimate of the ground state. By adjusting g, we can adjust the contribution of the double occupancy in $|\psi_0\rangle$ to the ground state energy estimate. From the definition of P_G, we see that when $g = 1$, it does not project out any double occupancy, but when $g = 0$, it projects out all of it. In principle, we could make the replacement $g \rightarrow g_i$. In practice, this is rarely done.

Frequently, we use some Hartree-Fock approximation to guide the selection of the orbitals defining $|\psi_0\rangle$. In Hartree-Fock, we replace the terms in the potential energy, which are quartic in the creation and destruction operators, by ones that are quadratic. For example, we could approximate (9.11) as

$$H \approx \sum_{k\sigma} \varepsilon_k n_{k\sigma} + U \sum_i \left[n_{i\uparrow} \langle n_{i\downarrow}\rangle + n_{i\downarrow} \langle n_{i\uparrow}\rangle\right] - U \sum_i \langle n_{i\uparrow}\rangle \langle n_{i\downarrow}\rangle. \qquad (9.14)$$

A Bogoliubov transformation (Blaizot and Ripka, 1986) formally diagonalizes the new quadratic form by defining creation and destruction operators that are linear combinations of the old ones. The new quadratic form, however, contains unknown scalar quantities, such as the $\langle n_{i\sigma}\rangle$ in the above formula. In deterministic uses of such approximations, these scalars constitute the set of parameters we determine self-consistently. In variational Monte Carlo calculations, instead of choosing their values self-consistently, we adjust them to minimize the energy.

We often make various assumptions to reduce the number of parameters. For example, if we assume that the $\langle n_{i\sigma}\rangle$ in (9.14) are independent of i and σ, we obtain the *restricted Hartree-Fock approximation*, which has the single unknown $\langle n\rangle$. In this case, the orbitals are plane waves, and for $|\psi_0\rangle$ we choose the Fermi sea $|FS\rangle$. To construct it, we define creation operators $\alpha_{k\sigma}^\dagger$ for putting an electron of spin σ into a plane-wave state of wave number k. These operators are simply the Fourier transform of the original operators:

$$\alpha_{k\sigma}^\dagger = \frac{1}{\sqrt{V}} \sum_\ell \exp(ik\ell)\, c_{\ell\sigma}^\dagger.$$

The *Fermi sea* is

$$|FS\rangle = \prod_k \alpha_{k\uparrow}^\dagger \prod_k \alpha_{k\downarrow} |0\rangle. \tag{9.15}$$

In restricted Hartee-Fock, the orbitals are independent of $\langle n \rangle$. To adjust $\langle n \rangle$, we simply need to cut off the products in (9.15) so the number of occupied lowest energy states equals the number of electrons. Our variational state then has g as the only adjustable parameter. In this case, the power of variational Monte Carlo is limited to estimating the energy without further approximation, avoiding, for example, the commonly used Gutzwiller approximation (Gutzwiller, 1965; Fazekas, 1999).

Making $\langle n_{i\sigma} \rangle$ independent of i, leaving $\langle n_\sigma \rangle$ unknown, yields the *unrestricted Hartree-Fock approximation*. The unrestricted Hartree-Fock approximation is useful for studying magnetic ground states. Such states have a well-defined wave number K signifying a ferromagnetic, antiferromagnetic, or incommensurate state of long-range magnetic order. The spin-density wave picture is in many respects more useful. For an antiferromagnetic spin-density-wave, for example, we take for $|\psi_0\rangle$ (Yokoyama and Shiba, 1987b)

$$|SDW\rangle = \prod_k d_{k\uparrow}^\dagger d_{k\downarrow}^\dagger |0\rangle, \tag{9.16}$$

where

$$d_{k\uparrow}^\dagger = \alpha_k c_{k\uparrow}^\dagger + \beta_k c_{k+K,\uparrow}^\dagger, \quad d_{k\downarrow}^\dagger = \alpha_k c_{k\downarrow}^\dagger - \beta_k c_{k+K,\downarrow}^\dagger \tag{9.17}$$

with

$$\alpha_k^2 = \frac{1}{2}\left(1 + \frac{\varepsilon_k}{E_k}\right), \quad \beta_k^2 = \frac{1}{2}\left(1 - \frac{\varepsilon_k}{E_k}\right), \tag{9.18}$$

and $E_k = \sqrt{\varepsilon_k^2 + \Delta^2}$ (Δ is a variational parameter). For a two-dimensional square lattice, $K = (\pi, \pi)$.

When we expect superconducting pairing correlations, we take guidance from still another type of Hartree-Fock approximation. Here, we appeal to the BCS theory of superconductivity and take for $|\psi_0\rangle$

$$|BCS\rangle = \prod_k \left(u_k + v_k c_{k\uparrow}^\dagger c_{-k\downarrow}^\dagger\right)|0\rangle, \tag{9.19}$$

where

$$u_k^2 = \frac{1}{2}\left(1 + \frac{\varepsilon_k}{E_k}\right), \quad v_k^2 = \frac{1}{2}\left(1 - \frac{\varepsilon_k}{E_k}\right), \tag{9.20}$$

and $E_k = \sqrt{\varepsilon_k^2 + \Delta_k^2}$. The parameterization of Δ_k is usually made simple with a k-dependence reflecting a particular partial wave. For example, for d-wave symmetry in two dimensions, $\Delta_k = \Delta(\cos k_x - \cos k_y)$.

We emphasize that in both (9.18) and (9.20), Δ is a variational parameter, not an order parameter as would be the case in Hartree-Fock theory. In these latter two trial states, Δ is a parameter that on adjustment hopefully helps to capture the ground state physics and makes the simulation more efficient. In contrast to Hartree-Fock, a nonzero value of Δ is not a signature of long-range order. Long-range order is signified by nonzero expectation values of the operators representing the order parameters.

Besides order parameters, the variational Monte Carlo method can be used to estimate a range of other physical quantities. For example, estimations of quasi-particle weights and dispersions (Paramekanti et al., 2001; Nave et al., 2006; Yang et al., 2007), Fermi surface deformations (Himeda and Ogata, 2000), stripe phases (Himeda et al., 2002), and spectral functions (Paramekanti et al., 2001; Yunoki et al., 2005) are possible. Edegger et al. (2007) provide a brief review of related work.

Besides including the essential physics, the trial wave function should also be easy to sample and easy to use for estimating the configuration energy and other observables. The sampling typically uses the Metropolis algorithm (Shiba, 1986; Yokoyama and Shiba, 1987a). A configuration state $|C\rangle$ is the state $|R\rangle = |r_{1\uparrow}, r_{1\downarrow}, r_{2\uparrow}, \ldots, r_{N\downarrow}\rangle = c_{r_{1\uparrow}}^\dagger c_{r_{1\downarrow}}^\dagger \cdots c_{r_{N\downarrow}}^\dagger |0\rangle$; that is, it is parametrized by the electron positions and spins. We accept or reject a proposed change either in position or in spin depending on whether

$$\langle \psi(R') | \psi(R') \rangle \geq \zeta \langle \psi(R) | \psi(R) \rangle,$$

where ζ is a uniform random number in $[0, 1]$. For the Hubbard model, we almost always use

$$\sum_{k\sigma} \varepsilon_k n_{k\sigma} = -t \sum_{\langle ij \rangle \sigma} \left(c_{i\sigma}^\dagger c_{j\sigma} + c_{j\sigma}^\dagger c_{i\sigma} \right), \tag{9.21}$$

which suggests the type of configuration changes we propose. For example, we can choose the electron position randomly and then select with equal probability whether we move the electron or flip its spin. We move the electron with equal probability to a neighboring site. If our trial state is (9.12), then (Yokoyama and Shiba, 1987a)

$$\frac{\langle \psi(R') | \psi(R') \rangle}{\langle \psi(R) | \psi(R) \rangle} = g^{2\Delta N_D} \left| \frac{\det D_\uparrow(R')}{\det D_\uparrow(R)} \right|^2 \left| \frac{\det D_\downarrow(R')}{\det D_\downarrow(R)} \right|^2,$$

where ΔN_D is the change in double occupancy, and if we move the electron with spin σ at site r_j to $r_j + a$, we have

$$E(R) = \frac{\langle R|H|\psi_T\rangle}{\langle R|\psi_T\rangle} = -t\sum_{j\sigma a} g^{2\Delta N_D} \frac{\det D_\sigma(\ldots, r_{j\sigma} + a, \ldots)}{\det D_\sigma(\ldots, r_{j\sigma}, \ldots)} + UN_D.$$

We evaluate other expectation values similarly to (9.7).

The computationally expensive step is computing the ratios of the determinants. This expense would be inhibiting if it were not for a trick that reduces the cost by a factor at least equal to the order of the matrices defining the determinants. This trick is the same as the one we used in the update procedure discussed for the determinantal methods in Chapter 7. (Also see Hammond et al. (1994), appendix B.) Briefly, the inverses of the matrices D_σ are computed and stored. Then,

$$\frac{\det(D_\sigma + \delta D_\sigma)}{\det D_\sigma} = \det\left(I + D_\sigma^{-1}\delta D_\sigma\right).$$

Because we know D_σ^{-1} and the Monte Carlo move is local, δD_σ has only one row and column that are nonzero. Hence the above determinant is effortlessly calculated. Then, if the move is accepted, we use the Sherman-Morrison-Woodbury formula given in Appendix G (Press et al., 2007, chapter 2) to update D_σ^{-1}.

As stated in (9.19), $|BCS\rangle$ lies outside of the algorithm just described because it lacks a fixed number of electrons and hence is inexpressible as a Slater determinant. Formally, we amend the situation by projecting the trial state to a fixed number of electrons

$$|\psi_T\rangle = P_N P_G |BCS\rangle.$$

Examining the issue more closely we note that

$$|BCS\rangle = \prod_k \left(u_k + v_k c_{k\uparrow}^\dagger c_{-k\uparrow}^\dagger\right)|0\rangle = \exp\left(\sum_k \phi_k c_{k\uparrow}^\dagger c_{-k\downarrow}^\dagger\right)|0\rangle$$

with $\phi_k = v_k/u_k$. Expanding the exponential and keeping only the term to the P-th power, where $P = N/2$ is the number of pairs (Bouchoud et al., 1988), we obtain

$$P_N|BCS\rangle = \left(\sum_k \phi_k c_{k\uparrow}^\dagger c_{-k\uparrow}^\dagger\right)^P |0\rangle$$

$$= \sum_{k_1 k_2 \ldots k_P} \phi_{k_1}\phi_{k_2}\cdots\phi_{k_P} c_{k_1\uparrow}^\dagger c_{k_2\uparrow}^\dagger \cdots c_{k_P\uparrow}^\dagger c_{-k_1\uparrow}^\dagger c_{-k_2\uparrow}^\dagger \cdots c_{-k_P\uparrow}^\dagger |0\rangle$$

$$= (b^\dagger)^P |0\rangle,$$

where

$$b^\dagger = \sum_k \phi_k c_{k\uparrow}^\dagger c_{-k\uparrow}^\dagger.$$

If $\phi_k = \exp(ik\ell)$, then

$$b_\ell^\dagger = \frac{1}{\sqrt{N}} \sum_i c_i^\dagger c_{i+\ell} e^{ik\ell},$$

which has the interpretation of a sum of valence bonds in the direction ℓ.

Two electrons may pair as a singlet or a triplet. Taking $\phi(r)$ to be the Fourier transform of ϕ_k, we can show for a singlet and a triplet with $S_z = 0$ that[2]

$$\langle R| P_N |BCS\rangle = \det \phi \left(r_{i\uparrow} - r_{j\downarrow} \right),$$

where $\phi(r_{i\uparrow} - r_{j\downarrow})$ is an element of a $P \times P$ matrix (Bouchoud et al., 1988; Gros, 1988).

The state

$$|RVB\rangle = P_N P_G^0 |BCS\rangle \qquad (9.22)$$

with $P_G^0 = \prod_i (1 - n_{i\uparrow} n_{i\downarrow})$, is the famous *resonant valence bond* state suggested by Anderson (1987). Anderson proposed that the resonant valence bond (RVB) state leads to superconductivity when preformed singlet pairs present in the parent insulating ground state of the Heisenberg model become mobile charged superconducting pairs when the system is hole doped. The study of such ground states for electronic (and quantum spin systems) in many respects has become as much a theory of matter as a suggestion for useful trial states (Edegger et al., 2007).

The trial states for electronic systems keep growing in sophistication with an accompanying increase in the number of variational parameters (Sorella, 2005; Tahara and Imada, 2008; Neuscamman et al., 2012). The overwhelming emphasis has been on unveiling the ground state of high-temperature superconductors. In high-temperature superconductors, the accompanying presence of antiferromagnetism has led to the increasing use of trial states that accommodate both types of long-range order. For example (Giamarchi and Lhuillier, 1990, 1991),

$$|\psi_T\rangle = P_N P_G^0 \prod_k \left(u_k + v_k d_{k\uparrow}^\dagger d_{-k\downarrow}^\dagger \right) |0\rangle$$

with the creation operators defined by (9.17) and the coefficients defined in (9.20). In practice a variety of additional projectors can prefix $|\psi_T\rangle$. Generally, these projectors introduce additional variational parameters. Examples are given in Sorella

---

[2] For a triplet with $|S_z| \neq 0$, the state is a *Pfaffian* and not a determinant (Bouchoud et al., 1988).

et al. (2002); Sorella (2005); Watanabe et al. (2006); Weber et al. (2006); Tahara and Imada (2008).

While many variational Monte Carlo calculations have been performed on the single-band Hubbard model, some have addressed other models such as the periodic Anderson model (Shiba, 1986; Shiba and Fazekas, 1990; McQueen and Wang, 1991; Oguri and Asahata, 1992; Yanagisawa, 1999; Watanabe and Ogata, 2009) and three-band Hubbard models (Oguri and Asahata, 1992; Oguri et al., 1994). As evidenced in the review by Edegger et al. (2007) close behind the Hubbard model as an object of interest is the *tJ* model, which we will discuss next. It bridges the topics of Gutzwiller projected states and RVB states, the subject of the next section. It was actually for the *tJ* model that Anderson (1987) proposed (9.22) as a ground state.

tJ model

When the repulsive on-site Coulomb interaction is large, instead of projecting all double occupancy from $|\psi_0\rangle$ with P_G^0, we can alternately replace the (Hubbard) Hamiltonian with one that is equivalent to it in the limit of $U \gg t$. By "equivalent," we mean a Hamiltonian that has the same eigenvalues and the same eigenstates, up to an overall unitary transformation, as the original one. We then perform our analysis with trial states appropriate for the physics of the equivalent Hamiltonian.

From (9.11) and (9.21), we can easily convince ourselves that for $t = 0$ the number of doubly occupied sites N_D is a good quantum number, and for a given electron density the states having the smallest N_D are the states of lowest energy. These states are highly degenerate, with the degeneracy being the greatest for the half-filled case that has one electron per lattice site. Restricting our discussion to electron densities of half-filling or less, we note that $N_D = 0$ marks the states of lowest energy. In constructing an equivalent Hamiltonian, we view the kinetic energy as a perturbation breaking the degeneracy among the $N_D = 0$ states and derive the equivalent Hamiltonian by performing second-order degenerate perturbation theory.

We can perform this perturbation theory most easily by making the unitary transformation

$$H' = e^{iS}He^{-iS} = H + \sum_{n=1}^{\infty} i^n \frac{[S_n, H]}{n!},$$

where S is some Hermitian operator and

$$S_n = [S, [S_{n-1}, H]], \quad S_1 = S.$$

Writing $H = H_0 + H_1$, we take for H_0 the Coulomb term and for H_1 the kinetic energy term. Now we expand this transformation as a power series in S,

$$H' = H_0 + H_1 + i[S, H_0] + i[S, H_1] - \tfrac{1}{2}[S, [S, H_0]] - \tfrac{1}{2}[S, [S, H_1]] + \cdots,$$

and by choosing S such that

$$H_1 + i[S, H_0] = 0,$$

we obtain an effective Hamiltonian

$$H' \approx H_0 + \tfrac{1}{2}i[S, H_1] + \cdots.$$

More generally, if we are perturbing states labeled by some quantum number n, then in a matrix representation, we would take

$$\langle n| S |n\rangle = 0,$$

$$\langle n| S |m\rangle = i\frac{\langle n| H_1 |m\rangle}{E_n - E_m},$$

where $m \neq n$. For the problem at hand, the states n are those with no sites doubly occupied, and the states m are those with one or more sites doubly occupied. Those states with just one doubly occupied site dominate. In a second-quantized representation,

$$S = -i\frac{t}{U} \sum_{\langle ij\rangle,\sigma} \left[\left(1 - n_{i,-\sigma}\right) c_{i\sigma}^{\dagger} c_{j\sigma} n_{j,-\sigma} + \left(1 - n_{j,-\sigma}\right) c_{j\sigma}^{\dagger} c_{i\sigma} n_{i,-\sigma} \right.$$

$$\left. -n_{j,-\sigma} c_{j\sigma}^{\dagger} c_{i\sigma} \left(1 - n_{i,-\sigma}\right) - n_{j,-\sigma} c_{i\sigma}^{\dagger} c_{j\sigma} \left(1 - n_{j,-\sigma}\right) \right].$$

After some algebra, we arrive at

$$H_{tJ} \equiv H' = P_G^0 \left(H_1 + H_2\right) P_G^0, \tag{9.23a}$$

with

$$H_2 = J \sum_{\langle ij\rangle} \left(\vec{S}_i \cdot \vec{S}_j - \tfrac{1}{4}n_i n_j\right) \tag{9.23b}$$

and

$$P_G^0 = \prod_i \left(1 - n_{i\uparrow} n_{i\downarrow}\right). \tag{9.23c}$$

Here

$$S_i^{\alpha} = \sum_{\sigma,\sigma'=\uparrow,\downarrow} c_{\sigma}^{\dagger} \sigma_{\sigma\sigma'}^{\alpha} c_{\sigma'}, \qquad \alpha = x, y, z,$$

where the σ^{α} are the Pauli spin matrices. The exchange interaction is $J = 4t^2/U^2$.

The effective Hamiltonian (9.23) is the tJ model. It operates only in a space where there is no double occupancy at any lattice site. As stated and used, a term dropped makes it only approximately equivalent to the Hubbard model in the large U limit. At half-filling, the lattice has an equal number of up and down electrons with one electron per site, and the tJ model reduces to the spin-$\frac{1}{2}$ antiferromagnetic Heisenberg model

$$H_{\text{Heisenberg}} = J \sum_{\langle ij \rangle} \left(\vec{S}_i \cdot \vec{S}_j - \tfrac{1}{4} \right), \quad J > 0. \tag{9.24}$$

Upon hole-doping away from half-filling, each site has zero or one electron (up or down) and the kinetic energy has the physical interpretation of holes hopping amid a background of antiferromagnetic fluctuations.

9.2.3 Valence bond states

With the introduction of the tJ model, valence bond states became prominent both as a representative of a new insulating ground state (spin liquid) and as a convenient variational basis for quantum Monte Carlo studies of Heisenberg models. Anderson suggested well prior to high-temperature superconductivity that such states are descriptors of a new type of insulating ground state (Anderson, 1973). Still prior to high-temperature superconductivity, Fazekas and Anderson suggested them as descriptors of the ground states of antiferromagnetic Heisenberg models on frustrated lattices (Fazekas and Anderson, 1974). In some sense these states have become a paradigm for novel ground states of strongly correlated electron systems that break spin-rotational and translational symmetries.

The novel ground state suggested by Anderson is the spin liquid state, a state with an energy gap and only short-range order among oppositely aligned electron spins. Some quantum spin models in one dimension have an RVB spin-liquid ground state (Majumdar and Ghosh, 1969a,b; Affleck et al., 1987, 1988), that is, a state that is a linear combination of valence bond states. Gapless spin liquids and RVB crystals are just a few of the additional proposals for novel RVB ground states. Often these states lack an order parameter. They seem especially relevant for models on frustrating lattices such as the triangular lattice. Here, we consider these states just as a convenient basis for variational Monte Carlo simulations.

In the valence bond basis, the trial state is a superposition of the valence bond states,

$$|\psi_T\rangle = \sum_V f_V |V\rangle. \tag{9.25}$$

A valence bond state is a tensor product of all possible pairs of electrons bound into singlets. These states are overcomplete, that is,

$$\sum_V |V\rangle\langle V| \neq 1, \quad \langle V|V'\rangle \neq \delta_{VV'},$$

and they number $N!/([(N/2)!]^2(N/2+1))$. Our main objective is detailing several facts about this basis that are useful in a variational Monte Carlo simulation.

A theorem of Marshall (1955) states that the ground state of the Hubbard and Heisenberg models on a bipartite lattice must have a total spin S of magnitude zero. For the Heisenberg model, the standard basis $\{|C\rangle\}$ is the set $\{|s_1, s_2, \ldots, s_n\rangle\}$ of 2^N configurations of possible eigenstates of the z-component of the spin-$\frac{1}{2}$ operator at each lattice site. It is easy to select states with total $S^z = 0$ by choosing states with half the sites having spin up and the other half having spin down, but the basis set with these states may place into the trial state expansion contributions of higher total spin. The valence bond basis is a basis with a given value of total spin. We focus only on constructing ones with $S = 0$.

We work with a tensor product state of the form

$$|(i_1, j_1)\rangle \otimes |(i_2, j_2)\rangle \otimes \cdots |(i_{N/2}, j_{N/2})\rangle = |(i_1, j_1)(i_2, j_2) \cdots (i_{N/2}, j_{N/2})\rangle,$$

where $|(i, j)\rangle$ is a *spin singlet* bond between lattices sites i and j. On a bipartite lattice, the convention is

$$|(i, j)\rangle = \frac{1}{\sqrt{2}}(|(i_\uparrow, j_\downarrow)\rangle - |(i_\downarrow, j_\uparrow)\rangle),$$

where i is an A-sublattice site and j is a B-sublattice site. This convention fixes the phase of the singlet. Without the phase convention we would have to keep track of the sign of the basis states as spins on the sites are exchanged. The convention also reduces the number of states in the basis to $(N/2)!$, but the basis is still overcomplete. The overcompleteness follows from the general relation

$$|(m, n)(k, l)\rangle + |(m, k)(l, n)\rangle + |(m, l)(n, k)\rangle = 0$$

for products of singlet states. This relation says that if the singlet pair bonds sites on the same sublattice, we can reexpress it as a sum of states where the singlets are bonding sites on different sublattices. It is straightforward to show that

$$\left(\vec{S}_i \cdot \vec{S}_j - \tfrac{1}{4}\right)|\cdots(i, j)\cdots\rangle = -|\cdots(i, j)\cdots\rangle,$$

$$\left(\vec{S}_i \cdot \vec{S}_j - \tfrac{1}{4}\right)|\cdots(i, k)\cdots(l, j)\cdots\rangle = -\tfrac{1}{2}|\cdots(i, j)\cdots(l, k)\cdots\rangle.$$

From the first equation, it follows that the magnitude of the total spin for a pair is zero, that is, $(\vec{S}_i + \vec{S}_j)^2|(i, j)\rangle = 0$.

284 Variational Monte Carlo

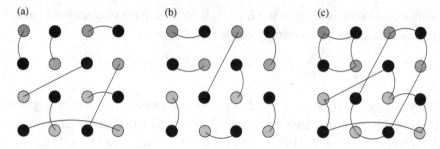

Figure 9.1 Two valence bond states (a) $|V\rangle$ and (b) $|V'\rangle$ and their overlap (c) $\langle V|V'\rangle$. The A-sublattice sites are the black dots. In the text the sites on the left edges are numbered 1, 5, 9, and 13 from the bottom up.

In Fig 9.1a and b we present two possible valence bond states for a 4×4 square lattice. The A-sublattice sites are the black dots. For convenience of discussion we will represent these states as[3]

$$|V\rangle = \begin{pmatrix} 1 & 3 & 6 & 8 & 9 & 11 & 14 & 16 \\ 4 & 12 & 2 & 7 & 13 & 5 & 10 & 15 \end{pmatrix} \tag{9.26}$$

and

$$|V'\rangle = \begin{pmatrix} 1 & 3 & 6 & 8 & 9 & 11 & 14 & 16 \\ 5 & 2 & 15 & 4 & 10 & 7 & 13 & 12 \end{pmatrix}. \tag{9.27}$$

On the top row of these arrays are the A-sublattice positions. On the bottom row are the B sites to which they are connected. This connection is the valence bond. The bond is not necessarily a physical bond. Given one member of the basis, say, $|V\rangle$, we can generate all others by permuting the $N/2$ numbers in its bottom row. With our choice of the normalization of each singlet pair,

$$\langle V|V\rangle = 1.$$

More generally,

$$\langle V|V'\rangle = 2^{N_{\text{loops}}}/2^{N/2}. \tag{9.28}$$

The latter expression says that for $\langle V|V\rangle$, $N_{\text{loops}} = N/2$. This expression also requires some explanation.

Figure 9.1c is the graph corresponding to $\langle V|V'\rangle$. It is the overlay of the graphs of Figs 9.1a and b. We see that the bonds form N_{loops} loops. In the present case $N_{\text{loops}} = 3$ and $N = 16$, so $\langle V|V'\rangle = 1/32$. If we were to define the adjoint of $|V\rangle$ by

---

[3] We leave it as an exercise for the readers to decipher the convention we used to number the lattice sites. (Hint: Start at the bottom left corner.)

$$\langle V| = \begin{pmatrix} 2 & 4 & 5 & 7 & 10 & 12 & 13 & 15 \\ 6 & 1 & 11 & 8 & 14 & 3 & 9 & 16 \end{pmatrix}, \tag{9.29}$$

where we now put the B sites on the top row, we could represent $\langle V|V'\rangle$ by interleaving (9.26) and (9.29)

$$\langle V|V'\rangle =$$

$$\begin{pmatrix} 1 & \underline{2} & 3 & \underline{4} & 5 & \underline{6} & 7 & \underline{8} & 9 & \underline{10} & 11 & \underline{12} & 13 & \underline{14} & 15 & \underline{16} \\ 5 & 6 & 2 & 1 & \underline{11} & 15 & 8 & 4 & \underline{10} & 14 & 7 & 3 & 9 & 13 & \underline{16} & 12 \end{pmatrix}.$$

In this representation the A site columns are the bonds for the ket, while the B site columns are those for the bra. Picking a site in the top row we find its bonding site in the bottom row. Taking this site back to the top row, we can find the site bonding to it. If we repeat this process until we point back to the starting site, we trace out one of the loops in the overlap graph. It is easy to demonstrate that there are only three loops in this particular case.

Besides for computing overlaps, identifying and counting loops is important because

$$\frac{\langle V| \vec{S}_i \cdot \vec{S}_j |V'\rangle}{\langle V \mid V'\rangle} = \begin{cases} 3/4, & \text{if } i,j \in \text{same loop and same sublattice,} \\ -3/4, & \text{if } i,j \in \text{same loop but different sublattices,} \\ 0, & \text{if } i,j \in \text{different loops.} \end{cases} \tag{9.30}$$

With this result we can compute the energy, the spin-spin correlation function, and the staggered magnetization, the main objects of interest for the antiferromagnetic spin-$\frac{1}{2}$ Heisenberg model.

In the valence bond basis a trial state has the form

$$|\psi_T\rangle = \sum_V f_V |V\rangle$$

$$= \sum_{\{i_1,j_1,i_2,\dots,i_{N/2}\}} f_{i_1,j_1,i_2,\dots,i_{N/2}} |(i_1,j_1)(i_2,j_2)\cdots(i_{N/2},j_{N/2})\rangle.$$

As suggested by Liang et al. (1988), we assume

$$f_{i_1,j_1,i_2,\dots,i_{N/2}} = h(i_1,j_1)h(i_2,j_2)\cdots h(i_{N/2},j_{N/2})$$

and then invoke another theorem of Marshall (1955) that for the Heisenberg model on any bipartite lattice we can take $f_{i_1,j_1,i_2,\dots,i_{N/2}} > 0$, and hence we can take $h(i,j) > 0$. We further assume that $h(i,j) = h(|i-j|)$. A variety of choices for h have been made. Liang et al. (1988), for example, optimized $|\psi_T\rangle$ by adjusting $h(i,j)$ over a few nearest neighbor distances and then described the remainder by adjusting the power α in a power law assumption for the function $h(i,j) \sim |i-j|^\alpha$. The important fact is that the trial state supports both short-range and long-range

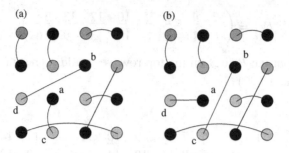

Figure 9.2 The proposed Monte Carlo move.

order depending on whether the decay is short or long range. This flexibility is one advantage of the valence bond basis.

Because the valence bond basis is nonorthogonal and overcomplete, we cannot directly use the variational Monte Carlo method described in Section 9.1.2. Instead, for those models where Marshall's positivity theorem applies, we use (9.1) and (9.25) and write

$$E_T = \frac{\sum_{VV'} f_V f_{V'} \langle V | H | V' \rangle}{\sum_{VV'} f_V f_{V'} \langle V | V' \rangle} = \sum_{VV'} H_{VV'} P_{VV'}$$

with

$$H_{VV'} = \frac{\langle V | H | V' \rangle}{\langle V | V' \rangle}, \quad P_{VV'} = \frac{f_V f_{V'} \langle V | V' \rangle}{\sum_{VV'} f_V f_{V'} \langle V | V' \rangle},$$

which suggests that we can use (9.30) to evaluate the configuration energy $H_{VV'}$ and use $f_V f_{V'} \langle V | V' \rangle$ as the sampling function. Following Liang et al., we employ the Metropolis method to do the sampling. Their suggestion was to select two A-sublattice sites, say, a and b, and then update either the $|V\rangle$ or $|V'\rangle$ associated with these sites. Let us choose $|V\rangle$. If its bond ends of the a and b sites are the sites c and d, the proposed move is creating a new state $|V''\rangle$ by interchanging the ends. We illustrate this move in Fig. 9.2. Using $P_{VV'}$, we accept the proposal according to

$$P_{\text{accept}} = \min\left\{1, \frac{f_{V''} \langle V'' | V' \rangle}{f_V \langle V | V' \rangle}\right\} = \min\left\{1, \frac{h(i_a, i_d) h(i_b, i_c) \langle V'' | V' \rangle}{h(i_a, i_c) h(i_b, i_d) \langle V | V' \rangle}\right\}.$$

We can reduce the ratio of the overlaps to

$$\frac{\langle V'' | V' \rangle}{\langle V | V' \rangle} = 2^{\Delta N_{\text{loops}}}$$

with a marked gain in efficiency. The proposed move either increases the number of loops by 1 if all sites are initially on the same loop or decreases this number by 1 if they are not. Determining which case occurs is a simple matter of a table lookup.

After accepting a move we need to update this structure. Sandvik and Evertz (2010) proposed another method with the promise of greater computational efficiency.

9.2.4 Tensor network states

We now very briefly discuss matrix product states (MPS) and projected entangled pair states (PEPS). We use the phrase *tensor network states* as an umbrella term, and a long growing list of related states is actively being developed mainly for extending the utility of the density matrix renormalization group (DMRG) method (White, 1992). A feature distinguishing these states from the ones discussed in the previous subsections is the emphasis on increasing the degree of entanglement in $|\psi_T\rangle$ within the construction of the estimate of the ground state $|\psi_0\rangle$, rather than on increasing it ex post facto by a projection operator acting on $|\psi_0\rangle$ as in, for example, a Gutzwiller state (9.12).

Until now, we have used as our basis expansion of a many-body trial wave function the form

$$|\psi_T\rangle = \sum_{i_1,i_2,...,i_N} c_{i_1,i_2,...,i_N} |\phi_{i_1}, \phi_{i_2}, ..., \phi_{i_N}\rangle, \qquad (9.31)$$

where the basis state $|\phi_{i_1}, \phi_{i_2}, ..., \phi_{i_N}\rangle$ is a tensor product state

$$|\phi_{i_1}, \phi_{i_2}, ..., \phi_{i_N}\rangle = |\phi_{i_1}\rangle \otimes |\phi_{i_2}\rangle \otimes \cdots \otimes |\phi_{i_N}\rangle.$$

With this ansatz, the space of our system is a composite of N subspaces associated with the different lattice sites. Each subspace is spanned by a basis $\{|\phi_{i_j}\rangle\}$. For quantum spin systems, for example, $|\phi_{i_j}\rangle$ was one of the two eigenstates, $|\uparrow\rangle$ and $|\downarrow\rangle$, of the S_i^z spin-$\frac{1}{2}$ operator. Generally, each subspace dimension is the same. For the spin-$\frac{1}{2}$ example, this dimension is 2. In what follows we will assume it is d.

Expansions of a physical state in such composite bases are common, in part because the system dynamics is often local. For lattice models, bases built this way have the added convenience of expanding naturally as we increase the number of lattice sites. However, as the number of sites increases, the number of basis states and hence the number of terms in the expansion (9.31) increases exponentially fast. Tensor network states also use a tensor product basis in the expansion (9.31). The new ingredient is the manner in which the amplitudes of these basis states, and hence the parametrization of the trial state, is fabricated.

Let us first make a point of contrast. Approaching the fabrication of a trial state from a generalized Hartree-Fock perspective leads to the replacement of the many-body problem by an effective one-body problem whose ground state in wave-number space is very compact:

$$|\psi_0\rangle \rightarrow |k_1, k_2, \ldots, k_{\text{Nelec}}\rangle .$$

This state has no entanglement. Various Jastrow factors and other projectors may induce entanglement, but do so only after the construction of $|\psi_0\rangle$. The new story is putting more entanglement into the ground state a priori by approximating $c_{i_1,i_2,\ldots,i_n}$ by something other than the mean-field construct $c_{i_1,i_2,\ldots,i_n} \approx c_{i_1} c_{i_2} \cdots c_{i_N}$.

We start with the MPS in one dimension. If our system has open boundary conditions, we define an MPS by

$$|\psi_T\rangle = \sum_{\substack{i_1,i_2,\ldots,i_N \\ \alpha_1,\alpha_2,\ldots,\alpha_N}} \left[A_1^{i_1}\right]_{\alpha_1} \left[A_2^{i_2}\right]_{\alpha_1\alpha_2} \cdots \left[A_{N-1}^{i_{N-1}}\right]_{\alpha_{N-2}\alpha_{N-1}} \left[A_N^{i_N}\right]_{\alpha_N} |\phi_{i_1}, \phi_{i_2}, \ldots, \phi_{i_N}\rangle.$$

More compactly,

$$|\psi_T\rangle = \sum_{i_1,i_2,\ldots,i_N} A_1^{i_1} A_2^{i_2} \cdots A_{N-1}^{i_{N-1}} A_N^{i_N} |\phi_{i_1}, \phi_{i_2}, \ldots, \phi_{i_N}\rangle. \tag{9.32}$$

To each state at each lattice site, we associate a different matrix, except for the boundary sites where we associate row and column vectors. The matrices differ at each lattice site because with open boundary conditions the properties of the system are not uniform. If the system has periodic boundary conditions, we place the same matrix at each site and define

$$|\psi_T\rangle = \sum_{i_1,i_2,\ldots,i_N} \text{Tr}\left(A^{i_1} A^{i_2} \cdots A^{i_{N-1}} A^{i_N}\right) |\phi_{i_1}, \phi_{i_2}, \ldots, \phi_{i_N}\rangle.$$

By exploiting the extra degrees of freedom associated with the A's at each lattice site, we can entangle the states on neighboring sites and coordinate this entanglement across many sites. The order of the vectors and matrices is usually denoted by D and called the *bond dimension*. This parameter, and the imposition of physical or assumed symmetries, controls the degree of parameterization of the trial state.

For MPS, as for other tensor network states, we construct the expansion coefficients $c_{i_1,i_2,\ldots,i_N}$ not only to reflect the symmetries of the lattice model but also to maximize quantum entanglement. We expect the structure of the expansion coefficients $c_{i_1,i_2,\ldots,i_N}$ to also reflect the locality of the dynamics, for example, by being small whenever the distance between any pair of sites exceeds a few lattice spacings. If so, the small expansion coefficient reduces the importance of the associated basis state. This expectation was the justification for the pairwise factorization of the expansion coefficients for the trial state in the RVB basis. In that basis, the subsystems were the spin states on pairs of lattice sites, as opposed to spin states at individual sites, and instead of involving all possible spin states, only spin singlet states are present. The RVB states are a particular class of tensor network states. Spin singlets are entangled states.

Matrix product states naturally emerge in one dimension as the fixed-point basis of the DMRG method (Takasaki et al., 1999). This deterministic ground state method works best for models with open boundary conditions and short-range interactions defined on one-dimensional lattices or on quasi-one-dimensional systems such as few-legged ladders and Cayley trees. There is some physical insight into why the method works outstandingly well for these geometries and this boundary condition. The insight centers around the *area law* (Verstraete et al., 2008; Cirac and Verstraete, 2009).

The area law applies to systems with a gap between the ground and first excited states and short-range interactions. It says that if we consider a block A of nearest-neighbor particles (say, spins), then the information theory entropy $S = -\text{Tr}\,\rho_A \ln \rho_A$ associated with the ground state's reduced density matrix for this block scales as the number of particles on the boundary of the block and not as the number inside. PEPS reduce to MPS. In both cases, we place an object at a lattice site whose number of subscripts equals its coordination number with nearest neighbor lattice sites. In two dimensions, lattice sites have a higher coordination than in one. Equally important is the fact that the number of required basis states needed scales polynomially with the size of the system. This scaling says that we in principle can construct an efficient basis from these states. Making them more efficient than they presently are is an active topic of research. In a number of cases the scaling, while polynomial, is still poor.

These basis states have also served to unveil the variational nature of the DMRG (Ostlund and Rommer, 1995; Dukelsky et al., 1998; Takasaki et al., 1999). It follows that they can be used as variational *Ansätze* in variational Monte Carlo calculations. Details of the DMRG are reviewed in Schöllwock (2005), while technical details about tensor product states and their connection to the DMRG are surveyed in McCulloch (2007), Verstraete et al. (2008), and Cirac and Verstraete (2009). Here, we will be more modest and first motivate the form of the matrix product basis and then very briefly discuss the use of tensor network states in variational Monte Carlo simulations. The use of MPS and PEPS in variational Monte Carlo simulations is a rapidly expanding topic.

To motivate the form of the MPS, we start by imagining that our system is a composite of two subsystems that have the orthonormal basis $\{|a_i\rangle\}$ spanning one subspace and the orthonormal basis $\{|b_j\rangle\}$ spanning the other. Then, creating a standard tensor product basis, we could write

$$|\psi_T\rangle = \sum_{ij} c_{ij} |a_i\rangle \otimes |b_j\rangle.$$

Schmidt's theorem (Nielsen and Chaung, 2000, chapter 2.5), however, says that a more compact representation exists: If a state $|\psi\rangle$ is in the composite of two subsystems, whose spaces are not necessarily of the same dimension, then there exists an orthonormal basis $\{|u_i\rangle\}$ and $\{|v_j\rangle\}$ for the subsystems such that

$$|\psi\rangle = \sum_i \lambda_i |u_i\rangle \otimes |v_i\rangle,$$

with $\sum_i \lambda_i^2 = 1$. The λ_i are nonnegative real numbers. The number of nonzero values of λ_i is called the *Schmidt number*. It equals the dimension of the smaller of the two subspaces. If the Schmidt number equals 1, $|\psi\rangle = |u\rangle \otimes |v\rangle$ is *unentangled* (a pure state). If the Schmidt number is greater than 1, then the state is *entangled*. Accordingly, the Schmidt number is a measure of the degree of entanglement in $|\psi\rangle$. The Schmidt decomposition works for only two subsystems at a time and is unique up to a unitary transformation of any subsystem basis.

We now apply the Schmidt decomposition to the tensor product expansion (9.31), first taking one subsystem spanned by the states $\{|\phi_{i_1}\rangle\}$ and the other by the states $\{|\phi_{i_2}, \phi_{i_3}, \ldots, \phi_{i_N}\rangle\}$. Invoking Schmidt's theorem, we thus have

$$|\psi_T\rangle = \sum_{\alpha_1} \lambda_{\alpha_1}^2 |\psi_{\alpha_1}\rangle |\psi_{\alpha_1}^{[2,N]}\rangle$$

Because the basis sets $\{|\phi_{i_1}\rangle\}$ and $\{|\psi_{\alpha_1}\rangle\}$ span the same space, we can transform back to our original basis $|\phi_{i_1}\rangle$ and thus write

$$|\psi_T\rangle = \sum_{\alpha_1} \sum_{i_1} \lambda_{\alpha_1}^2 \left[\Lambda_1^{i_1}\right]_{\alpha_1} |\phi_{i_1}\rangle |\psi_{\alpha_1}^{[2,N]}\rangle.$$

Next, we apply the Schmidt decomposition to $|\psi_{\alpha_1}^{[2,N]}\rangle$ and again transform the leading state so that

$$|\psi_T\rangle = \sum_{\alpha_1,\alpha_2} \sum_{i_1,i_2} \lambda_{\alpha_1}^2 \left[\Lambda_1^{i_1}\right]_{\alpha_1} \lambda_{\alpha_2}^2 \left[\Lambda_2^{i_2}\right]_{\alpha_1\alpha_2} |\phi_{i_1}\rangle |\phi_{i_2}\rangle |\psi_{\alpha_2}^{[3,N]}\rangle.$$

Repeating these steps until we have visited every site yields a messy looking expression but one identical in form to (9.32). These steps produce a parameterization of $c_{i_1,i_2,\ldots,i_n}$ in terms of coefficients in the matrix product expansion. Clearly, the parameterization can be rich.

When we use tensor network states in variational Monte Carlo calculations, the basic algorithm remains as previously described in Section 9.1.2, with

$$P(C) = P(\phi_{i_1}, \phi_{i_2}, \ldots, \phi_{i_N}) \propto \left|\text{Tr} A^{i_1} A^{i_2} \cdots A^{i_N}\right|^2.$$

One difference for their use is the computational cost of computing the probability ratio $P(C')/P(C)$ because of the overhead associated with the computation of the matrix products. It is important to exploit the cyclic nature of the trace operation, store partial matrix products, and update them as opposed to the entire string of products. Details about the efficient computation of matrix elements, their products, and expectation values of operators can be found in such articles as Sandvik and Vidal (2007), Schuch et al. (2008), Sfondrini et al. (2010), and Wang et al. (2011a). The field is still evolving, and the efficiency will likely increase further.

Efficiency is often achieved by exploiting the flexibility of the tensor network concept. There is no need to restrict entanglement to nearest neighbor pairs. We can entangle the spins in a nearest neighbor plaquette of spins. In this regard we can also factorize the expansion coefficients as a product of pairs (Changlani et al., 2009; Marti et al., 2010; Boguslawski et al., 2011; Neuscamman et al., 2011; Neuscamman et al., 2012), in the manner of Liang et al. (1988) but with regularity in the placement of singlet pairs, as products of plaquettes (Sandvik, 2008; Mezzacapo et al., 2009; Mezzacapo, 2011; Wang et al., 2011b), or as products of strings (Schuch et al., 2008; Sfondrini et al., 2010). We can include triplets (Neuscamman et al., 2012) as well as singlets. We can also bring projectors back into the game. Tensor network proposals exist for variational Monte Carlo simulations of Bosons, electrons (Corboz et al., 2010a,b; Neuscamman et al., 2012), quantum spin systems, and more recently electron systems in the continuum (Marti et al., 2010; Boguslawski et al., 2011; Neuscamman et al., 2012). While our presentation was mainly from the viewpoint of one dimension, most of the computational activity focuses on two dimensions.

With the increased parameterization of the wave function, the minimization of the trial energy or its energy variance becomes a very important part of the simulations (Schuch et al., 2008; Sfondrini et al., 2010; Wang et al., 2011b). In some instances, specific optimization techniques nicely accommodate specific tensor network *Ansätze*. In other cases, the optimizations follow procedures similar to the ones we will now discuss. The focus of this discussion is a recently proposed general-purpose trial state method, suitable for problems with both small and large parameter sets and with or without Jastrow-type projectors. The work of Neuscamman et al. (2012) used this method.

9.3 Trial-state optimization

The obvious method for trial-state optimization is to minimize the variational energy evaluated on a set of Monte Carlo samples. In fact, until about 1988, this was the commonly used method. Roughly speaking, the method was to start with a good trial state $|\psi_T(p)\rangle$ that depended on a set of parameters $p = (p_1, p_2, \ldots, p_{N_{opt}})$ and then use the modulus square of this state as the distribution function in a variational Monte Carlo sampling of a set of configurations $\{C_1, C_2, \ldots, C_{N_{config}}\}$

sufficiently large so that the sum $\sum_i |\psi_T(C_i,p)\rangle/N_{\text{config}}$ approximates the ground state reasonably well. Next, with the ensemble of configurations fixed, successive sets of new parameters were generated, each having small variations relative to the previous set. A "greedy" Monte Carlo was used to accept or reject a new state depending on whether the variational energy was raised or lowered.

The downside of this method is that the number of Monte Carlo samples in the ensemble of configurations must be many orders of magnitude larger than the number of parameters being optimized. Consequently, only a few parameters could be optimized. The reason why this imbalance is necessary is that a multiple parameter trial wave function has the flexibility to fit the data well, possibly leading to parameter sets for which the variational energy on the ensemble is lower than the true energy. However, when these parameters are used to evaluate the variational energy of a new independently drawn ensemble of configurations, a much higher energy might result. Minimizing the variance of the configuration energy (Bartlett et al., 1935; Coldwell, 1977; Umrigar et al., 1988) instead of minimizing the configurational energy was found to overcome this problem. Doing this variance minimization was the standard procedure in the time period between 1988 and 2001.

More sophisticated methods that efficiently optimize the energy or, better yet, an arbitrary linear combination of the energy and the variance of the local energy now exist (Lin et al., 2000; Nightingale and Melik-Alaverdian, 2001; Schautz and Filippi, 2004; Sorella, 2005; Umrigar and Filippi, 2005; Toulouse and Umrigar, 2007, 2008; Umrigar et al., 2007). We discuss here only the most efficient of these methods, originally proposed by Nightingale and Melik-Alaverdian (2001) for optimizing both ground and excited-state trial wave functions. Their trial wave functions have linear and nonlinear dependencies on the parameters. They optimize the linear parameters by minimizing the energy and optimize the nonlinear parameters by minimizing the variance. The Nightingale approach was extended to allow optimization of both linear and nonlinear parameters by energy minimization (Toulouse and Umrigar, 2007; Umrigar et al., 2007), and then further extended to optimize an arbitrary linear combination of the energy and the variance (Toulouse and Umrigar, 2008). We focus primarily on this last optimization method and refer to it as the "linear method" even though it optimizes linear and nonlinear parameters. The various methods just mentioned were developed in the context of many-electron problems in the continuum and hence for trial states of the Slater-Jastrow form. So in the literature some algorithmic detail may focus on optimally handling this form, but the basic methods are generally applicable and in fact have been applied to lattice problems (Changlani et al., 2009).

9.3.1 Linear method

At this point, it becomes convenient to adopt the compact notation of the seminal papers. In particular, we take our configuration basis $\{|C\rangle\}$ to be the basis set $\{|R\rangle\}$ of labeled positions of N particles $R = (r_1, r_2, \ldots, r_N)$:

$$E(C) = \frac{\langle C|H|\psi_T\rangle}{\langle C|\psi_T\rangle} \rightarrow E(R) = \frac{\langle R|H|\psi_T\rangle}{\langle R|\psi_T\rangle}.$$

We next use a functional notation for projections into this basis, e.g., $\psi_T(R) = \langle R|\psi_T\rangle$. Finally we drop the subscript T so that $\psi(R) = \psi_T(R)$. We note that, as in the case with the configuration energy, when $\psi(R)$ becomes an exact eigenstate, $E(R)$ is constant and equal to the energy of that state. Thus, it is possible to optimize the trial state by minimizing the variance of the configuration energy. In the continuum, the configuration energy is called the local energy.

At each optimization step, we expand the trial wave function to linear order in $\Delta p = p - p^0$ around the current set of parameters p_0,

$$\psi_{\text{lin}}(R, p) = \psi_0(R) + \sum_{i=1}^{N_{\text{opt}}} \psi_i(R)\,\Delta p_i, \qquad (9.33)$$

where $\psi_0(R) = \psi(R, p_0)$ is the current wave function (not yet the ground state) and

$$\psi_i = \partial\langle R|\psi(p)\rangle/\partial p_i|_{p=p_0} \equiv (\partial\psi(p)/\partial p_i)_{p=p_0}$$

are the N_{opt} derivatives of the wave function with respect to the parameters. This linearized wave function becomes the trial state used in the variational principle for the energy. With an infinite number of Monte Carlo samples, the optimal parameter variations Δp minimizing the energy calculated with the linearized wave function of (9.33) are the lowest eigenvalue solutions of the resulting generalized eigenvalue equation

$$H\Delta p = E\,S\Delta p, \qquad (9.34)$$

where H and S are the Hamiltonian and overlap matrices in the $(N_{\text{opt}} + 1)$-dimensional basis formed by the current wave function and its derivatives $\{\psi_0, \psi_1, \psi_2, \ldots, \psi_{N_{\text{opt}}}\}$. Here, $\Delta p_0 = 1$. Drawing a finite number of Monte Carlo samples, we estimate these matrices as

$$H_{ij} = \langle \psi_i | H | \psi_j \rangle = \int dR \, \frac{\langle \psi_i | R \rangle \, \langle R | H | \psi_j \rangle}{\langle \psi_0 | R \rangle \, \langle R | \psi_0 \rangle} \, \langle R | \psi_0 \rangle^2,$$

$$\equiv \left\langle \frac{\psi_i}{\psi_0} \frac{H \psi_j}{\psi_0} \right\rangle_{\psi_0^2}$$

$$= \begin{cases} \langle E \rangle_{\psi_0^2} & \text{if } i = j = 0, \\[2ex] \left\langle \frac{\psi_i}{\psi_0} \frac{\psi_j}{\psi_0} E \right\rangle_{\psi_0^2} + \left\langle \frac{\psi_i}{\psi_0} E_{,j} \right\rangle_{\psi_0^2} & \text{if } i \neq 0 \text{ and } j \neq 0, \end{cases}$$

$$S_{ij} = \langle \psi_i | \psi_j \rangle \equiv \left\langle \frac{\psi_i}{\psi_0} \frac{\psi_j}{\psi_0} \right\rangle_{\psi_0^2},$$

where

$$\left\langle \frac{\psi_i}{\psi_0} \frac{A \psi_j}{\psi_0} \right\rangle_{\psi_0^2} = \frac{1}{N_{\text{config}}} \sum_{k=1}^{N_{\text{config}}} \frac{\psi_i(R_k)}{\psi_0(R_k)} \frac{A \psi_j(R_k)}{\psi_0(R_k)},$$

and

$$E_{,j} = \frac{\partial}{\partial p_j} \left(\frac{H\psi}{\psi} \right) = \frac{H\psi_j}{\psi} - \frac{\psi_j}{\psi_0} E.$$

On an infinite sample $\langle j \rangle = 0$ and

$$\left\langle \frac{\psi_i}{\psi_0} E_{,i} \right\rangle_{\psi_0^2} = \left\langle \frac{\psi_j}{\psi_0} E_{,i} \right\rangle_{\psi_0^2},$$

so that H_{ij} is symmetric, but this is not so on a finite sample. Since the true Hamiltonian matrix is symmetric and because a nonsymmetric generalized eigenvalue problem can have complex eigenvalues, we might think it is desirable to symmetrize the matrix. This asymmetry is actually desirable because it creates a *strong zero variance property* (Nightingale and Melik-Alaverdian, 2001) whereby the variance of each component of Δp vanishes not only in the limit of $\psi_T(R, p)$ becoming exact but also in the limit of just $\psi_{\text{lin}}(R, p)$ becoming exact, that is, in the limit of the basis states spanning an invariant subspace of the Hamiltonian. In practice, the asymmetric Hamiltonian reduces the parameter fluctuations by one to two orders of magnitude.

Updating the parameters

Part of the optimization algorithm is now in place. At each step in the iteration we linearize ψ about p_0 and estimate H_{ij} and S_{ij} by Monte Carlo sampling. Then, using readily available software, we solve the generalized eigenvalue problem for the eigenvector Δp with the lowest eigenvalue. Having obtained the parameter variations Δp by solving (9.34), the next question is: How do we use them to update the parameters in the trial wave function?

The simplest procedure of incrementing the current parameters by Δp, $p_0 \leftarrow p_0 + \Delta p$ works for the linear parameters but will not work for the nonlinear parameters if the linear approximation of (9.33) is not good. Adjusting the parameters in ψ so it fits the optimized linear form is one possible strategy. Umrigar and coworkers suggest a simpler alternative that exploits the freedom in the normalization of the trial state. The arbitrariness in the choice of the normalization of ψ leads to an arbitrariness in the solution for the parameters. For linear parameters, the same physical wave function is obtained but with a different normalization, while for nonlinear parameters it changes the predicted wave function at each optimization step.

We now consider the differently normalized wave function $\overline{\psi}(R, p) = N(p)\psi(R, p)$ for which we require $\overline{\psi}(R, p_0) = \psi(R, p_0) \equiv \psi_0(R)$ and $N(p)$ depends only on the nonlinear parameters. Then, the derivatives of $\overline{\psi}(p)$ at $p = p_0$ are

$$\overline{\psi}_i = \psi_i + N_i \psi_0, \quad \text{where} \quad N_i = (\partial N(p)/\partial p_i)_{p=p_0}. \tag{9.35}$$

The linear approximation to $\overline{\psi}(p)$ is

$$\overline{\psi}_{\text{lin}} = \psi_0 + \sum_{i=1}^{N_{\text{opt}}} \Delta \bar{p}_i \, \overline{\psi}_i.$$

Since we obtain $\overline{\psi}_{\text{lin}}$ and ψ_{lin} by optimization in the same variational space, they are proportional to each other,

$$\overline{\psi}_{\text{lin}} = c \psi_{\text{lin}},$$

$$\psi_0 + \sum_{i=1}^{N_{\text{opt}}} \Delta \bar{p}_i \, \overline{\psi}_i = c \left(\psi_0 + \sum_{i=1}^{N_{\text{opt}}} \Delta p_i \, \psi_i \right) = c \left(\psi_0 + \sum_{i=1}^{N_{\text{opt}}} \Delta p_i \left(\overline{\psi}_i - N_i \psi_0 \right) \right)$$

$$= c \left(\left(1 - \sum_{i=1}^{N_{\text{opt}}} \Delta p_i N_i \right) \psi_0 + \sum_{i=1}^{N_{\text{opt}}} \Delta p_i \, \overline{\psi}_i \right),$$

so $c = 1/(1 - \sum_{i=1}^{N_{\text{opt}}} \Delta p_i N_i)$ and $\Delta \bar{p}$ is related to Δp by a uniform rescaling

$$\Delta \bar{p} = \frac{\Delta p}{1 - \sum_{i=1}^{N_{\text{opt}}} N_i \Delta p_i}. \tag{9.36}$$

Also, since the rescaling factor can be anywhere between $-\infty$ and ∞, the choice of normalization can affect not only the magnitude of the parameter changes but also the sign.

We have yet to discuss the choice of the parameter gradients of the normalization factors N_i. A good choice, employed by Sorella (2005) in a different optimization method, is to choose N_i such that the parameter derivatives of the trial state $\overline{\psi}_i$ are orthogonal to ψ_0. Umrigar and coworkers employ a more general choice: They choose N_i such that

$$\left\langle \xi \frac{\psi_0}{||\psi_0||} + (1 - \xi) \frac{\psi_{\text{lin}}}{||\psi_{\text{lin}}||} \middle| \overline{\psi}_i \right\rangle = 0,$$

where ξ is a constant between 0 and 1, resulting in

$$N_i = -\frac{\xi D S_{0i} + (1 - \xi)(S_{0i} + \sum_j^{\text{nonlin}} S_{ij} \Delta p_j)}{\xi D + (1 - \xi)(1 + \sum_j^{\text{nonlin}} S_{0j} \Delta p_j)}, \tag{9.37}$$

where $D = \sqrt{1 + 2 \sum_j^{\text{nonlin}} S_{0j} \Delta p_j + \sum_{jk}^{\text{nonlin}} S_{jk} \Delta p_j \Delta p_k}$ and the sums are over only the nonlinear parameters. The choice $\xi = 0$ is that of Sorella and the choice $\xi = 1/2$ imposes the normalization of $\overline{\psi}_{\text{lin}}$ to be the same as that of ψ_0.

In the above description, we followed the formulation of Umrigar et al. (2007) in that we calculate Δp by solving the generalized eigenvalue problem with the Hamiltonian and overlap matrices in the original basis and then calculated $\Delta\overline{p}$ using (9.36). Alternatively, we could have followed the formulation in Toulouse and Umrigar (2007) and constructed $\Delta\overline{p}$ directly from the Hamiltonian and overlap matrices in the semi-orthogonalized basis. These two alternatives lead to identical parameter changes. However, if stabilization is used (a topic that will be discussed later), these alternatives lead to somewhat different parameter changes.

9.3.2 Newton's method

Before we discuss using the linear method for minimizing the energy variance, we make an important digression that unveils the linear method as a stabilized Newton method that, in practice, often converges faster than the standard Newton method.

Newton's method is a commonly used multivariate minimizer (Press et al., 2007, chapter 9). It starts by considering the truncated Taylor series expansion of the function $E(p)$ about some point p_0,

$$E(p) \approx E(p_0) + \Delta p^T \cdot g + \tfrac{1}{2} \Delta p^T \cdot h \cdot \Delta p, \tag{9.38}$$

where g is the gradient vector of E,

$$g_i = \left.\frac{\partial E}{\partial p_i}\right|_{p_0},$$

and h is the Hessian matrix of E,

$$h_{ij} = \left.\frac{\partial^2 E}{\partial p_i \partial p_j}\right|_{p_0}.$$

Minimizing this quadratic expression, we find that the optimal parameters are given by

$$\Delta p = -h^{-1}g.$$

Because $E(p)$ is not in general a quadratic function of p, it is necessary to iterate[4]

1. Solve $h \cdot \Delta p = -g$ for Δp
2. $p_0 \leftarrow p_0 + \Delta p$,

until the norm of Δp satisfies some tolerance condition (Algorithm 30).

With respect to trial-state optimization, Umrigar and Filippi (2005) provide zero-variance expressions for the gradient g and a reduced variance expression for the Hessian h, using the observation that under certain circumstances the fluctuations of a covariance $\mathrm{cov}(X, Y) = \langle (X - \langle X\rangle)(Y - \langle Y\rangle)\rangle = \langle XY\rangle - \langle X\rangle\langle Y\rangle$ of two random variables X and Y are much smaller than those of a product $\langle XY\rangle$. These zero-variance expressions typically result in a gain in efficiency of two to three orders in magnitude.

It is apparent that we can use Newton's method to minimize not only the energy but also the variance of the local energy and an arbitrary linear combination of the

---

Algorithm 30 Newton's method.

Input: p_0 and g_{tol} and a procedure to calculating $g_k = \left.\frac{\partial E}{\partial p}\right|_{p_k}$ and $h_k = \left.\frac{\partial^2 E}{\partial p^2}\right|_{p_k}$.
$k = 0$;
repeat
    Solve $h_k \delta p_k = -g_k$ for δp_k;
    $p_{k+1} = p_k + \delta p_k$;
    $k \leftarrow k + 1$;
until $\|g_{k+1}\| < g_{\mathrm{tol}}$.
return p_{k+1}.

---

[4] For a multivariate quadratic form, Newton's method converges in one step. Hence it is highly efficient as it approaches the minimum. The challenge is getting near the minimum. When the eigenvalues of the Hessian range widely, the linear system of equations in Newton's method becomes difficult to solve accurately.

two. We just need to replace the expression in (9.38) with a quadratic approximation to the desired linear combination.

9.3.3 Connection between linear and Newton methods

We now establish the connection between the linear method and Newton's method (Toulouse and Umrigar, 2008). We start by reexpressing the energy (9.38) in a more revealing block matrix form

$$E(p) = \frac{\begin{pmatrix} 1 & \Delta p^{\mathrm{T}} \end{pmatrix} \begin{pmatrix} E_0 & g^{\mathrm{T}}/2 \\ g/2 & \bar{H} \end{pmatrix} \begin{pmatrix} 1 \\ \Delta p \end{pmatrix}}{\begin{pmatrix} 1 & \Delta p^{\mathrm{T}} \end{pmatrix} \begin{pmatrix} 1 & 0^{\mathrm{T}} \\ 0 & \bar{S} \end{pmatrix} \begin{pmatrix} 1 \\ \Delta p \end{pmatrix}},$$

where $\bar{S}$ is the overlap matrix and

$$\bar{H} = h/2 + E_0 \bar{S}$$

is the $N_{\mathrm{opt}} \times N_{\mathrm{opt}}$ Hamiltonian matrix in the semi-orthogonalized basis with $\xi = 0$. Minimizing this energy with respect to p yields a generalized eigenvalue equation,

$$\begin{pmatrix} E_0 & g^{\mathrm{T}}/2 \\ g/2 & \bar{H} \end{pmatrix} \begin{pmatrix} 1 \\ \Delta p \end{pmatrix} = E_{\mathrm{lin}} \begin{pmatrix} 1 & 0^{\mathrm{T}} \\ 0 & \bar{S} \end{pmatrix} \begin{pmatrix} 1 \\ \Delta p \end{pmatrix},$$

as is the case in the linear method, except that we obtain a symmetric Hamiltonian matrix.

9.3.4 Energy variance optimization

Having established the connection between the linear method and Newton's method, we now modify the linear method to optimize a linear combination of the energy and the variance of the local energy. In analogy with (9.38), all that is required is a quadratic approximation to the variance,

$$V = V_0 + g_V^{\mathrm{T}} \cdot \Delta p + \frac{1}{2} \Delta p^{\mathrm{T}} \cdot h_V \cdot \Delta p.$$

Minimization leads to the following generalized eigenvalue equation

$$\begin{pmatrix} V_0 & g_V^{\mathrm{T}}/2 \\ g_V/2 & h_V/2 + V_0 \bar{S} \end{pmatrix} \begin{pmatrix} 1 \\ \Delta p \end{pmatrix} = V_{\mathrm{min}} \begin{pmatrix} 1 & 0^{\mathrm{T}} \\ 0 & \bar{S} \end{pmatrix} \begin{pmatrix} 1 \\ \Delta p \end{pmatrix}.$$

Toulouse and Umrigar (2008) give explicit expressions for g_V and h_V.

9.3.5 Stabilization

Both in the linear method and in Newton's method, the updated parameters may be unstable if a quadratic approximation for the energy is too poor or if the number of Monte Carlo samples used to evaluate the H and S matrices is too small. We can stabilize both methods in the following manner: Newton's method is stabilized by adding a positive constant $a_{\text{diag}} \geq 0$ to the diagonal elements of the Hessian matrix, so that the proposed moves get smaller and rotate toward the steepest-descent direction. We achieve the same effect in the linear method by adding a_{diag} to the diagonal of the Hamiltonian matrix: $H_{ij} \rightarrow H_{ij} + a_{\text{diag}} \delta_{ij}(1 - \delta_{i0})$. As a_{diag} becomes larger, the parameter variations Δp become smaller and rotate. We can automatically adjust the value of a_{diag} at each optimization step as follows: Once we compute the matrices H and S, we use three values of a_{diag}, differing from each other by factors of 10, to predict three new wave functions. Using correlated sampling, we then perform a short Monte Carlo run to compute energy differences of these wave functions more accurately than the energy itself. With these correlated energies we calculate a near optimal value of a_{diag}. By *correlated sampling*, we mean that instead of doing independent Monte Carlo runs to estimate related quantities, we use the N_{config} Monte Carlo configurations from a single Monte Carlo run to estimate them. The advantage is that the relative errors in these quantities is much smaller than the absolute error of the individual quantities. In our example, the related quantities are the energies of the three wave functions. We discuss correlated sampling in Appendix N.

9.3.6 Summary of the linear and Newton's optimization methods

We now summarize the linear method (Algorithm 31) and Newton's method (Algorithm 32) in a manner that emphasizes their common features. Both methods are iterative and alternate between two Monte Carlo runs. In particular, in the

---

Algorithm 31 Linearized optimization method.

Input: parameters p and tolerance δ.

    repeat

        Do a Monte Carlo run to estimate the Hamiltonian H_{ij} and overlap matrices S_{ij} ;

        Solve the generalized eigenvalue problem with three values of a_{diag} to calculate three sets of rescaled parameter increments ;

        Do a short Monte Carlo run with correlated sampling to calculate the relative energies of these three wave functions and thereby find a near optimal a_{diag} ;

        Set $p = p_0 + \Delta \bar{p}$ using the near optimal a_{diag} ;

    until $|\Delta \bar{p}| < \delta$.

    return the optimized p.

Algorithm 32 Newton's optimization method.

Input: parameters p and tolerance δ.

   repeat

      Do a Monte Carlo run to estimate the Hessian matrix h_{ij} and the gradient vector g_i ;

      Solve the linear equation with three values of a_{diag} to calculate three sets of rescaled parameter increments ;

      Do a short Monte Carlo run with correlated sampling to calculate the relative energies of these three wave functions and thereby find a near optimal a_{diag} ;

      Set $p = p_0 + \Delta\bar{p}$ using the near optimal a_{diag} ;

   until $|\Delta\bar{p}| < \delta$.

   return the optimized p.

---

linear and Newton methods, we first use one kind of Monte Carlo run to estimate important matrices. In the linear method, we use Monte Carlo sampling to calculate the overlap and Hamiltonian matrices. With the Hamiltonian and overlap matrices, we then solve a generalized eigenvalue equation to obtain the proposed parameter changes. To obtain the proposed parameter changes in Newton's method, we use Monte Carlo sampling to calculate the gradient vector and the Hessian matrix, and then we solve a system of linear equations. We solve either the eigenvalue problem or the linear system for three different values of a_{diag} to obtain three sets of proposed new parameters. Then, we do the second kind of Monte Carlo run that employs correlated sampling (Appendix N) to obtain the differences in the objective function (a linear combination of energy and energy variance) to greater accuracy than the individual values. Using these computed differences, we obtain a near optimal value of a_{diag} that we then use to obtain the parameters for the next iteration cycle.

Suggested reading

C. Gros, "Physics of projected wavefunctions," *Ann. Phys.* **189**, 53 (1989).

B. L. Hammond, W. A. Lester, Jr., and P. J. Reynolds, *Monte Carlo Methods for ab initio Quantum Chemistry* (Singapore: World-Scientific, 1994), chapters 2 and 5, and appendix B.

J. Toulouse and C. J. Umrigar, "Full optimization of Slater-Jastrow wave functions with application to the first-row atoms and homonuclear diatomic atoms," *J. Chem. Phys.* **128**, 174101 (2008).

J. I. Cirac and F. Verstraete, "Renormalization and tensor product states in spin chains and lattices," *J. Phys. A: Math. Theor.* **42**, 504004 (2009).

S. Sorella, "Variational Monte Carlo and Markov chains for computational physics," in *Strongly Correlated Systems*, Springer Series in Solid-State Physics **176**, ed. A. Avella and F. Mancini (Heidelberg: Springer-Verlag, 2013).

Exercises

9.1 If a trial state is accurate to $\mathcal{O}(\varepsilon)$, show that the Rayleigh quotient produces an energy estimate accurate to $\mathcal{O}(\varepsilon^2)$.

9.2 If A is a Hermitian matrix of order N and B is a Hermitian matrix defined in block form by

$$B = \begin{pmatrix} A & a \\ a^H & \alpha \end{pmatrix},$$

where a is a vector with N components and α is a scalar, show that the eigenvalues $\lambda_1 \leq \lambda_2 \leq \cdots \leq \lambda_{N+1}$ of B are interleaved by the eigenvalues of A. If A is positive definite, what is the sufficient and necessary condition that implies that B is also?

9.3 Consider a two-site electronic system without periodic boundary conditions (an H_2 molecule) and as Hamiltonian the Heisenberg antiferromagnet, the tJ model, and the positive U Hubbard model, respectively.

1. In each case analytically find the model's eigenvalues and eigenvectors, for electron fillings of one per site and less as appropriate for the model.
2. In the large U limit, map the energies and states of the Hubbard model to those of the tJ model.
3. For half-filling, connect the energies and states of the tJ model to those of the Heisenberg model.

9.4 Verify the Gutzwiller projection relation (9.13).

9.5 Instead of projecting the BCS variational wave function to a fixed N, Yokoyama and Shiba (1988) achieved the same result by a change of variables:

$$d_k^\dagger = c_{-k\downarrow}, \quad d_k = c_{-k\downarrow}^\dagger.$$

With an appropriately defined vacuum $|\bar{0}\rangle$ for the c and d particles such that $c_k|\bar{0}\rangle = d_k|\bar{0}\rangle = 0$, one finds

$$|BCS\rangle = \prod_k \left(u_k d_k^\dagger + v_k c_k^\dagger \right) |\bar{0}\rangle.$$

How is $|\bar{0}\rangle$ related to the original vacuum $|0\rangle$?

9.6 Derive (9.28), the expression for the overlap of two valence bond states, and (9.30), the expectation value of the spin-spin correlation function.

10

Power methods

Quantum Monte Carlo versions of the power method for finding ground states come with different names, including the projector method, diffusion Monte Carlo, and Green's function Monte Carlo. These quantum Monte Carlo methods are primarily adoptions of methods developed for classical problems. We summarize the basics of deterministic power methods, detail key features of Monte Carlo power methods, and put these concepts into the context of quantum ground state calculations. We conclude the chapter by outlining power methods that allow the computation of a few excited states. In the computation of excited states, we encounter sign problems, which are discussed in more detail in Chapter 11. The methods and techniques of the present chapter are very general and most usefully applied to systems of Bosons and to systems of Fermions and quantum spins that do not suffer from a sign problem.

10.1 Deterministic direct and inverse power methods

The power method is over a century old. As a deterministic method, it originally was combined with the deflation technique (Meyer, 2000; Stewart, 2001b) as a method to compute all the eigenvalues and eigenvectors of small matrices. Today, its deterministic use is limited primarily to special applications involving very large, sparse matrices. Many of these applications are in the study of quantum lattice models as the corresponding Hamiltonian matrices are typically very sparse. In these applications, we can perhaps initially store in computer memory the relatively small number of nonzero matrix elements and all the vectors. Soon, because of what is undoubtedly the now familiar exponentially increasing number of basis states that determines the order of the Hamiltonian matrix and hence the size of our vectors, we reach the point where computer memory restrictions allow us to store only a few vectors and we need to compute the matrix elements on the fly. When we have problems for which we cannot even store all the components of one vector, the power method's expression as a Monte Carlo procedure is our only option. With it, we can compute a few of the extremal eigenvalues, that is, a few of the largest or smallest eigenvalues.

The mathematical basis for the method is simple. We already briefly discussed it in Section 2.4.2. If $\{(\lambda_i, |i\rangle), i = 1, 2, \ldots, N\}$ is the set of eigenpairs of A, the eigenvalues are ordered as $|\lambda_1| > |\lambda_2| \geq \cdots \geq |\lambda_N|$, and the state $|\phi\rangle$ is of the form $|\phi\rangle = \sum_i w_i^0 |i\rangle$, then for an operator or matrix A

$$A^n|\phi\rangle = \lambda_1^n \left[w_1^0|1\rangle + \sum_{i=2}^{N} \left(\frac{\lambda_i}{\lambda_1}\right)^n w_i^0|i\rangle \right]. \tag{10.1}$$

By making n large enough, we project the dominant state $|1\rangle$ out of $|\phi\rangle$ because $(\lambda_{i>1}/\lambda_1)^n \to 0$ as $n \to \infty$. Given an initial state $|\phi\rangle$ and an operator A, the power method (Meyer, 2000; Stewart, 2001b) reaches the limit of large n by iterating the following steps:

$$|\psi\rangle = A|\phi\rangle, \tag{10.2}$$
$$|\phi\rangle = |\psi\rangle/\|\psi\|.$$

On convergence, the dominant eigenpair $(\lambda_1, |1\rangle)$ of A is $(\|\psi\|, |\phi\rangle)$. Any choice of norm works. For example, we commonly use $\|\phi\|_\infty = \sum_i |w_i^0|$. The method assumes that the initial state overlaps with the dominant state, that is, $w_1^0 \neq 0$. The rate of convergence to the dominant state is controlled by $|\lambda_2|/|\lambda_1|$.

For eventual points of comparison with our Monte Carlo power methods, we state a more detailed deterministic method in Algorithm 33. It is common to shift

---

Algorithm 33 Shifted power method.

Input: matrix A, shift σ, convergence criterion ϵ, starting state $|\phi\rangle$.

  norm $= \|A\|$;
  $|\phi\rangle \leftarrow |\phi\rangle/\|\phi\|$;
  repeat
    $|\psi\rangle = A|\phi\rangle$;
    $\lambda = \langle\phi|\psi\rangle$;
    $|r\rangle = |\psi\rangle - \lambda|\phi\rangle$;
    $|\phi\rangle \leftarrow |\psi\rangle - \sigma|\phi\rangle$;
    $|\phi\rangle \leftarrow |\phi\rangle/\|\phi\|$;
  until $\|r\| \leq (\text{norm})\epsilon$.
  return the eigenpair $(\lambda, |\phi\rangle)$.

---

the origin of the eigenvalue spectrum and iterate with $A - \sigma I$ instead of A. Convergence is now to the eigenvalue farthest from σ, which is either the largest or smallest one. Choosing σ becomes equivalent to making a good guess for which of these two eigenvalues is sought. With the shift, convergence scales as $|\lambda_2 - \sigma|/|\lambda_1 - \sigma|$ or $|\lambda_{N-1} - \sigma|/|\lambda_N - \sigma|$. The closer the guess is to the exact answer, the more rapid is the convergence.

As stated, Algorithm 33 partially disguises the Rayleigh quotient (9.4) by adjusting the estimate of the dominant eigenvalue at each step and at each step reducing the norm of the residual state $|r\rangle = |\psi\rangle - \langle\psi|\phi\rangle|\phi\rangle$. The Rayleigh quotient gives a more precise eigenvalue estimate than the one based on the norm. In the variational Monte Carlo method (Chapter 9.1), we used this same quotient and residual to provide an upper bound of the eigenvalue. Here, we are driving the iteration not to a bound but to the exact answer.

A second power method is the *inverse power method* (Meyer, 2000; Stewart, 2001b). In this method, with some initial state $|\phi\rangle$ and a shifted matrix $A - \sigma I$, we iterate

$$|\psi\rangle = (A - \sigma I)^{-1}|\phi\rangle, \tag{10.3}$$
$$|\phi\rangle = |\psi\rangle/\|\psi\|.$$

As for the direct power method, the iteration converges to a dominant eigenpair, but this time it is to the dominant pair of $(A - \sigma I)^{-1}$, that is, the eigenpair of $A - \sigma I$ whose eigenvalue is closest to zero.

Since knowing the inverse of $A - \sigma I$ is akin to knowing the solution to the problem, we instead iterate the following steps:

$$\text{Solve } (A - \sigma I)|\psi\rangle = |\phi\rangle,$$
$$|\phi\rangle = |\psi\rangle/\|\psi\|. \tag{10.4}$$

This seems just as problematic because the first step of each iteration requires solving a linear system of equations. However, solving a linear system is more efficient and stable than inverting a matrix.

Algorithm 34 gives a detailed version of the inverse power method. It uses the same initial good estimate of the ground state energy in each iterative step. We typically solve the linear systems of equations with the use of Gaussian elimination with partial pivoting (Stewart, 2001a). The first step of this method, which also consumes the most computer time, is an LU factorization (Stewart, 2001a) of the matrix $A - \sigma I$. By fixing our guess σ, we need only factor the matrix once, and we can do this outside of the iterative loop. Then, in the loop, we solve the linear system by the less computationally intensive forward and back substitution methods (Stewart, 2001a).

Algorithm 34 Shifted inverse power method.

Input: matrix A, shift σ, convergence criterion ϵ, nonzero starting state $|\phi\rangle$.

  norm $= \|A\|$;

  Perform an LU factorization of $A - \sigma I$;

  repeat

    Solve $(A - \sigma I)|\psi\rangle = |\phi\rangle$;

    $|\varphi\rangle = |\psi\rangle / \|\psi\|$;

    $|\chi\rangle = |\phi\rangle / \|\psi\|$;

    $\mu = \langle\varphi|\chi\rangle$;

    $\lambda = \sigma + \mu$;

    $|r\rangle = |\chi\rangle - \mu|\varphi\rangle$;

    $|\phi\rangle = |\varphi\rangle$;

  until $\|r\| \leq$ (norm)ϵ.

  return the eigenpair $(\lambda, |\phi\rangle)$.

10.2 Monte Carlo power methods

The just concluded discussion of the power method described the standard deterministic implementation. What we represented as $|i\rangle$ could have been some classical unit vector or some quantum basis state. Similarly, A could have been a matrix or an operator.

The core computation in a deterministic power method is a conventional matrix-vector multiplication for dense or sparse matrices and dense vectors. A Monte Carlo power method is most useful for cases where we have a means to generate the hopefully sparse matrix elements on the fly. In these implementations, what we move from one iteration step to another is not all the vector components but a subset of them that we call walkers. A *walker* is a tuple of attributes that is at least (*weight, state*) where "state" is a label specifying one of the basis vectors. It is useful to view the subset of components as a stack of walkers. In other words, we let the stack represent a vector in such a way that the representation is asymptotically exact in the limit of many walkers. During the simulation, we pop walkers off one stack and push them onto a new one or back onto the old one. It becomes convenient to interpret the symbol w_i as more than just being the i-th component of a vector but also as the weight of walker i, which just happens to be in a state labeled by i. Using a stack data structure is convenient for the exposition but not necessary for the implementation.

The computational and memory bottleneck in a deterministic power method is the multiplication of a vector by a matrix. In a Monte Carlo power method, we replace the exact (deterministic) multiplications with a Monte Carlo estimate. Each

Monte Carlo step (one matrix-vector multiplication) generates a sample of the components of the resultant vector. From this sample the iteration then produces another sample in such a way that averaging the iteration over many steps and a sufficient number of walkers estimates the complete multiplication.

Replacing these multiplications by a Monte Carlo procedure requires replacing the estimation of the eigenvalue by estimators that differ from the ones we used deterministically and adding extra procedures to make the sampling efficient. In this section, we describe only ways to do the multiplications and to solve the linear system by Monte Carlo sampling. In the following section (Section 10.3), we address efficiency issues. When this discussion is complete, we jump to the basic types of quantum Monte Carlo power methods that employ the matrix-vector multiplication sampling (Section 10.4). We then complete the discussion of quantum power methods in the sections after this one by discussing estimators of the ground state energy, correlation functions, and exited states.

10.2.1 Monte Carlo direct power method

At this point, we recall our discussion of Markov chain Monte Carlo in Section 2.4.2. Sampling via a Markov chain is a Monte Carlo power method. In Markov chain Monte Carlo, we project to a stationary distribution that is the right eigenvector of the stochastic matrix defining the chain. Because the matrix is stochastic, we know that its dominant eigenvalue is 1. In a detailed balanced algorithm, where we specify a priori the stationary distribution, we are designing a stochastic matrix such that this distribution is its right-hand eigenvector. We are now concerned with problems where we cannot use detailed balance because we want to estimate an a priori unknown eigenvalue and concomitantly averages of other quantities with respect to an unknown stationary distribution, that is, from an unknown right eigenvector of our matrix. We still have to construct a transition probability matrix.

Constructing a transition probability matrix is straightforward if A has no negative matrix elements. For notational simplicity, we now absorb the shift σI into the definition of A. Matrix representations of A, even if all the elements A_{ij} of A are nonnegative, in general, do not lead to a stochastic matrix. We can, however, always define the elements P_{ij} of a stochastic matrix P related to the A_{ij} elements of A via[1]

$$A_{ij} = w_{ij} P_{ij},$$

where $\sum_i P_{ij} = 1$, $P_{ij} = 0$ only if $A_{ij} \geq 0$, and $w_{ij} > 0$.

---

[1] More generally, let us suppose we have a matrix A that is nonnegative and irreducible and the relation $Aa = \alpha a$, where the components of the vector αa are positive. Then, if D is a diagonal matrix such that $De = a$, where e is a vector with unit elements, $D^{-1}AD/\alpha$ is a stochastic matrix (Householder, 2006).

How do we choose P_{ij} and how do we sample from it? Common and effective choices[2] for P_{ij} and w_{ij} are[3]

$$P_{ij} = \frac{A_{ij}}{\sum_i A_{ij}}, \quad w_{ij} = \sum_{i'} A_{i'j} \text{ (independent of } i\text{).}$$

Often, w_{ij} is called the weight transfer multiplier. An obvious assumption is that from a given state $|j\rangle$, we can determine (or look up) all the possible values of P_{ij}. To sample P_{ij}, which in a direct power method we regard as the transition probability from state $|j\rangle$ to state $|i\rangle$, we use the sampling concepts from Section 2.3. For example, if from a given state $|j\rangle$, the number of possible states $|i\rangle$ is relatively small, we calculate the cumulative probabilities and use Algorithm 1 or 2 to select the new state $|i\rangle$. If the number of available states from any given state is large, other techniques such as the sewing algorithm (Booth and Gubernatis, 2009a) are potentially useful. Still another approach is expressing A as a product of sparse matrices, say, $A = B_L B_{L-1} \cdots B_1$, then sampling the transition matrix in L stages.

To estimate the matrix-vector multiplications

$$|\psi\rangle = A^n |\phi\rangle$$

by Monte Carlo sampling, we suppose the initial state is $|\phi\rangle = \sum_i w_i^0 |i\rangle$, where $w_i^0 > 0$ and $\sum_i w_i^0 = 1$. Then, after n multiplications,

$$|\psi\rangle = \sum_{i_1,\dots,i_n} A|i_n\rangle \langle i_n|A|i_{n-1}\rangle \cdots \langle i_2|A|i_1\rangle w_{i_1}^0 \tag{10.5}$$

so the i-th component of $|\Psi\rangle$, $\langle i|\psi\rangle = \sum_i \langle i|\psi\rangle |i\rangle = \sum_i w_i |i\rangle$, is simply

$$w_i = \sum_{i_1,i_1,\dots,i_n} A_{ii_n} A_{i_n i_{n-1}} \cdots A_{i_2 i_1} w_{i_1}^0 > 0. \tag{10.6}$$

This last equation we can rewrite as

$$w_i = \sum_{i_1,i_2,\dots,i_n} w_{ii_n} w_{i_n i_{n-1}} \cdots w_{i_2 i_1} w_{i_1}^0 P_{ii_n} P_{i_n i_{n-1}} \cdots P_{i_2 i_1}, \tag{10.7}$$

allowing us to interpret w_i as being the expectation value of

$$w_{ii_n} w_{i_n i_{n-1}} \cdots w_{i_2 i_1} w_{i_1}^0 \tag{10.8}$$

with respect to the joint probability

$$P(i, i_n, i_{n-1}, \dots, i_1) = P_{ii_n} P_{i_n i_{n-1}} \cdots P_{i_2 i_1}. \tag{10.9}$$

---

[2] We are not considering a case where P_{ij} depends on the iteration step or the current weight of the walker.
[3] With these choices w_{ij} is independent of i. For notational convenience it is customary and useful to retain this subscript

This joint distribution naturally enjoys a Markovian factorization. The positivity of the w_i leads one to refer to them not just as components of $|\psi\rangle$ in some basis $\{|i\rangle\}$ but also as weights.

If we were to consider the i_1-th component of the initial state as a random walker, then we could interpret the indices $\{i, i_n, i_{n-1}, \ldots, i_1\}$ of each term in the sum (10.7) as defining a weighted path γ of length n taken by this walker through probability space and interpret the summation as a summing over all possible paths, which we represent as

$$w_i = \langle w_i(\gamma) \rangle_\gamma,$$

where $w_i(\gamma)$ is (10.8), the product accumulation of weight transfer along a particular path γ.[4] The Monte Carlo task is to sample the important paths for the important walkers.

We sample the paths in steps with an ensemble of walkers. At the first step, we sample M random walkers from the distribution defined by w^0, possibly leading to a population of walkers where a state is represented more than once. Next, we process each walker in this sample, one at a time. If the walker is in state $|j\rangle$, whose current weight is w_j, then we sample P_{ij} for a new state $|i\rangle$ and transition the walker to state $|i\rangle$ with the weight $w_i = w_{ij} w_j$ (pushing the walker onto a new stack). After we process the entire population of walkers (all that were in our original stack), we have completed one step, and then we repeat this procedure on the new walker population (the new stack). By design, the number of walkers is $M \ll N$ where N is the dimension of the basis.

The two panels on the left side of Fig. 10.1 schematically represent the difference between a matrix-vector multiplication step in a deterministic power method and one that is performed by sampling. In the figure, a column of circles represents all possible states in the system. When the matrix-vector multiplication is performed exactly, the old states may contribute to multiple new ones, and a new one may receive contributions from multiple old ones. The details of the matrix determine the connections between old and new. When the matrix-vector multiplication step is performed by Monte Carlo, a sample of old states contributes to a new sample of states usually on a one-to-one basis. From the way they were defined, the nonzero elements of P and A have the same ij index pairs, but in a given step, we do not use all of them. The population sizes are orders of magnitude smaller than the number of available states. Because these sizes are small, two or more old states contributing to the same new one state is a rare event.

Algorithm 35 details one way to perform a sweep (one matrix-vector multiplication sampling step) in a Monte Carlo implementation of the matrix-vector

---

[4] The subscript on the expectation value is a reminder that the average is over the distribution of paths (10.9).

Algorithm 35 Monte Carlo matrix-vector multiplication.

Input: Stack W of walkers and procedures to calculate (or look up) the transition probabilities and weight transfer matrices P and w.

Initialize a new stack W' ;

repeat

    Pop a walker (with weight w_k in state $|k\rangle$) off W ;

    Sample i from P_{ik} ;

    Update the weight $w_i = w_{ik}w_k$;

    Push a walker in state $|i\rangle$ with weight w_i onto W' ;

until W is empty.

return W'.

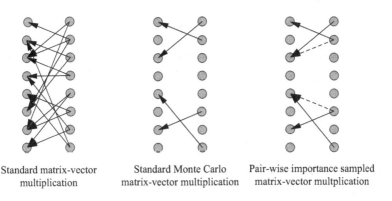

Standard matrix-vector multiplication   Standard Monte Carlo matrix-vector multiplication   Pair-wise importance sampled matrix-vector multplication

Figure 10.1 Matrix-vector multiplication. A column of open circles represents the possible states in the systems. When performed exactly, each old state contributes to multiple new ones, and a new state receives contributions from multiple old ones. When performed by Monte Carlo, a sample of old states contributes to a sample of new states, usually on a one-to-one basis. The right panel is referenced in Section 10.6.2.

multiplication in the direct power method. The other parts of a Monte Carlo power method are practically the same as in any other Monte Carlo method: We initialize and perform a number of steps until we equilibriate. In the equilibration stage, while stepping, we periodically make measurements. After we collect a sufficient number of measurements we stop.

We monitor convergence to stationarity in the Monte Carlo power method in the same way we would for any other Markov chain Monte Carlo simulation. In contrast to a deterministic power method, we do not explicitly monitor the decrease in the norm of the residual as a measure of convergence as we would in a deterministic method, because having only a sample of the vector components, we have insufficient information to compute the inner product between two different vectors and the subtraction of two vectors. We monitor the convergence of an estimator of the eigenvalue. We discuss estimators of the eigenvalue in Section 10.5. We note that the Monte Carlo algorithm we need to insert a new type of step in both the equilibration and measurements stages: *Occasionally stochastically reconfigure.* We will say more about this in Section 10.3.

10.2.2 Monte Carlo inverse power method

We now shift our attention to solving the linear system of equations in a Monte Carlo implementation of the inverse shifted power method. For notational simplicity, we again start by absorbing the σI term into the definition of A, but now we rewrite the first equation in (10.4) as

$$|\psi\rangle = (I - A)|\psi\rangle + |\phi\rangle.$$

With the extra definition of $B = I - A$ and the expression of the states and operator B in some basis, where, for example, $|\psi\rangle = \sum_i w_i |i\rangle$, the equation of interest is

$$w_i = w_i^0 + \sum_j B_{ij} w_j.$$

On iteration this equation becomes the Neumann series

$$w_i = w_i^0 + \sum_j B_{ij} w_j^0 + \sum_{jk} B_{ij} B_{jk} w_k^0 + \sum_{jkl} B_{ij} B_{jk} B_{kl} w_l^0 + \cdots, \qquad (10.10)$$

which we rewrite more compactly as

$$w_i = \sum_{k=1}^{\infty} \sum_{i_1} \cdots \sum_{i_k} \delta_{ii_k} B_{i_k i_{k-1}} \cdots B_{i_2 i_1} w_{i_1}^0. \qquad (10.11)$$

The task in Monte Carlo sampling (10.11) is more ambitious than that for (10.7). We can do the multiplications in a manner similar to the way we did them for (10.7), but we need to add a way to "sample" the number of terms in the power series. A stopping probability is the means to this end.

To proceed we note that the Neumann series converges if the matrix norm of B satisfies $\|B\| < 1$. Examples of *matrix norms* are (Meyer, 2000)

$$\|B\|_1 = \max_i \left(\sum_j |B_{ij}| \right), \quad \|B\|_\infty = \max_i \left(\sum_j |B_{ij}| \right). \qquad (10.12)$$

A matrix norm less than unity implies that the spectral radius of the matrix is less than unity. The *spectral radius* of a matrix defines a circle centered at the origin in the complex plane such that all eigenvalues of the matrix lie on or within this circle. If the spectral radius of B exceeds unity, then rescaling B is necessary.

Assuming $0 \leq B_{ij}$ and $0 < \sum_j B_{ij} < 1$, we next define

$$B_{ij} = w_{ij}P_{ij}, \qquad \sum_i P_{ij} = 1 - p_j,$$

and rewrite (10.11) as

$$w_i = \sum_{k=1}^{\infty} \sum_{i_1} \cdots \sum_{i_k} \delta_{ii_k} W_k p_{i_k} P_{i_k i_{k-1}} \cdots P_{i_2 i_1}, \tag{10.13}$$

where

$$W_k(\gamma) = w_{i_k i_{k-1}} \cdots w_{i_2 i_1} w_{i_1}^0 / p_{i_k}. \tag{10.14}$$

Here again, P_{ij} is the probability of a walker transitioning from state $|j\rangle$ to $|i\rangle$; however, its i-th row sum is less than one by an amount p_j. This amount is the probability of stopping at state $|j\rangle$.

Several things need noting. First, requiring all the elements of the B matrix to be nonnegative is an assumption that the problem under consideration has no sign problem. Assuming the spectral radius of this matrix is less than one so the series converges then means $\sum_i B_{ij} < 1$. Our choice of P_{ij} could then just as well be B_{ij} itself, in which case the definition of the weight transfer multiplier w_{ij} is unity. Instead of explicitly setting it to unity, we carry it along for hopeful insight in the structure of the procedures that evolve. Finally we now interpret the solution as being the following expectation value over all possible paths of the walker:

$$w_i = \langle W_i(\gamma) \rangle_\gamma. \tag{10.15}$$

With the above definitions we sample (10.13) in the following manner: We again sample the paths in steps. At the first step, we generate M random walkers from the distribution defined by w^0. Next, we process each walker in this sample, one at a time. If the walker is in state $|j\rangle$, whose current weight is w_j and $p_j < \xi$, then we sample P_{ij} for a new state $|i\rangle$, transition the walker to state $|i\rangle$, update its weight via $w_i = w_{ij}w_j$, and then repeat this step. If $p_j > \xi$, we update the walker's weight $w_j = w_j/p_j$, push the walker onto the new stack, and terminate its walk. Now, we pop a new walker off the original stack and restart the entire process. After we process the entire population of walkers (all that were in our original stack), we have completed one step, and then we repeat this procedure on the new walker population (the new stack).

The procedure just described performs the sampling in a Monte Carlo method that solves the linear system of equations in the inverse shifted power method. As for our discussion of the Monte Carlo direct shifted power method, we need to add steps that stochastically reconfigure the walkers and estimate the eigenvalue. We discuss these two topics in the following subsections.

Another way to sample (10.11) starts by rewriting it as

$$w_i = \sum_{k=1}^{\infty} \sum_{i_1} \cdots \sum_{i_k} \mathcal{W}_k p_{i_k} P_{i_k i_{k-1}} \cdots P_{i_2 i_1}, \qquad (10.16)$$

where

$$\mathcal{W}_k(\gamma) = \sum_{m=1}^{k} W_m \delta_{i i_m}. \qquad (10.17)$$

Now our solution becomes

$$w_i = \langle \mathcal{W}_i(\gamma) \rangle_\gamma.$$

Following Spanier and Gelbard (1969), we can show that this expectation value is equivalent to the previous one (10.15):

$$\langle \mathcal{W}_i(\gamma) \rangle_\gamma = \sum_{k=1}^{\infty} \sum_{i_1} \cdots \sum_{i_k} \left(\sum_{m=1}^{k} W_m \delta_{i i_m} \right) p_{i_k} P_{i_k i_{k-1}} \cdots P_{i_2 i_1}$$

$$= \sum_{m=1}^{\infty} \sum_{k_m}^{\infty} \sum_{i_1} \cdots \sum_{i_k} W_m \delta_{i i_m} p_{i_k} P_{i_k i_{k-1}} \cdots P_{i_2 i_1}$$

$$= \sum_{m=1}^{\infty} \sum_{i_1} \cdots \sum_{i_k} W_m \delta_{i i_m} P_{i_m i_{m-1}} \cdots P_{i_2 i_1}$$

$$\times \left\{ p_{i_m} + \sum_{i_{m+1}} p_{i_{m+1}} P_{i_{m+1} i_m} + \sum_{i_{m+2}} \sum_{i_{m+1}} p_{i_{m+2}} P_{i_{m+2} i_{m+1}} P_{i_{m+1} i_m} + \cdots \right\}.$$

Now we use the relation

$$p_{i_{m+1}} = 1 - \sum_{i_{m+2}} P_{i_{m+2} i_{m+1}}$$

to collapse the terms in the braces to $1 - \lim_{t \to \infty} P^t$ where P^t is the t-th power of the transition matrix. Because the spectral radius of P is less than one, $\lim_{t \to \infty} P^t = 0$ and we can then replace the term in the braces by one. The remainder of the expression is the same as (10.13).

What (10.16) and (10.17) say is that at each step we clone the walker, update its weight by dividing by the stopping probability for the current state, and push it onto the new stack. Then we let the walker proceed with the next step until an actual termination is sampled. In short, we accumulate contributions to the solution to the system of equations as the walker steps along. Sometimes, this way of sampling is more efficient than the previous one. Algorithm 36 details one way to execute this procedure.

Algorithm 36 Monte Carlo linear system solver.

Input: Stack W of walkers and the stopping and transition probabilities p and P.

  Initialize stack W' ;

  repeat

    Pop a walker (with weight w_k in state $|k\rangle$) off W ;

    Draw a ζ ;

    while $\zeta < 1 - p_k$ **do**

      Set $w \leftarrow w_k/p_k$;

      Push a walker in state $|k\rangle$ with w onto W' ;

      Sample m from P_{mk} ;

      Set $w_m = w_{mk}w_k$;

      Set $k = m$;

      Draw a ζ ;

    end while

  until W is empty.

  return W'.

Algorithm 35 performs the sampling in a second Monte Carlo method that solves the linear system of equations in the inverse shifted power method. As for our discussion of the direct shifted Monte Carlo power method and the first Monte Carlo method to solve a linear system of equations, we need to add steps that stochastically reconfigure the walkers and estimate the eigenvalue.

This second way of sampling is an illustration of the method of *expected values* (Spanier and Gelbard, 1969; Kalos and Whitlock, 1986). The method of expected values is a common procedure to reduce the variance of a Monte Carlo estimate of an average. It says that doing parts of the estimate exactly, instead of letting the Monte Carlo estimate it, under certain conditions reduces the variance of the entire result. In the present case, instead of recording the walker's contribution to the solution when it stops, we observe that at each step along the path it could have stopped. We record at each step along the path what we expect to happen on the average.

We remark that when using the power method deterministically, the requirement that A (or B for the linear system problem) has no negative matrix elements is unnecessary. In a Monte Carlo power method, direct or inverse, when an A_{ij} is negative, the absolute value of A_{ij} is used whenever we need to use it to define a transition probability, and its sign multiplies the weight transfer factor w_{ij}. If A has no negative matrix elements, then the *Perron-Frobenius theorem* (Section 2.4.2; Meyer, 2000) applies, saying that the dominant eigenvalue must be real and positive and the components of its eigenvector in this basis must be real and positive.

In our algorithms, we push walkers onto a stack without attempting to combine those that are in the same state. In general, it is a rare event if they are. If the walkers all have positive weight, the walkers would combine constructively. In some sense, there is not much to gain by combining them. If they have mixed signs, they would combine destructively. In this case, by not combining them, we can lose important cancellation effects. Generally, we must add an extra procedure to accommodate walkers of mixed sign. In Section 10.6, we discuss one such method, originally developed for the Green's function Monte Carlo method (Section 10.4), that has been successfully used in excited state calculations. *The sign problem is not just a Fermion problem or even just a quantum problem.*

10.3 Stochastic reconfiguration

The above Monte Carlo algorithms are incomplete until we define what we mean by *stochastic reconfiguration.*[5] There are several intertwined concepts in this phrase. One is ensuring ergodicity, another is promoting efficiency, and a third is computing conveniently.

The problem is that the successive updating of the walker weights generates a wide range in the values of these weights: Some walkers develop very large weights, other walkers very small ones. Having a few large-weighted ones leads to just a few states underrepresenting the many possible states. For the sampling to be ergodic, we need to ensure all important states are properly accessed. One way of doing this is limiting the weight accumulation of any one walker by occasionally splitting a large-weighted one into a number of smaller-weighted ones that share the original weight. Each of the smaller-weighted walkers subsequently takes a path different from the original one. The splitting and termination changes the size of the walker population. Thus, we also have to ensure that the population does not become too large or too small, which requires a procedure to control the population size. The control and efficiency tasks are often combined into one procedure. We call the combination of Monte Carlo techniques that help accomplish these goals *stochastic reconfiguration*. They change the mix of the walkers in the current configuration. Doing this reconfiguration stochastically, and occasionally, can reduce the potential for an unwanted bias in the resulting estimates.

There are two classes of stochastic reconfiguration procedures. One class operates on the walkers one at a time, the other operates on the entire population at once. For the direct power algorithm, it is natural to reconfigure the entire population

---

[5] Sorella (1998, 2000) uses the term "stochastic reconfiguration" to describe a method to stabilize the sign problem in the linear Green's function Monte Carlo method (Section 10.4.1) using the lattice fixed node approximation (Section 11.2).

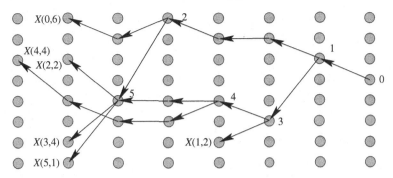

Figure 10.2 On-the-fly reconfiguration. A column of circles represents the possible states in the system, and each column represents the states after a power method step, moving right to left. Only one walker is being considered. This walker starts at the event labeled zero. Branching occurs at the nonzero event numbers, generating new walkers labeled by that event number. The bookkeeping symbol $X(m, n)$ marks the end of a walk: m is the event number of origin, and n is the number of steps from the origin to termination.

occasionally at the end of an iteration step. In the inverse power method, it is natural to follow a single walker until it terminates. Accordingly, walker control here is usually done individually and on the fly.

Figures 10.2 and 10.3 illustrate these two procedures. In Fig. 10.2, a single walker starts at the event labeled zero. Branching occurs at each other numbered event, generating new walkers labeled by that event number. The symbol $X(m, n)$ marks the end of a walk. Here, m is the event number of origin, and n is the number of steps from the event of origin to termination. Because branches are followed one at a time, walkers from different branches may occupy the same state, as they do in the state labeled event 5, but the occupancy is not simultaneous. When the initial walker and all the walkers it spawned terminate, the procedure moves to the next walker in the initial sample.

In Fig. 10.3 a sample of walkers starts simultaneously. For illustration purposes, we pretend we are reconfiguring at each step by splitting arbitrarily. At each step, the weights of the walkers change (this change is not illustrated in the figure), and as a result, some walkers terminate and others branch. During population reconfiguration, different walkers may occupy the same state simultaneously. If their weights are positive, the reconfiguration generally does not take this into consideration. If the weights are of mixed sign, it is advantageous to combine these walkers into one whose weight is the sum of the weights of those in the same state.

The iconic procedure for doing walker control on the fly is called *roulette*. It has a prespecified survival weight w_s and a cut-off weight $w_l < w_s$. If a walker's weight

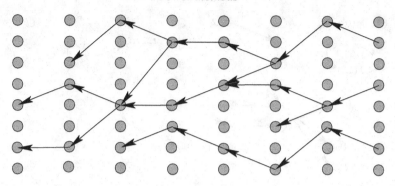

Figure 10.3 Population reconfiguration. A column of circles represents the possible states in the system, and each column represents the states after a power method step, moving right to left. For the purpose of illustration, reconfiguration occurs at each step. This procedure changes the weights of each walker in the population, leading to some branching and others terminating. Multiple walkers may occupy the same state.

w falls below w_l, then if $\zeta < w/w_s$, the walker is kept but its weight is increased to w_s. Otherwise, the walker is terminated.

The second commonly used procedure seems to lack a name, but we will call it *branching*. It addresses problems with weights being too large and too small. For this procedure, we calculate $m = \text{integer}(w + \zeta)$, where ζ is a random number in $[0, 1]$. The addition of ζ to weight reduces any bias the truncation to an integer might cause. If $m = 0$, we terminate the walker. If not, then we split it m times, with each replica assigned a weight w/m.

A third procedure, called *weight window*, offers better control at the expense of having several parameters to adjust (Booth, 2009). This method combines roulette and branching. The weight window has a lower, survival, and upper weight arranged so that $w_l < w_s < w_u$ with $w_u \geq 2w_l$. Typically, $w_u = 5w_l$ and $w_s = 3w_l$. As long as the walker's weight is in the weight window $[w_l, w_u]$, we don't do anything. If the walker's weight falls below w_l, we roulette it. If it rises above w_u, we branch it by finding the smallest integer m that places w/m within the weight window and split the walker m times. This method should generally be preferred over branching because it preserves the total weight, while branching introduces weight fluctuations and preserves only the total weight on the average. Algorithm 37 details the weight window algorithm.

In any of the on-the-fly procedures that split the walker into m walkers, $m - 1$ walkers with their current state and modified weights are pushed back onto the list of current walkers. The remaining walker, with its new weight, continues the walk. When its walk terminates, then a new walk starts with one of the waiting walkers. One iteration is completed when the stack of original and spawned

---

Algorithm 37 The weight window.

Input: A walker with weight w in state $|i\rangle$, a stack W of walkers plus the upper and lower cut-off w_u and w_l, survival weight w_s.

  if $w > w_u$ **then**

    Set $m = \text{floor}(w/w_u)$;

    $w = w_s$;

    Push $m - 1$ walkers with weight w in state $|i\rangle$ onto W ;

  else if $w < w_l$ **then**

    Set $p = w/w_s$;

    if $\zeta < p$ **then**

      $w = w_s$;

      Push a walker with weight w in state $|i\rangle$ onto W ;

    else

      $w = 0$;

    end if

  end if

  return the walker and W.

---

walkers depletes. An important point is that since the population of walkers sums the Neumann series, there is no need to terminate a walker arbitrarily on the basis of its weight falling below some prescribed value or exceeding a number of steps. Doing either of these terminations introduces a bias. The procedures just described stochastically terminate a walker in a way that at least preserves its weight on the average.

We could apply the on-the-fly techniques to the entire population of walkers in one swoop. Doing so however may generate a large sudden variation in population size and total weight. Instead, we use one of several convenient techniques for stochastically reconfiguring the entire population of walkers and resetting the population to a fixed size. The simplest procedure is *sampling with replacement*. Here, occasionally, but after an iteration step, we create the cumulative distribution function (CDF) of the walker weights, $C_i = \sum_{j=1}^{i} w_j$, and then from it sample with replacement M walkers with their weights and states tagging along. Those with small weights have a low probability of being selected, while those with large weights may get selected multiple times.

A second procedure is *stratified sampling* (Spanier and Gelbard, 1969; Kalos and Whitlock, 1986). Again we form the cumulative distribution of walker weights. After doing so, we divide its 0 to 1 range into a number of strata, for example, based on weight percentiles or fractions of walkers. Then we sample with replacement from each strata a prefixed number of walkers such that the total from all strata

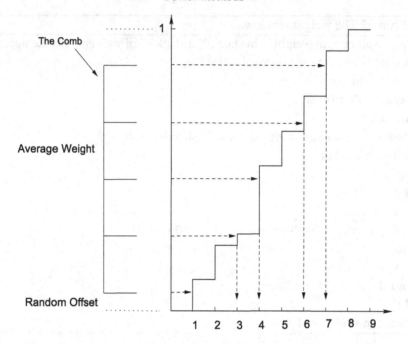

Figure 10.4   The comb. This figure illustrates the comb with equal spacing (average weight) between its teeth. The main figure is the cumulative distribution function (CDF), C_i, for nine walkers originally. The comb is randomly offset from the CDF by a random shift. Sampling with the comb assigns each walker the average weight per walker, the distance between the teeth. Here there are five walkers. The number of times a walker is selected is given by the number of teeth that lie in an interval of the CDF. In the present illustration, the original walkers 1, 3, 4, 6, and 7 are selected once. The others are not selected.

is the desired number. For example, if we want M samples, we could divide the $[0, 1]$ range of the cumulative distribution into M equal intervals and then draw one sample from each interval.

Still another procedure is the *comb* (e.g., Booth, 2009; Booth and Gubernatis, 2009b). It preserves the before and after total weight, eliminating a fluctuation, and uses only one random number (Fig. 10.4). Each selected walker has the same weight. If we have M walkers and want N, where M could be greater than or less than N, the value of the common weight is $w_{\text{average}} = \sum_{i=1}^{M} w_i/N$. Walker splitting and termination starts with the formation of the cumulative distribution C_i from the weights of M walkers. We point out, however, that we can compute the cumulative distribution on the fly, eliminating the need to store it. We detail the comb in Algorithm 38.

The three on-the-fly and the three population reconfiguration methods are *variance reduction methods*. Their individual effectiveness for this purpose, however,

---

Algorithm 38 The comb.

Input: Stack W on M walker and N, the size of the new stack.

    Initialize the new stack W' ;

    Set $C = 0$;

    Set $\delta = 1/M$;

    Compute the total weight w_{total} of walkers in W ;

    Set $w_{\text{average}} = w_{\text{total}}/N$;

    Draw a random number $\zeta \in [0, 1]$;

    Set $\Delta = \zeta\delta$;

    repeat

        Pop a walker (with weight w in state $|i\rangle$) off W ;

        $C \leftarrow C + w$;

        while $\Delta > C/w_{\text{total}}$ **do**

            Push walker of weight w_{average} in state $|i\rangle$ onto W' ;

            $\Delta \leftarrow \Delta + \delta$;

        end while

    until W is empty.

    return W'.

---

can vary widely. In variance reduction, we seek to get the biggest "bang for the buck," that is, the smallest variance for a given amount of computer time. In general, the objective of variance reduction is minimizing the size of the fluctuations of individual measurements about their averages. Its objective therefore differs from that of population control. Often the two intents are confused. Eliminating branching is not always a good idea from the point of view of variance reduction, although it is convenient for population control. As noted above, branching promotes ergodic sampling so its elimination might in fact be ill-advised. In a similar vein, resetting the population to a fixed size, even with branching included, is also convenient for population control, but whether the population is adjusted one walker at a time or all at once, the main objective should be bias and variance reduction obtained by setting the parameters of the simulation so a roughly constant population propagates from iteration to iteration. In each reconfiguration method described, observable averages before and after reconfiguration differ. This fluctuation adds to the variance. Thus, we want to make these changes as small and as infrequently as possible. One way of doing this is reducing fluctuations among the various walker weights; that is, we can try to keep them within a weight window. If we were reconfiguring by means of the weight window algorithm, for example, then we would virtually eliminate splittings and terminations if we were able to keep most of the weights within the window.

We comment that reconfiguring the entire population at once has two consequences. One is that the simulation is no longer Markovian. Reconfiguration introduces correlations between the walkers, since each walker's weight now depends on the others in the population. The correlations carry from one step to another. These correlations are a reason why we stated we reconfigure "occasionally." Reconfiguration should not be done after every iteration. Another consequence of reconfiguring the entire population at once is that a bias, which varies as the reciprocal of the number of walkers, generally develops in the estimates. Typically, the number of walkers is large, so this bias is small. If the computation time is affordable, the cure is simply increasing the number of walkers. Other procedures exist that help reduce the need for huge numbers of walkers. The series method of Brissenden and Garlick (1986) and Gelbard and Gu (1994) is easily applied to the block averages with noticeable positive consequences.

Reconfiguring on the fly does not eliminate these consequences. This type of reconfiguration changes both the total weight and the number of walkers. The overarching need is to maintain both at a nearly constant size throughout the simulation. Making the adjustment to do this can induce non-Markovian behavior and biases. In general, for both types of reconfiguration, these adverse effects of reconfiguration are small but should be noted. While the simulations may appear to implement what mathematically are Markovian and unbiased procedures and estimates, what occurs in practice can often be less than ideal.

10.4 Green's function Monte Carlo methods

In quantum ground state calculations, the power method projects to the dominant eigenpair of some operator A. For A, Green's function Monte Carlo methods use a function of the Hamiltonian H whose dominant eigenvalue maps to the subordinate eigenvalue of H. Three functions are used:

1. $A = I - \tau(H - \sigma I)$, which shifts and rescales the eigenspectrum of the Hamiltonian
2. $A = [I + \tau(H - \sigma I)]^{-1}$, which inverts a shifted and rescaled Hamiltonian
3. $A = \exp[-\tau(H - \sigma I)]$, which exponentiates the shifted Hamiltonian.

Because the first case uses a linear mapping of the eigenvalues, we call it the *linear method*. The Green's function method using the exponential is called *diffusion Monte Carlo*. The method using the inverse is called *exact Green's function Monte Carlo*. All three are often simply called Green's function Monte Carlo. Clearly, a lowest order power series expansion of the last two operators reduces both to the first. Indeed, as we discuss next, the three approaches have a common interpretation and share many techniques.

To develop this interpretation, we start with the time-dependent Schrödinger equation (Section 1.2)

$$i\frac{\partial}{\partial t}|\psi(t)\rangle = H|\psi(t)\rangle,$$

which after the substitution

$$|\psi(t)\rangle = |\psi\rangle e^{-itE}$$

becomes the eigenvalue equation

$$H|\psi\rangle = E|\psi\rangle, \tag{10.18}$$

called the time-independent Schrödinger equation. Associated with the inverse of the Hamiltonian is an operator G called the *time-independent* Green's function that by definition satisfies

$$HG = GH = I.$$

Multiplying (10.18) on the left with G yields the eigenvalue equation

$$|\psi\rangle = EG|\psi\rangle. \tag{10.19}$$

If we were to express this Green's function operator G in some basis $\{|C\rangle\}$, this equation would become the matrix-vector equation

$$\langle C|\psi\rangle = E\sum_{C'}\langle C|G|C'\rangle\langle C'|\psi\rangle.$$

The transformation $it \to \tau$ converts the time-dependent Schrödinger equation into the imaginary-time-dependent Schrödinger equation

$$\frac{\partial}{\partial \tau}|\psi(\tau)\rangle = -H|\psi(\tau)\rangle, \tag{10.20}$$

which upon formal integration creates an operator that enables the generation of the state at one imaginary time from another state at another time, that is,

$$|\psi(\tau_2)\rangle = e^{-H(\tau_2-\tau_1)}|\psi(\tau_1)\rangle = G(\tau_2-\tau_1)|\psi(\tau_1)\rangle.$$

This result defines the *imaginary-time-dependent* Green's function, which is sometimes called the *imaginary-time propagator*. Because of the implied time translation invariance, we rewrite the above as

$$|\psi(\tau_2)\rangle = \underbrace{G(\tau)G(\tau)\cdots G(\tau)}_{M \text{ factors}}|\psi(\tau_1)\rangle,$$

where $\tau_2 - \tau_1 = M\tau > 0$.

To obtain the state at one time in terms of a state at another time, we could also use a finite-difference approximation (Euler's method) to the imaginary-time derivative

$$\frac{|\psi(\tau_1 + \tau)\rangle - |\psi(\tau_1)\rangle}{\tau} = -H\,|\psi(\tau_1)\rangle$$

or

$$|\psi(\tau_1 + \tau)\rangle = [I - \tau H]\,|\psi(\tau_1)\rangle.$$

These considerations motivate calling the techniques "Green's function" methods, although this name is a bit deceptive. Our intent is computing the ground state properties and not solving the initial-value problem represented by the equation of motion (10.20). *Accordingly, for what we called the linear method, τ need not be small.* In general, τ is a parameter we use to rescale the eigenvalue spectrum of the shifted Hamiltonian so that the matrix G used in the power method has nonnegative matrix elements. For the exact Green's function method, we also use τ to scale the shifted Hamiltonian so that its spectral density is less than one, a necessary condition for the convergence of the Neumann series representing the inverse. τ is small in the diffusion Monte Carlo method because we use the Trotter approximation to evaluate the exponential. In short, when possible, we use a combination of τ and σ to create a nonnegative matrix representation of the operator. For some basis $\{|C\rangle\}$, necessary conditions for nonnegativity are

1. For $A = I - \tau(H - \sigma I)$: $\langle C|H - \sigma I|C'\rangle \leq 0$ for all C and C'
2. For $A = [I + \tau(H - \sigma I)]^{-1}$: $\langle C|H - \sigma I|C'\rangle \leq 0$ for all $C \neq C'$
3. For $\exp[-\tau A]$ with $A = (H - \sigma I)$: $\langle C|H - \sigma I|C'\rangle \leq 0$ for all $C \neq C'$.

Of the three Green's function methods, the exact Green's function method, which is based on the inverse power method, is the one most infrequently used. We have given its basic definition and next focus on just discussing extra details of the other two methods.

10.4.1 Linear method

The linear method is convenient and generally simple to implement. As just discussed, we want the matrix elements of $A = I - \tau(H - \sigma I)$ in some basis $\{|C\rangle\}$ to be nonnegative and can ensure this if $C \neq C'$ and $\langle C|H - \sigma I|C'\rangle < 0$. As an example of how to construct the power method, we take for H the one-dimensional quantum Ising model in a transverse field (5.6) and choose for our basis the spin configuration basis (5.5).

The Hamiltonian is

$$H = H_0 + H_1 \tag{10.21}$$

with

$$H_0 = -J^z \sum_{i=1}^{N} S_i^z S_{i+1}^z \quad \text{and} \quad H_1 = -H^x \sum_{i=1}^{N} S_i^x. \tag{10.22}$$

As noted before (Section 6.5.2), this model is exactly solvable, so we do not need quantum Monte Carlo simulations to understand its properties. What we do know from the exact solution is that the model possesses a *quantum critical point* (Sachdev, 1999); that is, as a function of the transverse field, it has a phase transition at zero temperature.

To motivate this transition, we regard J^z as setting the scale of energy so that the properties of the model become a function of the coupling constant $g = H^x/J^z$. If this constant is very small, we expect the model to behave like the zero-field Ising model whose ground state is the state of long-range magnetic order with all spins up or down. On the other hand, if g is very large, the natural basis functions are not the $|C\rangle = |s_1\rangle|s_2\rangle \cdots |s_N\rangle$ but rather $|C\rangle = |t_1\rangle|t_2\rangle \cdots |t_N\rangle$ where the $|t_i\rangle$ are eigenfunctions of S_i^x. These states are $|1\rangle = \frac{1}{\sqrt{2}}(|+1\rangle+|-1\rangle)$ and $|2\rangle = \frac{1}{\sqrt{2}}(|+1\rangle-|-1\rangle)$ with eigenvalues $\lambda_1 = +1$ and $\lambda_2 = -1$. In the large coupling constant limit, the S_i^z independently fluctuate between these two states, and we expect the ground state to be a linear combination of all spin configurations, a quantum paramagnet (Sachdev, 1999). Thus, as a function of g, the ground state changes. The critical point in fact is at $g_c = \frac{1}{2}$, that is, where $J^z = 2H^x$.

From the above discussion, we discover that our chosen basis is not the most appropriate one for a phase of the model. Accordingly, our basis choice is unlikely to capture the physics of that phase easily. We also expect difficulty near the critical point.

We first note that H_0 is diagonal in our basis ($s_{i,k} = \pm\frac{1}{2}$)

$$\langle C_i|H_0 - \sigma I|C_j\rangle = \left[-\sum_k \left(J^z s_{i,k} s_{i,k+1} + \sigma \right) \right] \delta_{ij}. \tag{10.23}$$

We next note that H_1 is strictly off-diagonal:

$$\langle C_i|H_1|C_j\rangle \equiv -H^x \sum_k \delta_{C_i^{jk},C_j}.$$

H_1 acting on some state $|C_j\rangle$ produces the sum of N states $|C_i^{jk}\rangle$ where each differs from $|C_j\rangle$ only at lattice site k where it has the flipped spin $-s_{j,k}$. Thus, for $H^x > 0$ all off-diagonal matrix elements are negative, and from (10.23), it is clear that we can make all diagonal elements non positive if $\sigma = J^z$. Accordingly,

$$\langle C_i|I - \tau\,(H - \sigma I)\,|C_j\rangle = \sum_k w_{jk} \left[P_{jk}\delta_{C_i,C_j} + \bar{P}_{jk}\delta_{C_i^{jk},C_j} \right],$$

with

$$P_{jk} = \left[1 + \tau J^z \left(s_{j,k} s_{j,k+1} + 1\right)\right] / w_{jk}, \quad \bar{P}_{jk} = \tau H^x / w_{jk},$$

and

$$w_{jk} = 1 + \tau J^z \left(s_{j,k} s_{j,k+1} + 1\right) + \tau H^x.$$

By unveiling possible transition probabilities P and weight transfer multipliers w, the result leads to a simple sampling strategy we can use in Algorithm 35 to generate a sweep: For a walker representing a spin configuration $|C_j\rangle$, we choose a lattice site k at random, update the walker's weight by $w \leftarrow w w_{jk}$, and then if $P_{jk} < \zeta$, where $\zeta \in [0, 1]$ we keep the walker in its current state or else we jump it to a new state formed from the old one by flipping the spin state at the lattice site k. In the present case, $\tau > 0$ is a free parameter for adjusting the efficiency of the sampling, for example, keeping the flipping and nonflipping rates roughly equal as the coupling constant is varied.

Algorithm 35 generates different states, each of which is a spin configuration; that is, it generates different populations of walkers. With them a variety of measurements are possible (Section 10.5). How well does this algorithm work? We do not know. We constructed it mainly for illustration purposes. In general, it is hard to predict the efficiency of an algorithm. During the implementation, usually an understanding emerges about the degree of efficiency and the location of bottlenecks. Since for this problem the exact answer is known, trying to reproduce it becomes a learning opportunity to experiment with the proposed algorithm and to improve it.

10.4.2 Diffusion Monte Carlo

To illustrate the diffusion Monte Carlo method, we again take the transverse-field Ising model as our example and continue using the spin configuration basis. This method differs from the linear method by requiring that we exponentiate the Hamiltonian. To do the exponentiation, we have to use a Trotter approximation. In the present case we need it because the spin operators S_i^x and S_j^z lack on-site $(i = j)$ commutivity, so that the spin-spin Hamiltonian H_0 does not commute with the spin-field Hamiltonian H_1. These two parts of H are thus not simultaneously diagonalizable. Accordingly, the states in (5.5) are not eigenstates of the Hamiltonian (10.21). However, in this basis both $\langle C' | \exp(-H_0 / kT) | C \rangle$ and $\langle C' | \exp(-H_1 / kT) | C \rangle$ are easily found. In contrast to the linear method, τ must be small so the Trotter approximation is accurate.

We also need slightly different techniques for sampling the matrix-vector multiplication, but we still have to ensure that in our chosen basis the matrix elements of the exponential are nonnegative. The condition to ensure nonnegativity is the same as for the linear method. However, this condition ensures the positivity of the exact exponentiation and not of the individual pieces of the Trotter approximation. The positivity of each piece has to be established separately. In any case, in what follows we take $\sigma = J^z$.

There are several different ways to construct the algorithm. Our choice is somewhat arbitrary. We first write

$$H - \sigma I = -\sum_{k=1}^{N} \left[J^z \left(S_k^z S_{k+1}^z + I \right) + H^x S_k^x \right],$$

define the terms $h_{k,k+1}^{(0)} = -J^z \left(S_k^z S_{k+1}^z + I \right)$ and $h_{k,k+1}^{(1)} = -H^x S_k^x$, and then use the following form of the Trotter approximation (5.13),

$$e^{-\tau H} \approx \prod_k e^{-\frac{1}{2}\tau h_{k,k+1}^{(0)}} e^{-\tau h_{k,k+1}^{(1)}} e^{-\frac{1}{2}\tau h_{k,k+1}^{(0)}}.$$

The matrix elements of exponentials of the individual factors are easily calculated:

$$\langle C_i | e^{-\frac{1}{2}\tau h_{k,k+1}^{(0)}} | C_j \rangle = e^{\frac{1}{2}\tau J^z \left(s_{j,k} s_{j,k+1} + 1 \right)} \delta_{C_i C_j},$$

$$\langle C_i | e^{-\tau h_{k,k+1}^{(1)}} | C_j \rangle = \cosh \left(\tfrac{1}{2}\tau H^x \right) \delta_{C_i C_j} + 2 \sinh \left(\tfrac{1}{2}\tau H^x \right) \delta_{C_i^{jk} C_j}.$$

Thus,

$$\langle C_i | e^{-\tau (H - \sigma I)} | C_j \rangle \approx \prod_k w_{jk} \left[P_{jk} \delta_{C_i C_j} + \bar{P}_{jk} \delta_{C_i^{jk} C_j} \right],$$

where

$$P_{jk} = e^{\tau J^z s_{j,k} s_{j,k+1}} \cosh \left(\tfrac{1}{2}\tau H^x \right) / w_{jk}, \quad \bar{P}_{jk} = 2\sinh \left(\tfrac{1}{2}\tau H^x \right) / w_{jk},$$

and

$$w_{jk} = e^{\frac{1}{2}\tau J^z s_{j,k} s_{j,k+1}} \cosh \left(\tfrac{1}{2}\tau H^x \right) + 2 \sinh \left(\tfrac{1}{2}\tau H^x \right).$$

A sampling strategy emerges that we can fit into Algorithm 35. For a walker of weight w representing a spin configuration $|C_j\rangle$, we choose the first lattice site $k = 1$, update the walker's weight by $w \leftarrow w w_{jk}$, and then if $P_{jk} < \zeta$, where $\zeta \in [0, 1]$, we keep the walker in its current state, updating its weight via $w \leftarrow w$ $\exp[-\frac{1}{2}\tau J^z s_{j,k} s_{j,k+1}]$, or else we jump it to a new state formed from the old one by flipping the spin at the lattice site $k = 1$. We then move to the next lattice site $(k = 2)$ and repeat the process until a sweep of the lattice is completed. We note that for very small τ, P_{jk}, $\bar{P}_{jk}$, and w_{jk} reduce to those in the linear method. The method here differs by successively updating the walkers' spin configuration through the entire lattice before moving to the next walker.

Using the above in Algorithm 35 is one way to generate configurations in the form of a population of walkers. With these configurations, we can make a variety of measurements. One difference in quantum uses of the power method is that they invariably modify the transition probability (and hence the weight transfer multiplier) by using what is called importance sampling. Proper importance sampling is a very important performance issue.

10.4.3 *Importance sampling*

Importance sampling is a common Monte Carlo technique to reduce variance. We first define it and then discuss its use in the diffusion Monte Carlo method. Its use in the linear and exact Green's function methods follows by analogy. It is difficult to find any quantum Monte Carlo power method that does not use importance sampling. We follow the standard route to the introduction of the technique.

Suppose we seek a Monte Carlo estimate of

$$\langle X \rangle = \int dx\, X(x) f(x), \tag{10.24}$$

where $f(x)$ is some probability distribution that is not especially convenient to sample. Also suppose $g(x)$ is a more convenient one. Then, instead of estimating (10.24) by sampling $f(x)$ to obtain $\langle X \rangle = \frac{1}{M} \sum_{i=1}^{M} X(x_i)$, we estimate

$$\langle X \rangle = \int dx\, \frac{X(x) f(x)}{g(x)} g(x) \tag{10.25}$$

by sampling $g(x)$ to obtain

$$\langle X \rangle \approx \frac{1}{M} \sum_{i=1}^{M} \frac{X(x_i) f(x_i)}{g(x_i)}.$$

The variance of X when $g(x)$ is sampled is

$$\sigma_X^2 = \int dx\, \left[\frac{X(x) f(x)}{g(x)} \right]^2 g(x) - \langle X \rangle^2.$$

If we want the $g(x)$ that minimizes this variance, subject to the normalization constraint $\int dx\, g(x) = 1$, then it is straightforward to show by substitution into the above that the optimal (zero variance) importance function is

$$g(x) = \frac{X(x) f(x)}{\langle X \rangle},$$

a comforting result that says if we know the answer, that is, $\langle X \rangle$, then the Monte Carlo measurement will produce it! The more practical inference is that a good importance function reduces the variance. A poor one, however, can increase it. Unfortunately, there are few good rules to guide the choice of this function other than the rule that $g(x)$ should "cover" $f(x)$ closely but have longer tails.

If we have a good starting state $|\phi\rangle$, which is more conventionally written as $|\psi_T\rangle$ for the power method, we can use this same state for the importance function $\langle C|\psi_G\rangle \equiv \langle C|\psi_T\rangle$. The starting states typically come from approximate theories or from the variational Monte Carlo method (Chapter 9). The importance function $|\psi_G\rangle$ is often called a *guiding function* as its use has the interpretation of guiding the walkers toward the important parts of configuration space. We assume that $|\psi_G\rangle = |\psi_T\rangle$. We also use this same $|\psi_T\rangle$ to estimate σ via $\sigma \equiv E_T = \langle \psi_T|H|\psi_T\rangle / \langle \psi_T|\psi_T\rangle$.

We perform importance sampling in a quantum Monte Carlo power method by transforming the basic iteration step (10.2)

$$\langle C|\psi\rangle = \sum_{C'} \langle C|A|C'\rangle \langle C'|\phi\rangle$$

to

$$\langle C|\psi_T\rangle \langle C|\psi\rangle = \sum_{C'} \frac{\langle C|\psi_T\rangle \langle C|A|C'\rangle}{\langle C'|\psi_T\rangle} \langle C'|\psi_T\rangle \langle C'|\phi\rangle.$$

This is a similarity transformation, $A' = SAS^{-1}$, of the original eigenvalue equation with a diagonal transformation matrix S whose nonzero elements are $\langle C|\psi_T\rangle$. The eigenvalues are unchanged but the eigenstates are not. What we now iterate is

$$\langle C|\bar{\psi}\rangle = \sum_{C'} \langle C|A|C'\rangle \frac{\langle C|\psi_T\rangle}{\langle C'|\psi_T\rangle} \langle C'|\bar{\phi}\rangle$$

where the ratio $\langle C|\psi_T\rangle / \langle C'|\psi_T\rangle$ is the analog of $f(x)/g(x)$ in (10.25) and becomes an extra factor contributing to the weight-transfer multiplier $w_{CC'}$. $\langle C|\bar{\psi}\rangle = \langle C|\psi_T\rangle \langle C|\psi\rangle$ and $\langle C|\bar{\phi}\rangle = \langle C|\psi_T\rangle \langle C|\phi\rangle$. The guiding function need not be the same as the starting state, but if the starting state is good, then it should be a good guiding function. One use of variational Monte Carlo is providing this function (Chapter 9).

10.5 Measurements

We now come to the promised discussion of measurements. In diffusion Monte Carlo the standard estimator of the eigenvalue is called a *mixed estimator* (Hammond et al., 1994),

$$E_0 \le E_m \equiv \frac{\langle \psi_T | H | \psi \rangle}{\langle \psi_T | \psi \rangle},$$

as it uses both the starting state $|\psi_T\rangle$ and the projected state $|\psi\rangle$. Clearly, this expression yields an exact estimate as $|\psi\rangle$ approaches the ground state. We use this same estimator for any observable that commutes with H. For other operators, we use a combination of the mixed and variational estimates called the *extrapolated estimate*

$$\langle X \rangle = 2\frac{\langle \psi_T | X | \psi \rangle}{\langle \psi_T | \psi \rangle} - \frac{\langle \psi_T | X | \psi_T \rangle}{\langle \psi_T | \psi_T \rangle} + \mathcal{O}\left(\| \psi_T - \psi \|^2 \right).$$

The diffusion Monte Carlo method often uses a special estimator of the ground state energy, the *growth estimator* (Hammond et al., 1994). As we previously remarked, in the power method the norm of the eigenstate estimates this eigenvalue. Writing the norm as $N(\psi) = \sum_C |\langle C | \psi \rangle|$, we note that if we are sampling from the ground state, which we assume to be real, then because this state is nodeless, $|\langle C | \psi \rangle| = \langle C | \psi \rangle$, and the norm of the new state produced by one iteration step in the ground state is

$$N(\phi) = \sum_C \langle C | \phi \rangle = e^{-\tau(E_0 - E_T)} N(\psi).$$

However, the energy over any τ step actually has a value E_τ that fluctuates about E_0, so we have

$$N(\phi) = \sum_C \langle C | \phi \rangle = e^{-\tau(E_\tau - E_T)} N(\psi).$$

Thus,

$$E_\tau = E_T - \frac{1}{\tau} \ln \frac{N(\phi)}{N(\psi)}$$

so that

$$E_0 \equiv \langle E_\tau \rangle.$$

Common and useful refinements of the measurement process are *forward walking* (Hammond et al., 1994) and *back propagation* (Zhang et al., 1997). To discuss back propagation, let us suppose we have propagated $|\psi_T\rangle$ in the stationary state by n steps to obtain $|\psi\rangle = A^n |\psi_T\rangle$ and have saved this state. Now suppose we take $m \ll n$ additional steps and save the m configurations leading to the new state $|\psi'\rangle$. Instead of computing an average via

$$\langle \mathcal{O} \rangle = \frac{\langle \psi_T | \mathcal{O} | \psi' \rangle}{\langle \psi_T | \psi' \rangle},$$

where $|\psi'\rangle = A^m |\psi\rangle$, we use

$$\langle \mathcal{O} \rangle = \frac{\langle \psi_T | A^m \mathcal{O} | \psi \rangle}{\langle \psi_T | A^m | \psi \rangle} = \frac{\langle \psi_T' | \mathcal{O} | \psi \rangle}{\langle \psi_T' | \psi \rangle},$$

that is, we do not make measurements with the configuration at the current $(m + n)$-th step, but we make it with the configuration of a past step at n and use the intervening configurations to project $|\psi_T\rangle$ m steps closer to the ground state. We saw something similar in the ground state determinant method (Appendix I). There, we made measurements around the $\beta/2$ point in the projection whose imaginary-time length was β.

Without importance sampling, the Monte Carlo calculation estimates

$$|\psi\rangle = \sum_C w_C |C\rangle,$$

and accordingly for any observable $\mathcal{O}$, its mixed estimate is

$$\langle \mathcal{O} \rangle = \frac{\sum_C w_C \langle \psi_T | \mathcal{O} | C \rangle}{\sum_C w_C \langle \psi_T | C \rangle}.$$

Importance sampling, however, returns a weight $\bar{w}_C = w_C \langle \psi_T | C \rangle$ instead of w_C. In terms of the importance sampled weights, the mixed estimate becomes

$$\langle \mathcal{O} \rangle = \frac{\sum_C w_C \langle \psi_T | C \rangle \frac{\langle \psi_T | \mathcal{O} | C \rangle}{\langle \psi_T | C \rangle}}{\sum_C w_C \langle \psi_T | C \rangle} = \frac{\sum_C \bar{w}_C \frac{\langle \psi_T | \mathcal{O} | C \rangle}{\langle \psi_T | C \rangle}}{\sum_C \bar{w}_C}.$$

10.6 Excited states

Excited state information is important, and with the power method a few such states are often accessible. Computing them accurately is usually difficult. Thus, their computation is infrequent. We now briefly discuss two methods for computing excited states. The first is a classic with a track record, and the other is new and less developed. The classic method (Ceperley and Bernu, 1988), called the *correlation function quantum Monte Carlo method*, starts with the standard variational method for computing excited states and adds features of the variational and diffusion Monte Carlo methods to it. The excited states are determined concurrently, not one at a time.

10.6.1 Correlation function Monte Carlo

In the deterministic variational method, if we want to compute M excited states, we need M trial states, $\{|\psi_{T,i}\rangle, i = 1, 2, \dots, M\}$. If we parametrize these states in terms of a linear combination of some set of basis states $\{|\phi_i\rangle, i = 1, 2, \ldots, N \geq M\}$, with an N not less than M, such that

$$|\psi_{T,i}\rangle = \sum_j |\phi_j\rangle N_{ji},$$

then minimizing the Rayleigh quotient with respect to the elements of the matrix N yields the generalized eigenvalue problem

$$A|\psi\rangle = vS|\psi\rangle,$$

where the Hamiltonian and overlap matrices are $N \times N$ matrices defined as

$$A_{ij} = \langle \phi_i|H|\phi_j\rangle, \quad S_{ij} = \langle \phi_i|\phi_j\rangle.$$

The computed eigenvalues v_i interlace the eigenvalues λ_i of A (9.5); that is, there is a set of the eigenvalues of A such that $\lambda_1 \leq v_1 \leq \lambda_2 \leq v_2 \cdots$ (Section 9.1).

In the correlation function quantum Monte Carlo method (Chapter 9), we first generate the best possible trial states from a variational Monte Carlo calculation or a mean-field theory and then use them in the generalized eigenvalue problem

$$\bar{A}|\phi\rangle = v\bar{S}|\phi\rangle,$$

where

$$\bar{A}_{ij}(\tau) = \langle \phi_i|Ae^{-\tau A}|\phi_j\rangle, \quad \bar{S}_{ij}(\tau) = \langle \phi_i|e^{-\tau A}|\phi_j\rangle.$$

Next we create a sampling density by adding a guiding function $|\psi_G\rangle$ and express these matrices in a form similar to those of the Hamiltonian and overlap matrices in the variational Monte Carlo method (Section 9.3),

$$\bar{A}_{ij}(\tau) = \sum_C \frac{\langle \phi_i|e^{-\tau A/2}|C\rangle}{\langle \psi_G|C\rangle} \frac{\langle C|Ae^{-\tau A/2}|\phi_j\rangle}{\langle C|\psi_G\rangle} \frac{\langle C|\psi_G\rangle^2}{\sum_C \langle C|\psi_G\rangle^2},$$

$$\bar{S}_{ij}(\tau) = \sum_C \frac{\langle \phi_i|e^{-\tau A/2}|C\rangle}{\langle \psi_G|C\rangle} \frac{\langle C|e^{-\tau A/2}|\phi_j\rangle}{\langle C|\psi_G\rangle} \frac{\langle C|\psi_G\rangle^2}{\sum_C \langle C|\psi_G\rangle^2}.$$

We need to sample a configuration from the distribution defined by the guiding function. Then we need to project all basis states and compute their overlap for the sampled configuration. We do the projection with diffusion Monte Carlo. As $\tau \to \infty$, the eigenvalues from the generalized eigenvalue problem using the Monte Carlo average of $\bar{A}$ and $\bar{S}$ approach those of the excited states overlapped by the trial states.

In choosing a basis for the variational trial states and also in choosing a guiding state, we must make these states general enough to avoid inadvertently missing an excited state with a symmetry of interest. Of course, if we know the symmetries of interest, we tailor the basis and guiding function to incorporate them.

The original papers give satisfying detail about the validity of the method and useful information about its implementation. However, as we discussed in the variational Monte Carlo chapter (Section 9.3), newer methods for optimizing the trial state now exist, and it has been understood that while the Hamiltonian and overlap matrices in the generalized eigenvalue problem should be symmetric, symmetrizing the expression that arises upon sampling destroys the zero-variance property of the variational method.

10.6.2 Modified power method

The second method for computing excited states is quite different. The components of the excited eigenvectors must have mixed signs. The correlation function quantum Monte Carlo method never directly samples the excited states. The new method does and hence has to solve a sign problem. We call this new method the *modified power method* (Gubernatis and Booth, 2008; Booth and Gubernatis, 2009b). As with the correlation function quantum Monte Carlo method, it needs as many starting states as the number of eigenvalues sought and determines several eigenpairs concurrently. To simplify the discussion we assume we are seeking just two eigenpairs. The generalization to more than two is fairly obvious.

An important feature of this new method is that it converges to the extremal states, say, the ground state and first excited state, as $(\lambda_3/\lambda_1)^n$ for the ground state and as $(\lambda_3/\lambda_2)^n$ for the first excited state. (Recall (10.1).) Convergence is faster to the ground state and first excited state because during convergence the excited state is subtracted out of the ground state and vice versa as opposed to simply being reduced ("powered out") by the iteration. The modified power method explicitly projects to both states. The other state are "powered out." Details are given in Booth (2003a,b), Gubernatis and Booth (2008), and Booth and Gubernatis (2009b).

If the starting states are $|\phi_1\rangle$ and $|\phi_2\rangle$, the modified power method iterates

$$|\psi_1\rangle = A\,|\phi_1\rangle,$$
$$|\psi_2\rangle = A\,|\phi_2\rangle,$$
$$|\phi_1\rangle = |\psi_1\rangle + \eta_1\,|\psi_2\rangle,$$
$$|\phi_2\rangle = |\psi_1\rangle + \eta_2\,|\psi_2\rangle,$$
$$|\phi_1\rangle \leftarrow |\phi_1\rangle / \|\phi_1\|,$$
$$|\phi_2\rangle \leftarrow |\phi_2\rangle / \|\phi_2\|.$$

The middle two equations prevent the method from being two independent power methods. If the iteration were two independent power methods, then both starting states would project to the same ground state. The middle two equations make the projections dependent in such a way that if one state, say, $|\psi_1\rangle$, converges to the dominant state, the other converges to the subdominant state. It is through these two equations that $|\psi_2\rangle$ gets subtracted out of $|\psi_1\rangle$ and vice versa. The claim is that there exist two numbers, η_1 and η_2, that make this happen.

In the middle two equations, we want to form an eventual eigenstate from the linear combination of two states. Clearly the values of η_1 and η_2 cannot be arbitrary. To find them, we observe that if λ is an eigenvalue of A and the components of the associated eigenstate are ψ_i, then for any nonzero ψ_i[6]

$$\lambda = \frac{\sum_j A_{ij}\psi_j}{\psi_i}.$$

If this is true for one component, then a similar relation is true for a sum over any subset of components.

There are many ways of choosing such a subset, which we call a *region*. For any two regions R_1 and R_2, we must have

$$\lambda = \frac{\sum_{j\in R_1} A_{ij}\psi_j}{\sum_{i\in R_1}\psi_i} = \frac{\sum_{j\in R_2} A_{ij}\psi_j}{\sum_{i\in R_2}\psi_i}. \tag{10.26}$$

Now if $|\psi\rangle = |\psi_1\rangle + \eta|\psi_2\rangle$ is an eigenvector, and we substitute this state into the above equation and cross-multiply the numerators and denominators on opposite sides of the equal sign, we find that η must satisfy a quadratic equation

$$a\eta^2 + b\eta + c = 0,$$

whose coefficients are sums of products of the region sums:

$$a = \sum_{i\in R_2}\psi_{1i}\sum_{i\in R_1}\sum_j A_{ij}\psi_{2j} - \sum_{i\in R_1}\psi_{2i}\sum_{i\in R_2}\sum_j A_{ij}\psi_{1j},$$

$$b = \sum_{i\in R_2}\psi_{2i}\sum_{i\in R_1}\sum_j A_{ij}\psi_{1j} - \sum_{i\in R_1}\psi_{2i}\sum_{i\in R_2}\sum_j A_{ij}\psi_{1j}$$

$$+ \sum_{i\in R_2}\psi_{1i}\sum_{i\in R_1}\sum_j A_{ij}\psi_{2j} - \sum_{i\in R_1}\psi_{1i}\sum_{i\in R_2}\sum_j A_{ij}\psi_{2j},$$

$$c = \sum_{i\in R_2}\psi_{1i}\sum_{i\in R_1}\sum_j A_{ij}\psi_{1j} - \sum_{i\in R_1}\psi_{1i}\sum_{i\in R_2}\sum_j A_{ij}\psi_{1j}.$$

---

[6] Until a deterministic power method converges, what we really should define is $\lambda(i) = \frac{\sum_j A_{ij}\psi_j}{\psi_i}$, an estimate of the eigenvalue that varies from component to component. As the deterministic power method converges, these estimates for different i converge to the same value. In a Monte Carlo simulation, while the global eigenvalue estimator converges, this local estimate still shows small variations.

At each step of the iteration, we compute these coefficients and solve the quadratic equation for its two roots. When its roots are both real,[7] we use each root in $|\psi\rangle = |\psi_1\rangle + \eta|\psi_2\rangle$, compute its eigenvalue estimate (10.26) using one of the regions, and then associate one root η_1 with the largest eigenvalue estimate $|\psi_1\rangle$ and the other root η_2 with the subdominant estimate. Doing this consistently causes the modified power method to converge to the dominant and subdominant states as claimed. By shifting the spectrum of A, we can enforce convergence to the ground state and first excited states.

For a deterministic implementation, whether we are seeking a few dominant eigenvalues of a matrix or a continuous operator, just about any choice of regions works. Additionally, the mixed signs of the elements of the matrix or of the components of its eigenvectors cause few problems because we do the matrix-vector multiplication exactly. A Monte Carlo implementation, which uses a sample of just a few components of the vector, misses important sign cancellations, and we are thus confronted with a sign problem.

To address the sign problem, we start by associating with each walker a state and two weights representing the components of the two eigenstates. If the matrix is nonnegative, one weight is positive while the other may have either sign. As the number of walkers becomes minuscule with respect to the number of available states, the likelihood that the Monte Carlo algorithm puts two walkers of opposite sign into the same state is small. We therefore add to the Monte Carlo sampling some procedures that promote these events. Several ways to do this have been developed (Booth and Gubernatis, 2009b; Rubenstein et al., 2010; Gubernatis, 2012). One way was actually developed for quantum Monte Carlo problems in the continuum (Arnow et al., 1982). We now describe it.

We are using a Monte Carlo algorithm to sample the sum $\sum_j A_{ij}x_j$. We start by expressing this product as a sum $\sum_{j\in\{k\}} A_{ij}x_j$ over the set of indices $\{k\}$ representing the current population of M walkers, which of course is only a small subset of the available states. Instead of sampling an i for a given j in $\{k\}$, we involve the entire population in the following way:

$$\sum_{j\in\{k\}} A_{ij}w_j = \sum_{j\in\{k\}} \frac{A_{ij}w_j}{\sum_{j\in\{k\}} P_{ij}} \sum_{j\in\{k\}} P_{ij}$$

$$= \sum_{j\in\{k\}} \frac{A_{ij}w_j}{\sum_{j\in\{k\}} P_{ij}} \left[P_{ij_1} + P_{ij_2} + \cdots + P_{ij_M} \right].$$

---

[7] We could simply use the complex roots if we were to use complex arithmetic.

From the terms in the square brackets, we now sample an i_1 from P_{ij_1} to obtain a walker in state $|i_1\rangle$ with weight

$$\sum_{j\in\{k\}} A_{i_1j}w_j \bigg/ \sum_{j\in\{k\}} P_{i_1j} \qquad (10.27)$$

and i_2 from P_{ij_2} to obtain a walker in state $|i_2\rangle$ with weight

$$\sum_{j\in\{k\}} A_{i_2j}w_j \bigg/ \sum_{j\in\{k\}} P_{i_2j}\,, \qquad (10.28)$$

and so on. Consequently, each of the current walkers contributes multiple new walkers, instead of the usual one per walker. Each new walker typically receives contributions from multiple old walkers. This increase in population increases the likelihood of new walkers with opposite sign sharing the same state. It is easy to combine walkers in the same state and thus cancel signs that otherwise would be missed.

This importance sampling procedure scales as M^2 and proves computationally expensive. Instead of using the entire population at once, we can take all pairs of walkers, or all triplets, and so on. Generally there is little gain in grouping beyond a certain number. Or we may choose to use just subsets of the population most likely to have walkers of mixed sign. We illustrate the pair-wise sampling in the right-hand panel of Fig. 10.1. In this figure, the solid lines connecting the old and new states are the connections we sample with the standard Monte Carlo power method sampling. This sampling defines the new population. Then, for a given pair of old states, the dotted lines represent additional possible contributions to the new pair of states. We modify the weights of the new state according to (10.27).

The modified power method with proactive sign cancellation has been successfully applied to a number of classical problems and is just being explored for quantum problems. More details are given in a series of papers (Gubernatis and Booth, 2008; Booth and Gubernatis, 2009a; Rubenstein et al., 2010; Gubernatis, 2012). To date, its success surprisingly has been without the use of importance sampling. Much work remains before its full potential for quantum Monte Carlo is known.

Booth has proposed power methods for computing one excited state at a time without the knowledge of other eigenstates and methods to assess convergence (Booth, 2003a,b,c, 2010, 2011a,b,c).

10.7 Comments

As mentioned at the beginning of this chapter, most Monte Carlo techniques described here were originally developed for classical problems. In fact, we can easily argue that the problem of neutron transport through fissile material, the original application that inspired the development of the Monte Carlo method, is the father of ground state quantum Monte Carlo methods in the continuum. The materials in these nuclear engineering problems are typically piecewise homogeneous. Each piece is characterized by scattering cross sections that determine the mean free path of a neutron and the probability of it being exited, absorbed, scattered, and fissioned. There are two classes of neutron transport problems, called *fixed source* and *criticality*. Fixed-source problems map onto solving a linear system of equations whose right-hand side (the known) is the source of the radiation. The criticality problem maps onto an eigenvalue problem. When the material produces as many neutrons by fission as those lost by absorption and escape, the system is critical with a dominant eigenvalue of unity. In a subcritical system, this eigenvalue is less than unity, while in a supercritical system, it is greater than unity.

In these transport problems, the neutron needs to move from one spatial point in a continuum to another. This movement is relatively easy to sample. Sampled is the distance to the next collision whose probability is an exponential that has the path lengths weighted by the reciprocal of the mean free path for the pieces of material transversed.

Diffusion Monte Carlo in the continuum borrowed many of its sampling methods from the neutron transport problem. A point of difference is the sampling of the movement of a particle (walker) through the continuum. In the quantum problem the Hamiltonian is exponentiated. The Trotter approximation expresses this exponential as a product of the exponential of the kinetic energy and the exponential of the potential energy (Section 7.1.1). In a position basis, the latter is diagonal. The former moves the particle and requires the computation of terms such as $e^{D\nabla^2}\psi(r)$. We can execute this operation via a Hubbard-Stratonovich transformation in the following manner (Zhang et al., 1997):

$$e^{D\nabla^2}\psi(r) = \frac{1}{\sqrt{2\pi}}\int_{-\infty}^{+\infty} dx\, e^{-\frac{1}{2}x^2} e^{-i\sqrt{D}x\cdot\nabla}\psi(r)$$

$$= \frac{1}{\sqrt{2\pi}}\int_{-\infty}^{+\infty} dx\, e^{-\frac{1}{2}x^2}\psi\left(r + \sqrt{D}x\right).$$

Sampling x from the Gaussian produces movement from r to $r + \sqrt{D}x$, a diffusive motion. Accordingly, for problems in the continuum, diffusion Monte Carlo provides a convenient and almost necessary means to sample particle movement.

A curious difference between the quantum and nuclear engineering applications is the way importance sampling, in the sense of (10.4.3), is used. In the engineering problems different importance functions may be used for specific processes in different pieces of the material. The overall problem is too complex to have one importance function for the entire problem. Accordingly, the emphasis has been on the generation of variance reduction methods for the difficult pieces. In general, multiple techniques are used in multiple parts of the material.

The stochastic reconfiguration methods we discussed owe their existence to the neutron transport problem. In the quantum problems, however, we have means to generate a global importance function, for example, by the variational Monte Carlo method or a mean field theory. As a consequence, there has been less emphasis on exploring a range of methods for variance reduction. Usually, one stochastic reconfiguration method is used for the entire problem, making its intent more about controlling the size of the population. Compared with the spectrum of techniques used in the neutron transport simulations, the spectrum of techniques used in quantum Monte Carlo power method simulations appears somewhat monochromatic.

Ground state quantum Monte Carlo on a lattice in turn borrowed many techniques from ground state quantum Monte Carlo simulations in the continuum. In lattice models, the walkers generally move from one discrete state to another, whether or not we are using a discrete Hubbard-Stratonovich transformation (Section 7.1.1). Often the states available from any one state are very limited in number. To some extent this feature of lattice problems is underexploited. While diffusion Monte Carlo maintains advantages, the disadvantages of using the linear method and the exact Green's function method in the continuum are mitigated. Designing a ground state quantum Monte Carlo solution for lattice problems has more options. The elegance of the exact Green's function method becomes harder to keep bypassing.

The methods in this chapter bring us back to where Fermi started us in Chapter 1. We have illustrated that the imaginary-time Schrödinger equation has a Monte Carlo realization as a collection of walkers (particles) in which each independently performs a random walk while at the same time being subjected to multiplication of their weights at their locations in configuration space by parts of the Hamiltonian. In the Monte Carlo simulation the weight of the walkers decays exponentially in imaginary time at a rate controlled by the eigenvalue E_0 and the distribution of walkers $\langle C | \psi \rangle$. This eigenpair corresponds to the ground state solution of the Hamiltonian.

Suggested reading

G. Goertzel and M. H. Kalos, "Monte Carlo methods in transport problems," in *Progress in Nuclear Energy*, vol. 2 (New York: Pergamon Press, 1958), p. 315.

J. Spanier and E. M. Gelbard, *Monte Carlo Principles and Neutron Transport Problems* (Reading, MA: Addison-Wesley, 1969), chapter 2.

T. E. Booth, "Particle transport applications," in *Rare Event Simulation Using Monte Carlo Methods*, ed. G. Rubino and B. Tuffin (Chichester: John Wiley, 2009), p. 215.

Exercises

10.1 For a homogeneous, real, symmetric, tridiagonal matrix of order N whose nonzero elements are positive, find analytically the eigenpair associated with the largest and smallest eigenvalue.

10.2 For the matrix chosen in Exercise 10.1, use Algorithm 33 to find the eigenpair associated with the largest and smallest eigenvalue for several different values of N.

10.3 For the matrix chosen in Exercise 10.1, use Algorithm 35 to find the eigenpair associated with the largest and smallest eigenvalue for several different values of N. Use branching for stochastic reconfiguration.

10.4 Propose a quantum Monte Carlo algorithm for the transverse-field Ising model by adapting the steps for the linear method in Section 10.4.1 using the basis in which H_1 is diagonal. Do the analogous development for the transverse field Ising model for the diffusion Monte Carlo method described in Section 10.4.2.

10.5 For the matrix chosen in Exercise 10.1, use the modified power method deterministically to find the two eigenpairs associated with the two largest and the two smallest eigenvalues for several different values of N.

10.6 Propose a Monte Carlo method to solve the linear system of equations $\sum_j A_{ij} x_j = b_i$.

11

Fermion ground state methods

In this chapter, we discuss the fixed-node and constrained-path Monte Carlo methods for computing the ground state properties of systems of interacting electrons. These methods are arguably the two most powerful ones presently available for doing such calculations, but they are approximate. By sacrificing exactness, they avoid the exponential scaling of the Monte Carlo errors with system size that typically accompanies the simulation of systems of interacting electrons. This exponential scaling is called the *Fermion sign problem*. After a few general comments about the sign problem, we outline both methods, noting points of similarity and difference, plus points of strength and weakness. We also discuss the constrained-phase method, an extension of the constrained-path method, which controls the phase problem that develops when the ground state wave function cannot be real.

11.1 Sign problem

The "sign problem" refers to the exponential increase of the Monte Carlo errors with increasing system size or decreasing temperature (e.g., Loh et al., 1990, 2005; Gubernatis and Zhang, 1994) that often accompanies a Markov chain simulation whose limiting distribution is not everywhere positive. Such a case generally arises in simulations of Fermion and frustrated quantum-spin systems. It seems so inherent to Monte Carlo simulations of Fermion systems that the phrase "the sign problem" to many seems almost synonymous with the phrase "the Fermion sign problem."

Explanations for the cause of the sign problem vary and are still debated. In this chapter, we choose to summarize two explanations that seem to connect the causes in ground state Fermion simulations in the continuum and on the lattice. The sign problem, of course, is not limited to ground state calculations or even to Fermion simulations. While the cause we discuss in Sections 11.2 and 11.3 focuses on the low-lying states of diffusion-like operators, several topological pictures have been proposed (Muramatsu et al., 1992; Samson, 1993; Gubernatis and Zhang, 1994; Samson, 1995). Some of these discussions are done in the context of the zero- and finite-temperature determinant methods (Muramatsu et al., 1992; Gubernatis and Zhang, 1994). Others are done more analytically from a Feynman path-integral point of view (Samson, 1993, 1995). Some are for particles with statistics other than Fermions. The presentation of the sign problem in this chapter is appropriate for the Monte Carlo methods discussed in this chapter.

It is often difficult to know a priori whether a given Fermion or quantum-spin Monte Carlo simulation suffers from a sign problem. Experience, however, has shown that the simulations for certain classes of Hamiltonians are likely to have such a problem. For example, quantum-spin systems with antiferromagnetic exchange interactions are a good bet, unless the lattice is bipartite. In the latter case, the Marshall transformation (Marshall, 1955) might convert the simulation of Hamiltonians with nearest-neighbor exchange interactions to a sign-free one with a ground state wave function that must be positive (Section 5.2.4). For the same Hamiltonian on a "frustrated" lattice, such as the triangular or Kagome lattice, the transformation ceases to remove the sign problem. Simulations of spin models with nearest and next-nearest neighbor exchange interactions of opposite sign typically have a sign problem for any lattice structure.

For interacting electrons on a lattice, sign-problem-free situations are even rarer. Proofs exist that special Hamiltonians lack a sign problem (e.g., Wu and Zhang, 2005). Unfortunately, these Hamiltonians are unphysical. Hamiltonians with particle-hole symmetry at certain electron fillings usually lack a sign problem (Hirsch, 1985), as do most Hamiltonians in one dimension, if the hopping is nearest neighbor. In the latter case, the restricted motion along the line allows an unambiguous relabeling of the electrons that eliminates the exchange of electrons and the source of the sign problem. Certain types of spinless Fermion models can be simulated with the auxiliary field method after switching to a Majorana Fermion representation (Li et al., 2014). The idea here is that the formulation in terms of Majorana Fermions doubles the number of determinants, which eliminates the sign problem in a manner similar to the case of spinful Fermion models with particle-hole symmetry.

If the sign problem cannot be trumped by the choice of the Hamiltonian or its expression in an alternative basis, then the burden of dealing with the sign problem falls upon the algorithm. In the continuum, there are several Green's function methods that have no Fermion sign problem (Zhang and Kalos, 1991; Kalos and Pederiva, 2000), but unfortunately these methods do not appear to be suitable for general efficient use. The fixed-node and constrained-path methods fill a void by being well-benchmarked approximate methods that do not suffer from a sign problem. Accepting the approximation enables us to perform simulations for modestly sized systems instead of struggling with an exact simulation for small systems.

In principle, the strategy for sampling from a mixed-sign distribution is straightforward. In regions where the distribution is positive, we sample normally. In regions where it is negative, we sample normally from the distribution's absolute value and change the sign of the observable. Expectation values become sign-weighted averages. More explicitly, if $p(x)$ is the distribution of interest and its sign function is $s(x) = p(x)/|p(x)|$, then the *sign-weighted average* is

$$\langle A \rangle = \frac{\int dx s(x) A(x) \, |p(x)|}{\int dx s(x) \, |p(x)|} = \frac{\langle sA \rangle_{|p|}}{\langle s \rangle_{|p|}}.$$

The subscript on the angular brackets indicates that the average is with respect to the distribution $|p(x)|$.

For many Markov chain Monte Carlo simulations, and in particular for the Monte Carlo power methods, the limiting distribution is unknown and typically not nonnegative. Particularly problematic is the situation when we do not know the location of the nodal surface that delineates the regions of phase space differing in sign, because this limits the utility of using sign-weighted averages. Without this knowledge, the weights of the configurations have mixed sign, and eventually about half of the configuration population has a positive sign and the other half a negative sign. This almost balanced mixture causes $\langle sA \rangle \approx 0$ and $\langle s \rangle \approx 0$, and the sign-weighted average becomes indeterminate (the error bar becomes much larger than the value of the observable). Various heuristic arguments (Section 5.4) and the experience from countless simulations demonstrate that these sign-weighted averages approach zero exponentially fast as the number of lattice sites or the inverse temperature increases. The associated relative errors also grow exponentially fast, leading to an algorithm that quickly becomes unusable. The irony of the Fermion sign problem is that in principle we do not need sign-weighted averages. We only need to solve for the wave function in the region corresponding to one sign. From this solution, we can use an antisymmetrization procedure to generate the remainder of the solution explicitly. The problem is that we do not know the boundaries of this region.

As discussed in the previous chapter, Monte Carlo power methods define a transition probability $P(C'|C)$ for generating the Markov chain from a Green's function $\langle C'|G|C \rangle = G(C',C) \equiv w(C',C)P(C'|C)$. While we can always define nonnegative transition probabilities, we cannot always at the same time define nonnegative weights $w(C',C)$. As also discussed in the previous chapter, these power methods usually importance sample via a guiding function $\psi_G(C) \equiv \langle C|\psi_G \rangle$ so the weights and transition probabilities actually used are defined relative to $\bar{G}(C',C) \equiv \psi_G(C')G(C',C)/\psi_G(C)$. For systems of frustrated quantum spins, a typical situation is that the guiding function is positive, while the matrix elements of the Green's function have mixed signs. For systems of Fermions, the situation is often reversed with $G(C', C)$ being nonnegative but the guiding function at C' and C having opposite signs. On a transition from $|C\rangle$ to $|C'\rangle$, as defined by a positive $G(C', C)$, the importance-sampled Green's function $\bar{G}(C', C)$ can change sign because the transition produces an odd number of electron exchanges from $\psi_G(C)$ to $\psi_G(C')$.

The fixed-node and constrained-path methods typically scale as the cube of the system size. They eliminate the exponential error scaling by preventing walker movement across an approximate nodal surface of the sampling distribution. We first discuss the basics of the fixed-node method and then the basics of the constrained-path method. The details of the methods have important implications with respect to the estimation of expectation values. We discuss both methods simultaneously for mixed estimators and highlight the differences between the forward-walking estimator in the fixed-node method and the back-propagation estimator in the constrained-path method that often replaces the mixed estimator. After the discussion of mixed estimators we discuss the constrained-phase method, an extension of the constrained-path method, that controls the phase problem that develops when the ground state wave function cannot be real.

11.2 Fixed-node method

The fixed-node method was developed decades ago for diffusion Monte Carlo (Section 10.4.2) simulations of Fermions in the continuum (Anderson, 1975, 1976; Moskowitz et al., 1982; Reynolds et al., 1982). Associated with it is an interesting physical picture of what happens to the wave function when a sign problem exists. In this power method, a Monte Carlo algorithm evolves a configuration $C = (r_1, r_2, \ldots, r_N)$ of N labeled electron positions through phase space. When importance sampling is used (Section 10.4.3 and Section 11.1), the Monte Carlo sampling generates empirically a distribution function $f(C, \tau) = \psi(C, \tau)\psi_G(C)$ where $\psi_G(C)$ is the guiding (importance) function representing the ground state.[1] One can show (Hammond et al., 1994) that this distribution solves a partial-differential equation that defines the electron distribution function's deterministic imaginary-time dynamics

$$\frac{\partial f}{\partial \tau} = D\nabla^2 f - D\nabla \cdot \left[f F_Q \right] + [E_T - E_L]f,$$

---

[1] For simplicity, we temporarily drop the spin of each electron that is associated with each spatial position r_i.

where the "diffusion constant" is $D = \hbar^2/2m$, the local energy is (Section 9.3.1)

$$E_L = \langle C|H|\psi \rangle / \langle C|\psi \rangle \equiv H\psi(C)/\psi(C),$$

and the vector field F_Q is given by $F_Q = \nabla \ln |\psi|^2$. This field, called the *quantum force*, is a drift term appearing when the dynamics is importance sampled, that is, when the weight $f(C, \tau) = \psi(C, \tau)\psi_G(C)$.

The importance of this differential equation, which is not one that we are interested in solving numerically, is that it is well known that the spatial extremal eigenfunction of a diffusive differential equation is nodeless. The nodeless solution maximizes the effect of the Laplacian ∇^2 that is driving the diffusion. In continuum diffusion Monte Carlo simulations, we call the nodeless solution Bosonic. The first excited state, as all other eigenstates, must have nodes.[2] This first excited state is the Fermionic state we want. For large systems, the gap between the Bosonic and Fermionic eigenvalues becomes exponentially small and eventually lost in the noise of the simulation. Hence, in the continuum, the collapse of the Fermionic solution to the Bosonic solution causes the sign problem.

The continuum fixed-node method prevents the Bosonic collapse by imposing an infinite potential energy barrier at the Fermionic nodal surface, preventing any random walker from crossing it and thereby generating walkers of mixed sign. With these added potentials, the original Hamiltonian becomes some H_{eff}. The method exactly solves the Schrödinger equation for the effective Hamiltonian for the ground state $(E_{\text{eff}}, \psi_{\text{eff}})$ within this nodal interior. Because the location of the nodal surface is unknown, the method approximates the exact surface by using the nodal surface of a good approximation to the Fermionic ground state. If the method used the exact surface, it would generate the exact answer. The Fermionic guiding state $|\psi_G\rangle$ used in the importance sampling hopefully closely approximates the exact ground state and serves a dual purpose by also fixing the nodal surface. *The Monte Carlo solution inherits its Fermionic character from this guiding state.*

This sign problem picture, which is supported by continuum simulations, is enlightening but not as appropriate on a lattice as it is in the continuum. On a lattice, the picture changes as the connection with a diffusion-like equation of motion is less clear. Not surprisingly, the implementation of the fixed-node method changes with respect to defining a nodal surface in a discrete space and to the details of the effective Hamiltonian acting when a random walker tries to cross the surface.

In the *lattice fixed-node method* (Bemmel et al., 1994; ten Haaf et al., 1995), we use what we called the linear Green's function Monte Carlo method (Section 10.4.1). Sometimes the diffusion Monte Carlo method is used (Section 10.4.2).

---

[2] We recall that the lowest eigenpair of H is the maximal pair of $\exp(-\tau H)$.

We recall that the linear method has no Trotter approximation and iterates

$$\left|\psi'\right\rangle = [I - \tau \, (H - E_T)] \left|\psi\right\rangle .$$

In this iteration we represent the ground state $\left|\psi_0\right\rangle$ in a complete orthonormal basis of labeled electrons as

$$\left|\psi_0\right\rangle = \sum_C |C\rangle\langle C|\psi_0\rangle = \sum_C \psi_0(C)|C\rangle, \quad \psi_0(C) > 0, \tag{11.1}$$

and after we project the iteration onto the labeled-configuration basis, we obtain

$$\psi'(C') = \sum_C \langle C'|I - \tau \, (H - E_T)|C\rangle\psi(C) = \sum_C G(C', C)\psi(C).$$

When the iteration becomes stationary, that is, when the condition $\psi' = \psi_0$ is fulfilled, the Monte Carlo simulation is sampling from

$$P(C) = \frac{\psi_0(C)}{\sum_C \psi_0(C)}.$$

With importance sampling (Section 10.4.3)

$$\bar{\psi}'(C') = \sum_C \bar{G}(C', C)\bar{\psi}(C),$$

where $\bar{\psi}(C) = \psi_G(C)\psi(C)$. Importance sampling thus changes the transition probability from which we draw our samples. Almost always we take $\psi_G(C) = \psi_T(C)$. Sampling only in the region where $\psi_T(C)$ is positive resolves the issue. Making the matrix elements of $\bar{G}$ positive yields the transition probability for the Markov chain. In the continuum, G is generally easily made positive.[3]

A nodal surface for lattice problems cannot be precisely located. It is something not approachable by making successively smaller and smaller spatial steps that avoid crossing the surface. On a lattice, the sign problem morphs into something associated with *sign flips* as electrons hop between lattice points. Sign flips occur for several reasons. As in the continuum, the guiding state can change sign. In contrast to the continuum, where the kinetic energy is always negative, the hopping part of many lattice Hamiltonians has amplitudes of mixed signs. More generally, the Hamiltonian almost always connects lattice sites in different nodal regions.

We also recall that in the linear Green's function Monte Carlo method (Section 10.4) the condition for no sign problem was $\langle C|H - \sigma I|C'\rangle \leq 0$ for all C and C'.[4] If so, then a sign change occurs only if the walker moves from one nodal region

---

[3] In the continuum making the elements of G positive is a matter of making its diagonal elements negative.
[4] In what follows $\sigma = E_T$, and for notational simplicity we replace $H - \sigma I$ by H.

to another. More generally, when configuration changes cause $\langle C'|H|C\rangle \psi_T(C)$ $\psi_T(C') > 0$, the lattice fixed-node method (Bemmel et al., 1994; ten Haaf et al., 1995) prevents sign flips by replacing the original Hamiltonian with an effective one created from the original by truncating the off-diagonal terms and adding terms to the diagonal.[5] The off-diagonal elements of this effective Hamiltonian are

$$\langle C'|H_{\text{eff}}|C\rangle = \begin{cases} \langle C'|H|C\rangle & \text{if } \langle C'|H|C\rangle \psi_T(C') \psi_T(C) < 0 \\ 0 & \text{otherwise,} \end{cases} \tag{11.2}$$

while the diagonal elements are

$$\langle C|H_{\text{eff}}|C\rangle = \langle C|H|C\rangle + \langle C|V_{\text{sf}}|C\rangle. \tag{11.3}$$

The last term in this equation is a spin-flip potential that partially corrects for the contribution of the transitions left of H_{eff}. This potential has only diagonal elements and is defined by

$$\langle C|V_{\text{sf}}|C\rangle = \sum_{C'}{}^{\text{sf}} \langle C|H|C'\rangle \psi_T(C')/\psi_T(C). \tag{11.4}$$

For the diagonal elements, the superscript "sf" on the summation sign signifies that the summation over C' is over all configurations C for which $\langle C'|H|C\rangle \psi_T(C)$ $\psi_T(C') > 0$. On the diagonal of the effective Hamiltonian, a positive potential energy replaces a hop (an off-diagonal term) that would flip a sign. The Monte Carlo power method exactly solves for the ground state $(E_{\text{eff}}, \psi_{\text{eff}})$ of this effective Hamiltonian. An important feature of this method is that the ground state of H_{eff} is, as we will show, an upper bound of the exact ground state energy.

Several key characteristics of the lattice fixed-node method are: (1) It accommodates sign problems other than the Fermion sign problem; (2) the exact ground state energy is obtained only if $\psi_T(C) = \psi_0(C)$ is exact (Section 11.4); and (3) the effective ground state energy is variational, that is, $E_{\text{eff}} \geq E_0$.

Whether there is a sign problem or not, efficient use of Green's function methods requires trial states we can easily evaluate for different configurations. This is possible for a number of interesting many-body wave functions such as those discussed in Section 9.2. The ability to use a variety of trial states is a strength of this method. Best results are obtained if the variational Monte Carlo method, a good approximate theory, or some other wave function optimization method first refines this trial state. We note that refining a trial state to produce a truly excellent estimate of the ground state energy may produce a trial state less effective than desired in estimating other observables.

---

[5] Recall from Section 10.4 the conditions for the positivity of $G = I - \tau(H - E_T I)$.

11.3 Constrained-path method

In the constrained-path method (Zhang et al., 1995, 1997; Carlson et al., 1999), we represent the ground state $|\psi_0\rangle$ as a sum of Slater determinants $|\psi_0\rangle = \sum_\phi c_\phi |\phi\rangle$ with $c_\phi > 0$ instead of as a sum over spatial configurations (11.1). This difference and its implications distinguish the constrained-path method from the diffusion Monte Carlo fixed-node method (Section 10.4.2). In many other respects, they are formally the same.

The constrained-path method was developed for projection problems where the Trotter approximation and Hubbard-Stratonovich transformation are used on the exponential of the Hamiltonian to enable the projection of one Slater determinant onto another (Appendix F). The resulting random walk is in the space of Slater determinants $|\phi\rangle$, and the Monte Carlo sampling from the distribution of the Hubbard-Stratonovich fields generates a distribution $f(\phi, \tau)$ that estimates the imaginary-time projection to the ground state.

Analogous to the continuum problem, a deterministic diffusion-like partial differential equation

$$\frac{\partial f}{\partial \tau} = \frac{1}{2} D(\phi) f + \left[\nabla_\phi V_1(\phi)\right] \cdot \nabla_\phi f + V_2(\phi) f \qquad (11.5)$$

for the weight $f(\phi, \tau)$ exists when the Hubbard-Stratonovich fields are Gaussian distributed (Fahy and Hamann, 1991). Here the diffusion operator D and the drift V_1 and branching V_2 operators have explicit representations in terms of the operators of the Hamiltonian.

What is special about this equation is its invariance under the Slater determinant parity transformation $|\phi\rangle \rightarrow -|\phi\rangle$ (Fahy and Hamann, 1990, 1991; Hamann and Fahy, 1990). This symmetry enables the classification of its Slater determinant eigenfunctions as having an odd or even parity. The lowest eigenstate of this partial differential equation has even parity, that is, $f_0(\phi, \tau) = f_0(-\phi, \tau)$. As it is a sum of Slater determinants, it is naturally Fermionic in configuration space.

In terms of this distribution, the imaginary-time-dependent wave function is

$$|\psi(\tau)\rangle = \sum_\phi f(\phi, \tau)|\phi\rangle, \qquad (11.6)$$

where the sum is over all normalized Slater determinants, and expectation values become

$$\langle A(\tau)\rangle = \frac{\langle \psi(\tau)|A|\psi(\tau)\rangle}{\langle \psi(\tau)|\psi(\tau)\rangle} = \frac{\sum_\phi \sum_{\phi'} f(\phi, \tau) f(\phi', \tau)\langle \phi|A|\phi'\rangle}{\sum_\phi \sum_{\phi'} f(\phi, \tau) f(\phi', \tau)\langle \phi|\phi'\rangle}. \qquad (11.7)$$

Fahy and Hamann (1990, 1991) showed that the lowest eigenstate of (11.5) has even parity. As we have done for other discussions of projection methods, let us make the eigenfunction expansion

$$f(\phi, \tau) = \sum_i c_i f_i(\phi) \exp(-E_i \tau).$$

As $\tau \to \infty$, we see that the even-parity lowest eigenstate dominates this expansion and also dominates the integrands of the numerator and denominator in (11.7). In this limit, because $|\psi\rangle$ and $-|\psi\rangle$ are different points in the manifold of Slater determinants, the even-parity $f(\phi, \tau)$ contributions in (11.6) must cancel out to yield the lowest odd-parity eigenstate. Thus, the integrals in (11.7) are dominated by the lowest odd-parity eigenfunction. A statistical evaluation of these integrals ultimately lacks sufficient accuracy to capture the difference in the different parity contributions.

This "diffusion in the space of Slater determinants" approach was developed when the main ground state method for lattices was the determinant method discussed in Appendix I. The approach presents another interesting perspective on the sign problem. It was also trying to reconcile the similar nature of the sign problem in diffusion Monte Carlo in the continuum and the sign problem in the determinant method. The continuum method was at the time believed not to be using Hubbard-Stratonovich fields. We now know that the continuum diffusion Monte Carlo method is an auxiliary-field method with a Gaussian Hubbard-Stratonovich transformation applied to the exponential of the kinetic energy (Zhang et al., 1997; Section 10.7). In the determinant method, the transformation is applied to the exponential of the potential energy. The main difference between the methods is that one executes a random walk in configuration space and the other executes one in Slater determinant space.

The analysis leading to the diffusion equation (11.5) assumes that the auxiliary fields are Gaussian distributed. The details of the diffusion equation depend on the distribution of Hubbard-Stratonovich fields, and generally, we use discrete fields. In any case, the advantage of this formal analysis was insight and not an efficient algorithm. However, the analysis did lead to a proposal in Slater determinant space called the *positive projection method* (Fahy and Hamann, 1990) as an analog to the fixed-node approximation in configuration space. The positive projection method became the springboard for the constrained-path method. If $\psi < 0$, the positive projection method assumes $f(\psi) = 0$, which is equivalent to assuming $V_2(\psi)$ in (11.5) is infinity. Under these conditions the lowest eigenstate has odd parity.

At its core the constrained-path method is a power method of the diffusion Monte Carlo type. The key operator in this method is $\exp(-\Delta \tau H)$, which is the same key operator as in the zero- and finite-temperature determinant methods (Chapter 7 and Appendix I). We recall that in these chapters an operator B represented $\exp(-\Delta \tau H)$ after the Trotter approximation and the Hubbard-Stratonovich transformation, and the propagation was by a noninteracting Hamiltonian that depended on a time-dependent auxiliary field. We can represent one step in this propagation as

$$|\psi'\rangle = \sum_x P(x)B(x)|\psi\rangle.$$

Here, B is an operator and $P(x)$ is the distribution of the auxiliary field x.

In the constrained-path method, each walker $|\phi\rangle$ is a Slater determinant: $|\phi\rangle = a^\dagger_{N_\sigma} \cdots a^\dagger_2 a^\dagger_1 |0\rangle$, where $a^\dagger_i = \sum_j c^\dagger_j \Phi_{ji}$ is defined by the $N \times N^\sigma$ matrix Φ. By Thouless's theorem (Appendix F), the propagation of this state $e^{-\Delta\tau c^\dagger Mc}|\phi\rangle$ via a noninteracting Fermion Hamiltonian is to a new Slater determinant $|\phi'\rangle$, where $|\phi'\rangle = b^\dagger_{N_\sigma} \cdots b^\dagger_2 b^\dagger_1 |0\rangle$, and $b^\dagger_i = \sum_j c^\dagger_j \Phi'_{ji}$ is defined by the $N \times N^\sigma$ matrix $\Phi' = \exp(-\Delta\tau M)\Phi = B\Phi$. Here B is a matrix.[6]

To define a way to implement the positive projection method, we first recall that in the ground state determinant method (Appendix I) our expectation values were of the form

$$\langle A \rangle = \frac{\sum_x P(x)\langle \psi_T | B_L(x_L) \cdots B_{L/2-1}(x_{L/2-1}) A B_{L/2}(x_{L/2}) \cdots B_1(x_1)|\psi_T\rangle}{\sum_x P(x)\langle \psi_T | B_L(x_L) B_{L-1}(x_{L-1}) \cdots B_1(x_1)|\psi_T\rangle}.$$

Here $x = (x_1, x_2, \ldots, x_L)$ represents a configuration of Hubbard-Stratonovich fields, and $P(x)$ is the joint distribution function associated with these fields (Section 7.1.1 and Appendix G). Constraining the sign of the terms in the denominator creates the sign of the numerator. This constraint creates a nodal surface equal to the hyperplane perpendicular to some constraining wave function $|\psi_C\rangle$, which might be one or a sum of a few Slater determinants. The better $|\psi_C\rangle$ approximates the ground state, the closer the hyperplane becomes to the exact nodal surface of f. The positive projection method imposes the L conditions

$$\langle \psi_C | B_1 |\psi_T\rangle > 0,$$
$$\langle \psi_C | B_2 B_1 |\psi_T\rangle > 0,$$
$$\vdots$$
$$\langle \psi_C | B_{L/2} B_{l-1} \cdots B_1 |\psi_T\rangle > 0,$$
$$\langle \psi_T | B_L B_{L-1} \cdots B_{L/2+1} |\psi_C\rangle > 0,$$
$$\vdots$$
$$\langle \psi_T | B_L |\psi_C\rangle > 0.$$

Only configurations of Hubbard-Stratonovich fields that satisfy all L constraints are used. For Gaussian-distributed Hubbard-Stratonovich fields these conditions create an impenetrable barrier for the random walkers.

---

[6] We are overloading notation to curb the proliferation of notation. The context clearly determines whether B is an operator or a matrix.

The constrained-path method (Zhang et al., 1995, 1997; Carlson et al., 1999) makes the positive projection suggestion more practical by two modifications. The first makes the projection open ended,

$$\langle \psi_C | B_1 | \psi_T \rangle > 0,$$
$$\langle \psi_C | B_2 B_1 | \psi_T \rangle > 0,$$
$$\vdots$$

that is, it becomes the power method instead of the fixed-length projection in the determinant method. The second modification eliminates any random walker violating $\langle \phi | \psi(\tau) \rangle > 0$; that is, it does not look at what might happen in the future. In practice, we choose the constraining state $| \psi_C \rangle$ to be the guiding state $| \psi_G \rangle$, which in turn is just $| \psi_T \rangle$. The constraint is now something easily imposed.

Asymptotically, the constrained-path method samples from the imaginary-time-independent distribution $\pi(\phi) = c_\phi / \sum_\phi c_\phi$ with $c_\phi > 0$. The decomposition of $| \psi_0 \rangle$ in terms of a set of Slater determinants Ω is not unique. We could have just as well written $| \psi_0 \rangle = \sum_{\Omega'} d_\phi | \phi \rangle$ with $d_\phi > 0$ or $| \psi_0 \rangle = \sum_{\Omega''} f_\phi | \phi \rangle$ with $f_\phi > 0$, where Ω' and Ω'' are different sets of Slater determinants. This flexibility arises because the set of Slater determinants is not orthonormal, $\langle \phi' | \phi \rangle \neq \delta_{\phi\phi'}$, and is overcomplete, $\sum_\phi | \phi \rangle \langle \phi | \neq I$.

Because the random walks in the constrained-path method occur in a different basis than the random walks in the fixed-node method (Fig. 11.1), we need a different type of nodal constraint. In this different basis, transitions from a region where the overlap $\langle \psi_T | \phi(\tau) \rangle$ is positive to one where it is negative cause the sign problem. The regions with overlaps of different signs are not physically distinguishable, and the propagation thus mixes degenerate bases caused by the plus-minus symmetry. We need a procedure to prevent this mixing from happening indiscriminately. Breaking the plus-minus symmetry by constraining each walker to the region where $\langle \psi_T | \phi(\tau) \rangle > 0$ prevents such mixing.

Instead of applying the constraint to the entire population of walkers at once, we apply the constraint to one random walker at a time: If $| \psi(\tau) \rangle = \sum_\phi c_\phi | \phi(\tau) \rangle$ with $c_\phi > 0$, then $\langle \psi_T | \psi(\tau) \rangle > 0$ if $\langle \psi_T | \phi(\tau) \rangle > 0$ for all $| \phi(\tau) \rangle$. For a given step, $| \psi' \rangle = \sum_x P(x) B(x) | \psi \rangle$, so we sample an x from $P(x)$ and then generate a new walker $| \phi' \rangle = B(x) | \phi \rangle$. If $\langle \psi_T | \phi' \rangle > 0$, we keep the walker. If not, we eliminate it.

The constrained-path approximation is not a fixed-node approximation. While the chosen $| \psi_T \rangle$ may constrain a $| \phi(\tau) \rangle$ to be a member of a "positive" set, this $| \phi(\tau) \rangle$ may overlap a member of a "negative" set. In contrast to the fixed-node method, the constrained-path method thus does not separate the basis into orthogonal sets. One consequence is that the constrained-path method can sometimes produce the exact solution even if the constraining wave function is approximate

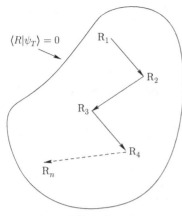

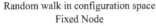

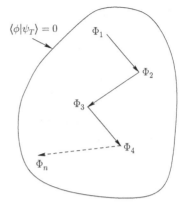

Random walk in configuration space   Random walk in Slater determinant space
                Fixed Node   Constrained Path

Figure 11.1 Schematic representation of the difference between the random walk in the fixed-node and constrained-path methods. In the fixed-node method, the walk is from point to point $R = (r_1, r_2, \ldots, r_n)$ in configuration space and is within the nodal surface $\langle R|\psi_T \rangle = 0$. In the constrained-path method the walk is from point to point in the space of Slater determinants and is within the nodal surface $\langle \phi|\psi_T \rangle = 0$.

and at any given imaginary time has the incorrect nodal surface in configuration space. The eigenpair $(E_{\text{eff}}, \psi_{\text{eff}})$ satisfies

$$E_v = E_{\text{eff}} = \frac{\langle \psi_{\text{eff}}|H|\psi_{\text{eff}} \rangle}{\langle \psi_{\text{eff}}|\psi_{\text{eff}} \rangle}.$$

Some key characteristics of the constrained-path method are (1) the nodal surface of $\langle \psi_T|\phi(\tau) \rangle = 0$ is not the same as that of $\langle \psi_{\text{eff}}|\phi(\tau) \rangle = 0$; (2) sometimes an approximate nodal surface $\langle \psi_T|\phi(\tau) \rangle = 0$ produces the exact ground state energy; and (3) the exact solution is found in the noninteracting case.

11.4 Estimators

11.4.1 Mixed estimator

In the chapter on the power method (Section 10.5), we discussed three estimators of the energy: the variational estimate E_v, the mixed estimate E_m, and the growth estimate E_g. In the fixed-node method, the three are equivalent and provide an upper bound to the ground state energy E_0. In the constrained-path method, the relation of the three among themselves is unclear (Carlson et al., 1999). Only the variational estimator is always an upper bound.

In the fixed-node method, because $\psi(C, \tau) = \bar{\psi}(C, \tau)/\psi_T(C)$, we clearly have

$$E_v = \frac{\sum_C \psi(C, \tau)(H\psi(C, \tau))}{\sum_C \psi^2(C, \tau)} \geq E_0.$$

At large τ, the walkers are distributed with a probability density $\psi_T(C)\psi(C)$, and both $\psi_T(C)$ and $\psi(C)$ go to zero linearly near the nodal surface. Because the Hamiltonian and the constraint are local operators, this implies that the growth and mixed estimators are equal:

$$E_g = E_m = \frac{\sum_C \psi_T(C)E_L(C)\psi(C)}{\sum_C \psi_T(C)\psi(C)}.$$

The constrained propagator is identical to the exact one except near the nodal surface, and the constraint discards any contributions to the ground state that are orthogonal to $|\psi_T\rangle$ and $|\psi\rangle$. Accordingly, the nodal region gives no net contribution to either $\langle \psi_T|H|\psi\rangle$ or $\langle \psi|H|\psi\rangle$. Therefore the variational energy is identical to the growth and mixed estimators. The mixed energy is the one more efficiently computed.

We can demonstrate that the variational energy defined by the effective fixed-node Hamiltonian is an upper bound to the ground state energy E_0 (ten Haaf et al., 1995). To do this, we need to show that the ground state of H_{eff} is an upper bound to the true ground state energy. First, we write the effective and actual Hamiltonians as

$$H_{\text{eff}} = H_{\text{tr}} + V_{\text{sf}},$$

$$H = H_{\text{tr}} + H_{\text{sf}},$$

and then define the configuration space matrix elements of H_{tr}, H_{sf}.

To define these elements, we first use the first of the above equations and note that (11.3) implies that the diagonal elements of H_{tr} equal the diagonal elements of H_{eff}. Then, from (11.2), we see that the off-diagonal elements of H_{eff} equal the off-diagonal elements of H. Hence, because V_{sf} is diagonal, the off-diagonal elements of H_{tr} equal those of H. Because the diagonal elements of the truncated Hamiltonian H_{tr} are $\langle C|H|C\rangle$, it follows from the second of the aforementioned equations that the diagonal elements of H_{sf} must be zero. Finally, because the off-diagonal elements of H_{tr} are zero, this same equation implies that the off-diagonal elements of H_{sf} equal those of H.

Now we compute the energy difference between H and H_{eff}, relative to an arbitrary state $|\psi\rangle$. We find

$$\Delta E = \langle \psi| (H_{\text{eff}} - H) |\psi\rangle = \langle \psi| (V_{\text{sf}} - H_{\text{sf}}) |\psi\rangle,$$

which on using the matrix elements of V_{sf} and H_{sf} equals

$$\Delta E = \sum_C \psi(C) \left[\langle C| V_{\text{sf}} |C\rangle \, \psi(C) - \sum_{C'} \langle C| H_{\text{sf}} |C'\rangle \psi(C') \right].$$

We next rewrite this expression in terms of matrix elements of H,

$$\Delta E = \sum_C \psi(C) \left[\sum_{C'}^{\text{sf}} \langle C|H|C'\rangle \frac{\psi_T(C')}{\psi_T(C)} \psi(C) - \sum_{C'}^{\text{sf}} \langle C|H|C'\rangle \psi(C') \right]. \quad (11.8)$$

Denoting $s(C, C')$ as the sign of the matrix element $\langle C|H|C'\rangle$ and noting that all terms in the summations satisfy the condition

$$\langle C|H|C'\rangle \psi_T(C)\psi_T(C') < 0,$$

we obtain

$$\Delta E = \sum_{C,C'}^{\text{sf}} |\langle C|H|C'\rangle| \left| \psi(C)\sqrt{\left|\frac{\psi_T(C')}{\psi_T(C)}\right|} - s(C, C')\psi(C')\sqrt{\left|\frac{\psi_T(C)}{\psi_T(C')}\right|} \right|^2,$$

an expression that is obviously nonnegative for any state $|\psi\rangle$. The ground state energy of H_{eff} is therefore an upper bound for the ground state energy of the original Hamiltonian H. Hence, $E_{\text{eff}} \geq E_0$.

The lattice fixed-node method has several other important properties. First, we can easily show that $H|\psi_T\rangle = H_{\text{eff}}|\psi_T\rangle$. This equality ensures that the fixed-node Green's function Monte Carlo procedure improves the energy of the trial state:

$$\frac{\langle \psi_T| H |\psi_T\rangle}{\langle \psi_T|\psi_T\rangle} = \frac{\langle \psi_T| H_{\text{eff}} |\psi_T\rangle}{\langle \psi_T|\psi_T\rangle} \geq E_{\text{eff}} = \frac{\langle \psi_{\text{eff}}|H|\psi_{\text{eff}}\rangle}{\langle \psi_{\text{eff}}|\psi_{\text{eff}}\rangle} \geq E_0.$$

Next, we can show that if the exact ground state is used for the trial state, then the ground state energy of the effective Hamiltonian equals the true ground state energy. Thus, by varying the trial state, we can reach the true ground state energy.

There is an important difference between the lattice and continuum fixed-node methods. In the continuum, only the sign of the trial state matters. If the nodal surface is correct, we obtain the exact energy regardless of the magnitude of the trial state. On the lattice, it follows from (11.8) that when $\Delta E = 0$, we obtain the exact energy if

$$\frac{\psi(C')}{\psi(C)} = \frac{\psi_0(C')}{\psi_0(C)}$$

for all sign-flipping configuration pairs (C, C'). Thus, the sign and relative magnitude of the trial state in configurations connected by sign flips must be correct.

In the constrained-path method, while the variational energy constructed from $|\psi(\tau)\rangle$ satisfies

$$E_v = \frac{\langle \psi_{\text{eff}}(\tau)|H|\psi_{\text{eff}}(\tau)\rangle}{\langle \psi_{\text{eff}}(\tau)|\psi_{\text{eff}}(\tau)\rangle},$$

the relative ranking of E_v, E_m, and E_g is unclear because the constraint discards configurations that are orthogonal to $|\psi_T\rangle$, but these discarded configurations are not necessarily orthogonal to $|\psi\rangle$ (Carlson et al., 1999).

On general grounds, if the constrained-path method defines a Markov process with a stationary distribution such that

$$H_{\text{eff}}|\psi(\tau)\rangle = E_g|\psi_{\text{eff}}(\tau)\rangle,$$

where $H_{\text{eff}} = H + \delta H$, then

$$\frac{\langle \psi(\tau)|H|\psi(\tau)\rangle}{\langle \psi(\tau)|\psi(\tau)\rangle} = E_m + \delta E_m \geq E_0,$$

with

$$\delta E_m = \frac{\langle \psi_T|\delta H|\psi(\tau)\rangle}{\langle \psi_T|\psi(\tau)\rangle} - \frac{\langle \psi(\tau)|\delta H|\psi(\tau)\rangle}{\langle \psi(\tau)|\psi(\tau)\rangle}.$$

Only when $\delta E_m \leq 0$ is E_m an upper bound to the ground state energy. The mixed estimator for the energy is generally not an upper bound to the ground state energy. In general, the growth and mixed estimators are not equal and only the variational estimator is an upper bound.[7]

11.4.2 *Forward walking and back propagation*

The mixed estimator provides accurate estimates of observables that commute with the Hamiltonian. For those that do not, it can provide poor estimates. For example, the mixed estimator predicts the total energy well but generally gives poor estimates for the kinetic and the potential energies. For observables that do not commute with the Hamiltonian we need to use other estimators. For these cases the simplest estimator is the *extrapolated estimator*

$$\langle A \rangle \approx \langle A \rangle_{\text{ex}} = 2\langle A \rangle_m - \langle A \rangle_v. \tag{11.9}$$

If the deviation of the trial state from the true ground state is significant, then this estimator also provides poor estimates. Fortunately, more reliable estimators exist. We touched upon these in Section 10.5. For the fixed-node method we discuss only

---

[7] We note that other estimators exist so that the constrained-path method can yield an upper bound to the ground state energy (Carlson et al., 1999).

what is called the *forward (or future) walking method* (Hammond et al., 1994). In discussing it, we distinguish between cases where the observable is diagonal in the simulation basis and cases where it is not (Nightingale, 1999).

We start with the *diagonal* case. We want

$$\langle A \rangle = \frac{\langle \psi_0 | A | \psi_0 \rangle}{\langle \psi_0 | \psi_0 \rangle} = \frac{\sum_C A(C) \psi_0^2(C)}{\sum_C \psi_0^2(C)}.$$

The problem is that when the Markov chain is stationary, it samples from $\psi_T(C)\psi_0(C)$ instead of $\psi_0^2(C)$. To sample from the ground state distribution the strategy is to estimate a weight $w_0(C)$ so that

$$\langle A \rangle = \frac{\sum_C A(C)\psi_0^2(C)}{\sum_C \psi_0^2(C)} = \frac{\sum_C w_0(C) A(C) \psi_T(C) \psi_0(C)}{\sum_C w_0(C) \psi_T(C) \psi_0(C)}.$$

Clearly, we need $w_0(C) \propto \psi_0(C)/\psi_T(C)$.

To estimate this weight we first note that for large L

$$\langle A \rangle = \frac{\langle \psi_T | G^L A | \psi_0 \rangle}{\langle \psi_T | G^L | \psi_0 \rangle}.$$

With multiple insertions of unity, we transform this expression into

$$\langle A \rangle = \frac{\sum_{C_0, C_1, \dots, C_L} \psi_T(C_L) G(C_L, C_{L-1}) \cdots G(C_1, C_0) A(C_0) \psi_0(C_0)}{\sum_{C_0, C_1, \dots, C_L} \psi_T(C_L) G(C_L, C_{L-1}) \cdots G(C_1, C_0) \psi_0(C_0)}.$$

The next step is writing the relevant quantities in terms of their importance sampled values. Doing so we obtain

$$\langle A \rangle = \frac{\sum_{C_0, C_1, \dots, C_L} \bar{G}(C_L, C_{L-1}) \cdots \bar{G}(C_1, C_0) A(C_0) \bar{\psi}_0(C_0)}{\sum_{C_0, C_1, \dots, C_L} \bar{G}(C_L, C_{L-1}) \cdots \bar{G}(C_1, C_0) \bar{\psi}_0(C_0)},$$

where

$$\bar{G}(C', C) = \frac{\langle C' | \psi_T \rangle \langle C' | G | C \rangle}{\langle C | \psi_T \rangle} \tag{11.10}$$

is the modified transfer matrix (because of importance sampling). We now have a useful result. We know that the simulation eventually reaches a step 0 at which it starts to sample from the stationary importance sampled distribution $\bar{\psi}_0(C) = \psi_T(C)\psi_0(C)$. At this step we record the value of $A(C_0)$ plus the weight and state of each walker. We also know that for each walker at 0 the Markov chain of length L generates a population of descendants that sample a stationary distribution approximately equal to $\psi_0(C)$. The weights of the descendants thus give us the needed estimate of $w_0(C)$.

Some walkers at 0, however, have no descendants at L because of the killings in the stochastic reconfigurations (Section 10.3). Because of the branching, others might have multiple descendants. Consequently, to estimate $\langle A \rangle$, we need to record the identity of a walker's parents at each step along the L-segments. We obtain the contribution of a walker at 0 to the numerator by multiplying its weight by its value of A at 0 and by the weights of its descendants at L. Not all walkers at 0 contribute to the value of the numerator. We accumulate the estimate of the denominator similarly.

Estimates of *nondiagonal* operators are more involved and potentially very difficult (Nightingale, 1999). For starters, we now have

$$\langle A \rangle = \frac{\sum_{C_0,C_1,\ldots,C_{L+1}} \bar{G}(C_{L+1},C_L)\cdots\bar{G}(C_2,C_1)A(C_1,C_0)\bar{\psi}_0(C_0)}{\sum_{C_0,C_1,\ldots,C_{L+1}} \bar{G}(C_{L+1},C_L)\cdots\bar{G}(C_1,C_0)\bar{\psi}_0(C_0)}, \qquad (11.11)$$

which we can immediately write more conveniently as

$$\langle A \rangle = \frac{\sum_{C_0,C_1,\ldots,C_{L+1}} \bar{G}(C_{L+1},C_L)\cdots\bar{G}(C_2,C_1)\bar{G}(C_1,C_0)\dfrac{A(C_1,C_0)}{\bar{G}(C_1,C_0)}\bar{\psi}_0(C_0)}{\sum_{C_0,C_1,\ldots,C_{L+1}} \bar{G}(C_{L+1},C_L)\cdots\bar{G}(C_1,C_0)\bar{\psi}_0(C_0)}.$$

The fraction in the numerator of the last equation replaces the $A(C_0)$ in (11.11), and we execute the sampling procedure remembering a walker's parent just as we did in the diagonal case. Things work well as long as $A(C_1,C_0)$ is zero when $\bar{G}(C_1,C_0)$ is and vice versa. Generally this is not the case, and other techniques, which we do not describe, are needed (Nightingale, 1999). These techniques are much more involved than the previously mentioned.

For the constrained-path method, the mixed and extrapolated estimators have the same range of utility as they do in the fixed-node method. We move beyond these estimators with the same strategy we used for the fixed-node method, employing what is called the *back-propagation method* (Zhang et al., 1997) as the analog of the forward-walking method. In contrast to the forward-walking method, we do not need separate strategies for the diagonal and off-diagonal observables. Here, we calculate expectation values as sums of products of Green's functions in the same manner as we do for the determinant methods (Section 7.4 and Appendix I). As a consequence, we enjoy here the same ease and flexibility in the implementation of measurements as we do there. This ease is a positive feature of the constrained-path method.

In back-propagation, at step 0, we save the walkers (weights and Slater determinants) and then propagate the population via the B matrices according to the standard algorithm, noting the identity of parents and additionally saving the Hubbard-Stratonovich fields for each walker at each step. For each walker reaching L, we now back-propagate $\langle \psi_T |$ via

$$((((\langle \psi_T | B_L) B_{L-1}) B_{L-2}\cdots)$$

until we arrive back at step 0, say, with walker $\langle \phi' |$. For each step of the back-propagation, we do not sample new Hubbard-Stratonovich fields, but rather we reuse those recorded in the forward steps but in reverse order. More explicitly, at the first back-propagation step, we use the fields for the last forward-propagation step; at the second back-propagation step, we use the fields for the next-to-last forward propagation step; etc. In the backward direction, we perform matrix stabilizations, but we do not apply the constrained-path condition or stochastically reconfigure. Each walker arriving at L has already passed the constrained-path test and the entire population was likely reconfigured several times. For each returning walker $\langle \phi' |$ we calculate $\langle \phi' | A | \phi \rangle$ for it and its parent $| \phi \rangle$ and then multiply this value times the product of the parent's weight at 0 and the descendant's weight at L. Not all parents contribute, only those with descendants do. Monte Carlo measurements are generally made periodically (Section 3.9). We can conveniently choose L to be a multiple of this period. After we complete the back-propagation measurement, we return to the population at L and use it to make the next L steps.

11.5 The algorithms

An implementation of the lattice fixed-node method follows the basic steps of an implementation of a standard power method. We refer the reader to the chapter on power methods (Chapter 10) for details. The key difference from a standard power method is that the fixed-node method uses H_{eff} instead of the original Hamiltonian H to avoid the sign problem. An implementation of the constrained-path method also uses the basic steps of a power method algorithm. However, it has additional Fermion-specific steps that resemble those in the zero-temperature determinant method (Appendix I). The way in which the sign problem is eliminated resembles the positive projection procedure.

Algorithm 39 gives the details of an implementation of the constrained-path method (Zhang et al., 1997). We simplify the discussion by excluding spin. As we noted in our discussion of the determinant methods (Chapter 7), we add spin by requiring the overlaps and Slater determinants to be products of a spin-up and a spin-down part. Other adjustments are fairly obvious.

In the algorithm, we suppose we have a stack of random walkers, each characterized by a weight w, a Slater determinant state $| \phi \rangle$, and an overlap integral $\mathcal{O}_T(\phi) = \langle \psi_T | \phi \rangle$. We represent the Slater determinant by an $N_{\text{states}} \times N_{\text{electrons}}$ matrix Φ (Appendix I). N_{states} is the number of orbitals and $N_{\text{electrons}} = N_{\text{up}} + N_{\text{down}}$ is the number of electrons. For the single-band Hubbard model N_{states} is the number of lattice sites, while for the periodic Anderson model, it is twice the number of lattice sites. It follows from Appendices F and H that $\mathcal{O}_T(\phi) = \det(\Phi_T^T \Phi)$.

Algorithm 39 Constrained-path Monte Carlo method.

Input: A trial state $|\psi_T\rangle$, described by the matrix Ψ_T, and an estimate of the ground
state energy E_T.

Initialize a stack of N walkers described by the matrix $\Phi = \Psi_T$, a unit weight w,
and a unit overlap integral $\mathcal{O}_T$;

repeat

  repeat

    Pop a walker off the stack ;

    Update $w \leftarrow we^{-\Delta\tau E_T}$;

    Compute $\Phi' = B_{K/2}\Phi$;

    Compute $\mathcal{O}'_T = \mathcal{O}_T(\Phi')$ and $[\Psi_T^T\Phi']^{-1}$;

    if $\mathcal{O}'_T < 0$ **then**

      Cycle to the next walker ;

    end if

    Update $\Phi \leftarrow \Phi', w \leftarrow w\mathcal{O}'_T/\mathcal{O}_T, \mathcal{O}_T \leftarrow \mathcal{O}'_T$;

    Use Algorithm 40 to propagate the walker via B_V ;

    if $\mathcal{O}'_T < 0$ **then**

      Cycle to the next walker ;

    end if

    Compute $\Phi' = B_{K/2}\Phi$;

    Compute $\mathcal{O}'_T = \mathcal{O}_T(\phi')$ and $[\Psi_T^T\Phi']^{-1}$;

    if $\mathcal{O}'_T < 0$ **then**

      Cycle to the next walker ;

    end if

    Update $\Phi \leftarrow \Phi', w \leftarrow w\mathcal{O}'_T/\mathcal{O}_T, \mathcal{O}_T \leftarrow \mathcal{O}'_T$;

    Push the walker onto the new stack ;

  until the stack is empty.

  If equilibrated, periodically estimate physical quantities ;

  Periodically re-orthogonalize the columns of Φ ;

  Periodically stochastically reconfigure the new stack to restore population size
  to N (Section 10.3) ;

  Make the new stack the old one ;

until an adequate number of measurements is collected.

Compute final averages and estimate their statistical errors ;

return the averages and their error estimates.

The main task in the algorithm is repeating the step that projects the population
of walkers onto a distribution estimating the ground state. In Algorithm 39 we are
using a higher-order symmetric Trotter approximation (5.13) to do this projection,

$$\exp[-\Delta\tau H] \approx \exp[-\tfrac{1}{2}\Delta\tau K]\exp[-\Delta\tau V]\exp[-\tfrac{1}{2}\Delta\tau K],$$

$$\Phi' = B_{K/2}B_V B_{K/2}\Phi.$$

In the above, $H = K + V$ and the matrices $B_{K/2}$ and B_V represent the half-step and full-step imaginary-time propagators for the hopping and potential energies (Section 7.2). In practice, we modify Algorithm 39 slightly to reduce the number of matrix-matrix multiplications by noting that

$$\Phi' = \left(B_{K/2}B_V B_{K/2}\right)\left(B_{K/2}B_V B_{K/2}\right)\cdots\left(B_{K/2}B_V B_{K/2}\right)\Phi$$

$$= B_{K/2}B_V\, B_K B_V \cdots B_K B_V B_{K/2}\Phi$$

and adjusting the algorithm so that when we start a new measurement period, we first multiply Φ by $B_{K/2}$ and then again at the end of the period, just before we make measurements.

We execute the propagation by B_K and $B_{K/2}$ in procedures that do several things. Each does the matrix-matrix multiplication. With the resulting matrix Φ', each calculates the overlap matrix $\Psi_T^T\Phi'$ where the superscript T denotes transposition. Next, each uses standard LU-factorization[8] software to compute $\mathcal{O}_T = \det\Psi_T^T\Phi'$ and the inverse $[\Psi_T^T\Phi']^{-1}$. Then, they test if $\mathcal{O}_T < 0$. If so, that walker is not placed on the new stack. For each walker, we also compute the inverse of its overlap matrix because each multiplication by B_V follows a multiplication by $B_{K/2}$ or B_K. As we now discuss, we need this inverse at the start of the procedure that does the B_V propagation.

Algorithm 40 describes the propagation of a walker by B_V. This procedure is the one most computationally intensive. It borrows some of the algebra developed for the zero-temperature determinant method (Appendix I). There, as here, the Hubbard-Stratonovich transformation casts B_V into the form $\prod_i P(x_i)b_V(x_i)$, that is, the propagation factors into propagations for each Hubbard-Stratonovich field. As in the determinant method, the matrix representing $b_V(x_i)$ is a unit matrix except for the i-th diagonal element: $x_i = \pm 1$ and $P(x_i) = \tfrac{1}{2}$ (Section 7.2).

For a given i, we want to produce a Φ' from Φ. To do so, we importance sample a Hubbard-Stratonovich field x'. With importance sampling (Section 10.4.3), we select x' from $\mathcal{O}_T(\Phi')P(x)/\mathcal{O}_T(\Phi)$, but we do not yet know Φ'. We note that we can write

$$\Phi' = \sum_x \mathcal{O}_T\left[\Phi(x)\right]P(x)/\mathcal{O}_T(\Phi)b_V(x)\Phi \qquad (11.12)$$

$$= N\left[p_1 b_V(+1) + p_2 b_V(-1)\right]\Phi, \qquad (11.13)$$

---

[8] This factorization expresses a matrix as a product of a unit lower-triangular matrix L and an upper-triangular matrix U. The determinant of the given matrix is simply the product of the diagonal elements of U.

Algorithm 40 Constrained-path Monte Carlo method: propagation by B_V.

Input: A walker and the inverse of its overlap matrix $[\Psi_T^T \Phi]^{-1}$.

    for $i = 1$ to N_{states} **do**

        Compute $G_{ii} = 1 - [\Phi[\Psi_T^T \Phi]^{-1}\Psi_T^T]_{ii}$;

        Using (7.43) compute $C(1) = \mathcal{O}_T(1)/\mathcal{O}_T$ and $C(-1) = \mathcal{O}_T(-1)/\mathcal{O}_T$;

        Using (11.14) and (11.16) compute p_1, p_2 ;

        $p_1 = \max(0, p_1)$;

        $p_2 = \max(0, p_2)$;

        $N = p_1 + p_1$;

        if $N = 0$ **then**

            Set $\mathcal{O}_T < 0$;

            return the negatively weighted walker.

        end if

        if $p_1/N < \xi$ **then**

            $x' = 1$;

        else

            $x' = -1$;

        end if

        Update $w \leftarrow Nw$, $\mathcal{O}_T(\phi) \leftarrow [\mathcal{O}_T(x')/\mathcal{O}_T(\phi)]\mathcal{O}_T(\phi)$, and $\Phi \leftarrow b_V(x')\Phi$;

        Using (7.45), update $[\Psi_T^T \Phi]^{-1} \leftarrow [\Psi_T^T \Phi']^{-1}$;

        Save the Hubbard-Stratonovich field for this walker, this step, and this site ;

    end for

    Record the identity of the walker's parent ;

    return the updated walker.

---

where

$$p_1 = \frac{P(1)}{N}\frac{\mathcal{O}_T(1)}{\mathcal{O}_T}, \quad p_2 = \frac{P(-1)}{N}\frac{\mathcal{O}_T(-1)}{\mathcal{O}_T}, \tag{11.14}$$

with

$$\mathcal{O}_T(1) = \det[\Psi_T^T b_V(1)\Phi], \quad \mathcal{O}_T(-1) = \det[\Psi_T^T b_V(-1)\Phi] \tag{11.15}$$

and

$$N = P(1)\frac{\mathcal{O}_T(1)}{\mathcal{O}_T} + P(-1)\frac{\mathcal{O}_T(-1)}{\mathcal{O}_T}. \tag{11.16}$$

If either p_1 or p_2 is not positive, we set it equal to zero. If both are not positive, we set $\mathcal{O}_T < 0$ and return from the procedure. Otherwise, we choose x' equal to $+1$ or -1 with probability p_1 or p_2. With this x' we then update the walker

$$w \leftarrow wN, \quad O_T \leftarrow O_T(x'), \quad \Phi \leftarrow b_V(x')\Phi.$$

We also update $[\Psi_T^T \Phi]^{-1} \leftarrow [\Psi_T^T \Phi']^{-1}$.

We could have simplified the expressions for p_1 and p_2, but we deliberately wrote them with explicit ratios of determinants. We recall from the chapter on the determinant methods (Chapter 7) that when updating a change in the Hubbard-Stratonovich fields x_i, we needed to compute the ratio of two determinants, a computation that simplifies significantly if we know the i-th diagonal element of the zero-temperature Green's function $G = [I - \Phi(\Psi_T^T \Phi)^{-1} \Psi_T]^T$. Knowing this element requires knowing the inverse of the overlap matrix $\Psi_T^T \Phi$. The important point is that in the power method, we do not need to know all spatial elements of the Green's function, just the one for the site whose auxiliary field we are updating. By inputting the determinant and inverse of the overlap matrix, plus the state of the walker, we can easily perform the updating of this first field:

$$G_{ii} = 1 - \sum_{jk} \Phi_{ij} \left[\left(\Psi_T^T \Phi \right)^{-1} \right]_{jk} [\Psi_T]_{jk}.$$

To cycle to the next field, we need to update those quantities necessary to compute the diagonal element of the Green's function corresponding to this field. These quantities are the value of the new determinant, the new state of the walker, and the inverse of the new overlap matrix. We can get the new overlap determinant from the ratio $\mathcal{O}'_T/\mathcal{O}_T$ and the old overlap $\mathcal{O}_T$ by simple multiplication. To get the new overlap matrix, we first need to update the state of the walker, which we do by the multiplication of the matrix representing the old state by the matrix $b_V(x')$ that propagates it. To get the new inverse, we do not form $\Psi_T^T \Phi'$ and invert it, but as we did in the determinant method and in the continuous-time impurity algorithms (Chapters 7 and 8), we exploit the fact that Φ changes by only a few elements (here, just one row) and that we know the original inverse. If we call A^{-1} the known inverse, u the i-th row of Ψ_T, and v the i-th row of Φ, then the new inverse is (Sherman-Morrison formula; Meyer, 2000; Press et al., 2007)

$$\left(A + uv^T \right)^{-1} = A^{-1} - \frac{A^{-1} uv^T A^{-1}}{1 + uA^{-1}v^T}.$$

Thus, we can update the inverse of the overlap matrix by using (7.45) (see also Section 8.4.2) after notationally substituting G_{ij} with $A_{ij} = [\Psi_T^T \Phi']_{ij}$. This more efficient way of computing is the reason we compute the inverse of the overlap matrix at the end of the procedure that computes $B_{K/2}$ and B_K, and also while propagating by B_V, the reason we update this inverse after each selection of the new Hubbard-Stratonovich fields. At the end of a sweep we do not have the values of the Green's function matrix needed for doing measurements. Generally, we measure every few sweeps or so and simply compute the correct Green's function when it is time to do measurements.

As a power method the constrained-path method benefits from stochastic reconfiguration (Section 10.3). One advantage is the standard variance reduction. A second is a replenishment of the walker population. As the constraint removes negatively signed walkers, the population would drop to zero if not replenished. Here, the comb is a useful procedure (Algorithm 38).

The repeated matrix-matrix multiplications inherent to the method accumulate numerical errors. The matrix-product stabilization procedures are simpler than the one for the finite-temperature determinant method (Chapter 7), but the same as for the zero-temperature method (Appendix I). The stabilization reduces to periodically orthonormalizing the columns of the propagation matrix Φ. Again, the modified Gram-Schmidt method works well (Section 7.4). In Appendix I, we give a pseu-docode (Algorithm 50) for performing this orthonormalization. The justification for just needing this part of the matrix factorization follows the same arguments as given there for the zero-temperature determinant method.

As with the finite- and zero-temperature determinant methods, with the constrained-path method we can compute the expectation value of any observable from sums of products of Green's functions. Almost all quantities of interest require the use of back-propagation. As discussed in the previous sections, back-propagation replaces $\langle \psi_T |$ in the mixed estimate with $\langle \psi_B | = \langle \psi_T | B_L B_{L-1} \cdots B_1$ where $B_i = B_{K/2} B_V(i) B_{K/2}$ and $B_V(i)$ is the B_V matrix for the i-th term in the product. The Hubbard-Stratonovich fields used in the left-to-right multiplication are the ones used in the right-to-left multiplication $B_L B_{L-1} \cdots B_1 | \phi \rangle$. As we also noted the walkers now have to carry the identification of their parent over these L steps, and to execute the back-propagation we have to store the Hubbard-Stratonovich fields generated for each walker in the propagation $B_V = \prod b_V(x)$. This bookkeeping is most conveniently done as the last step in the B_V procedure. When we stochastically reconfigure, we need to maintain ancestry.

Recently, Shi and Zhang (2013) have illustrated how using symmetries can increase the accuracy and efficiency for ground state simulations that use auxiliary fields.

11.6 Constrained-phase method

The *constrained-phase method*, sometimes called the phaseless method, extends the constrained-path method to zero-temperature Fermion simulations that have complex weights (Ortiz et al., 1997; Schmidt and Fantoni, 1999; Schmidt et al., 2001, 2003; Sarsa et al., 2003; Zhang and Krakauer, 2003; Zhang, 2013). Complex weights arise for multiple reasons. The prime one is the system being simulated has broken time-reversal symmetry. Typically, this symmetry is broken by an external magnetic field, making the wave function complex. In lattice models, the magnetic field enters via the hopping matrix element t_{ij}, which becomes $t_{ij} \exp[i \int A \cdot d\ell]$ because of a vector potential A acting along the path ℓ connecting lattice positions i and j. A net phase through a closed path is a measure of the magnetic flux threading the loop. The Hamiltonian matrix is now Hermitian as opposed to being simply symmetric. Strictly Hermitian Hamiltonians also result when they include a spin-orbit interaction, a situation of great current interest in many-electron physics. If the Hubbard-Stratonovich transformation is not chosen to be real, complex weights can develop (Appendix G). Such a choice is convenient for electron-electron interactions whose form is more general than the simple on-site Hubbard interaction.

An artificial but useful way to obtain a complex wave function is to add a phase shift to the periodic boundary conditions, for example, in one dimension by using

$$\psi(x + L) = \exp(i\varphi)\,\psi(x). \tag{11.17}$$

This boundary condition breaks the symmetries that were creating the degeneracies among the eigenvalues and is the basis for the *twisted-phase boundary condition method* (Gammel et al., 1993; Lin et al., 2001). This method averages the properties of the system over many different values of φ uniform randomly chosen over the interval $[0, 2\pi]$. The degeneracy splitting and the averaging remove closed-shell effects in the total energy as a function of electron density and create a density dependence closer to that of an infinite system. For small finite systems, *closed-shell effects* appear as discontinuities in the slope of the total energy at densities where the up and down electrons each have a gap between the energy level of the highest occupied electron state and the next unoccupied state. Typically, the highest occupied energy level of each spin is degenerate because of a point-group symmetry of the Brillouin zone for the bands of the noninteracting electrons. Interactions in small systems often only sparsely fill the gaps that were present in the noninteracting problem.

The constrained-phase method is the analog of the fixed-phase method (Ortiz et al., 1993) that was developed for continuum problems in the presence of a magnetic field. The latter method substitutes the wave function expressed as

$$\psi(C) = |\psi(C)| \exp(i\varphi(C))$$

into Schrödinger's equation to generate two coupled partial differential equations, one for the amplitude $|\psi(C)|$ and the other for the phase $\varphi(C)$. Then, in the equation for $|\psi(C)|$, it approximates $\varphi(C)$ with the phase $\varphi_T(C)$ of the trial state. The diffusion Monte Carlo method now solves the equation for $|\psi(C)|$. No sign problem exists as this function is by construction everywhere nonnegative.

For lattice Fermion problems, where we propagate one Slater determinant into another (Appendix F), we generate a constrained-phase approximation in a manner similar to the way we created the constrained-path approximation. There, for real $\langle \psi_T | \psi(\tau) \rangle$, we achieve positivity by requiring $\langle \psi_T | \psi(\tau) \rangle > 0$ and by applying the constraint to each walker at each step of the imaginary-time propagation. Here, for complex $\langle \psi_T | \psi(\tau) \rangle$, we achieve reality and positivity by requiring either $\langle \psi_T | \psi(\tau) \rangle = \text{Re}\{\langle \psi_T | \psi(\tau) \rangle\} > 0$ or $\langle \psi_T | \psi(\tau) \rangle = |\langle \psi_T | \psi(\tau) \rangle| > 0$. As before, we apply the constraint to one walker at a time. For either option we define

$$\langle \phi | \psi(\tau) \rangle = e^{i\varphi_\tau} |\langle \phi | \psi(\tau) \rangle|$$

and fix the phase by taking $\varphi_\tau(\phi) = \varphi_T(\phi) = \text{Im}\{\ln\langle \phi | \psi_T \rangle\}$. Both options produce the exact result if the trial state is the exact ground state.

The basic step is to propagate $|\phi\rangle$ to $|\phi'\rangle$, that is, $\exp[-\Delta\tau H]|\phi\rangle = |\phi'\rangle$, and replace $|\phi'\rangle$ with $|\tilde\phi\rangle$ in one of two ways. For $\text{Re}\{\langle \psi_T | \psi(\tau) \rangle\} > 0$,

$$|\tilde\phi\rangle = \cos\varphi_T(\phi')e^{-i\varphi_T(\phi')}|\phi'\rangle,$$

while for $|\langle \psi_T | \psi(\tau) \rangle| > 0$,

$$|\tilde\phi\rangle = e^{-i\varphi_T(\phi')}|\phi'\rangle.$$

These replacements differ only in subtle ways, but the first replacement sometimes leads to energy estimates with smaller variance. However, testing of the fixed-phase method has been limited to date.

With either option, $\mathcal{O}_T(\phi)$ is real but $\mathcal{O}_T(\phi')$ is not. The replacement of $|\phi'\rangle$ with $|\tilde\phi\rangle$ creates a $\mathcal{O}_T(\tilde\phi)$ that is real. As defined above, $|\tilde\phi\rangle$ is of the form $a\det[\Phi']$, which equals $\det[a^{1/N_{\text{elec}}}\Phi']$, which in turn defines $\tilde\Phi$ to be $a^{1/N_{\text{elec}}}\Phi'$.

The conversion of the constrained-path algorithm (Algorithms 39 and 40) to the constrained-phase one involves just a few steps. The first step is switching to complex arithmetic and variables where necessary. The next steps are replacing in the procedures that propagate by $B_{K/2}$ and B_V the testing whether $\mathcal{O}_T(\phi') < 0$ and the updating of the walker weight with $w \leftarrow w\,|\mathcal{O}_T(\phi')|$. Then, in the procedure that executes B_V, we replace the $\mathcal{O}_T(i)$ in the computation of the p_i with $|\mathcal{O}_T(i)|/\mathcal{O}_T(\phi)$. Finally, after the execution of $B_{K/2}B_V B_{K/2}$, the phase is constrained. Here the first step is computing $\varphi_T(\phi')$ and testing whether its absolute value is less than $\pi/2$. If so, then $|\phi\rangle$ is updated with $|\tilde\phi\rangle$ and $\mathcal{O}_T(\phi)$ is updated with $\exp[-i\varphi_T(\phi')]\mathcal{O}_T(\tilde\phi)$. If not, then the weight of the walker is set to zero.

Suggested reading

S. Fahy and D. R. Hamann, "Diffusive behavior of states in the Hubbard-Stratonovich transformation," *Phys. Rev. B* **43**, 765 (1991).

B. L. Hammond, W. A. Lester, Jr., and P. J. Reynolds, *Monte Carlo Methods in ab initio Quantum Chemistry* (Singapore: World-Scientific, 1994), chapters 3 and 4.

Exercises

11.1 For the lattice fixed-node method, prove $H|\psi_T\rangle = H_{\text{eff}}|\psi_T\rangle$.

11.2 Derive (11.9).

11.3 For a nearest-neighbor tight-binding model on a square lattice with periodic boundary conditions, derive analytically the energy dispersion relation. For a succession of increasing lattice sizes, plot $E(k)$ versus k and document the scaling of the gap between the highest occupied state and the lowest unoccupied state at half-filling. Now derive the dispersion for the phased boundary condition, using (11.17):

$$\psi(x, y) = \psi(x + L, y)\, e^{i\phi} = \psi(x, y + L)\, e^{i\phi}.$$

For a succession of increasing lattice sizes and an increasing number of randomly chosen values of ϕ for each size, plot $E(k)$ versus k. Compare and contrast the behavior of the gaps, particularly near half-filling.

11.4 For a nearest-neighbor tight-binding model on a hexagonal lattice with periodic boundary conditions, derive analytically the energy dispersion relation. For a succession of increasing lattice sizes, plot $E(k)$ versus k and document the scaling of the gap between the highest occupied state and the zero mode and/or the lowest unoccupied state and zero mode at half-filling. Now derive the dispersion for the phased boundary condition, using (11.17) in each dimension. For a succession of increasing lattice sizes and an increasing number of randomly chosen values of ϕ for each size, plot $E(k)$ versus k. Compare and contrast the behavior of the gaps, particularly near half-filling.

11.5 If the Slater determinants $|\phi\rangle$ and $|\phi'\rangle$ are defined by the matrices Φ and Φ', show that the overlap integral equals $\det(\Phi^T \Phi')$.

Part IV

Other topics

12

Analytic continuation

The presence of dynamical information is a feature distinguishing a finite-temperature quantum Monte Carlo simulation from a classical one. We now discuss numerical methods for extracting this information that use techniques and concepts borrowed from an area of probability theory called Bayesian statistical inference. The use of these techniques and concepts provided a solution to the very difficult problem of analytically continuing imaginary-time Green's functions, estimated by a quantum Monte Carlo simulation, to the real-time axis. Baym and Mermin (1961) proved that a unique mapping between these functions exists. However, executing this mapping numerically, with a simulation's incomplete and noisy data, transforms the problem into one without a unique solution and thus into a problem of finding a "best" solution according to some reasonable criterion. Instead of executing the analytic continuation between imaginary- and real-time Green's functions, thereby obtaining real-time dynamics, we instead estimate the experimentally relevant spectral density function these Green's functions share. We present three "best" solutions and emphasize that making the simulation data consistent with the assumptions of the numerical approach is a key step toward finding any of these best solutions.

12.1 Preliminary comments

The title of this chapter, "Analytic Continuation," is unusual in the sense that it describes the task we wish to accomplish instead of the method we use to accomplish it. If we used the name of the method, the title would be something like "Bayesian Statistical Inference Using an Entropic Prior." A shorter title would be "The Maximum Entropy Method." We hope by the end of the chapter the reader will agree that using the short title is perhaps too glib and the longer one has meaningful content.

The task we want to accomplish is solving the integral equation

$$G(\tau) = \int_{-\infty}^{\infty} d\omega\, K(\tau,\omega) A(\omega), \quad K(\tau,\omega) = \frac{e^{-\tau\omega}}{1 \pm e^{-\beta\omega}} \tag{12.1}$$

for $A(\omega)$ using quantum Monte Carlo estimates of $G(\tau)$. We review the origin of this equation in the next section (Section 12.2). For now, it suffices to say that $G(\tau)$ is some dynamical (imaginary-time) many-body correlation function, and the equation is a standard result from many-body theory relating this correlation function to $A(\omega)$, which is called the *spectral density*. We also discuss in the next section that solving this equation is a surrogate for analytically continuing the dynamical correlations from $\tau \rightarrow it$.

Solving this linear integral equation seems simple, as discretizing τ and ω reduces it to a linear system of equations

$$G_i = \sum_j K_{ij} A_j.$$

As the number of knowns does not necessarily equal the number of unknowns, a least-squares solution for the A_i has to be found. Such a calculation is easily implemented. The problem is that this approach to the analytic continuation problem almost always fails to produce an acceptable solution.

Similar tasks have for decades been known to be major computational challenges. The exponential character of the kernel $K(\tau,\omega)$ makes solving the analytic continuation problem akin to parameterizing a two-nuclei radioactive decay problem

$$n(t) = a_1 e^{-\lambda_1 t} + a_2 e^{-\lambda_2 t},$$

where $n(t)$ is the number of decays as a function of time t. As discussed passionately by Acton (1970), this is an ill-posed problem as small changes in even very precise, closely spaced measurements produce large changes in the fitted values of a_1, a_2, λ_1, and λ_2. On the other hand, if we know λ_1 and λ_2 (the decay rates of the nuclei), good estimates of a_1 and a_2 (the relative proportions of the nuclei) are easily obtained.

A common way to get reasonable solutions to the radioactive decay and similar problems is to regularize the least-squares solution. In mathematics and statistics, particularly for machine learning and inverse problems, regularization refers to adding additional information to the solution to help solve an ill-posed problem or to avoid overfitting a least-squares problem. In the present case, we are dealing with both types of problems, and we take "regularization" to mean adding one or more constraints to the fit of the data. If we add one constraint, we maximize

$$\chi^2(A) = \lambda R(A) - \frac{1}{2} \sum_i \left(G_i - \sum_j K_{ij} A_j \right)^2,$$

where λ is a Lagrange multiplier and $R(A)$ is some function of our unknown.

One generally tries to use functions R that are consistent with known properties of the expected solution. Doing this forces some of the prior knowledge into the fit. For example, we might know that $A(\omega)$ is smooth, and thus we might want to constrain the solution so that its first derivative with respect to ω is continuous. As we discuss in the next section (Section 12.2), in the analytic continuation problem, we know at least that $A(\omega)$ is nonnegative, and its integral over ω is finite (that is, we know the integral satisfies a physical sum rule). For a physical solution, we need to take into account both of these pieces of prior information. The magnitude of λ controls how strongly this prior information is imposed on the solution. Fixing the numerical value of λ such that it does not undesirably bias the solution is often a problem in itself.

We choose to solve (12.1) and to use the prior information we know about $A(\omega)$ from the point of view of Bayesian statistical inference. We discuss this statistical approach in Section 12.3. As we will see, to use this approach, we need to recast our problem into the language of probability theory. In the present case, this recasting has one possible method of solution, maximizing an unnormalized probability function that looks like

$$\exp\left(\lambda R(A) - \frac{1}{2}\sum_i\left(G_i - \sum_j K_{ij}A_j\right)^2\right),$$

which of course is equivalent to maximizing the argument of the exponential. Hence, at this level of analysis, the Bayesian approach is equivalent to a regularized least-squares method of solution. Finding the maximum (the mode) of a probability function, that is, finding a regularized least-squares solution, is unfortunately often inadequate. Ultimately, we present a method for estimating a mean solution relative to this probability.

A component in the Bayesian inference approach is choosing something called a *prior probability*, which replaces the regularizer. It represents our belief about the solution prior to (before) the data. We use an entropic prior: The nonnegativity and boundedness of the spectral density, our minimal prior information, allows us to pretend it is a probability density, and consequently it has an associated information theory entropy (Section 2.7). Additionally, the character this quasi-regularizer imposes on the solution is something well documented. For instance, in the absence of data, varying any one A_i does not require any specific one or more of the remaining A_i to change, apart from the changes required to maintain the sum rule (Section 12.3). In effect, this prior probability induces structure in the solution only if that structure is enforced by the data or by other prior knowledge. We can even go beyond the standard regularization and develop a method to determine the

Lagrange multiplier (Section 12.4). The Bayesian-based analysis also unveils that the kernel of the integral "reveals" to the data only a handful of effective (latent) parameters. These parameters are not the A_i, but rather the A_i as functions of these effective parameters (Section 12.5).

We begin by discussing general properties of dynamical correlation functions and the kernel that connects them with the spectral density and the spectral densities with sum rules. We then discuss the basic principles of Bayesian statistical inference. We have a nonlinear optimization problem, but because we need to determine only a handful of latent variables, we can use a method that finds the solution to this optimization problem with negligible computational cost. Appendix O details this method.

12.2 Dynamical correlation functions

The fluctuations of a system in thermal equilibrium are characterized by time-correlation functions of the type $\langle C(t)B(0)\rangle$, where B and C are operators. Linear response theory (Negele and Orland, 1988) tells us that the dynamical response of the system to these operators is described by a retarded Green's function (with $\hbar = 1$)

$$iG_R(t > 0) = \left\langle \left[C(t), B(0)\right]_{\pm}\right\rangle, \tag{12.2}$$

where the angular brackets denote thermal averaging and the operators $C(t)$ and $B(0)$ are in the Heisenberg representation. The sign on the commutator is determined by whether the operators B and C (in the Schrödinger representation) satisfy Fermionic $(+)$ or Bosonic $(-)$ commutation relations.

The spectral density $A(\omega)$ associated with this Green's function satisfies (Negele and Orland, 1988)

$$G_R(\omega + i\eta) = \int_{-\infty}^{\infty} d\omega' \, \frac{A(\omega')}{\omega - \omega' + i\eta}, \quad 0 < \eta \ll 1, \tag{12.3}$$

where the frequency Fourier transform of $G_R(t)$ is defined by

$$G_R(\omega) = \frac{1}{2\pi} \int_0^{\infty} dt \, e^{i\omega t} G_R(t). \tag{12.4}$$

Knowing $A(\omega)$ thus yields the real-time and frequency-dependent retarded Green's functions: Substituting $A(\omega)$ into (12.3) gives $G_R(\omega)$, and then taking the inverse Fourier transform yields

$$G_R(t) = \int_{\infty}^{\infty} d\omega \, e^{-i\omega t} G_R(\omega).$$

Finite-temperature Monte Carlo simulations compute the imaginary-time Green's functions[1]

$$G(\tau) = \langle C(\tau)B(0)\rangle, \quad 0 \le \tau < \beta. \tag{12.5}$$

Because this type of Green's function is antiperiodic (Fermions) or periodic (Bosons) in imaginary time, that is, $G(\tau) = \mp G(\tau + \beta)$, where β is the inverse temperature, the Fourier transform of this correlation function is

$$\hat{G}(i\omega_n) = \int_0^\beta d\tau \, e^{i\omega_n \tau} G(\tau), \tag{12.6}$$

where ω_n is a Matsubara frequency equal to $(2n + 1)\pi/\beta$ for Fermion and $2n\pi/\beta$ for Boson operators.

Knowing $A(\omega)$ yields the imaginary-time and the Matsubara frequency-dependent Green's functions: Substituting $A(\omega)$ into

$$\hat{G}(i\omega_n) = \int_{-\infty}^{\infty} d\omega' \, \frac{A(\omega')}{i\omega_n - \omega'} \tag{12.7}$$

gives $\hat{G}(i\omega_n)$. Then, the inverse transform

$$G(\tau) = \frac{1}{\beta} \sum_n e^{-i\omega_n \tau} \hat{G}(i\omega_n)$$

yields $G(\tau)$. We see that $\hat{G}(i\omega_n)$ in (12.7) and $G_R(\omega + i\eta)$ in (12.3) are the analytic continuations of each other: $i\omega_n \leftrightarrow \omega + i\eta$. This continuation connects the real- and imaginary-time Green's functions.

If (12.7) is substituted into the inverse transform, it yields

$$G(\tau) = \int_{-\infty}^{\infty} d\omega \, \frac{e^{-\tau\omega}}{1 \pm e^{-\beta\omega}} A(\omega). \tag{12.8}$$

This is the basic equation of this chapter. Given a Monte Carlo estimate of $G(\tau)$ from a finite-temperature simulation, we wish to solve this equation for $A(\omega)$. With $A(\omega)$, we then know $G_R(t)$ (Bonča and Gubernatis, 1993b).

Most often we do not go all the way to $G_R(t)$ but instead stop with $A(\omega)$. We stop with $A(\omega)$ because experimentally interesting properties of the system, such as the optical conductivity, NMR relaxation time, dynamic spin structure factors, and the like are related to various spectral densities. These functions give a picture of how interactions change the nature of the eigenstates of the Hamiltonian. Often, they give direct information about the existence of quasi-particles and collective modes

---

[1] Here, we use the phase convention of (7.22).

(e.g., Kawashima et al., 1996). A tool to extract this type of information from a simulation clearly enhances the value of the simulation.

Solving (12.1) for $A(\omega)$, given $G(\tau)$, is extremely difficult. The difficulty is that the kernel of the integral equation,

$$K(\tau, \omega) = \frac{e^{-\tau\omega}}{1 \pm e^{-\beta\omega}}, \tag{12.9}$$

becomes exponentially small at large positive and negative frequencies. In the forward problem, "Given A, what is G?," this behavior suppresses the sensitivity of $G(\tau)$ to the large-$|\omega|$ features of $A(\omega)$. In the inverse problem, "Given G, what is A?," the large-$|\omega|$ features of $A(\omega)$ depend on subtle features of $G(\tau)$, which are compromised by noise and incompleteness. In short, with incomplete and noisy data we are trying to extract features of A to which the data are insensitive. Many different spectral densities fit the data equally well.

Typically, a Hamiltonian has several interesting spectral densities. For illustration, we now discuss two such densities for the single-impurity (spin-degenerate) Anderson model

$$H = \sum_{k\sigma} \epsilon_k n_{k\sigma} + \sum_{k\sigma} \left(V_k c_{k\sigma}^\dagger d_\sigma + V_k^* d_\sigma^\dagger c_{k\sigma} \right) + \varepsilon_d \sum_\sigma d_\sigma^\dagger d_\sigma + U n_{d\uparrow} n_{d\downarrow},$$

where ϵ_k is the conduction band energy of an electron in momentum state k, V_k is the strength of the hybridization of the impurity orbital and the band, and U is the strength of the Coulomb repulsion between two electrons if both occupy the impurity orbital.

Of particular interest are spectral densities associated with the impurity orbital. If $G^\sigma(\tau) = \langle \mathcal{T} d_\sigma^\dagger(\tau) d_\sigma(0) \rangle$ for an electron with spin σ and $A_\sigma(\omega)$ is the associated spectral density, it is convenient to define $G(\tau) = \frac{1}{2}\sum_\sigma G^\sigma(\tau)$ and $A(\omega) = \frac{1}{2}\sum_\sigma A_\sigma(\omega)$. Then, because $d_\sigma^\dagger$ and d_σ satisfy Fermionic anticommutation relations,

$$G(\tau) = \int_{-\infty}^{\infty} d\omega \frac{e^{-\tau\omega}}{1 + e^{-\beta\omega}} A(\omega).$$

We can also show that $A(\omega) \geq 0$ and obeys the sum rule (Negele and Orland, 1988)

$$\int_{-\infty}^{\infty} d\omega\, A(\omega) = 1.$$

The symmetric Anderson impurity model describes a situation where the energy bands are symmetric around $E_{\text{Fermi}} = 0$ and $\varepsilon_d = -\frac{1}{2}U$. For this case, $A(\omega) = A(-\omega)$, and our integral equation becomes ($0 \leq \tau < \beta$)

$$G(\tau) = \frac{1}{2} \int_0^{\infty} d\omega \frac{e^{-\tau\omega} + e^{-(\beta-\tau)\omega}}{1 + e^{-\beta\omega}} A(\omega),$$

where the new kernel is

$$K(\tau, \omega) = \frac{1}{2} \frac{e^{-\tau\omega} + e^{-(\beta-\tau)\omega}}{1 + e^{-\beta\omega}}$$

and the sum rule becomes

$$\int_0^\infty d\omega\, A(\omega) = \frac{1}{2}.$$

The new kernel is an even function of ω.

Also of interest is the two-particle Green's function $(0 \le \tau < \beta)$

$$\chi(\tau) = \left\langle d_\uparrow^\dagger(\tau) d_\downarrow(\tau) d_\uparrow^\dagger(0) d_\downarrow(0) \right\rangle = \frac{1}{\pi} \int_{-\infty}^\infty d\omega\, \frac{e^{-\tau\omega}}{1 - e^{-\beta\omega}} \operatorname{Im}\chi(\omega).$$

Because $d_\uparrow^\dagger d_\downarrow$ commutes with itself, the kernel is Bosonic. For this Green's function, the spectral density $\operatorname{sign}(\omega)\operatorname{Im}\chi(\omega) \ge 0$ is the transverse magnetic susceptibility, which satisfies the sum rule

$$\frac{1}{\pi} \int_{-\infty}^\infty d\omega\, \frac{\operatorname{Im}\chi(\omega)}{\omega} = \chi(T).$$

$\chi(T)$ is the magnetic susceptibility at temperature T. Another quantity of interest is

$$\frac{1}{T_1 T} = K \lim_{\omega \to 0} \frac{\operatorname{Im}\chi(\omega)}{\omega},$$

where K is some constant that depends on the details of the coupling between the impurity nucleus and the d-electron spin. T_1 is the nuclear magnetic relaxation time. Because $\operatorname{Im}\chi(\omega)$ is an odd function of frequency, working with the antisymmetrized kernel

$$K(\tau, \omega) = \frac{1}{2} \frac{e^{-\tau\omega} - e^{-(\beta-\tau)\omega}}{1 - e^{-\beta\omega}}$$

restricts the problem to just the domain of positive frequencies.

As we soon discuss, our procedure for selecting a "best" solution of the integral equation (12.1) is information theory based. Increasing the amount of embodied information generally increases the quality of the solution. A symmetry is precise information. Symmetrizing the kernel as was done in the previous examples is important when using the numerical methods we eventually describe.

12.3 Bayesian statistical inference

Central to a Bayesian method is the use of probability theory and, concomitantly, the use of Bayes's theorem. We discussed this theorem in Section 2.1. To review, if we have two sets of events, $X = (X_1, X_2, \ldots, X_m)$ and $Y = (Y_1, Y_2, \ldots, Y_n)$, to which we have assigned probabilities, then Bayes's theorem says (Section 2.1)

$$P(X|Y) = \frac{P(Y|X)P(X)}{P(Y)}.$$

We use this theorem in the following manner: A given Hamiltonian fixes $A(\omega)$. The $G(\tau)$ data produced by the simulation, which we now denote as $\bar{G}(\tau)$, are thus conditioned on $A(\omega)$. However, we want $A(\omega)$ conditioned on the data $\bar{G}(\tau)$. Accordingly,

$$P(A|\bar{G}) = \frac{P(\bar{G}|A)P(A)}{P(\bar{G})}. \tag{12.10}$$

From this Bayesian perspective, the probability of the spectral density given the data, that is, $P(A|\bar{G})$, is the solution to this problem. As with any other probability function, we reduce its vast amount of information to fewer characteristic metrics, such as modes, means, variances, and the like. Obtaining the probability of the spectral density is constructive, because faced with an infinite number of possible solutions, we have quantitatively assigned degrees of belief about our options. We can now investigate them and decide how to reasonably designate something as the "best" solution. For example, if the probability $P(A|\bar{G})$ has a single sharp peak in the space of functions $A(\omega)$, then we can reasonably take this most probable $A(\omega)$, the mode of $P(A|\bar{G})$, as our solution. In fact, as we increase the amount of data, we a priori expect this situation to occur. If we have a single peak, but it is skewed and broad, then selecting an average spectral density, such as $\int \mathcal{D}A\, A\, P(A|\bar{G})$, would seem reasonable. On the other hand, a multiply-peaked $P(A|\bar{G})$ would require a situation-specific analysis.

   The various probabilities appearing in Bayes's theorem have names. The probability of A, that is, $P(A)$, is called the *prior probability*. It represents the probability of A prior (logically, not temporally) to the data. It is the task of the data, and other information we may add to the problem, to pull A away from this prior knowledge. The probability of the data given the spectral density, that is, $P(\bar{G}|A)$, is called the *likelihood function*, and the probability of the spectral density given the data, $P(A|\bar{G})$, is called the *posterior probability*. Finally, the probability of the data, $P(\bar{G})$, is called the *evidence*. The evidence normalizes the posterior probability. To show this, let us functionally integrate

$$P(A|\bar{G})P(\bar{G}) = P(\bar{G}|A)P(A)$$

over A on both sides of the equation. Because $\int \mathcal{D}A\, P(A|\bar{G}) = 1$,

$$P(\bar{G}) = \int \mathcal{D}A\, P(\bar{G}|A)P(A).$$

Comparing this result with (12.10) establishes the evidence as the normalization of the posterior probability when we construct the posterior probability from the product of the likelihood function and prior probability, which is what we do. Thus, the evidence, as well as the posterior probability, depends on the likelihood function and prior probability.

We now begin defining our choices for the prior probability and the likelihood function. In defining the prior probability, we appeal to the *theory of most probable distributions* and the *principle of maximum entropy*, so we first discuss these two important concepts.

12.3.1 Principle of maximum entropy

Schrödinger championed the theory of most probable distributions as a simple and unified approach to generate the distribution functions at the foundation of statistical mechanics (Schrödinger, 1952). To illustrate his point, he considered finding the distribution of an energy E over an ensemble of N identical independent systems, each in one of many possible energy states ε_i. With n_i defined as the number of systems in state i, the number of possible states having the set of occupation numbers $(n_1, n_2, \ldots, n_i, \ldots)$ is

$$\Omega(n_1, n_2, \ldots, n_i, \ldots) = \frac{N!}{n_1! n_2! n_3! \cdots n_i! \cdots}.$$

Given that the energy of the ensemble is E, the numbers n_i must satisfy the constraints $\sum_i n_i = N$ and $\sum_i \varepsilon_i n_i = E$. Seeking the maximum of

$$\ln \Omega - \lambda \sum_i n_i - \mu \sum_i \varepsilon_i n_i,$$

where λ and μ are Lagrange multipliers, he assumed that N and the n_i are large, and then after using Stirling's formula

$$\ln n! \approx n \ln n - n, \tag{12.11}$$

he showed that the probability p_i to be in state i is

$$p_i = \frac{n_i}{N} = \frac{e^{-\mu \varepsilon_i}}{\sum_i e^{-\mu \varepsilon_i}}.$$

Thermodynamic consistency requires that $\mu = 1/kT$. In the large N and large n_i limits, the microcanonical entropy $S = k \ln \Omega$ becomes

$$S = -kN \sum_i p_i \ln p_i. \tag{12.12}$$

Hence, the entropy per system is $S/N = -k \sum_i p_i \ln p_i$.

The information theory approach to assigning probability densities, the principle of maximum entropy, maximizes a constrained entropy. It is similar to Schrödinger's use of the theory of most probable distributions. It, however, does not appeal to counting states and to the law of large numbers to define an entropy, but rather appeals to a small set of axioms (Shore and Johnson, 1980). The functional

$$S = - \int dx\, p(x) \ln \left(\frac{p(x)}{m(x)} \right), \qquad (12.13)$$

is shown to satisfy them, up to an overall constant. These axioms are that the entropy functional should

- be unique
- have coordinate independence
- have system independence
- have subset independence.

They define the character of the entropy. Coordinate independence means invariance under a change of variables. System independence says it should not matter whether one accounts for independent information about independent systems in terms of separate distributions or in terms of a joint distribution. Finally, subset independence says it should not matter whether one treats an independent subset of systems in terms of a conditional density or in terms of the full system density.

For *discrete probabilities*, the entropy axioms of information theory say that up to an overall positive constant, the entropy expression is

$$S = - \sum_i p_i \log \left(\frac{p_i}{m_i} \right). \qquad (12.14)$$

Seemingly this expression is the natural discretization of (12.13). In fact, a fifth axiom is required (Shore and Johnson, 1983). The entropy functional should

- be logically consistent,

that is, in the absence of additional information, maximizing the entropy for a discrete probability (12.14) must yield $p_i = m_i$.

Analogous results exist for a continuous distribution. In the entropy functional (12.13), $p(x)$ is a probability distribution and $m(x)$ is a measure necessary for invariance under a change of variables.[2] If we were to maximize (12.13), subject to the constraint that $\int dx\, p(x) = 1$, we would start with

---

[2] The invariance is easily seen by making the standard change of variables to the integration and recalling from Section 2.1 how probability distributions transform under a change of variables. The various Jacobians of the change of variables cancel.

$$Q = \lambda_0 \left[\int dx\, p(x) - 1 \right] - \int dx\, p(x) \ln \left(\frac{p(x)}{m(x)} \right).$$

Then, requiring the variations δQ in Q due to arbitrary variations δp in p to be zero,

$$\delta Q = -\int dx\, \delta p(x) \left[\ln \left(\frac{p(x)}{m(x)} \right) + 1 - \lambda_0 \right] = 0,$$

implies $p(x) = m(x)e^{-(1-\lambda_0)}$. Integrating both sides of the equation over x, using the normalization condition on $p(x)$, and assuming one for $m(x)$, for convenience, shows that $\exp[-(1-\lambda_0)] = 1$, that is, $\lambda_0 = 1$. Therefore, $p(x) = m(x)$. Thus, we can view $m(x)$ as representing our prior knowledge of $p(x)$, that is, prior to the use of the data. In Bayesian analysis, this Lebesgue measure is sometimes called the *default model*. The last axiom says that in the absence of additional information we must recover our prior information.

Insight about m_i (and $m(x)$) comes from revisiting Schrödinger's problem, but now restricting the total number of energies to be M and assigning a probability m_i to each. Now, for a set of measures $(m_1, m_2, \ldots, m_M)$, which for convenience we assume are normalized,

$$1 = (m_1 + m_2 + \cdots + m_M)^N = \sum_{\substack{m_1, m_2, \ldots, m_M \\ m_1 + m_2 + \cdots + m_M = 1}} \frac{N!}{n_1! n_2! \cdots n_M!} m_1^{n_1} m_2^{n_2} \cdots m_M^{n_M}.$$

$$(12.15)$$

Accordingly, the probability of a given set $(n_1, n_2, \ldots, n_M)$ of occupation numbers, conditioned on $(m_1, m_2, \ldots, m_M)$, is

$$P(n_1, n_2, \ldots, n_M | m_1, m_2, \ldots, m_M) = \frac{N!}{n_1! n_2! \cdots n_M!} m_1^{n_1} m_2^{n_2} \cdots m_M^{n_M}. \qquad (12.16)$$

For N and n_i large,

$$\ln P = -N \sum_i p_i \ln \left(\frac{p_i}{m_i} \right) = S, \qquad (12.17)$$

where $p_i = n_i/N$. We note that p_i must be zero wherever m_i is zero.

The *principle of maximum entropy* says that to assign probabilities on the basis of partial information, we maximize the entropy, constrained by whatever information we know about the probability. What is produced is the least informative probability consistent with the constraints.

An iconic use of this principle is predicting the joint probability of a set of variables $x = (x_1, x_2, \ldots, x_N)$, each ranging from $-\infty$ to ∞, whose means and covariance matrix (Section 3.24) are

$$\langle x_k \rangle = \int dx \, x_k \, p(x),$$

$$C_{ij} = \langle (x_i - \langle x_i \rangle) (x_j - \langle x_j \rangle) \rangle.$$

Straightforward analysis yields

$$p(x|\{\langle x \rangle\}, C) = e^{-\frac{1}{2}\delta x^T C^{-1} \delta x} / \sqrt{\det(2\pi C)},$$

where the vector of deviations δx is $(x_1 - \langle x_1 \rangle, \ldots, x_N - \langle x_N \rangle)^T$. We express the result as a conditional probability, as logically the derivation is conditional on knowing the means and covariance matrix beforehand. We comment that if the variables did not range from $-\infty$ to ∞, we would need to compute the normalization constant by performing the integrations numerically. While the predicted probabilities would have the functional form of a Gaussian, they would rather be functions proportional to Gaussians.

In general, the principle of maximum entropy is useful for suggesting the functional forms of probabilities. In the next subsection, we adopt this multivariate Gaussian as a likelihood function, but we do not invoke the principle of maximum entropy. Entropy is part of our choice for a prior probability. The solutions to our integral equations involve competitions between maximizing the log-likelihood and maximizing the entropy.

12.3.2 The likelihood function and prior probability

We now discuss our choices for the likelihood function and prior probability. For the likelihood function, we start very generally with

$$P(\bar{G}|A) = e^{-\mathcal{L}(\bar{G},A)} / Z_{\mathcal{L}},$$

where $\mathcal{L}(\bar{G}, A)$ is some positive function and $Z_{\mathcal{L}}$ is the normalization constant. We now need to be more specific about the functional form of $\mathcal{L}(\bar{G}, A)$.

We want our choice of $\mathcal{L}$ to be compatible with the data the simulation generates. If $\bar{G}_i^{(j)}$ is the value of G_i for the j-th configuration of the simulation, then the simulation eventually gives an estimate of the mean

$$\bar{G}_i = \frac{1}{M} \sum_{j=1}^{M} \bar{G}_i^{(j)} \tag{12.18}$$

and the covariance matrix (3.24)

$$C_{ik} = \frac{1}{M(M-1)} \sum_{j=1}^{M} \left(\bar{G}_i^{(j)} - \bar{G}_i \right) \left(\bar{G}_k^{(j)} - \bar{G}_k \right). \tag{12.19}$$

With the data represented by the mean and covariance, it is natural and convenient to take

$$\mathcal{L}(\bar{G}, A) = \tfrac{1}{2}\chi^2(\bar{G}, A),$$ (12.20)

with

$$\chi^2(\bar{G}, A) = \sum_{i,j=1}^{L} (\bar{G}_i - G_i) \left[C^{-1}\right]_{ij} (\bar{G}_i - G_i),$$ (12.21)

where G_i is the exact value of $G(\tau)$ at τ_i for a given A. With this choice of $\mathcal{L}$, the normalization constant $Z_{\mathcal{L}}$ is $(2\pi)^{N/2}\sqrt{\det C}$.

This choice is the same as the one made for simple and regularized least-squares problems. The discussion there implicitly assumed that some Gaussian process has generated the data and hence defines the mean and the covariance. In our case, a Monte Carlo simulation generates the data from which we compute the mean and covariance. If we computed them with a sufficiently large amount of statistically independent information, then by the central limit theorem the means are distributed by a Gaussian whose width is defined by the covariance matrix.

Equation (12.17) hints at our choice of the prior probability: It says that in the absence of other information, the probability of a given set of occupation numbers is proportional to the exponential of the entropy. Instead of appealing to combinatorial arguments, we can appeal to the information theory arguments for the functional form of the entropy and use the same exponential form as a prior probability. In fact, in statistical inference, when assigning probabilities and needing to maintain positivity, the most common choice for the prior probability is the entropic prior

$$P(A) = e^{\alpha S(A)}/Z_S(\alpha),$$ (12.22)

where α sets the scale of the entropy's contribution and $Z_S(\alpha)$ is the normalization constant. In general, α is unknown a priori.

There are philosophical reasons for choosing an entropic prior. Our main reason, however, is practical: It is a convenient choice for maintaining the positivity and the normalization of the solution. Another reason is the choice that has value added: We constrain our solutions with the known characteristics prescribed by the axioms that establish the form of the entropy. In effect, because of the additive nature of the entropy, this prior probability induces structure in the solution only if that structure is enforced by the data or by other prior knowledge. Finally, we like the fact that the default model is an explicit mechanism for putting our prior information into the solution. For most physics problems, various approximations or exact results exist that we can use as default model.

These choices of the likelihood and prior enable us to state the joint distribution

$$P(A, \bar{G}) = P(\bar{G}|A)P(A) = \frac{e^{\alpha S - \mathcal{L}}}{Z_{\mathcal{L}} Z_S(\alpha)}. \tag{12.23}$$

The two remaining probabilities in Bayes's theorem are the posterior probability and the evidence. Our choices for the likelihood function and prior probability set both. Our task now is discussing how these choices affect our choice of a "best" solution to the integral equation.

12.3.3 The "best" solutions

We discuss three approaches to finding a solution. With our choices of the likelihood function (12.20) and prior probability (12.22), we can write the posterior probability in terms of the evidence as

$$P(A|\bar{G}) = \frac{P(A, \bar{G})}{P(\bar{G})} = \frac{e^{Q(A)}}{Z_{\mathcal{L}} Z_S(\alpha)}, \tag{12.24}$$

with

$$Q(A) = \alpha S(A) - \tfrac{1}{2} \chi^2(A). \tag{12.25}$$

With this posterior probability we now state our first best solution: We ignore the evidence, which does not depend on A, and take as our solution the most probable spectral density A given the data $\bar{G}$. We find this by maximizing $Q(A)$, which, in turn, maximizes the posterior probability. We call this the *constrained fit*, as it is simply a regularized least-squares fit with the regularizing function being the entropy. This solution depends on α. We choose α so that $\chi^2 = N$, where N is the number of $\bar{G}_i$. This way of choosing α is basically ad hoc and usually tends to under-fit. The resulting procedure is commonly called the *historic maximum entropy method*.

At this level of analysis, we clearly see a duality between fitting and inference. We can regard the constrained fit as minimizing χ^2, which corresponds to choosing the A_i parameters to fit the data as closely as possible, subject to the entropy constraints. Or we can regard this solution as maximizing the entropy, which puts the least amount of information into the solution, subject to the constraints of the data.

An alternative approach to choosing α is to use Bayesian analysis to guide the choice. We call this second approach to solving the integral equation the *Bayesian constrained fit*. More commonly, it is called the *classic maximum entropy method*.

In starting the development of this alternative approach, we first note that our prior probability is conditional on α and we write (12.23) as

$$P(A, \bar{G}|\alpha) = P(\bar{G}|A)P(A|\alpha) = \frac{e^Q}{Z_{\mathcal{L}} Z_S(\alpha)}. \tag{12.26}$$

Note that the likelihood function in (12.26) is determined by the physical dynamics and hence is not conditional on α, which was introduced to scale the contribution of the entropy to the entropic prior. We then define a new joint probability

$$P(A, \bar{G}, \alpha) = P(A, \bar{G}|\alpha)P(\alpha) = P(\alpha)\frac{e^Q}{Z_{\mathcal{L}}Z_S(\alpha)} \qquad (12.27)$$

and a new posterior probability

$$P(A, \alpha|\bar{G}) = \frac{P(\alpha)}{P(\bar{G})}\frac{e^Q}{Z_{\mathcal{L}}Z_S(\alpha)}.$$

Introduced is a new probability, $P(\alpha)$, the probability of α. This number is usually chosen to be a constant or what is called Jeffery's prior, $P(\alpha) \propto 1/\alpha$ (Sivia and Skilling, 2006). Jeffery argued that the probability assigned to a scale-setting parameter should be done so that $p(x)dx = p(cx)d(cx)$. Since $d(cx) = cdx$, this requirement reduces to $p(x) = cp(cx)$, which is satisfied only by $p(x) \propto 1/x$. Normalization requires an accompanying restriction on the range of x. In practice, a good solution is reasonably insensitive to choosing $P(\alpha)$ as either a constant or $1/\alpha$. For a fixed α, we note that the maximum of the new posterior probability still occurs at the maximum of Q.

To move toward finding the "best" value of α, we write

$$P(A|\bar{G}) = \int d\alpha \, P(A, \alpha|\bar{G}) = \int d\alpha \, P(A|\alpha, \bar{G})P(\alpha|\bar{G}).$$

If the number of data points is large and well connected to the inference problem, it seems reasonable to expect that the many data restrict the possible values of the single parameter α significantly, making $P(\alpha|\bar{G})$ sharply peaked at some value $\alpha = \hat{\alpha}$; that is, $P(\alpha|\bar{G}) \approx \delta(\alpha - \hat{\alpha})$. The posterior probability becomes

$$P(A|\bar{G}) \approx P(A|\bar{G}, \hat{\alpha}).$$

For our second solution of the integral equation, we find $\hat{\alpha}$, the value of α at which $P(\alpha|\bar{G})$ peaks, and *then* for this value of α, we find the A that maximizes Q as the solution. These two steps specify the Bayesian constrained fit. We note that this solution is not equivalent to solving

$$\frac{\partial P(A, \alpha|\bar{G})}{\partial A} = 0, \quad \frac{\partial P(A, \alpha|\bar{G})}{\partial \alpha} = 0.$$

These equations fix A and α simultaneously.

In Section 12.4, we discuss how to estimate $P(\alpha|\bar{G})$ and find $\hat{\alpha}$. Here we derive the form of this probability. We start by marginalizing A from the joint probability (12.27):

$$P(\alpha, \bar{G}) = \int \mathcal{D}A\, P(A, \alpha, \bar{G}).$$

Then, using $P(\alpha, \bar{G}) = P(\alpha|\bar{G})P(\bar{G})$ and (12.27), we find that

$$P(\alpha|\bar{G}) = \frac{P(\alpha)}{P(\bar{G})} \int \mathcal{D}A\, \frac{e^Q}{Z_{\mathcal{L}}Z_S}. \qquad (12.28)$$

The parameter α is called a nuisance parameter, and a nuisance parameter generally is best handled by integrating it out of the problem (marginalizing it) and working with $P(A|\bar{G})$ instead of $P(A, \alpha|\bar{G})$. In fact, our third solution to the problem does precisely this. We call this approach the *average spectrum method*. In the past, it was called *Bryan's method* (Bryan, 1990) to acknowledge its source. The switch in name emphasizes the nature of the solution being an average instead of being a mode. This nature has often been overlooked.

For this third method, we first find $\hat{A}_\alpha$ from

$$\left.\frac{\partial Q}{\partial A}\right|_{A=\hat{A}_\alpha} = 0,$$

that is, we obtain the constrained fit as a function of α, and then we choose as the solution the average defined by

$$\langle A \rangle = \int d\alpha\, \hat{A}_\alpha P(\alpha|\bar{G}),$$

where $P(\alpha|\bar{G})$ is given by (12.28). We need this solution because $P(\alpha|\bar{G})$ is sometimes not sharply peaked but broadly peaked and skewed.

All three methods require finding the A that maximizes $Q(A)$ for a fixed value of α, and two require knowing something about $P(\alpha|\bar{G})$. In the next section, we discuss the nature of the maximum of $Q(A)$ in detail and present an approximation for $P(\alpha|\bar{G})$. Before moving to that section, we first will say a few words about the evidence.

There are two kinds of evidence present in our methods of solution, $P(\bar{G})$ and $P(\bar{G}|\alpha)$. We now derive formal expressions for both. First, using

$$P(\bar{G}|\alpha) = \frac{P(\alpha|\bar{G})P(\bar{G})}{P(\alpha)}$$

and (12.28), we immediately find that

$$P(\bar{G}|\alpha) = \int \mathcal{D}A\, \frac{e^Q}{Z_{\mathcal{L}}Z_S}. \qquad (12.29)$$

Next, integrating both sides of (12.28) over α yields

$$P(\bar{G}) = \int d\alpha\, P(\alpha) \int \mathcal{D}A\, \frac{e^Q}{Z_{\mathcal{L}} Z_S}.$$

From these two results we see that the conditional evidence $P(\bar{G}|\alpha)$ is central to both the evidence $P(\bar{G})$ and to $P(\alpha|\bar{G})$. Fortunately, the evidence $P(\bar{G})$ for our posterior probability $P(A|\bar{G})$ is the expectation value of the conditional evidence $P(\bar{G}|\alpha)$ with respect to the probability $P(\alpha)$. Germane to this point are a few remarks about how to know when the evidence is significant.

If we accept as our solution a quantity that depends on the mode of the posterior probability, the evidence $P(\bar{G})$, being its normalization factor, plays no role in this solution as it has no explicit dependence on the fitting parameters α and the values of A. This situation is the case for the constrained and the Bayesian constrained fits. The evidence does, however, play an essential role if our solution is an average over A and α. The conditional evidence plays a role in the Bayesian constrained fit and average spectrum method.

12.4 Analysis details and the Ockham factor

Finding the maximum of Q (12.25) as a function of A for a fixed α, which is central to all three of our approaches, is simple in principle because both the likelihood function and the entropic prior are concave functions of the A_i and hence a unique maximum exists. In practice, finding the maximum can be difficult because it may not be sharp. The concavity of the likelihood function is familiar: The covariance matrix is positive definite, so we can always transform the data into the coordinate system where this matrix is diagonal. Then, in the likelihood function, the curvature tensor is diagonal with positive matrix elements equal to the positive eigenvalues of the covariance matrix. The concavity of the entropic contribution to Q is also easily established. By direct calculation,

$$\frac{\partial^2 S}{\partial A_i \partial A_j} = -\frac{\delta_{ij}}{A_i} = -\frac{\delta_{ij}}{\sqrt{A_i A_j}}.$$

Hence, the curvature of the entropy equals $\sqrt{A_i A_j}\delta_{ij}$.

The nature of Q in the vicinity of this maximum determines how easy it is to find it. To describe the maximum, it is convenient to transform the deviations from it, that is, the $\delta\hat{A}_i = A_i - \hat{A}_i$, to the space of new variables X_i in which the entropy curvature is flat. The new coordinate system satisfies

$$\frac{\partial A_j}{\partial X_i} = \sqrt{A_j}\delta_{ij}. \tag{12.30}$$

With this change of variables,

$$Q(A, \alpha) \approx Q(\hat{A}_\alpha) - \tfrac{1}{2} \sum_{ij} \delta X_i \Gamma_{ij} \delta X_j,$$

where Γ is a positive-definite matrix $\Gamma_{ij} = \alpha \delta_{ij} + \Lambda_{ij}$ with

$$\Lambda_{ij} = \left[\sqrt{A_i} \frac{\partial^2 \mathcal{L}}{\partial A_i \partial A_j} \sqrt{A_j} \right]_{A = \hat{A}_\alpha}$$

and

$$\frac{\partial^2 \mathcal{L}}{\partial A_i \partial A_j} = \left[K^T \cdot C^{-1} \cdot K \right]_{ij} = \sum_{kl} K_{ki} \left[C^{-1} \right]_{kl} K_{lj}.$$

Here, K_{ij} is the time-frequency discretization of the kernel of the integral equation. In the new coordinate system, Γ controls the curvature of Q around the maximum. If the eigenvalues of Γ are small, the curvature is flat. A flat curvature complicates finding the maximum and leads to considerable uncertainty in the result. The covariance of the X_i is a measure of the uncertainty,

$$\langle \delta X_i \delta X_j \rangle = \int \mathcal{D}X \, \delta X_i \delta X_j P(A|\bar{G}) \approx \left[\Gamma^{-1} \right]_{ij}.$$

In a coordinate system where Γ is diagonal, $\langle \delta X_i \delta X_j \rangle \rightarrow \langle \delta X_i^2 \rangle \delta_{ij} = \delta_{ij}/\gamma_i$, where the γ_i are the eigenvalues of Γ. Similarly, we find that

$$\langle \delta A_i \delta A_j \rangle \approx \left[\sqrt{\hat{A}} \, \Gamma^{-1} \sqrt{\hat{A}} \right]_{ij} = \sqrt{\hat{A}_i} \left[\Gamma^{-1} \right]_{ij} \sqrt{\hat{A}_j}. \qquad (12.31)$$

Thus, a small eigenvalue of Γ leads to a large variance in the result. We also note that taking $\sqrt{A_i}$ as the metric for the problem (12.30) yields $\mathcal{D}A = \prod_i dA_i/\sqrt{A_i}$, and with the change in variables, $\mathcal{D}A \rightarrow \mathcal{D}X = \prod_i dX_i$.

We now make explicit the normalization factors needed in the analysis. Physically, as $A(\omega)$ ranges between 0 and 1, the values of the Green's function for the different imaginary times range between another set of bounds. For example, if $G(\tau)$ describes Fermions of a particular spin, they range between 0 and 1. Hence, the normalization integrals are something that at first glance we need to do numerically. However, if the error in the data is sufficiently small to make the exponential of the likelihood function sharply peaked, which is to say the exponential dies off over the range of integration, we can extend the limits of integration over all space and then evaluate the integral analytically: If

$$Z_{\mathcal{L}} = \int \mathcal{D}\bar{G} \, e^{-\frac{1}{2} x^2} = \int \prod_i d\bar{G}_i e^{-\frac{1}{2} x^2}$$

with

$$\chi^2 = \sum_{i,j=1}^{L} (\tilde{G}_i - G_i) \left[C^{-1}\right]_{ij} (\tilde{G}_j - G_j),$$

then when the limits of the integrations are extended over all space, we find

$$Z_{\mathcal{L}} = (2\pi)^{L/2} \sqrt{\det C}.$$

Our likelihood function is thus the normalized Gaussian we assumed.

This redundant analysis illustrates an important point about the central limit theorem: While the range of random variables for a given distribution, such as the uniform distribution over $[0, 1]$, is bounded, the allowed range of the average of a large number of these variables is over $-\infty$ to ∞. As the number in the sum increases, the probability of the average lying outside the original range becomes vanishingly small.

For the normalization factor of the prior probability, that is, $Z_S(\alpha)$, the situation is different. We need to derive an approximate expression and do so by a similar analysis. We start with

$$Z_S(\alpha) = \int \mathcal{D}A \, e^{\alpha S} = \prod_i \int \frac{dA_i}{\sqrt{A_i}} e^{\alpha S_i},$$

where $S_i = A_i \ln(A_i/m_i)$. Next we approximate the exponential by a Gaussian centered at $A_i = m_i$, extend the range of the integration to be over all space, and find

$$\int \frac{dA_i}{\sqrt{A_i}} e^{\alpha S_i} \approx \int dX e^{-\alpha X^2/2} = \sqrt{2\pi/\alpha}.$$

Consequently,

$$Z_S(\alpha) \approx \left(\frac{2\pi}{\alpha}\right)^{N/2} = \frac{(2\pi)^{N/2}}{\sqrt{\det \alpha I}}, \tag{12.32}$$

where I is an $N \times N$ identity matrix. It is definitely true that in the vicinity of its maximum, the entropy is well approximated by a quadratic function. The width around that maximum is controlled by the size of $\alpha > 0$, and the prior is most sharply peaked when α is large.

We now derive an approximate expression for the conditional evidence

$$P(\tilde{G}|\alpha) = \int \mathcal{D}A \, \frac{e^Q}{Z_{\mathcal{L}} Z_S(\alpha)}.$$

With it, the expressions we need for $P(\alpha|\bar{G})$ and $P(\bar{G})$ follow readily. We simply approximate the integrand by a Gaussian form centered at the maximum of Q:

$$P(\bar{G}|\alpha) \approx \frac{e^{Q(\hat{A}_\alpha)}}{Z_{\mathcal{L}}Z_S(\alpha)} \int \mathcal{D}X\, e^{-\frac{1}{2}\delta X^T \cdot (\alpha I + \Lambda) \cdot \delta X}$$

$$= \frac{e^{Q(\hat{A}_\alpha)}}{Z_{\mathcal{L}}Z_S(\alpha)} \frac{(2\pi)^{N/2}}{\sqrt{\det[\alpha I + \Lambda(\hat{A}_\alpha)]}}. \qquad (12.33)$$

If the peak around Q is sufficiently sharp, this approximation should be sufficiently good.

For the Bayesian constrained fit, we need to maximize $P(\alpha|\bar{G})$ with respect to α. This function is simply $P(\alpha)P(\bar{G}|\alpha)/P(\bar{G})$. As $P(\bar{G})$ is independent of α, we find the maximum of $P(\alpha|\bar{G})$ from the maximum of $P(\alpha)P(\bar{G}|\alpha)$. In general, the common choices of $P(\alpha)$ are featureless. What we effectively need is the maximum of $P(\bar{G}|\alpha)$ as a function of α. With (12.33), the condition $\partial \ln P(\bar{G}|\alpha)/\partial \alpha = 0$ leads to

$$-2\hat{\alpha}S(\hat{A}_{\hat{\alpha}}) = \mathrm{Tr}\left[\Lambda(\alpha I + \Lambda)^{-1}\right] - \mathrm{Tr}\left[\frac{d \ln \Lambda}{d \ln \alpha}\Lambda(\alpha I + \Lambda)^{-1}\right].$$

The logarithmic derivative is expected to be small. In any case, we drop it from this expression. The defining equation for the Bayesian constrained fit becomes

$$N_{\mathrm{good}} \equiv -2\hat{\alpha}S(\hat{A}_{\hat{\alpha}}) = \mathrm{Tr}\left[\Lambda(\alpha I + \Lambda)^{-1}\right] = \sum_i \frac{\lambda_i}{\alpha + \lambda_i}, \qquad (12.34)$$

where the λ_i are the eigenvalues of Λ. The quantity $-2\hat{\alpha}S(\hat{A}_{\hat{\alpha}})$, called N_{good}, is a measure of the shift of the solution away from the default model (if the solution were the default model, the entropy would be zero). When a λ_i is much greater than α, it contributes a value of unity to N_{good}. When it is much smaller, it contributes zero. Thus, N_{good} measures the amount of good information in the solution and is a convenient indicator of the amount of structure in the solution.

When N_{good} is large, one expects $P(\alpha|\bar{G})$ to be sharply peaked. Unfortunately, the extremely ill-posed nature of the analytic continuation fitting problem is typically characterized by $N_{\mathrm{good}} \ll N$ with N_{good} often being between 5 and 10. When this figure of merit is small, we can expect to parameterize the locations and widths of only a few peaks.

While the Bayesian constrained fit often gives an acceptable result, the average spectrum method is generally more consistent with the information in the data. Once again, this solution is

$$\langle A \rangle = \int \mathcal{D}A\, d\alpha\, A(\alpha)P(A, \alpha|\bar{G}) \approx \int d\alpha\, \hat{A}_\alpha P(\alpha|\bar{G}), \qquad (12.35)$$

where the integral over α is done numerically. The average spectrum solution repro-duces the Bayesian constrained fit if $P(\alpha|\bar{G})$ is sharply peaked and returns the model in the absence of data. To obtain the average spectrum result, we need a normalized $P(\alpha|\bar{G})$. What is done is to evaluate $P(\alpha)P(\bar{G}|\alpha)$, using (12.33) for $P(\bar{G}|\alpha)$, for a number of discrete values of α and then compute the area under this curve. The area is the normalization factor. Dividing the values of $P(\alpha)P(\bar{G}|\alpha)$ by this factor gives $P(\alpha|\bar{G})$.

We concluded the last subsection with a few remarks about when the evidence is significant. We conclude this subsection with a few remarks about the significance of the evidence. We do so by first discussing the relationship between the condi-tional evidence and what is called the Ockham factor. With (12.32), the conditional evidence

$$P(\bar{G}|\alpha) = \frac{e^{Q(\hat{A}_\alpha)}}{Z_{\mathcal{L}}} \frac{(2\pi)^{N/2}}{Z_S(\alpha)\sqrt{\det[\alpha I + \Lambda]}} \approx e^{\alpha S(\hat{A})} \underbrace{\frac{e^{-\mathcal{L}(\hat{A}_\alpha)}}{Z_{\mathcal{L}}}}_{\text{best fit}} \underbrace{\sqrt{\frac{\det[\alpha I]}{\det[\alpha I + \Lambda]}}}_{\text{Ockham factor}} \quad (12.36)$$

is approximately the product of two factors. One comes from the mode of the posterior probability, which represents the constrained least-squares fitting, and the other modifies the fit. From this perspective the maximum for a given α results from a competition between fitting the values of the spectral density to good data, which is the tendency for small α, and because of the Ockham factor, defaulting to the model, which is the tendency for large α. When the data are closely fitted, the mode represents a solution with many parameters, while the Ockham factor favors fewer parameters. As we showed, the evidence is the expectation value of the conditional evidence, and as the common choices of $P(\alpha)$ are featureless, if not flat, the full evidence carries with it the Ockham character. The evidence helps establish a balance between the accuracy of the fit to the data and the number of parameters being fitted.

In Fig. 12.1, we depict the Ockham factor schematically. Before we have data, our knowledge of the solution is expressed by the prior probability. It admits a possible solution over some volume ΔA in the space of parameters and hence roughly equals $1/\Delta A$. With data, the likelihood function restricts the solution to some smaller volume δA centered around the mode. In general, if our prior information admits solutions over a wide range of parameter space, only a small portion of the prior contributes to the evidence. Typically, as we increase the number of parameters, we make δA smaller while making ΔA larger. The Ockham factor, $\delta A/\Delta A$, expresses a penalty for using too many parameters for the sake of getting a good fit. In short, the evidence contains an Ockham factor that favors a simpler physical model than the least-squares solution in the spirit of Ockham's

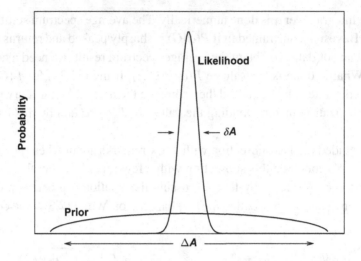

Figure 12.1 Schematic representation of the Ockham factor. This factor penalizes results for "wasting" volume in parameter space. The evidence $P(\bar{G}) =$
$\int P(\bar{G}|A)P(A)\mathcal{D}A \approx P(\bar{G}|\hat{A})\delta A/\Delta A =$ maximum likelihood $\times$ Ockham factor.

centuries-old principle that one should always opt for an explanation in terms of
the smallest possible number of causes, factors, or variables. The evidence helps
avoid the "with enough parameters you can fit anything" syndrome.

12.5 Practical considerations

We now address several practical considerations associated with the use of our
Bayesian entropic method to generate a solution to the integral equation (12.1).
Each of the three basic steps to the method requires special considerations. These
steps are:

1. Verifying that the data are consistent with the assumptions of the likelihood
   function
2. Solving the integral equation
3. Assessing the acceptance of the solution.

Jarrell and Gubernatis (1996) presented a detailed case study of these steps for the
infinite-dimensional periodic Anderson model. A slightly updated version is given
by Jarrell et al. (2008). Here, we highlight the main points of these studies.

   1. Verifying that the data are consistent with the assumptions of the likelihood
function has two parts. The first is establishing that the data used to calculate
the covariance matrix are statistically independent. Here, we need to remove the
correlations in the data that exist from one Monte Carlo step to another. The second

addresses removing the correlations that exist between the data at a given Monte Carlo step. Here, we remove these correlations by diagonalizing the covariance matrix.

The data consist of two types of measurements: the mean values (12.18) and the covariance matrix (12.19). As with other Monte Carlo estimates of means, the estimates of the $\bar{G}_i$ are unaffected by correlations existing between successive Monte Carlo steps. On the other hand, as with other Monte Carlo estimates of uncertainty, the estimates of the elements of the covariance matrix are affected. To generate the assumed statistically independent measurements needed to estimate the covariance properly, we use the method of blocked means, focusing on the *diagonal elements* of the covariance matrix. As explained in Section 3.4 of Chapter 3, we break up the data stream into larger and larger blocks until the block averages are statistically independent.

One difficulty in "verifying the data" lies with a number of estimates of the $\bar{G}_i$ being inclined to have a nonzero skewness and kurtosis. For example, the fluctuations in the Green's function $G(\tau)$ associated with the spectral density of the single-impurity Anderson model are bounded above by 1 when τ is close to 0 and bounded below by zero when τ is close to $\beta/2$ and β is large. Clearly, reducing the fluctuations in the block averages of correlations for these Green's function elements is more challenging than for others. If we recall the discussion in the previous subsection about the central limit theorem, we can begin to understand why. The central limit theorem locates the Gaussian at the mean value of the random variable. If that variable is bounded and its mean is close to the boundary, more reduction of the variance is needed before the distribution about the mean has a strongly Gaussian shape. In the present case, increasing the block sizes eventually promotes most of the *diagonal elements* of the covariance matrix to exhibit features of statistical independence and a Gaussian distribution.

Focusing on the diagonal elements of C to establish statistical independence is a practical approach more so than an insightful statistical one. It, however, is not the same as throwing away the off-diagonal elements of C. Throwing away removes the correlations among measurements at different τ values but leads to poor estimates of the errors of independent information as measured by C.

After the block size becomes reasonably established, increasing the number of blocks promotes the proper calculation of the positive-definite covariance matrix. Here, we are interested in diagonalizing this matrix so we can transform χ^2 into the standard estimate of the error,

$$\chi^2 = \sum_{i,j=1}^{L} (\bar{G}_i - G_i)[C^{-1}]_{ij}(\bar{G}_j - G_j) \rightarrow \sum_{i=1}^{L} (\bar{G}'_i - G'_i)^2/\sigma_i^2, \qquad (12.37)$$

for uncorrelated measurements. If C is diagonalized by the similarity transformation $C = S\Sigma S^T$, then $G' = SG$ and $\bar{G}' = S\bar{G}'$. If the number of statistically independent blocks is insufficient, the eigenvalues of C, that is, the σ_i^2, when indexed from high to low, fall precipitously at some value of the index. The small eigenvalues are less accurate than the large ones. To prevent this break, it seems necessary that $N_{blocks} > 2L$. Producing a reasonably large number of sufficiently large blocks defines the numerical task of the quantum Monte Carlo method.

2. With the data qualified, we turn to the second step of our method and now seek a solution of the integral equation. We first need to discuss its discretization. Although our problem (12.1) is one of assigning a quasi-probability density $A(\omega)$, when we discretize the problem, we convert it to one of assigning probabilities. At times, a nonuniform discretization is advantageous to focus computational effort around peaks and not in smoothly varying featureless regions of frequency. When discretizing, assigning a mean value to the density over an interval is akin to assigning each interval a different probability. We need to do this in such a way that if we change, for example, from a uniform grid to a logarithmic one, we leave our entropy form invariant. (Recall the second axiom.)

After discretization, we have a system of linear equations to solve for the values of $A(\omega)$ at a set $(\omega_1, \omega_2, \ldots, \omega_N)$ of discrete values of ω using the values of $\bar{G}(\tau)$ at a set $(\tau_1, \tau_2, \ldots, \tau_L)$ of discrete values of τ. The system of equations is

$$\bar{G}_i = \sum_{j=1}^{N} K_{ij} A_j, \quad i = 1, \ldots, M, \tag{12.38}$$

where $\bar{G}_i = \bar{G}(\tau_i)$, $K_{ij} = K(\tau_i, \omega_j)$, and $A_i = A(\omega_i)\Delta\omega_i$. Similarly, we define $m_i = m(\omega_i)\Delta\omega_i$. With the latter two definitions, A_i represents the *probability* of being in the interval $(\omega_i, \omega_i + \Delta\omega_i)$, and the problem properly shifts from finding a probability density to assigning probabilities.

Each of the three solutions discussed in Section 12.3.3 requires maximizing

$$Q(A) = Q(A_1, A_2, \ldots, A_N)$$

$$= \alpha S(A_1, A_2, \ldots, A_N) - \frac{1}{2}\sum_{i=1}^{L}\left(\frac{\bar{G}_i - \sum_{j=1}^{N} K_{ij}A_j}{\sigma_i}\right)^2 \tag{12.39}$$

for a given value of α. (For notational convenience, we dropped the primes on the $\bar{G}_i$.) The core strategy for each of these solutions is illustrated in Fig. 12.2. The contours with solid lines are values of the χ^2 misfit function, and those with dotted lines are isoentropic values. For large values of α, the solution is dominated by the default model, with a likely poor fit to the data, while for small values it is dominated by the least-squares fit, with likely a tight fit. Starting with a large value

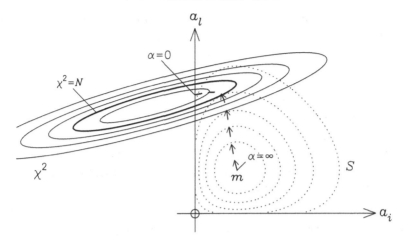

Figure 12.2 Schematic illustration of a solution trajectory. It starts with a default model m and takes small steps in α in such a way to reduce χ^2 while keeping the entropy S as large as possible (from Sivia and Skilling (2006)).

of α, we seek to take small steps reducing the misfit to the data and increasing the entropy (that is, moving it away from the default model). The concavity of both the misfit function and entropy means the solution is unique. We do the maximization by using a Levenberg-Marquardt algorithm (Golub and Loan, 1989; Press et al., 2007) after recasting Q to a form that reduces the complexity of the problem. We give the details of this algorithm in Appendix O. Here we summarize the main points of the overall procedure.

On the one hand, finding the maximum of Q is simple because it is unique, but on the other hand, it is touchy because the maximum is broad. This broadness is a reflection of $N_{\text{good}} \ll N$. The maximization method, however, is not directly applied to the original parameters A_i but rather to a smaller number of new parameters u_i that emerge after the problem is transformed into what is called the *dominant subspace*.

Normally, there may be a thousand or so values of $\bar{G}_i$ and a few times that for A_i. If we were to do a singular value composition (Golub and Loan, 1989; Press et al., 2007) of the transpose of the kernel, that is, $K^T = U \cdot D \cdot V^T$, where U is an $L \times N$ orthogonal matrix, D is an $N \times N$ diagonal matrix, and V is an $N \times N$ orthogonal matrix, then we would find that most of the diagonal elements of D, the singular values d_i^2 ordered from largest to smallest, lose all or almost all numerical accuracy. If s is the number of surviving accurate singular values, then the standard procedure is to use the factorization $K^T = \bar{U} \cdot \bar{D} \cdot \bar{V}^T$ where $\bar{U}$ is an $L \times s$ matrix obtained by retaining only the first s columns of the original U matrix, $\bar{D}$ is an $s \times s$ diagonal matrix, obtained by retaining only the first s columns and rows of the original D, and $\bar{V}$ is an $N \times s$ matrix, obtained by retaining only the first s columns of the original V. The value of s is typically smaller than 10.

With this decomposition,

$$\bar{G} = K \cdot A \rightarrow (\bar{U}^T \cdot \bar{G}) = \bar{D} \cdot (\bar{V}^T \cdot A). \tag{12.40}$$

While this expression does not represent the procedure, it does illustrate that the severe ill-conditioned nature of the kernel reduces the effective dimension of the search space, the space of the latent parameters for the global maximum, to s or smaller. This is the dominant subspace. It also illustrates that the number of "good" parameters in this space is also s, that is, the number of elements of $(V^T \cdot A)$. The Levenberg-Marquardt method, a variant of Newton's method, is applied in this much smaller space whose dimension is principally set by the dimension of the dominant subspace of the kernel. Following Bryan (1990), we adopt a spectral function parameterized as

$$A_i = m_i \exp\left(\sum_{j=1}^{s} \bar{U}_{ij} u_j \right) \tag{12.41}$$

and Q is maximized with respect to the s parameters u_i. Because of the small dimension of the dominant subspace, for a given value of α, the maximization executes rapidly.

We can use this numerical procedure for finding the maximum of Q for a given α to obtain the constrained fit (historic maximum entropy solution) by applying a bisection technique to $\frac{1}{2}\chi(\alpha) - N$. We simply start with acceptable values of α small and large and with its value at the midpoint, and successively halve the intervals, until we locate the one containing $\frac{1}{2}\chi^2 - N = 0$. Convergence is quick. The Bayesian constrained fit (classic maximum entropy) requires the maximum of Q at the value of α that maximizes $P(\alpha|\bar{G})$ (12.33). Coupling a line search (Press et al., 2007) for this maximum with the procedure for maximizing Q is straightforward. The average spectrum method requires the numerical computation of an integral. Simpson's rule works well.

For the purpose of discussion, in Algorithm 41 we give a simpler description of the computational procedures. It is simpler in the sense that it does not compute explicitly any of the three solutions but rather computes the information needed to compute them. The information generated is the $A(\alpha)$ that maximize $Q(\alpha)$ for a given value of α, the misfit statistic $\chi^2(\alpha)$, and the probability $P(\alpha|\bar{G})$ associated with $A(\alpha)$. A graph of $\chi^2(\alpha)$ versus α enables a simple estimate of the value of α for which $\chi^2 \approx N$. Similarly, a graph of $P(\alpha|\bar{G})$ versus α enables an estimate of the $\hat{\alpha}$ that maximizes the conditional probability. Then, the constrained Bayesian solution $A(\hat{\alpha})$ is read from this table. With the tables, estimating the integral (12.35) is also

Algorithm 41 Core analytic continuation.

Input: Input $\bar{G}_i$, C_{ij}, K_{ij}, A_i, m_i, and ω_i. Specify $P(\alpha)$.

Diagonalize C and transform $\bar{G}$ and K^T to this basis ;

Perform a singular value decomposition on K and determine the size of the dominant subspace. Transform $\bar{G}$ to this basis (12.40) ;

Solve (12.41) for a set of starting parameters u_i ;

Choose a large starting α, a decrement $\Delta\alpha$, and number N of α values ;

for $i = 1$ to N **do**

    Find the u_i that maximizes $Q(\alpha)$ (Appendix O) ;

    From (12.41) compute $A_i(\alpha)$;

    Compute $\chi^2(\alpha)$;

    Compute $P(\alpha|\bar{G})$;

    $\alpha \leftarrow \alpha - \Delta\alpha$;

end for

return $A_i(\alpha)$, $\chi^2(\alpha)$, and $P(\alpha|\bar{G})$ as a function of α.

straightforward. However, this simpler approach to the solution is no substitute for the procedures mentioned in the previous paragraph.

In the algorithm, we see that first steps are inputting information generated in the "qualifying the data" step of the analysis and transforming it for use in the minimization procedure. We first define a grid of τ and ω values to discretize the integral equation (12.38), taking care of possible symmetries in the kernel. This same grid discretizes $A(\omega)$ and $m(\omega)$ as probabilities over intervals on this grid. We also need to choose the prior probability $P(\alpha)$, usually the flat or Jeffery's prior.

The key part of qualifying the data is accurately estimating the covariance matrix C. The first step in the maximization algorithm is diagonalizing C, and then transforming the Green's function data and the kernel to this space. Next, the transpose of the transformed kernel is factorized by a singular value decomposition, and the dimension s of the dominant subspace is determined. The matrix $\bar{U}$ from this factorization (used in (12.41)), with an initial guess for A and the choice of the default model, yields a starting point u for the algorithm that maximizes $Q(\alpha)$ for a sequence of decreasing values of α. Appendix O describes a specialized form of the Levenberg-Marquardt method that does this. It is similar to the Newton method's use for optimizing trial wave functions in the variational Monte Carlo method (Section 9.3) but is more involved as it exploits the fact that the ill-posed nature of the problem makes the effective parameter space much smaller than the number of parameters being fitted. Once a u is found that maximizes Q, it is a simple matter to find the corresponding A and to compute $\chi^2(\alpha)$ and $P(\alpha|\bar{G})$. The value of u is used as the starting point for the solution for a smaller value of α. The algorithm returns the saved values of α, $A_i(\alpha)$, $\chi^2(\alpha)$, and $P(\alpha|\bar{G})$.

Computing all three solutions generates a comparative basis for confidence in the average spectrum result. There are several other things that we can do. One is studying how the solutions change if less or more statistically independent measurements are used. Another is reducing the number of τ values used. The imaginary-time Green's functions are very smooth, U-shaped curves. This behavior means successive τ values are correlated. The natural tendency is to improve the solution by increasing the number of τ values. However, because of the correlations, increasing the number of τ values only slowly adds new information to the solution.

3. With the solution stable relative to the quality of the numerical input, the third and final step of our procedure is to assess the acceptance of the solution with respect to its sensitivity to the default model and to estimate its errors. Does the result have a feature not present in the default model or does it lack such a feature? How does the situation change if more data or less data are used? If the result and the model have the same features, how does the result change if more or less data are used? Good solutions exhibit relative independence to the choice of the model. How does the result change if we change the model? For example, if the model were a Gaussian centered at a location where a single peak in the spectrum may be expected, we can ask how much the result would vary if we were to change the peak position and width of the Gaussian. A flat default model is the least commitment we can make about our prior knowledge. If we switch from a more physically motivated model to a flat one, what are the differences? Do we believe they are significant? While the default model can serve to represent our state of prior knowledge about the solution, one that is too informative (for example, the exact solution) can be counterproductive. In the absence of data the solution defaults to the model. Data pull the solution from the model toward one of many possible results consistent with the data. If the data do not support details of the model (or the exact solution), the result might prove difficult to accept.

We can perform a limited form of error estimation. We need to forgo the concept of an "error bar." We cannot estimate the error associated with individual values of A_i, because we do not know the amount of correlations between points,[3] but rather we can estimate the error in functions $f(\omega)A(\omega)$ integrated over a frequency interval. We recall the definitions of the posterior probability (12.24) and the covariance of the spectral density (12.31). With this covariance, we compute the error associated with the measurement

$$F = \int d\omega f(\omega)A(\omega)$$

---

[3] The computed $A(\omega)$ is smooth and hence correlations exist among neighboring values of ω.

of some function of frequency $f(\omega)$ via

$$\langle F^2 \rangle = \iint d\omega d\omega' f(\omega) f(\omega') \langle \delta A(\omega) \delta A(\omega') \rangle . \qquad (12.42)$$

This follows from the quadratic approximation, $P\left(A|\bar{G}\right) \propto e^{-\frac{1}{2}\delta A^T \cdot \nabla \nabla Q \cdot \delta A}$. We can readily approximate the expectation $\langle \delta A \delta A \rangle$ from the inverse of the Hessian (Appendix O, (O.3)) after the Levenberg-Marquardt maximization converges. If the function $f(\omega)$ is unity, then we can estimate the errors associated with regions of $A(\omega)$. Doing so is useful for establishing confidence in peaks and shoulders in the solution.

12.6 Comments

Admittedly, our discussion has been abstract. As we noted, several reviews give a detailed case study of the methods (Jarrell and Gubernatis, 1996; Jarrell et al., 2008). Our intent was to provide more details of the basis of the methods and to highlight the assumptions and approximations.

In our presentation we opted to describe the three Bayesian-based solutions by names that differ from their original presentations. Often, the phrase "the maximum entropy method" is used for all three, obscuring their differences and leaving vague which one is actually being used. The historical development of these three solutions is a progressive sequence refining what to do about the parameter α. From one point of view α is a Lagrange multiplier, and fixing such multipliers is a challenge in almost all constrained least-squares problems. Our use of the term "average spectrum method" is similar in spirit but different in detail from the average spectrum method proposed by several authors, for example, Syljuåsen (2008), whose solution is

$$\langle A \rangle = \frac{\int \mathcal{D}A \, A P(A|\bar{G})}{\int \mathcal{D}A \, P(A|\bar{G})}.$$

Typically, the averages in these proposals are computed by a Monte Carlo evaluation of the integrals, and different strategies, sometimes ad hoc ones, are used to fix α. A number of these issues have been reviewed and discussed by Fuchs et al. (2010). *In many respects, the broader issue is better estimating the evidence.* We approximated it. Doing this estimate via Monte Carlo (or other means) is an active research topic in Bayesian analysis (Friel and Wyse, 2012). The method of nested sampling has become widely used within certain communities (Sivia and Skilling, 2006, chapter 9).

Still other variations and alternatives to the three solutions described can be found in the literature. Limited space here prevents their review and assessment. In general, these publications address the perceived complexity of the procedure, its tendency to produce smooth results, and its difficulty with having multiple peak structures and peaks at high frequencies. Clearly, the described procedures are not absolute, but care is needed in distinguishing between them and alternatives targeted for more specific problems or burdened with shortcuts and misconceptions. Definitely, the solution procedures are not black boxes; that is, piping the quantum Monte Carlo output into any of them will not automatically produce an acceptable result. In some sense, the three procedures are hypotheses, testable by refining the input (the quantum Monte Carlo data and prior information), approximations, and assumptions. Hopefully, the just completed discussion imparts the understanding and insight to evaluate the past and promote future advances.

The presented analysis makes Gaussian approximations to multiple integrands, rendering exact expressions for the resulting integrals. The main purpose was to develop insight into what makes the solution difficult. A consequence was the ability to develop an algorithm that executes in a trivial amount of computer time. Proposed Monte Carlo approaches to analytical continuation are devoid of these approximations but require computer time that is a significant fraction of the time needed to generate the quantum Monte Carlo data. If the Gaussian approximations become invalid, then the numerical algorithm can yield unreliable results, and Monte Carlo solutions are the only hope. We note that in another field, for a quite different problem, the breakdown of these approximations has been reported (Skilling, 1998).

The Bayesian approach and the principle of maximum entropy have been used in a variety of quantum Monte Carlo contexts for purposes other than finding a spectral density. We note the use of this approach for extracting thermodynamics (Huscroft et al., 2000), ground state gaps (Cafferel and Ceperley, 1992), and excited states (Blume et al., 1997, 1998). Analytic continuation in the presence of a sign problem is discussed in Jarrell et al. (2008).

Suggested reading

M. Jarrell and J. E. Gubernatis, "Bayesian inference and the analytic continuation of imaginary-time quantum Monte Carlo data," *Phys. Rept.* **269**, 133 (1996).

D. S. Sivia and J. Skilling, *Data Analysis: A Bayesian Tutorial* (Oxford University Press, 2006).

Exercises

12.1 What distribution function results from maximizing the entropy subject to the constraints of normalization and knowledge of the mean? Recall that a Gaussian results if the constraints also include the variance.

12.2 Assign a probability p_i to each face of a die by maximizing the entropy $S = -\sum_{i=1}^{6} p_i \log p_i$ subject to the constraints $\sum_{i=1}^{6} p_i = 1$ and $\sum_{i=1}^{6} i p_i = 3.5$. The latter is the average face value if each of the six faces is equally likely.

1. How does the solution change if $3.5 \rightarrow 3.5 \pm 0.5$?
2. Will any prior knowledge of the mean give roughly $1/6$ for all p_i?

12.3 Construct a discrete probability $p = (p_1, p_2, \ldots, p_N)$ that has two prominent peaks, and set the default model $m = (m_1, \ldots, m_N)$ equal to it. Keeping m fixed, permute the indices of p and study how values of $S = -\sum_i p_i \ln(p_i/m_i)$ change as the peak positions in the permuted p move relative to those in m. Note that $S' = -\sum_i p_i \ln p_i$ is invariant to the permutations and that S takes a minimum when $p = m$.

12.4 Consider the inverse problem

$$\begin{pmatrix} 11 \\ 2 \end{pmatrix} = \begin{pmatrix} 10 & 1 \\ 1 & 0.1 \end{pmatrix} \begin{pmatrix} x_1 \\ x_2 \end{pmatrix},$$

where x_1 and x_2 are known to be positive, but not normalized, and the value of x_1 has a standard deviation of 0.01 and x_2 has one of 1. Using a flat default model, compare the exact values of x_1 and x_2 with the ones that maximize $Q = \alpha S - \frac{1}{2}\chi^2$ as a function of α obtained using a quadratic approximation to S. The quadratic approximation reduces the maximization problem to a linear one. Using a quadratic prior probability is called *Tikhonov regularization*.

12.5 Repeat the above problem using the exact expression for S.

12.6 Show that the alternative expression for the entropy,

$$S(A) = \sum_i \left[A_i - m_i - A_i \log\left(\frac{A_i}{m_i}\right) \right], \qquad (12.43)$$

allows the normalization conditions on A_i and m_i to be relaxed.

12.7 Show that the entropy of a multivariate Gaussian is $\frac{1}{2}\ln[(2\pi e)^N \det C]$.

12.8 A third of all kangaroos have blue eyes and a quarter of all kangaroos are left-handed. On the basis of this information alone, what proportion are both blue-eyed and left-handed (Gull and Skilling, 1984)? This information alone will not allow a unique answer. Select an answer based on constraining the problem with (1) $-\sum_i p_i$ and (2) $-\sum_i p_i^2$, subject to the constraints of the available information. We have no prior knowledge that the handedness and eye color are correlated, so it is reasonable to expect that for a given eye color, left- and right-handed kangaroos are equally likely. Which functional form does not inject correlations that are not part of the prior information? Now constrain with (1) $\sum_i \log p_i$ and (2) $\sum_i \sqrt{p_i}$. How do the correlations change?

13

Parallelization

Recent trends in computer hardware have made modern computers parallel computers; that is, the power of today's computers depends on the number of processors.[1] In order to use these powerful machines, we must split the algorithmic tasks into pieces and assign them to a large number of processors. In many cases, this requires nontrivial modifications of the algorithm. In this chapter, we discuss several basic concepts about parallelizing an algorithm and illustrate them in the context of loop/cluster identification.

13.1 Parallel architectures

The key issue in parallel computation is the distribution of computer memory. In other words, how much memory is available and at what access speed. Memories are organized in a hierarchical structure: L1 cache, L2 cache, main memory, and so on, all with different access speeds. The access speed also depends on the physical distance between the computing unit and the memory block.[2]

Discussing how to fine-tune computer programs, taking all machine details into account, is clearly beyond the scope of this book. Therefore, in the following, we focus our discussion of parallel computers on two common types of architectures (Fig. 13.1): shared memory and distributed memory. In either case, we assume that the parallel computer has N_p processors.[3]

In most parallel computers available today, each local memory block is directly accessible by only a small number of processors in the local processor block which is physically closest to it. To access a remote block of memory, a processor

---

[1] In 2016, it is of the order of 10^6 for the most powerful machines.

[2] If a computer operates with the clock rate of 1 GHz, a photon can travel only 30 cm in a clock cycle. Therefore, if two processors are separated by 10 meters, which is not unrealistic for big parallel machines, they have to wait at least 30 clock cycles for a message to travel from one to the other, no matter what technology is used.

[3] Here a "processor" means the smallest computing unit. In the case where a central processing unit (CPU) consists of multiple processors, that is, *cores*, our use of "processor" means a core rather than a CPU.

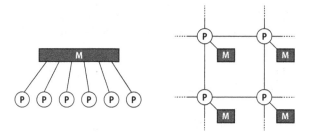

Figure 13.1 The shared-memory model (left) and the distributed-memory model (right). Circles with "P" represent processors, and rectangles with "M" the memory.

must communicate with another processor that has direct access to this block. In some sense, the shared-memory architecture is a model for a local processor block. Alternatively, it can be regarded as a model of the whole computer system in which the communication cost is negligible. In this model, we do not have to account for the process of communicating between the processors. We simply assume that they are all reading and writing to the same block of memory, and each processor immediately knows when something is written to memory by another. The distributed-memory architecture, on the other hand, is a model in which every processor monopolizes the access to the block of memory that it possesses, and for a processor to get the information written on another processor's memory, the owner of the information must explicitly send it the content of this memory. Because today's high-performance computers have aspects of both architectures, two different parallelization approaches might be used at the same time, a process called *hybrid* parallelization.

13.2 Single-spin update on a shared-memory computer

Let us start with a simple problem: parallelizing the single-spin update algorithm for an Ising chain (Chapter 4.2) on a shared-memory computer. As for the size of the memory, we assume that the available memory is infinitely large, which is not unrealistic because Monte Carlo simulations are often computation limited, not memory limited. We may thus assume that the information about the spin configuration of the whole lattice fits into the shared memory.

The most natural way to split the task is to divide the set of spins into N_p subsets $\Omega_1, \Omega_2, \ldots, \Omega_{N_p}$ and assign a subset to each processor (Algorithm 42). Let M be the size of Ω_i, that is, $M \equiv N/N_p$, where N is the total number of spins. The p-th processor performs Algorithm 6 with some fixed order of the spins in Ω_p instead of random picking. The probability of the new spin state relative to the state with the chosen spin s_i satisfies the detailed balance condition.

Algorithm 42 Single-spin update of the Ising model: the p-th processor's task on a shared-memory computer. (Error prone on a parallel computer.)

Input: Spin configuration in block p, $\{s_i = \pm 1\}$ ($i \in \Omega_p$), and values of neighboring spins.

    for $k = 1, 2, \ldots, M (\equiv N/N_p)$ **do**

        Let i be the k-th member of Ω_p ;

        Change s_i to the new state chosen probabilistically ;

    end for

    return the updated spin configuration in block p.

Algorithm 42 yields correct results (up to the statistical error) as long as the operations do not interfere with each other. However, since this independency condition does not hold in general, naive parallelization may yield erroneous results. When a processor, say, P_1, tries to update spin s_i in its domain, it needs the information about the spins surrounding s_i. However, some neighboring spin, say, s_j, may belong to a different processor, say, P_2. The processor P_1 obtains the information about s_j at the beginning of the update cycle described in Algorithm 42. The problem is that this information is not updated until the cycle completes on all processors and they communicate with each other. Therefore, the value of s_j used by P_1 for updating s_i may have been changed by P_2 when P_1 starts to work on s_i. As a simple example, let us consider a small system with only two ferromagnetically coupled Ising spins. Suppose we assign one spin to each of the two processors, and apply Algorithm 42 with the Metropolis update as in Algorithm 6, which always flips the spins whenever this flip lowers the energy. What will happen? Once these spins become antiparallel at some point of time, they will remain antiparallel forever![4]

To avoid this problem,[5] we must carefully order the examination of spins in Ω_p and ensure that the processors are appropriately synchronized. In the case of the one-dimensional Ising model, we replace the somewhat generic algorithm just described by Algorithm 43, illustrated in Fig. 13.2. This new procedure ensures that neighboring spins are never updated simultaneously. In the figure, the update order of the solid circles covered by the same processor might affect the efficiency, but it should not cause systematic errors. At the end of each step, we synchronize all processors to avoid mixing open and solid circles; that is, we let the proces-

---

[4] At the beginning of the cycle, each processor obtains the information that the neighboring spin is antiparallel, and therefore judges that flipping its own spin will lower the energy. Because both processors flip their spins, the spin configuration will stay antiparallel.

[5] This particular type of problem is caused by what is called a *race condition*, as the result of the operation is determined by a race between two or more processors.

Algorithm 43 Single-spin update of the one-dimensional Ising model: the p-th processor's task on a shared-memory computer. (Free from error even when $M > 1$.)

Input: Spin configuration in block p, $s_i = \pm 1$ ($M(p-1)+1 \leq i \leq Mp$), and value of neighboring spins.

  for $k = 1, 2, 3, \ldots, M/2$ **do**

    $i \leftarrow M(p-1)+k$;

    Update s_i ;   ▷ e.g., by the Metropolis algorithm

  end for

  Synchronize ;   ▷ Wait for the other processors to finish their tasks

  for $k = M/2+1, M/2+2, \ldots, M$ **do**

    $i \leftarrow M(p-1)+k$;

    Update s_i ;

  end for

  Synchronize ;

  return the updated spin configuration in block p.

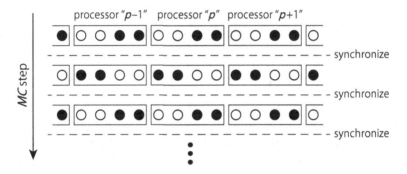

Figure 13.2 The parallelized single-spin update. Each processor takes care of four spins in this example, that is, $M = 4$. In each step, only the spins represented by open circles are updated so that each processor can carry out its task independently. At the end of each step, all the processors are synchronized, and in the case of the distributed memory machine, the value of the boundary spin is passed to the neighboring processor.

sors wait until all reach the same point in the algorithm.[6] Extensions to higher dimensions are obvious: We split each processor's domain into a few subdomains (domain decomposition) in such a way that two neighboring spins are not examined simultaneously.

---

[6] While synchronization resolves the issue in the present example, the use of this procedure becomes less and less efficient as the computer has more and more processors.

Algorithm 44 Single-spin update of the one-dimensional Ising model: the p-th processor's task on a distributed-memory computer.

Input: Spin configuration in block p, $s_i = \pm 1$ $(M(p-1)+1 \leq i \leq Mp)$.

    Send s_{Mp} to the $(p+1)$-th processor and receive $s_{M(p-1)}$ from the $(p-1)$-th processor ;

    for $k = 1,2,3,\ldots, M/2$ **do**

        $i \leftarrow M(p-1)+k$;

        Update s_i ;

    end for

    Synchronize ;

    Send $s_{M(p-1)+1}$ to the $(p-1)$-th processor and receive s_{Mp+1} from the $(p+1)$-th processor ;

    for $k = M/2+1, M/2+2,\ldots, M$ **do**

        $i \leftarrow M(p-1)+k$;

        Update s_i ;

    end for

    Synchronize ;

    return the updated spin configuration in block p.

13.3 Single-spin update on a distributed-memory computer

In the distributed-memory computing model, a processor might not have all the information necessary for executing its task. Therefore, it must ask other processors, usually its neighbors, for the missing information. In the Ising model example, a processor lacks necessary information when it tries to update a spin on the boundary of its domain. Therefore, it must issue an inquiry to its neighbor before it can continue with its task. The same is true for all other processors. As is evident from Algorithm 44, these inquiries generate frequent exchanges of information.

Accordingly, we now have to pay special attention to the cost of the communication. Roughly speaking, if the time needed for interprocessor communication is shorter than the time required for each processor to execute its operations, we say that the process is well balanced, because we will benefit from increasing the number of processors as long as this condition remains true.[7]

Let us consider the performance of the computation when the problem size (that is, the number of spins N in the present case) increases while the number of

---

[7] In such a case, we can even effectively eliminate the communication time by so-called *communication-latency hiding*. Namely, by exploiting the fact that the calculating unit and the communicating unit can work simultaneously and independently from each other, a processor can update spins while sending/receiving the information to/from other processors.

processors remains fixed, making the size of the subsystem assigned to each processor proportional to the problem size. In the Ising chain example, the communication time is always $\mathcal{O}(1)$ as each processor sends and receives only two spins. The number of operations executed by each processor, on the other hand, is proportional to $M = N/N_p$. Therefore, the ratio of the interprocessor communication time to the operation time within each processor is $R = \mathcal{O}(M^{-1})$. It is easy to generalize this estimate to higher dimensions. Now, M is the length scale of each processor's domain, which is a d-dimensional hypercube, for example. Then, each processor must send and receive the values of the spins on the boundary, whose area scales as M^{d-1}. The communication time is therefore proportional to M^{d-1}. The operation time is again proportional to the number of spins covered by a processor, which is M^d. Therefore, the communication to operation ratio is

$$R = \mathcal{O}(M^{-1})$$

regardless of the dimension. From this we can conclude that the interprocessor communication time is negligible, as long as the decomposition is sufficiently coarse.[8] If, on the contrary, the size of each domain is too small, the relative communication cost can be large and we cannot benefit from increasing the number of processors. Simply put, *large computers are efficient only for large problems*. This statement also applies to other parallelized algorithms based on domain decomposition.

13.4 Loop/cluster update and union-find algorithm

We now consider the loop/cluster algorithm. We recall that it consists of two steps: graph assignment and cluster flips. Our aim is to split space-time into N_p regions and assign a processor to each region. Each processor therefore has the complete information about the region assigned to it, such as the space-time positions of the vertices, the local state of a given segment delimited by kinks, and the state at the boundary of the region.

We first consider the parallelization of the graph assignment step. If the space-time decomposition is only temporal and no boundary separates two regions spatially, interprocessor communication is absent because both ends of a graph element, which is local in time, always belong to the same region. On the other hand, when the decomposition is not purely temporal, we may have to place a graph element across the boundary between two regions, say, A and B. In this case, we must transfer information about the local state near the boundary of region B to the processor in charge of A. The A processor then computes the temporal positions

---

[8] Whether we can make the space decomposition sufficiently coarse in this sense, of course, depends on whether the original problem size is large enough for the number of processors available.

Problem   Solution

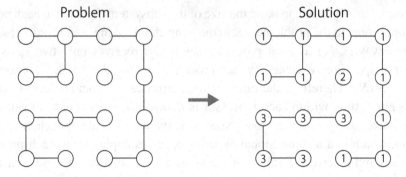

Figure 13.3 The cluster identification problem.

of the new vertices and sends this information back to the B processor. Therefore, the amount of interprocessor communication is proportional to the number of the vertices between A and B, which in turn is proportional to the length of the spatial boundary separating them.

Now we consider the parallelization of the loop/cluster spin flips, which is much less straightforward because of the nonlocal nature of the cluster identification. Because we will use graph manipulation algorithms for this identification, we adopt a graph theoretical terminology. Since the essence of the difficulty does not depend on whether it is a classical model or a quantum model, let us consider the classical model for the sake of clarity of the description. In this condition, a vertex is a point and an edge is a line connecting two points.

To update the spins we need to identify clusters. To be more specific, our task is to obtain an array, say, p, that defines a mapping from the vertex number to the cluster number. For example, $p(3) = 5$ means that the vertex 3 belongs to the cluster 5. The cluster number is such that $p(v_a) = p(v_b)$ if and only if the vertices v_a and v_b are connected by a sequence of edges (Fig. 13.3). In the cluster algorithm for the Ising model, once we obtain such an array, we update the spin variables simply by generating one-bit random numbers, $b(1), b(2), \ldots, b(n_c)$, that represent the spin values of the n_c clusters and assign $b(p(v))$ to each vertex v as its new spin value.

We first discuss the serial version of this union-find algorithm (Hoshen and Kopelman, 1976). To construct the array $p(v)$ for a given list of edges, we grow a forest of trees[9] such that each tree corresponds to a cluster to be identified. Every component of the tree (the root, a branching point, or a leaf) corresponds to a vertex. During the grafting operations we discuss below, $p(v)$ is always a pointer to the "parent" vertex, that is, the vertex at the next step along the path toward the root. We suppose the vertices are sequentially numbered, and their number is fixed.

---

[9] If a graph has N_v vertices, it is a tree if and only if it has $N_v - 1$ edges. For a tree the path between any two vertices is unique. In other words, a tree does not contain any loops.

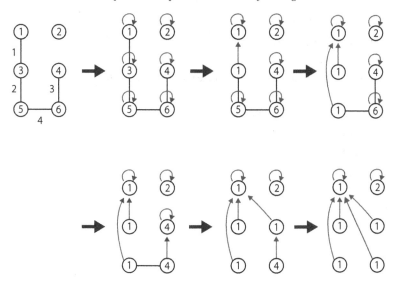

Figure 13.4 The "grafting" algorithm for the union-find problem (Algorithm 45). The numbers in the circles are $p(v)$ in the algorithm. The edges are numbered in the order in which they are examined.

In the initial state, every vertex is the root of a tree (second diagram in Fig. 13.4). In other words, we have N trees of height zero. We grow the trees by feeding them information about edges, one by one.

Suppose an edge e, the first in the list of unchecked edges, connects two vertices v_a and v_b, that is, $e = (v_a, v_b)$. Then we start from v_a and trace the branches until we reach the root of the tree, say, r_a. We do the same starting from v_b, reaching the root r_b. If r_a and r_b differ, v_a and v_b are on different trees. In this case, one tree must be grafted onto the other. If the vertex number of r_a is smaller than that of r_b, we graft tree b onto tree a as a branch connecting b directly to the root of a. Otherwise, we graft tree a onto tree b in the same manner. As is obvious from this procedure, after examining all edges, there must be a one-to-one correspondence between the trees and the clusters. In other words, any two vertices that are connected by a sequence of edges belong to the same tree. Once all trees are completed, we start from each vertex and trace the branches to the root and then assign the root's vertex number to the starting vertex as its cluster number. This procedure is summarized in Algorithm 45.

The theoretical upper bound for the number of operations required to execute the procedures of Algorithm 45 is $\mathcal{O}(N_E^2)$, where N_E is the number of edges. The worst case is an edge list of the type $(100, 99), (99, 98), (98, 97), \ldots, (2, 1)$. Here the order is important: If the edges are examined in this order, the content of the array $p(v)$ at the end of the grafting phase is $p(1) = 1, p(2) = 1, p(3) = 2, p(4) = 3, \ldots,$ $p(100) = 99$. Then, in the height reduction phase, we need to trace this sequence

---

Algorithm 45 Hoshen-Kopelman union-find algorithm (serial).

Input: List of edges $\{(v_a, v_b)\}$

    for every vertex v **do**

        $p(v) \leftarrow v$;   ▷ Initially, every vertex points to itself.

    end for

    for every edge $e = (v_a, v_b)$ **do**

        $r_a \leftarrow \text{Root}(v_a)$;

        $r_b \leftarrow \text{Root}(v_b)$;

        if $r_a < r_b$ **then**

            $p(r_b) \leftarrow r_a$;

        else

            $p(r_a) \leftarrow r_b$;

        end if

    end for

    for every v **do**

        $r \leftarrow \text{Root}(v)$;

        $p(v) \leftarrow r$;

    end for

    return p.

    function Root(v)

    $u \leftarrow v$;

    while $p(u) \neq u$ **do**

        $u \leftarrow p(u)$;

    end while

    return u.

---

to its root. Depending on the order in which the vertices are examined, the total number of operations varies. The worst case occurs when the vertices are examined in decending order, that is, 100, 99, ..., 2, 1. When we start from vertex 100, we need 99 replacement operations to find that the root of this vertex is 1. Then, $p(100)$ is set to 1. Next we start from vertex 99, and after 98 replacements we finish the task. The computational complexity for these operations will be $\mathcal{O}(N_V N_E)$ with N_V being the number of vertices. Because $O(N_V) = O(N_E)$ in most applications in condensed matter physics, we obtain a computational complexity of $\mathcal{O}(N_E^2)$.[10] This cost may completely invalidate the advantage of the loop/cluster procedure.

---

[10] When we examine the vertices in ascending order, we can always reach the root within two steps, which reduces the computational complexity to $\mathcal{O}(N_E)$. However, in a parallel implementation it may not be possible to control the order of the vertices.

Algorithm 46 Improved union-find algorithm (serial).

Input: Edge list $\{(v_a, v_b)\}$.

  for every vertex v **do**

    $p(v) \leftarrow v$;

    $w(v) \leftarrow 1$;

  end for

  for every edge $e = (v_a, v_b)$ **do**

    $r_a \leftarrow \text{Root}(v_a)$;

    $r_b \leftarrow \text{Root}(v_b)$;

    if $w(r_a) \geq w(r_b)$ **then**

      $r_0 \leftarrow r_a$;

      $r_1 \leftarrow r_b$;

    else

      $r_0 \leftarrow r_b$;

      $r_1 \leftarrow r_a$;

    end if

    $w(r_0) \leftarrow w(r_0) + w(r_1)$;   ▷ Updating the weights

    $p(r_1) \leftarrow r_0$;   ▷ The small tree grafted on the large

    $\text{Compress}(v_a, r_0)$;   ▷ Compressing one tree

    $\text{Compress}(v_b, r_0)$;   ▷ Compressing the other

  end for

  for every v **do**

    $r \leftarrow \text{Root}(v)$;

    $\text{Compress}(v, r)$;

  end for

  return p.

  function $\text{Root}(v)$

  $u \leftarrow v$;

  while $p(u) \neq u$ **do**

    $u \leftarrow p(u)$;

  end while

  return u.

  function $\text{Compress}(v, r)$

  $u \leftarrow v$;

  while $p(u) \neq u$ **do**

    $s \leftarrow p(u)$;

    $p(u) \leftarrow r$;

    $u \leftarrow s$;

  end while

  return

To reduce the computational cost, we make two improvements (Aho et al., 1983). One is to introduce a weight $w(v)$ for the tree defined as the number of vertices included in the tree, and when grafting always grafts the smaller weighted tree on the larger one. The other is to reduce the height of the trees frequently. The improved algorithm is Algorithm 46. Its complexity is $\mathcal{O}((N_E+N)\lg N)$ (Aho et al., 1983) where $\lg X$ is the inverse Ackerman function, which increases very slowly. Roughly speaking, it is the number of times we have to take $\log_2$ of X to obtain 1, for example, $\lg 1 = 0$, $\lg 2 = 1$, $\lg 2^2 = 2$, $\lg 2^{2^2} = 3$, $\lg 2^{2^{2^2}} = 4$, etc. For practical purposes, we can regard it as a constant.

13.5 Union-find algorithm for shared-memory computers

We now discuss the shared-memory parallelization of the union-find algorithm using the prescription proposed in Todo et al. (2012). It breaks the set of edges into smaller subsets and assigns a subset to each processor. First, let us imagine that each processor simply performs its task with Algorithm 45 or Algorithm 46. Obviously, we do not have control of the order in which the edges are examined. At first glance, this does not seem to matter, since the final outcome (clustering) should not depend on the order as long as all edges are taken into account.[11] It is, however, not trivial to ensure that all edges are properly accounted for.

To see this, let us consider a graph consisting of six vertices. Suppose the roots of the vertices 4, 5, and 6 are 1, 2, and 3 (Fig. 13.5) and that processors A and B are about to handle edges $(4,6)$ and $(5,6)$, respectively. We might expect the following sequence of events:

(A1) Processor A finds the root of 6 to be 3.
(A2) Processor A finds the root of 4 to be 1.
(A3) Processor A changes $p(3)$ to 1.
(B1) Processor B finds the root of 6 to be 1.
(B2) Processor B finds the root of 5 to be 2.
(B3) Processor B changes $p(2)$ to 1.

The result is a single cluster, as illustrated in the left-hand side of Fig. 13.5. However, the order of the events need not necessarily be this. The following is equally likely to happen:

---

[11] The order of edges may affect the shape of the tree in the final state, but any two vertices must belong to the same tree as long as all the edges are properly examined, as in Fig. 13.5.

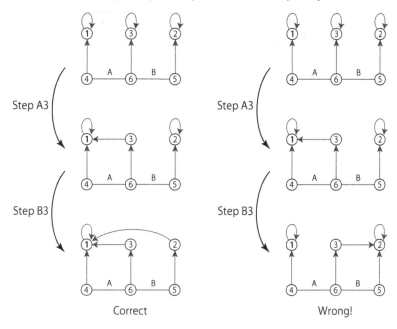

Figure 13.5 Nonthread-safe operations in the union-find algorithm.

(A1) Processor A finds the root of 6 to be 3.
(B1) Processor B finds the root of 6 to be 3.
(A2) Processor A finds the root of 4 to be 1.
(B2) Processor B finds the root of 5 to be 2.
(A3) Processor A changes $p(3)$ to 1.
(B3) Processor B changes $p(3)$ to 2.

Obviously, the information about the edge $(4, 6)$ is lost, as shown in the right-hand
side of Fig. 13.5. It happens since, at (B3) in the latter process, processor B uses
obsolete information about the tree with vertex 6 when it grafts the tree. If no other
edge reconnects the two clusters later, the result will be wrong.

To prevent such an information loss, whenever a processor tries to graft a root
r_a to another root r_b, it must first lock them so that no other processor can change
$p(r_a)$ or $p(r_b)$ before the first processor finishes grafting. One way of modifying
the algorithm is illustrated in Algorithm 47, in which a locked vertex v is encoded
by a negative value of $p(v)$. The function "Lock" must be implemented so that
the operations in the if-clause beginning with "**if** $p(v) = v$ **then**" are executed as a
single step. If not, an interruption from other processors might cause $p(v)$ in the first
operation (comparison, $p(v) = v$) and $p(v)$ in the second operation (substitution,
$p(v) \leftarrow -v$) to differ from each other, thus leading to an error. While the strategy
to avoid such an interruption may depend on the machine, we can do it with a

Algorithm 47 The task of a processor in union-find on a shared-memory computer. The omitted parts are the same as in Algorithm 46.

Input: Sublist of the edge list $E_p \equiv \{(v_a, v_b)\}$.

$\cdots$ *the initialization part omitted* $\cdots$

for every edge $e \in E_p$ **do**

   $r_a \leftarrow \text{Root}(v_a)$;

   $r_b \leftarrow \text{Root}(v_b)$;

   loop

      $q_a \leftarrow \text{Lock}(r_a)$;

      $q_b \leftarrow \text{Lock}(r_b)$;

      if $q_a = $ "success" **and** $q_b = $ "success" **then**

         Break ;

      else

         if $q_a = $ "success" **then**

            $\text{Unlock}(r_a)$;

         end if

         if $q_b = $ "success" **then**

            $\text{Unlock}(r_b)$;

         end if

         $r_a \leftarrow \text{Root}(r_a)$;

         $r_b \leftarrow \text{Root}(r_b)$;

      end if

   end loop

   $\cdots$ *the grafting part omitted* $\cdots$

   $\text{Unlock}(r_a)$;

   $\text{Unlock}(r_b)$;

end for

$\cdots$ *the tree-compression part omitted* $\cdots$

return p

function Lock(v)

   ▷ The following if-clause must be done as a single operation.

if $p(v) = v$ **then**

   $p(v) \leftarrow -v$;

   return "success" ;   ▷ Lock succeeded

end if

return "failure" ;   ▷ Lock failed

function Unlock(v)

$p(v) \leftarrow -p(v)$;

   return

compare-and-swap atomic (noninterruptable) instruction that compares and swaps in a single instruction (Todo et al., 2012).

13.6 Union-find algorithm for distributed-memory computers

We now discuss the parallelization of the union-find algorithm for a distributed-memory computer. Whereas most graph manipulation algorithms assume no particular spatial structure, the finiteness of the space dimension is crucial to the algorithm discussed below. In this sense, the algorithm is not for all union-find applications. We slightly generalize the ideas of Todo et al. (2012), and also describe the identification of clusters, instead of loops.

In this algorithm, space-time is split into domains and a processor is assigned to each space-time domain. A processor's task is to obtain the cluster number, $p(v)$, for all v in its domain, say, Ω_0. It starts the cluster identification using only the information of edges in Ω_0 by one of the algorithms discussed above. When this is done, the cluster identification of the domain Ω_0 is complete within Ω_0. (We say "the cluster identification of the domain A is complete within B" when the cluster numbers assigned to all the vertices in A would be correct if there were no other edges than those within region B.) The processor then takes into account edges within a neighboring region, say, Ω_1, and also edges on the boundary between Ω_0 and Ω_1, and updates the cluster number table. The result is the cluster number table of Ω_0 that is complete within $\Omega_0 \cup \Omega_1$. In this way, the processor keeps doubling the domain for edges until the cluster number table of Ω_0 is complete within Ω_{whole}, the whole space.

Each processor possesses information about the vertices in its initial domain, the vertices on the boundary of its current domain, and the root of trees that include one or more of the boundary vertices. To be more specific, the local memory associated with a processor stores $p(v)$ and $w(v)$ for the vertices in $\Omega_0 \cup \partial\Omega \cup \mathcal{R}(\partial\Omega)$, where Ω_0 is the processor's original domain, Ω and $\partial\Omega$ are the current domain and its boundary, and $\mathcal{R}(\partial\Omega)$ is the set of the roots of the trees that have nonvanishing overlap with $\partial\Omega$. The information on the edges that connect the current domain to others is also stored in memory, as well as information on the edges within the domain. The latter, however, is necessary only for the first iteration. Each processor uses these pieces of information in Algorithms 45, 46, or 47.

Initially, each processor uses the information of edges and vertices within Ω_0 to identify clusters by applying Algorithm 46.[12] However, this cluster identification is clearly incomplete, since two vertices that are disjoint within the domain might con-

---

[12] If each processor has many cores and is actually a shared-memory parallel machine itself, we execute hybrid parallelization by letting each processor use Algorithm 47 instead of the algorithm for the serial computer.

nect via a path that runs outside of the domain. To account for such outside paths, the processor, referred to as P_0 hereafter, must obtain information from others. To do this, it picks one of its nearest-neighbor processors, say, P_1, and asks for the information. It does not, however, need all of the information that P_1 has. P_0's task is to find the cluster numbers of vertices only in Ω_0, and any outside path must run through one of the boundary vertices. Therefore, for P_0, it suffices to obtain the pointers $p(v)$ and the weights $w(v)$ of vertices on the domain boundary between P_0's domain and P_1's and those of their roots. Once P_0 has obtained this information, it examines the edges that connect the two domains and grafts trees by the prescription of Algorithm 46. When finished, the cluster identification of Ω_0 is complete within $\Omega_0 \cup \Omega_1$ where Ω_1 is P_1's original domain. We repeat this procedure in a hierarchical fashion until the domain covered by each processor becomes the whole system.

There are several ways of organizing this hierarchy. Here, we consider one represented by an "escargot" pattern (Fig. 13.6). A processor, P_0, first completes the cluster identification within Ω_0 (the darkest square in the figure). Likewise, all the other processors do their job. Then, P_0 asks its nearest neighbor, P_1, in the x-direction for the cluster numbers of the boundary vertices. Using this information, P_0 completes its cluster identification within the domain whose area is now twice as large (the union of the darkest and the second darkest square). Again, all the others do the same. Next, P_0 asks for the information from its nearest neighbor in the y-direction, P_2. This time, the neighbor has the boundary information of the doubled domain (the horizontal rectangle above the darkest square). As a result, P_0

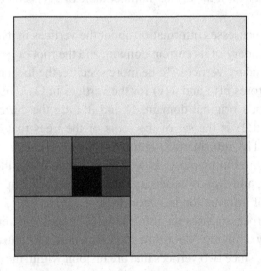

Figure 13.6 The domain covered by a processor. The darkest and smallest square represents the domain that it covers at the beginning. As the operation proceeds by one step, the domain is doubled, and the cluster identification is completed within this domain.

completes the cluster identification for a square twice as large as the initial square in both directions. In the third step, P_0 gets information from its neighbor, P_3, located at position $(-2, 0)$ relative to the original square in units of the original square size. After this, it repeats this procedure with the other processors located at $(0, -2)$, $(4, 0)$, $(0, 4)$, $(-8, 0)$, $(0, -8)$, and so on. In general, in d dimensions, this sequence would be $e_1, e_2, \ldots, e_d, -2e_1, -2e_2, \ldots, -2e_d, 4e_1, 4e_2, \ldots, 4e_d, \ldots$ with e_μ being the unit lattice vector in the μ-th direction.

In this "escargot" arrangement, if the clusters to be identified are small, a processor completes its task before its domain covers the whole system, because as soon as the processor covers all the clusters overlapping with Ω_0, the cluster numbers of vertices in Ω_0 are assigned correctly, and no further change occurs even if the processor keeps searching. This advantage can be significant when the typical cluster size is much smaller than the whole system, which is often the case in the disordered state. Specifically, if the maximum cluster size is ξ, all processors stop as soon as the size of the domain reaches 4ξ.

13.7 Back to the future

In Chapter 1 we recounted that in what is generally regarded as the first publication to use the phrase "Monte Carlo method," Metropolis and Ulam (1949) described the philosophy of this new method and noted its ability to do what we today say is "to break the curse of dimensionality" that often limits the effective numerical solution of partial differential equations for complex problems. Generally unnoticed in this landmark publication is a remark that the Monte Carlo method is ideal for parallel computing, an observation made decades before other scientists started to think about parallel computers and only two years after the now standard von Neumann architecture was proposed. The remark likely concerns the simplest, but unarguably the most efficient parallelization, the natural parallelization whereby we give each processor its own input (at least its own random number seed) and the same executable. Each processor runs the executable independently, and one collects the individual outputs when all processors are finished. The executable may be any of the Monte Carlo algorithms discussed in this book: single-spin update, cluster update, worm update, determinant method, variational Monte Carlo, and the rest. We omitted the discussion of the natural parallelization in this chapter, mainly because it is so natural. However, we emphasize here that natural parallelization is one of the big advantages of the Monte Carlo technique.[13]

---

[13] Instead of being called "naturally parallel," Monte Carlo methods are often called "embarrassingly parallel" because they are an embarrassment to methods such as molecular dynamics or finite element methods that are not naturally parallel.

Ulam and Metropolis were speaking about the Monte Carlo method that Ulam and von Neumann proposed just a few years prior for simulating the diffusion-collision transport of radiation through fissile material. As we also remarked in Chapter 1, Ulam and von Neumann used a Markov chain to predict the distribution of radiation that required sampling from a stationary distribution that is unknown beforehand, and in 1953, Metropolis et al. proposed the Metropolis algorithm, their now famous other form of Monte Carlo that samples from a stationary distribution that is known beforehand. In his recollections about the development of the Metropolis algorithm, Marshall Rosenbluth (2003) relates that they were studying the melting of a solid, and the most natural way to simulate it was to follow the dynamics of the interacting particles under the action of Newton's laws. However, the memory and speed of the computer available to them, Metropolis's MANIAC, were too limited to follow an adequate number of particles for a large number of small time steps, so they had to think of another way. As they were interested only in equilibrium properties, they decided to take advantage of statistical mechanics and ensemble averages instead of following the detailed kinematics. If Metropolis's computer, one of the earliest built with a von Neumann architecture, had been more advanced, they would have invented molecular dynamics instead of the Metropolis algorithm!

Over time, the basic characteristic of Monte Carlo algorithms has not changed as much as the computer hardware on which we use them. Indeed, in the future the Monte Carlo method will still break the curse of dimensionality, will still be naturally parallelizable, and will still require relatively modest core memory. We also know that it usually samples phase space by taking relatively local steps that in many cases promotes processor asynchronization. It needs to pass few messages, and these passings are often local.

If the main obstacle is the statistical noise in the Monte Carlo simulation, and this noise problem is not exponentially hard, the natural parallelization removes it and is the method one should use. However, in some cases, for example, if the computational autocorrelation time is too long to wait for the system to relax to its equilibrium distribution or if the system is too large to store a whole configuration in the memory of a single processor, then we need something that goes beyond natural parallelization.[14] In this chapter, we gave an example of such a nontrivial parallelization of a very important task for a popular class of classical and quantum Monte Carlo algorithms. In particular, the final approach to parallelization of the loop/cluster algorithm discussed in this chapter is a step up from natural parallelization. The parallelization of other types of algorithms is currently an active

---

[14] Another important example is the Fourier transformation.

research topic (see, e.g., Masaki-Kato et al. (2014) for the parallelization of the worm algorithm).

As we remarked in the text, big computers are efficient only for big problems. The visions for the next generation of big computers are ones with orders of magnitude more processors each being not much faster than today's processors. Writing parallel codes for these computers is a challenge for computational physics, but the inherent advantages of the Monte Carlo approach, as well as the development of nontrivial parallelization algorithms, poises Monte Carlo for extensive use in the dawn of exascale computing. How we use the Monte Carlo method will evolve. How we parallelize it will change. What will always remain important is identifying suitable problems that can be tackled successfully with improved algorithms or more powerful computing platforms.

Appendix A
Alias method

The methods to sample from a discrete probability for n events, presented in Algorithms 1 and 2, become inefficient when n is large. To devise an efficient algorithm for large n, we must distinguish the case where the list of probabilities is constantly changing from the case where it remains unchanged. In the latter case, there are several ways to boost efficiency by performing a modestly expensive operation once and then using more efficient operations in subsequent samplings. One such approach is to sort the list of probabilities, an operation generally requiring $\mathcal{O}(n \log n)$ operations, and use a bisection method, requiring $\mathcal{O}(\log n)$ operations, to select the events from the list. What we really want, however, is a method requiring only $\mathcal{O}(1)$ operations. Walker's alias algorithm (Walker, 1977) has this property.

Suppose we need to repeatedly choose one of five colors randomly – red, blue, yellow, pink, and green – with weights 6, 1, 3, 2, and 8 (Fig. A.1). We first generate five sticks with lengths 6, 1, 3, 2, and 8 and paint each stick with the color that it represents. Next, we define a linked list of the sticks that are longer than the average (which is 4), and another for the sticks that are shorter than the average. In the present case, the long-stick list is ("red" $\rightarrow$ "green"), and the short-stick list is ("blue" $\rightarrow$ "yellow" $\rightarrow$ "pink"). We pick the first item from each list and cut the longer (red) stick in two pieces in such a way that if we join one of the two to the shorter (blue) stick, we obtain an average-length stick. As a result, we are left with a red stick of length 3 and a joint stick of length 4. Since the original red stick was made shorter than the average length, we remove it from the long-stick list and append it to the short-stick list. On the other hand, the original blue stick has become an average-length stick, so we remove it from the short-stick list. Then, we pick the first item from each list and repeat the same operations again and again. When finished, we have five sticks of average length, some with a single color and others with two colors.

At each step at least one of the lists is shortened by one. Because the length of the list is finite, the whole procedure must come to an end within a finite number

Alias method

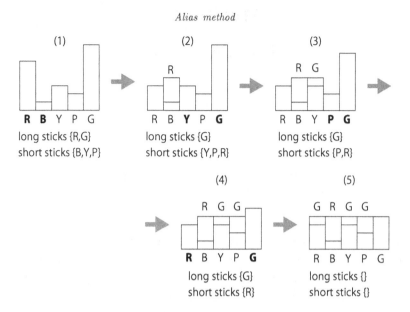

Figure A.1 Constructing aliases in Walker's method.

of operations. While constructing the joint stick-lists takes $\mathcal{O}(n)$ operations, this scaling does not matter in most cases, because once we have constructed the lists we can use them many times with little additional cost.

Now, to use these lists efficiently in the present case, we generate a random number $r \in [0, 20]$ (20 is the sum of the weights). Suppose it is 9.6. Instead of accumulating and comparing as in Algorithm 1, we simply divide r by the average stick length, that is, 4. This number means that our tentative choice is the third stick. But this time "third" does not necessarily mean "yellow." It can be an alias of "green." In the present case, the third stick, which was originally yellow, is now a composite of a yellow piece of length 3 and a green piece of length 1 so we have to choose among these two colors with probabilities 3/4 and 1/4. What is important is that this operation is $\mathcal{O}(1)$.

Appendix B

Rejection method

Rejection methods sample a variety of probability densities in a quite different way from the methods discussed in Chapter 2. The methods in Section 2.3 always returned a sample for each random number drawn. In a rejection method some draws do not return a sample: They are rejected. There are a number of rejection methods, some of which target specific densities. Of the general methods, von Neumann's is the most famous. It is quite general purposed, works in more than one dimension, and does not require the distribution of interest $f(x)$ to be normalized.

Suppose $f(x)$ equals $cp(x)$ where $p(x)$ is a probability density and c is the unknown normalization of $f(x)$, and $g(x)$ is another probability density. The procedure described in Algorithm 48 enables the sampling of x from $p(x)$. The loop generates an x until it is accepted. The other samples are rejected.

It follows directly from the **until** statement that because ζ is drawn from a uniform distribution over $[0, 1]$, the probability of accepting an x is

$$P(\text{accept}|x) = P\left(\zeta < \frac{f(x)}{Cg(x)}\right) = \frac{cp(x)}{Cg(x)},$$

where C is a constant such that $Cg(x) \geq f(x)$.

Algorithm 48 Rejection method.

Input: Function $f(x)$ to be sampled, some density $g(x)$, and a constant C such that $Cg(x) \geq f(x)$.

  repeat

    Sample x from $g(x)$;

    Compute the ratio $R(x) = \frac{f(x)}{Cg(x)}$ (≤ 1) ;

    Draw a uniform random number $\zeta \in [0, 1]$;

  until $\zeta < R(x)$.

  return x.

Next, we have

$$P(\text{accept}) = \int P(\text{accept}|x)g(x)dx$$

$$= \int \frac{cp(x)}{Cg(x)}g(x)dx = c/C.$$

We have $P(\text{accept}|x)$ but want $P(x|\text{accept})$. To find it we use Bayes's theorem:

$$P(x|\text{accept}) = \frac{P(x,\text{accept})}{P(\text{accept})}. \tag{B.1}$$

Because x is sampled independently of ζ, the two events in the joint probability are statistically independent, and we can write

$$P(x|\text{accept}) = \frac{P(\text{accept}|x)g(x)}{P(\text{accept})}$$

$$= \frac{[cp(x)/Cg(x)]g(x)}{c/C} = p(x),$$

which proves the validity of the algorithm.

The efficiency of the algorithm depends on the ratio c/C and the choice of $g(x)$. The first condition follows from (B.1): A small value of c/C implies low acceptance. The second is illustrated in Fig. B.1. Here, we assumed $f(x) = p(x)$ and $g(x)$ is uniform over the sampling interval (a, b). The algorithm is a variant of blindly throwing darts at a target and taking the ratio of the number of target hits relative to the number of times the bounding box is hit. For efficiency, $g(x)$ should "cover" $f(x)$ tightly and cost little to evaluate.

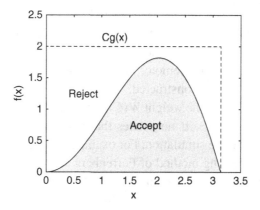

Figure B.1 The rejection algorithm. The function to be sampled is $f(x) = x\sin(x)$. The covering function $g(x)$ is uniform.

Appendix C

Extended-ensemble methods

While extending the state space with auxiliary graph variables, as done by the loop/cluster and worm algorithms, can be a powerful way of accelerating the simulation, there are also other strategies. The approach discussed here is a form of importance sampling (Section 10.4.3). In some cases, we can accelerate a slow relaxation by using a surrogate target weight $\tilde{W}(C)$ instead of using the given target weight $W(C)$. At finite temperatures, for example, we simply carry out Markov-chain Monte Carlo simulations with the transition probabilities satisfying the detailed balance condition for $\tilde{W}(C)$ instead of $W(C)$. To obtain the proper thermodynamic average with respect to the original weight, we calculate

$$\langle Q \rangle = \left\langle \frac{W(C)}{\tilde{W}(C)} Q(C) \right\rangle_{\tilde{W}} \Big/ \left\langle \frac{W(C)}{\tilde{W}(C)} \right\rangle_{\tilde{W}},$$

where $\langle \cdots \rangle_{\tilde{W}}$ is the Monte Carlo average with respect to the surrogate $\tilde{W}$. At finite temperatures, simulation techniques based on this idea are called *extended-ensemble Monte Carlo* methods. Into this category fall various algorithms such as the reweighting method (Ferrenberg and Swendsen, 1988), multicanonical Monte Carlo (Berg and Neuhaus, 1991), the Wang-Landau method (Wang and Landau, 2001), simulated tempering (Marinari and Parisi, 1992), replica Monte Carlo (Geyer, 1991; Hukushima and Nemoto, 1996), and the like. They differ in how the importance function $\tilde{W}(C)$ is constructed.

For an efficient sampling, the weight $\tilde{W}(C)$ must have a sufficiently large overlap with the target weight $W(C)$. If not, states that dominate the ensemble of $W(C)$ may seldomly appear in the simulation. For example, when applied to some classical systems, the reweighting method of Ferrenberg and Swendsen (1988) uses for $\tilde{W}(C)$ the Boltzmann weight at a temperature different from the one at which the thermodynamic average is needed, that is, $\tilde{W}(C) \propto e^{-E(C)/T'}$ and $W(C) \propto e^{-E(C)/T}$. In this case, $|T' - T|$ must be smaller than some energy scale proportional to $1/\sqrt{V}$ with V being the volume of the system. The proportionality has this form because the width of the energy distribution, or the density of states profile, is proportional to $\sqrt{V}$, whereas the distance between the mean values of the two distributions is proportional to $cV|T' - T|$ when T' and T are close to each other. Here, c is the specific heat, which is of the order of unity unless the system is not close to a phase transition.

In most extended-ensemble methods, we do not fix $\tilde{W}(C)$ but instead use it throughout the simulation as an adjustable function and gradually optimize it. One of the earliest successful examples of this adaptive strategy is the multicanonical method (Berg and Neuhaus, 1991, 1992). In the multicanonical method, the weight depends on the configuration C only through the energy $E(C)$. Therefore, we have

$$\tilde{W}(C) = \tilde{w}(E(C))$$

and may adjust $\tilde{w}(E)$ to make the frequency of the event $E(C) = E$ independent of E. This adjustment ensures that the resulting distribution function has a significant overlap with the Boltzmann distribution at any temperature. In this way, we are able to obtain information for a wide range of temperatures with only a single Monte Carlo simulation. For a number of problems, the energy takes only a discrete set of values, while for others, it takes a continuum of values, at least bounded from below. In the continuum case, we bin the energy; that is, we effectively discretize it. The Monte Carlo sampling we now describe is a random walk on the discrete values (or bins) of the energy E.

In an extended ensemble method, we perform several sets of samplings. The weight $\tilde{w}$ is fixed during each set, and it is updated at the end of each set for use in the next set. For the first set, we make an initial guess of the appropriate weight, most often taking $\tilde{w}(E) = \text{const}$. In every set, the histogram $h(E)$ is recorded; that is, every time a new state C is generated, we update the histogram (which is initially set to $h(E) = 0$ for all E) via

$$h(E(C)) \leftarrow h(E(C)) + 1. \tag{C.1}$$

We update C by Metropolis sampling with the weight $\tilde{W}(C) \equiv \tilde{w}(E(C))$. At the end of each set, $\tilde{w}(E)$ is updated using

$$\tilde{w}(E) \leftarrow \tilde{w}(E)/h(E),$$

and then the histogram is reset ($h(E) \leftarrow 0$).

The next set with this new weight starts with the last configuration of the current set.

When properly normalized, the histogram must converge to $h^*(E) \propto g(E)\tilde{w}(E)$ in the limit of infinite simulation length. Here

$$g(E) \equiv \sum_C \delta_{E,E(C)} = e^{S(E)/k_B}$$

is the density of states and $S(E) = k_B \ln g(E)$ is the microcanonical entropy. Therefore, if the simulation is sufficiently long, $h(E)/\tilde{w}(E)$ is a good approximation to $g(E)$ apart from a normalization constant. With $g(E)$ thus obtained, we compute the canonical average of an arbitrary quantity Q at an arbitrary inverse temperature β as

$$Q(\beta) \equiv \sum_E g(E)e^{-\beta E}Q(E) \Big/ \sum_E g(E)e^{-\beta E},$$

In this expression, $Q(E)$ is the microcanonical average of Q at discretized values of the energy E. We compute $Q(E)$ during the last set of samplings by classifying the sampled configurations according to the energy, and taking the average of each "microcanonical" set of configurations. In other words,

$$Q(E) \approx \langle Q(C)\delta_{E,E(C)}\rangle_{\mathrm{MC}} \Big/ \langle \delta_{E,E(C)}\rangle_{\mathrm{MC}}.$$

While this procedure is quite useful in studying various systems with slow dynamics, there is a drawback: The range of E visited by the random walk in energy space widens very slowly, partially due to the poor initial guess of the weight $\tilde{w}(E)$ and partially due to the diffusive nature of the random walk.

The Wang-Landau variant of the method (Wang and Landau, 2001) improves the situation. This method updates $\tilde{w}$ via

$$\tilde{w}(E) \leftarrow r\tilde{w}(E) \tag{C.2}$$

every time a new state is generated. However, the whole simulation may still be organized as multiple sets of samplings, and the reduction factor $0 < r < 1$ is fixed for each set. At the beginning of each sampling, the histogram is reset, that is, $h(E) = 0$ for all E; the reduction factor is updated as $r \leftarrow r^\alpha$ where $0 < \alpha < 1$; while the weight $\tilde{w}(E)$ is kept unchanged. Each set of samplings does not have a preset duration but is terminated when the histogram becomes approximately flat. Since the dynamic update (C.2) strongly penalizes random walkers staying at the same value of the energy, the visited region widens much faster than in the ordinary multicanonical Monte Carlo method. The efficiency of the algorithm depends on the choice of the initial value of r, the value of α, and the condition for terminating each set of samplings. The original work (Wang and Landau, 2001) suggested $r = 1/e$, $\alpha = 1/2$, and each set was terminated when the maximum deviation from the average in the histogram became less than 20% of the average. Obviously, there is room for further optimization of the updating scheme for these parameters and the termination condition.

It is natural to consider the combination of the extended-ensemble method with the loop/cluster algorithms. However, in a typical quantum Monte Carlo method, the weight $\tilde{W}(C)$ is not a function of the energy, and there is no reason to assign a special role to the energy. In addition, the value of the energy is not discretized or binned as conveniently. Janke and Kappler (1995), Yamaguchi and Kawashima (2002), Yamaguchi et al. (2002), and Troyer et al. (2003) used the extended-ensemble method in the loop algorithm and in the directed-loop algorithm by histogramming the number of vertices (or graph elements). In other words, they replaced the factor β^n in the high-temperature series expansion by a function $f(n)$,

$$Z = \sum_{n=0}^{\infty} \frac{\beta^n}{n!} \mathrm{Tr}(-H)^n \rightarrow \sum_{n=0}^{\infty} \frac{f(n)}{n!} \mathrm{Tr}(-H)^n.$$

This replacement changes the vertex assignment in the directed-loop algorithm and the graph assignment in the loop/cluster algorithm. For the directed-loop algorithm, the starting expression (6.37) becomes

$$Z_M = \sum_C \sum_G \frac{f(n)}{M^n} \prod_{k=1}^{M} \left\langle \psi(k+1) \left| \prod_{l=1}^{N_l} (-H_l)^{\gamma_{kl}} \right| \psi(k) \right\rangle. \tag{C.3}$$

Here, $n = \sum_{kl} \gamma_{kl}$ is the number of vertices in $G \equiv \{\gamma_{kl}\}$ and $1/M^n$ replaces $(M - n)!/M!$. The replacement of β^n by $f(n)$ affects the weight only as a function of the graph G. It does not change the weight as a function of $C \equiv \{\psi(k)\}$. Correspondingly, in the resulting extended-weight directed-loop algorithm, the worm updating cycle is exactly the same as before with only the graph assignment process modified.

With respect to the graph assignment, we need to modify the first two **for**-loops in Algorithm 20 to update the vertices for each uniform time-interval terminated by kinks at both ends. Let I be one such interval. From (C.3) it follows that the probability of having ν vertices in this interval is proportional to

$$\frac{f(n_0 + \nu)}{M^\nu} \langle \psi | - H_l | \psi \rangle^\nu \binom{I/(\Delta\tau)}{\nu},$$

where n_0 is the total number of vertices in the rest of the system. The binomial coefficient is the number of ways to assign ν vertices to the $I/(\Delta\tau)$ cells. Taking the limit of $\Delta\tau = \beta/M \rightarrow 0$, we obtain the unnormalized probability of having ν vertices as

$$w_I(\nu) = f(n_0 + \nu) \frac{(I\beta^{-1}\langle -H_l \rangle_I)^\nu}{\nu!}.$$

Therefore, to assign vertices in I, we could in principle generate a random integer ν from this distribution and assign ν vertices to positions sampled from a uniform distribution over I. However, doing this is often unpractical, as computing the normalization constant is generally expensive. An alternative is to first choose with probability $\frac{1}{2}$ whether we propose to increase or decrease the number of vertices by one, and then update the number of vertices by the Metropolis algorithm; namely, we accept the proposal with the probability $\min(1, w_I(v \pm 1)/w_I(v))$, with v being the current number of the vertices in I.

Extended-ensemble methods are especially useful in the computation of the quantities directly related to the density of states, such as the free energy, entropy, and specific heat, which normally would require a series of simulations at different temperatures or the numerical integration over temperature if other methods are adopted. An example of the computation of these quantities can be found in Troyer et al. (2003). Yamaguchi et al. (2002) proposed further improvements by employing the broad-histogram method (de Oliveira et al., 1998a,b; Wang et al., 1999). This method exploits the exact relation between the expectation value of the transition matrix and the density of states,

$$g(E)\langle N(S, E \rightarrow E')\rangle = g(E')\langle N(S, E' \rightarrow E)\rangle,$$

where $N(S, E \rightarrow E')$ is the number of possible paths to reach any of the states with energy E' when we are currently in a state with energy E. Generally, N is a macroscopic quantity proportional to the system's volume. This means that by using the above equation we can estimate $g(E)$ only from the expectation values of macroscopic quantities, which we can estimate with higher precision than each entry of the histogram itself.

Appendix D

Loop/cluster algorithms: SU(N) model

As mentioned at the beginning of Section 6.2.6, in some cases the loop algorithm (Algorithm 17) for $S = \frac{1}{2}$ is applicable to spin systems with $S > \frac{1}{2}$ without splitting spins. This is always possible if the Hamiltonian decomposes graphically. For example, let us consider the bilinear-biquadratic interaction model with $S = 1$ (Harada and Kawashima, 2001, 2002),

$$H = J \sum_{\langle ij \rangle} [(\cos \theta) \vec{S}_i \cdot \vec{S}_j + (\sin \theta)(\vec{S}_i \cdot \vec{S}_j)^2]. \tag{D.1}$$

The model (D.1) obviously has SU(2) symmetry. At $\theta = \pm\pi/2$ and $\theta = \pm\pi/4$, however, it possesses a higher symmetry. Consider the case of $\theta = -\pi/2$ for which

$$H = -J \sum_{\langle ij \rangle} [(\vec{S}_i \cdot \vec{S}_j)^2 - 1], \tag{D.2}$$

where the constant -1 has been added for convenience. The Hamiltonian (D.2), as well as the Hamiltonian at other special values of θ, has SU(3) symmetry. In the ordinary S^z-basis, $S_i^z |s_i\rangle = s_i |s_i\rangle$, the matrix elements of the pair Hamiltonian are

$$\langle s_i' s_j' | -H_{ij} | s_i s_j \rangle = J\sigma(S_p) \Delta(S_p, g_{\mathrm{H}}). \tag{D.3}$$

The function $\sigma(S_p) = \pm 1$ carries a sign: It is -1 if and only if either the initial state (s_i, s_j) is $(0, 0)$ and the final state (s_i', s_j') is $(\pm 1, \mp 1)$ or vice versa. It is easy to see that this sign is irrelevant since the negative signs always occur an even number of times, leaving the sign of the whole system positive. Therefore, in (D.3) we neglect σ. The function $\Delta(S_p, g_{\mathrm{H}})$ is

$$\Delta(S_p, g_{\mathrm{H}}) \equiv \begin{cases} 1, & \text{if } s_i + s_j = s_i' + s_j' = 0 \\ 0, & \text{otherwise.} \end{cases}$$

This quantity is the horizontal graph operator in Table 6.2, generalized to the present three-state model.

With the pair Hamiltonian expressed in terms of the graph operators, it is straightforward to apply the general prescription discussed in Section 6.2.2. The only difference between the resulting $S = 1$ algorithm and the $S = \frac{1}{2}$ case is that the density of the graph elements is J, not $J/2$, because of the larger diagonal matrix element and that the variables have three values, $s_i = \{-1, 0, 1\}$, instead of two. Accordingly, loop-flipping is not a coin toss, but proceeds by choosing one of the three values randomly with equal probability.

We can construct a similar algorithm for cases with lower symmetry, for example, SU(2) symmetry, for the parameter region $-3\pi/4 \le \theta \le -\pi/2$. The graphical decomposition of the pair Hamiltonian in this case is (Harada and Kawashima, 2001)

$$- H_{ij} = J\left[(-\sin\theta + \cos\theta)D_{ij}(g_{\mathrm{h}}) - (\cos\theta)D_{ij}(g_{\mathrm{d}})\right], \qquad (\mathrm{D.4})$$

where an irrelevant sign and an additive constant have been omitted. The symbol $D_{ij}(g_{\mathrm{d}})$ corresponds to the diagonal graph in Fig. 6.2. Its matrix elements are

$$\langle s_i's_j'| - D_{ij}(g_{\mathrm{d}})|s_is_j\rangle = \Delta(S_p, g_{\mathrm{d}}) \equiv \begin{cases} 1, & \text{if } s_i = s_i' \text{ and } s_j = s_j' \\ 0, & \text{otherwise.} \end{cases}$$

The loop construction and the loop flipping are done as before.

The special case of the SU(N) bilinear-biquadratic model that possesses SU(3) symmetry generalizes to a SU(N) symmetric model. The local spin variable of the SU(N) Heisenberg model takes N possible values, that is, $\sigma_i = \{1, 2, \ldots, N\}$. Expressed in terms of the generators of the SU(N) algebra,

$$H \equiv \frac{J}{N} \sum_{\langle ij \rangle} \sum_{\mu=1}^{N} \sum_{\nu=1}^{N} S_i^{\mu\nu} \bar{S}_i^{\nu\mu}. \qquad (\mathrm{D.5})$$

Here, $S_i^{\mu\nu}$, $\bar{S}_j^{\nu\mu}$ satisfy

$$[S_i^{\mu\nu}, S_j^{\sigma\tau}] = \delta_{ij}(\delta_{\mu\tau}S_i^{\sigma\nu} - \delta_{\sigma\nu}S_j^{\mu\tau}) \quad (\mu, \nu, \sigma, \tau = 1, 2, \ldots, N).$$

The operators $\bar{S}_j^{\mu\nu}$ satisfy analogous conditions. We consider the model defined on a bipartite lattice. On one sublattice (say, the sublattice A) we adopt the fundamental representation that corresponds to the Young tableau with a single box. On the other sublattice (B), we adopt the conjugate representation. The bar on the operator $\bar{S}_i^{\mu\nu}$ is a reminder that we are using different representations for the spins on sublattices A and B.

More explicitly, the matrix element of $S_i^{\mu\nu}$ on the sublattice A is $\langle \alpha |S_i^{\mu\nu}| \beta \rangle = \delta_{\alpha\mu}\delta_{\beta\nu} - \frac{1}{N}\delta_{\alpha\beta}$, while on B $\langle \alpha |\bar{S}_j^{\mu\nu}| \beta \rangle = -\delta_{\alpha\nu}\delta_{\beta\mu} + \frac{1}{N}\delta_{\alpha\beta}$. Therefore, we are able to rewrite (D.5) as

$$H = -J \sum_{\langle ij \rangle} \frac{1}{N} \sum_{\mu\nu} S_i^{\mu\nu} S_j^{\mu\nu}.$$

If we define the operator P_{ij} by its matrix element as

$$\langle \alpha', \beta' | P_{ij} | \alpha, \beta \rangle \equiv \frac{1}{N} \delta_{\alpha'\beta'} \delta_{\alpha\beta},$$

it is the projection operator to the color singlet-state on the bond $\langle i,j \rangle$,

$$|\text{singlet on } (ij)\rangle \equiv \frac{1}{\sqrt{N}} \sum_{\alpha} |\alpha, \alpha\rangle .$$

To convince ourselves that this is indeed a singlet state, we have to recall that we are using the conjugate representation to the fundamental one for one of the two sublattices. In other words, one of the two αs in $|\alpha, \alpha\rangle$ does not represent the same state as the other α, but it is the product of all the states but α. (It is easy to confirm that this state is invariant under P_{ij}.) Using the SU(N) generators, we express the projection operator as

$$P_{ij} = \frac{1}{N} \sum_{\mu\nu} S_i^{\mu\nu} S_j^{\mu\nu} + \frac{1}{N^2}.$$

Therefore, apart from an additive constant,

$$H = -J \sum_{\langle ij \rangle} P_{ij}.$$

Note also that the matrix elements of the projection operator are the product of Kronecker deltas, so they can be represented by a graph element, g_h, that is,

$$P_{ij} = \frac{1}{N} D_{ij}(g_h).$$

The two spins bound by this graph element have the same value, whereas they were antiparallel previously. But this difference is removable by a gauge transformation and is not essential. Thus, we have the graphical decomposition of the SU(N) Heisenberg model Hamiltonian, from which a loop algorithm follows. The algorithm is almost identical to the one discussed in Section 6.2.3 for the $S = \frac{1}{2}$ antiferromagnet. The only difference is that each local variable takes N values instead of 2.

Appendix E

Long-range interactions

Most examples discussed in this book are models with on-site or nearest-neighbor interactions. Physical systems, however, often have interactions reaching beyond one lattice constant, and there are cases where such long-range interactions alter the essential properties of the system. A classic example is the dipolar interaction in magnets. Because of the long-range dipolar interaction, magnetic domains form, reducing the total magnetization. Another more recent example is the effect of dipolar interactions among ultra-cold atoms in an optical lattice. Because of the much larger lattice constant in such systems, the magnitude of the dipolar interaction relative to the nearest-neighbor interaction is enhanced compared with a conventional solid. Frustrated systems are other good examples, since for these the short-range interactions sometimes cancel, making the middle/long-range interactions more relevant.

From the computational point of view, long-range interactions generally increase the computational complexity of Monte Carlo simulations. Consider, for example, the Ising model

$$H = -\sum_{ij} J(R_{ij}) s_i s_j,$$

where $J(R_{ij})$ is some decreasing function of the distance between two sites R_{ij}. If we adopt the single-spin update, every time we attempt flipping a spin we must compute the molecular field produced by all the other spins. This requires $\mathcal{O}(N)$ operations for a lattice of N sites. Therefore, every Monte Carlo sweep takes a computational time proportional to N^2 instead of N, as in the case of the short-range interaction models. A neat trick exists to reduce this N^2 dependence to NM, where M is the effective number of interacting spins defined by

$$M \equiv a^{-d} \int d^d R\, J(R).$$

If M is bounded from above by some constant that is independent of the system size, this trick reduces the computational complexity back to $\mathcal{O}(N)$. Even if M diverges in the thermodynamic limit, reducing the complexity from N^2 to NM can still be a major gain. For example, if the coupling constant $J(R)$ is inversely proportional to R, M is proportional to L^{d-1} if L is the linear size and the dimension is $d \geq 3$. In this case, $NM \propto L^{2d-1}$, which is less than N^2 by a factor L.

As a first method and specific example, let us discuss the approach originally proposed by Luijten and Blöte (1995), which is based on a look-up table. We consider the Swendsen-Wang cluster algorithm for an Ising model with long-range interactions. The system has N sites. In the following, we first compute the probability for placing a bond assuming that all the spins are parallel, and then, before we actually place a bond, we examine whether the spins are parallel. (If not, we do not place a bond, of course.) The probability of having a bond between the p-th ($p = 1, 2, \ldots, N(N-1)/2$) pair of spins can be written as

$$P_{\text{bond}}(p) = 1 - e^{-2K_p}, \tag{E.1}$$

where K_p describes the interaction strength for the pair p. For example, if p is the pair (i,j), $K_p \equiv \beta J_{ij}$. Now consider a sequence of $N_{\text{pair}} \equiv N(N-1)/2$ boxes, each representing a pair of sites, and suppose that a ball is placed in each box with the corresponding probability (E.1). The probability that no ball is put in any box between the m-th and the n-th becomes

$$P_{\text{no bond}}(m; n) = \prod_{p=m+1}^{n-1} e^{-2K_p} = \exp\left[-2 \sum_{p=m+1}^{n-1} K_p\right]. \tag{E.2}$$

If we define an array ("look-up table") A of length N_{pair} with elements

$$A_0 = 1, \quad A_n = \prod_{p=1,\ldots,n} e^{-2K_p} \quad (n = 1, \ldots, N_{\text{pair}}), \tag{E.3}$$

then (E.2) can be written as

$$P_{\text{no bond}}(m; n) = \frac{A_{n-1}}{A_m}.$$

The pairs that are to be connected can then be calculated as follows: A random number $\zeta_1 \in [0, 1)$ is chosen and the array A is searched for the first index, say, p_1, such that $A_{p_1} \leq \zeta_1$. For the calculation of the next bond, the values of A must be divided by A_{p_1}, or equivalently, the random number $\zeta_2 \in [0, 1)$ multiplied by A_{p_1}. The site number p_2 of the second connected spin is the first index, such that $A_{p_2} \leq A_{p_1}\zeta_2$. This process is continued until $A_{p_{k-1}}\zeta_k < A_{N_{\text{pair}}}$. As was already mentioned, the outlined procedure assumes that all spins are aligned. Before actually inserting

---

Algorithm 49 Cluster algorithm with long-range interactions: look-up table approach.

Input: N_{pair}, array A defined in (E.3). l, an initially empty list of connected pairs, $p = 0$.

   loop

      Draw random number $\zeta \in [0, 1)$;

      $r = A_p \zeta$;

      if $r < A_{N_{\text{pair}}}$ **then**

         break ;

      end if

      Search the lowest index p such that $A_p \leq \zeta$;

      If the spins connected by pair p are parallel, add p to the list l ;

   end loop

   return the list l of connected pairs.

---

a proposed bond, we have to check whether the two spins are parallel. Because $A_0 = 1$, the above algorithm corresponds to the pseudocode in Algorithm 49.

If the bisection method is used to search for the lowest index such that $A_p \leq \zeta$, the calculation time is of order $\mathcal{O}(\log N_{\text{pair}}) = \mathcal{O}(\log N)$ for each proposed bond. Since the number of bonds connecting to a given site is of order M, and we have N sites, the total calculation time for the bond placing is $O(NM \log N)$.

We can even get rid of the $\log N$ factor, although it may not matter in practical applications. To do this, let us regard a missing bond on a pair of interacting sites as a "pair survival." If the survival probability of a pair p is $e^{-\Gamma_p}$, we can interpret it as surviving the decay rate of unity for the duration Γ_p. ($\Gamma_p = 2\beta J_p$ in the Swendsen-Wang algorithm.) In the case of many pairs, we have many intervals with various durations Γ_p. Let us now consider a line segment of length $\Gamma \equiv \sum_{p=1}^{N_{\text{pairs}}} \Gamma_p$. Instead of judging for each individual interval whether pairs survive or not, we distribute uniformly and randomly decay events over this line segment with unit density. Specifically, we generate the total number of decay events by drawing an integer n from the Poisson distribution of mean Γ and distribute n events uniformly and randomly on the line. This procedure is like throwing n darts onto the line. As a result, some darts may or may not fall into the interval Γ_p. If we have no dart in Γ_p, the pair p survives (no bond assigned); otherwise, it is killed (a bond assigned). Given that the expectation value of n is proportional to Γ, the computational cost for generating n bonds is of order Γ. In the case of the Swendsen-Wang algorithm this cost is proportional to $\sum_{ij} J_{ij}$ and hence $O(NM)$.

Finding the intervals hit by any dart is less trivial. We carry out the task n times (for each of the randomly distributed darts). Generating a uniform random number is easy and costs $\mathcal{O}(1)$ for each dart thrown. The difficult part is finding the index p of the interval hit. Using the technique discussed in Appendix A we can do this in a time $\mathcal{O}(1)$ (Knuth, 1997; Walker, 1977; Fukui and Todo, 2009). As a result, throwing darts and judging survival takes a computational time of $\mathcal{O}(\Gamma) \sim \mathcal{O}(NM)$ in total.

The application of these ideas to quantum simulations is rather straightforward. For example, in the directed-loop algorithm for a long-range interaction model, we generate the total number of vertices n by the Poisson distribution with mean Γ, defined in this case as $\Gamma = \sum_l \Gamma_l$ with

$$\Gamma_l \equiv \beta \max_{\psi} \langle \psi | - H_l | \psi \rangle. \tag{E.4}$$

In the placement of the vertices, the algorithm, however, differs from the general procedure described above. There, the number of darts n_l on the interval Γ_l did not matter as long as $n_l \neq 0$. Here, it matters since n_l is the number of the vertices we assign to the interacting spins involved in H_l. For each of the n_l vertices, we generate an imaginary time in the interval $\tau \in [0, \beta]$ and with probability

$$P_{\text{accept}} = \Gamma_l^{-1} \beta \langle \psi(\tau) | - H_l | \psi(\tau) \rangle$$

decide whether we place the vertex there. In this way, we compensate for the overestimate of the graph assignment density made in (E.4) by using the maximum value.

Appendix F

Thouless's theorem

In this appendix, we prove *Thouless's theorem* (Thouless, 1961; Blaizot and Ripka, 1986). It says that if the state $|\phi\rangle$ is the Slater determinant $|\phi\rangle = a_{N_\sigma}^\dagger \cdots a_2^\dagger a_1^\dagger |0\rangle$, where $a_i^\dagger = \sum_j c_j^\dagger \Phi_{ji}$ is defined by the $N \times N^\sigma$ matrix Φ, then the propagation of this state $e^{-\Delta\tau c^\dagger M c} |\phi\rangle$ via a noninteracting Fermion Hamiltonian is to a new Slater determinant $|\phi'\rangle$, where $|\phi'\rangle = b_{N_\sigma}^\dagger \cdots b_2^\dagger b_1^\dagger |0\rangle$, and $b_i^\dagger = \sum_j c_j^\dagger \Phi_{ji}'$ is defined by the $N \times N^\sigma$ matrix $\Phi' = \exp(-\Delta\tau M)\Phi$. In other words, it says that if the Hamiltonian is noninteracting, a state of noninteracting Fermions evolves into another state of noninteracting Fermions and tells us how to compute the new state via the multiplication of two matrices.

This theorem underlies implicitly or explicitly many Fermion algorithms. Our use of the Hubbard-Stratonovich transformation replaces the exponential of the Hamiltonian of a system of interacting Fermions by the exponential of a Hamiltonian of Fermions that interact only with imaginary-time-dependent auxiliary fields. With this transformation, we were able to reduce the many-body formalism to numerical methods whose core computations are matrix-matrix multiplications. The role of the theorem is most apparent in the constrained path and phase algorithms (Chapter 11) where the projection to the ground state is cast as an open-ended imaginary-time propagation where each discrete imaginary-time step evolves one Slater determinant into another. It is implicit in the finite-temperature determinant method (Chapter 7). There, building the algorithm upon a generalization of the partition function of noninteracting Fermions was most natural. Our language was that of imaginary-time propagators, and our proof of the determinant form of basic single-particle propagators evoked the theorem. While we developed the zero-temperature determinant method by analogy to the finite-temperature one, we could have developed it from the point of view of the constrained path and phase ones by simply capping the length of the projection.

In fact, the code for any one of these four methods can be easily converted to one of the other three.

The proof of the theorem is simple and mainly uses the anticommutation relation $\{c_i, c_j^\dagger\} = \delta_{ij}$. To evaluate $e^{-\Delta \tau c^\dagger M c} |\phi\rangle$, we commute $e^{-\Delta \tau c^\dagger M c}$ and $a_{N_\sigma}^\dagger$,

$$e^{-\Delta \tau c^\dagger M c} a_{N_\sigma}^\dagger = \left[\sum_{ij} M c_i^\dagger \left[e^{-\Delta \tau M} \right]_{ij} \Phi_{jN_\sigma} \right] e^{-\Delta \tau c^\dagger M c},$$

so

$$e^{-\Delta \tau c^\dagger M c} a_{N_\sigma}^\dagger \cdots a_2^\dagger a_1^\dagger |0\rangle = \left[\sum_{ij} c_i^\dagger \left[e^{-\Delta \tau M} \right]_{ij} \Phi_{jN_\sigma} \right] e^{-\Delta \tau c^\dagger M c} a_{N_\sigma-1}^\dagger \cdots a_2^\dagger a_1^\dagger |0\rangle .$$

Repeating this process until the operator exponential acts on the vacuum state yields

$$e^{-\Delta \tau c^\dagger M c} a_{N_\sigma}^\dagger \cdots a_2^\dagger a_1^\dagger |0\rangle = \left[\sum_{ij} c_i^\dagger \left[e^{-\Delta \tau M} \right]_{ij} \Phi_{jN_\sigma} \right] \left[\sum_{ij} c_i^\dagger \left[e^{-\Delta \tau M} \right]_{ij} \Phi_{j,N_\sigma-1} \right]$$

$$\cdots \left[\sum_{ij} c_i^\dagger \left[e^{-\Delta \tau M} \right]_{ij} \Phi_{j1} \right] e^{-\Delta \tau c^\dagger M c} |0\rangle .$$

Next, we note that $e^{-\Delta \tau c^\dagger M c} |0\rangle = \left[I - \Delta \tau c^\dagger M c + \frac{1}{2} \left(\Delta \tau c^\dagger M c \right)^2 + \cdots \right] |0\rangle = |0\rangle$ and hence

$$e^{-\Delta \tau c^\dagger M c} a_{N_\sigma}^\dagger \cdots a_2^\dagger a_1^\dagger |0\rangle = \left[\sum_{ij} c_i^\dagger \left[e^{-\Delta \tau M} \right]_{ij} \Phi_{jN_\sigma} \right] \left[\sum_{ij} c_i^\dagger \left[e^{-\Delta \tau M} \right]_{ij} \Phi_{j,N_\sigma-1} \right]$$

$$\cdots \left[\sum_{ij} c_i^\dagger \left[e^{-\Delta \tau M} \right]_{ij} \Phi_{j1} \right] |0\rangle .$$

We could stop here, but we will add multiple propagator steps because they are required for the Trotterized path integrals in several algorithms. Each propagator step inherits the Slater determinant from the previous step. Multiple steps simply use multiple applications of the theorem.

Repeating this multiplication for each exponential produces

$$e^{-\Delta\tau c^\dagger M_{N_\tau} c} \cdots e^{-\Delta\tau c^\dagger M_2 c} e^{-\Delta\tau c^\dagger M_1 c} a^\dagger_{N_\sigma} \cdots a^\dagger_2 a^\dagger_1 |0\rangle =$$

$$\left[\sum_{ij} c^\dagger_i \left[e^{-\Delta\tau M_{N_\tau}} \cdots e^{-\Delta\tau M_2} e^{-\Delta\tau M_1} \right]_{ij} \Phi_{jN_\sigma} \right] \cdots$$

$$\cdots \left[\sum_{ij} c^\dagger_i \left[e^{-\Delta\tau M_{N_\tau}} \cdots e^{-\Delta\tau M_2} e^{-\Delta\tau M_1} \right]_{ij} \Phi_{j2} \right]$$

$$\times \left[\sum_{ij} c^\dagger_i \left[e^{-\Delta\tau M_{N_\tau}} \cdots e^{-\Delta\tau M_2} e^{-\Delta\tau M_1} \right]_{ij} \Phi_{j1} \right] |0\rangle .$$

After defining $b^\dagger_i = \sum_{jk} c^\dagger_j \left[e^{-\Delta\tau M_{N_\tau}} \cdots e^{-\Delta\tau M_2} e^{-\Delta\tau M_1} \right]_{jk} \Phi_{ki}$, we get

$$e^{-\Delta\tau c^\dagger M_{N_\tau} c} \cdots e^{-\Delta\tau c^\dagger M_2 c} e^{-\Delta\tau c^\dagger M_1 c} |\phi\rangle = b^\dagger_{N_\sigma} \cdots b^\dagger_2 b^\dagger_1 |0\rangle \equiv |\phi'\rangle . \qquad \text{(F.1)}$$

Appendix G

Hubbard-Stratonovich transformations

The original Hubbard-Stratonovich transformation converts the exponential of a two-body operator into a sum of exponentials of one-body operators. Each term in the sum depends on an auxiliary field. The models typically simulated have a small number of orbitals per lattice site and density-density Coulomb interactions acting only on site. This simplicity leads to the number of auxiliary fields being at most a few times the number of lattice sites. In these cases, Hirsch's discrete transformation (Hirsch, 1983) is more efficient than the standard continuous transformation. Here, we consider simulations of more complex models and discuss Hubbard-Stratonovich transformations for models with extended-range interactions and more general Coulomb interactions. More specifically, we discuss the use of the continuous Hubbard-Stratonovich transformation when the interactions are long ranged and present a novel discrete Hubbard-Stratonovich transformation for interactions that are not simply products of number operators. In both cases, the auxiliary fields are real. We also mention in passing other Hubbard-Stratonovich transformations used in special applications. These transformations include ones that preserve the SU(2) symmetry (Assaad, 1998) and couple the auxiliary field to fluctuations in the pairing field (Batrouni and Scalettar, 1990).

G.1 Extended interactions

We start by considering the exponentiation of the interaction

$$V = \sum_{ij} V_{ij} n_i n_j, \tag{G.1}$$

where the subscript i represents a combination of electron spin, orbital quantum numbers, and lattice position identifiers. The interactions, described by the elements V_{ij} of a symmetric $N \times N$ matrix, are not restricted to being on site or being between nearest neighbors. We have $n_i^2 = n_i$.

With $n = (n_1, n_2, \ldots, n_N)^T$ being a N-vector of occupation numbers, the operative Hubbard-Stratonovich transformation is

$$e^{-\frac{1}{2}\Delta\tau n^T V n} = \frac{1}{\sqrt{\Delta\tau \det V}} \int \prod_i \left(\frac{dy_i}{\sqrt{2\pi}}\right) e^{-\frac{1}{2}\Delta\tau y^T V y} e^{-\Delta\tau y^T n}, \qquad (G.2)$$

where $y = (y_1, y_2, \ldots, y_N)^T$ is a vector of auxiliary fields. The convergence of this multi-variable Gaussian integral requires V to be positive definite. If it is not, we write

$$e^{-\frac{1}{2}\Delta\tau n^T V n} = e^{\Delta\tau V_0 n^T n - \frac{1}{2}\Delta\tau n^T V n} e^{-\Delta\tau V_0 n^T n} \equiv e^{-\frac{1}{2}\Delta\tau n^T V' n} e^{-\Delta\tau V_0 n^T n}$$

and choose V_0 such that the matrix V' obtained by shifting the diagonal elements of V by $-2V_0$ is positive definite. Since $n^T n = \sum_i n_i = \sum_i c_i^\dagger c_i$, the additional exponential we introduce is that of a one-body interaction. We accommodate it by absorption into another one-body part of the Hamiltonian.

The generalized Hubbard-Stratonovich transformation (G.2) has had the same effect as the simpler transformations in Section 7.1.1, namely, converting the interacting problem into a noninteracting one. We can thus use it in a Monte Carlo simulation, instead of the others. What are the auxiliary fields and how do we sample them?

With V positive definite, a Cholesky factorization (Meyer, 2000) exists such that $V = R^T R$, where R is an upper triangular matrix with positive diagonal elements. Because of the factorization we can write $n^T V n = y^T y$ where $y = Rn$. After generating a vector $\zeta^T = (\zeta_1, \zeta_2, \ldots, \zeta_N)$ of Gaussian random variables of unit variance and zero mean, we then obtain a sample of the vector y by using the back-substitution method to solve the linear system of equations $\zeta = Ry$. Note that we need only N auxiliary fields.

A general two-body interaction can be put into the form

$$V = \sum_{ijkl} V_{ijkl} c_i^\dagger c_j c_k^\dagger c_l. \qquad (G.3)$$

With the definition of the operator $\rho_{ij} = c_i^\dagger c_j$, and the mapping of the subscript pairs ij and kl to the integers m and n, (G.3) takes the quadratic form

$$V = \sum_{mn} V_{mn} \rho_m \rho_n.$$

After ensuring the V_{mn} are elements of a positive-definite matrix, we sample the auxiliary field y in the same manner as we did for (G.1).

The Gaussian Hubbard-Stratonovich transformation is useful for a variety of applications other than those with long-range interactions. An important way to improve the efficiency in many of these simulations is to subtract background terms prior to making the transformation, leading to sampling from a shifted Gaussian (Purwanto and Zhang, 2004).

G.2 General discrete transformations

We ease into the general discrete Hubbard-Stratonovich transformation for interactions that are not simply products of number operators by first revisiting Hirsch's transformation from a perspective different from those that are. The new perspective is an expression of the exponential of the operators as a quadratic function of operators.

With $V_{ij} > 0$ and the relation $n_i^2 = n_i$, Hirsch's transformation for the exponential of each term in the potential energy (G.1) is

$$e^{-\Delta \tau V_{ij} n_i n_j} = \frac{1}{2} \sum_{\sigma = \pm 1} e^{-\sigma \Delta \tau J_{ij} (n_i - n_j) - \frac{1}{2} \Delta \tau V_{ij} (n_i + n_j)},$$

with $\cosh(\Delta \tau J_{ij}) = \exp(\frac{1}{2} \Delta \tau V_{ij})$. If expanded as a power series, the left-hand side of this equation becomes

$$e^{-\Delta \tau V_{ij} n_i n_j} = 1 + \lambda_{ij} n_i n_j,$$

where we defined

$$\lambda_{ij} = e^{-\Delta \tau V_{ij}} - 1. \tag{G.4}$$

From these results follows the trivial identity

$$e^{-\Delta \tau V_{ij} n_i n_j} = \frac{1}{2} \sum_{\sigma = \pm 1} \left(1 + \sigma \alpha_{ij} n_i + \sigma \beta_{ij} n_j + \alpha_{ij} \beta_{ij} n_i n_j \right) \tag{G.5}$$

if the product of α_{ij} and β_{ij} equals λ_{ij}. Noting that

$$e^{-\sigma \Delta \tau J_{ij} (n_i - n_j) + \frac{1}{2} \Delta \tau V_{ij} (n_i + n_j)} \neq 1 + \sigma \alpha_{ij} n_i + \sigma \beta_{ij} n_j + \alpha_{ij} \beta_{ij} n_i n_j$$

unless $\alpha_{ij} = -\beta_{ij}$, we choose[1]

$$\alpha_{ij} = \sqrt{|\lambda_{ij}|} \quad \text{and} \quad \beta_{ij} = \text{sign}\left(\lambda_{ij}\right) \sqrt{|\lambda_{ij}|}. \tag{G.6}$$

Next, we comment that because the action of $\exp[-\sigma \Delta \tau J_{ij} \left(n_i - n_j \right) + \frac{1}{2} \Delta \tau V_{ij} \left(n_i + n_j \right)]$ on a Slater determinant produces another one (Appendix F), then so must the action of the operator $1 + \sigma \alpha_{ij} n_i + \sigma \beta_{ij} n_j + \alpha_{ij} \beta_{ij} n_i n_j$. If

$$|\psi_A\rangle = a_1^\dagger a_2^\dagger \cdots a_M^\dagger |0\rangle,$$

---

[1] If $V_{ij} < 0$, we would take $\alpha_{ij} = \beta_{ij}$.

where

$$a_j^\dagger = \sum_i c_i^\dagger A_{ij},$$

the consequence of the exponential operators acting on $|\psi_A\rangle$ is $|\psi_{A'}\rangle$, where $A' = \exp(\chi^+)\exp(\chi^-)A$, χ^+ is an $N \times N$ matrix that is null except for its i-th diagonal element equaling $-\Delta\tau(\frac{1}{2}V_{ij}-\sigma J_{ij})$, and χ^- is an $N \times N$ matrix that is null except for its j-th diagonal element equaling $-\Delta\tau(\frac{1}{2}V_{ij}+\sigma J_{ij})$. The effect of $\exp(\chi^-)$ on A is to multiply the j-th row of A by $\exp[-\Delta\tau(\frac{1}{2}V_{ij}+\sigma J_{ij})]$, and the effect of $\exp(\chi^+)$ on this matrix product is to multiply its i-th row by $\exp[-\Delta\tau(\frac{1}{2}V_{ij}-\sigma J_{ij})]$. It is straightforward to show that these matrix-matrix multiplications are equivalent to

$$A' = \left(I + \sigma\alpha_{ij}I_{\bullet i}I_{i\bullet}^T + \sigma\beta_{ij}I_{\bullet j}I_{j\bullet}^T\right)A,$$

where α_{ij} and β_{ij} are as defined by (G.4) and (G.6). For an arbitrary matrix X, $X_{\bullet i}$ is an N-vector formed from the i-th column of X, and $X_{i\bullet}$ is an N-vector formed from its i-th row. In particular, the outer product $I_{\bullet i}I_{i\bullet}^T$ is an $N \times N$ null matrix except for its i-th diagonal element equaling unity. Hence, the two different-looking Hubbard-Stratonovich transformations, one using an exponential of a product of operators and the other using a quadratic polynomial of operators, are equivalent.

We now write the two-body potential energy in still another form,[2]

$$V = \sum_{ijkl} V_{ijkl}c_i^\dagger c_j^\dagger c_l c_k = \sum_{kl} q_{kl},$$

with

$$q_{kl} = \left(\sum_{ij} V_{ijkl}c_i^\dagger c_j^\dagger\right) c_l c_k.$$

We can show that

$$q_{kl}^2 = \lambda_{kl}q_{kl},$$

where $\lambda_{kl} = V_{kkll} - V_{klkl}$. Hence

$$e^{-\Delta\tau q_{kl}} = 1 + \gamma_{kl}q_{kl}$$

and

$$\gamma_{kl} = \begin{cases} \dfrac{e^{-\Delta\tau\lambda_{kl}}-1}{\lambda_{kl}} & \text{if } \lambda_{kl} \neq 0 \\ -\Delta\tau & \text{if } \lambda_{kl} = 0. \end{cases}$$

---

[2] The V_{ijkl} here are not necessarily the same as those in (G.3).

$$e^{-\Delta\tau q_{kl}} = \frac{1}{2} \sum_{\sigma=\pm 1} \sum_{ij} \frac{|V_{ijkl}|}{\Lambda_{kl}} \left(1 + \sigma\alpha_{kl}c_i^\dagger c_k + \sigma\beta_{kl}c_j^\dagger c_l + \alpha_{kl}\beta_{kl}c_i^\dagger c_j^\dagger c_l c_k\right), \quad (G.7)$$

and choose

$$\Lambda_{kl} = \sum_{ij} |V_{ijkl}|, \quad \alpha_{kl} = \sqrt{\Lambda_{kl}|\gamma_{kl}|}, \quad \beta_{kl} = \text{sign}\left(\gamma_{kl}V_{ijkl}\right)\sqrt{\Lambda_{kl}|\gamma_{kl}|} \quad (G.8)$$

and then ask, Do the σ-dependent operators in (G.7) yield a single Slater determinant $|\psi\rangle_{A'}$ if they act on $|\psi\rangle_A$?

Rombouts et al. (1999a) showed that the answer to this question is yes. More specifically, if $|\psi\rangle_A$ is a Slater determinant defined by the $N \times M$ matrix A, then the operator $I + \alpha c_i^\dagger c_l + \beta c_j^\dagger c_k + \alpha\beta c_i^\dagger c_j^\dagger c_l c_k$ acting on $|\psi_A\rangle$ produces a Slater determinant $|\psi_{A'}\rangle$ defined by the $N \times M$ matrix

$$A' = \left(I + \alpha I_{\bullet i}I_{k\bullet}^T + \beta I_{\bullet j}I_{l\bullet}^T\right)A. \quad (G.9)$$

Equations (G.7) and (G.8) define a discrete Hubbard-Stratonovich transformation for the general two-body interaction (G.2). After the Monte Carlo procedure selects an auxiliary field σ, (G.9) defines a simple procedure for updating the Slater determinant.

We now express this updating in a more familiar form. We start by letting $\Delta = \alpha I_{\bullet i}I_{k\bullet}^T + \beta I_{\bullet j}I_{l\bullet}^T$. If we are interested in the ratio of the overlap integrals of two Slater determinants after and before a proposed change, we have

$$\frac{\langle\psi_L|I + \alpha c_i^\dagger c_k + \beta c_j^\dagger c_l + \alpha\beta c_i^\dagger c_j^\dagger c_l c_k|\psi_R\rangle}{\langle\psi_L|\psi_R\rangle} = \det\left(I + G\Delta\right), \quad (G.10)$$

where $G = R(L^T R)^{-1}L^T$ is the single-particle Green's function matrix (Appendix I). Because of the sparsity of Δ, (G.10) reduces to a determinant of a 2×2 matrix

$$\det \begin{pmatrix} 1 + \alpha G_{ki} & \beta G_{li} \\ \alpha G_{kj} & 1 + \beta G_{lj} \end{pmatrix} = 1 + \alpha G_{ki} + \beta G_{lj} + \alpha\beta\left(G_{ki}G_{lj} - G_{li}G_{kj}\right),$$

a result we could have obtained from (G.10) by using Wick's theorem. Using the Sherman-Morrison formula (Meyer, 2000; Press et al., 2007)

$$\left(A + c d^T\right)^{-1} = A^{-1} - \frac{A^{-1}c d^T A^{-1}}{1 + d^T A^{-1}c}$$

twice yields an efficient procedure for updating the Green's function: First, we execute

$$B^{-1} \equiv \left(G + \beta I_{\bullet j} I_{l\bullet}^T\right)^{-1} = G^{-1} - \beta \frac{[G^{-1}]_{\bullet j} [G^{-1}]_{l\bullet}^T}{1 + \beta [G^{-1}]_{lj}}$$

and then

$$G' = \left(A + \alpha I_{\bullet i} I_{k\bullet}^T + \beta I_{\bullet j} I_{l\bullet}^T\right)^{-1} = \left(B + \alpha I_{\bullet i} I_{k\bullet}^T\right)^{-1} = B^{-1} - \alpha \frac{[B^{-1}]_{\bullet i} [B^{-1}]_{k\bullet}^T}{1 + \alpha [B^{-1}]_{ki}}.$$

Appendix H

Multi-electron propagator

In this appendix, we derive the expressions for the multi-electron propagator heavily used in Chapters 7 and 11 and in Appendix I. We choose to give the details for the derivations to introduce various techniques that are useful for other analyses and to engender a keener appreciation of the physical content of the multipropagator relation.

Our main objective is proving the identity

$$
\left\langle 0 \left| c_{i_1} c_{i_2} \cdots c_{i_{N_\sigma}} \left[\mathcal{T} \exp\left(-\int_{\tau_1}^{\tau_2} H(\tau)\, d\tau \right) \right] c_{j_{N_\sigma}}^\dagger \cdots c_{j_2}^\dagger c_{j_1}^\dagger \right| 0 \right\rangle =
$$

$$
\det \begin{pmatrix}
B_{i_1 j_1}(\tau_2, \tau_1) & B_{i_1 j_2}(\tau_2, \tau_1) & \cdots & B_{i_1 j_{N_\sigma}}(\tau_2, \tau_1) \\
B_{i_2 j_1}(\tau_2, \tau_1) & B_{i_2 j_2}(\tau_2, \tau_1) & \cdots & B_{i_2 j_{N_\sigma}}(\tau_2, \tau_1) \\
\vdots & \vdots & \ddots & \vdots \\
B_{i_{N_\sigma} j_1}(\tau_2, \tau_1) & B_{i_{N_\sigma} j_2}(\tau_2, \tau_1) & \cdots & B_{i_{N_\sigma} j_{N_\sigma}}(\tau_2, \tau_1)
\end{pmatrix}, \tag{H.1}
$$

which expresses the multielectron propagator as a determinant of a matrix of single-electron propagators

$$
B_{ij}(\tau_2, \tau_1) = \left\langle 0 \left| c_i \left[\mathcal{T} \exp\left(-\int_{\tau_1}^{\tau_2} H(\tau)\, d\tau \right) \right] c_j^\dagger \right| 0 \right\rangle. \tag{H.2}
$$

As a preliminary step, we prove a related result for electrons in nonoverlapping states, whose creation and destruction operators satisfy

$$
\{c_i^\dagger, c_j\} = \mathcal{B}_{ij}, \tag{H.3}
$$

where $\mathcal{B}_{ij} \equiv \langle 0 | c_i c_j^\dagger | 0 \rangle$ is the overlap. The result we first prove is

$$
\langle 0 | c_{i_1} c_{i_2} \cdots c_{i_{N_\sigma}} c_{j_{N_\sigma}}^\dagger \cdots c_{j_2}^\dagger c_{j_1}^\dagger | 0 \rangle = \det \begin{pmatrix}
\mathcal{B}_{i_1 j_1} & \mathcal{B}_{i_1 j_2} & \cdots & \mathcal{B}_{i_1 j_{N_\sigma}} \\
\mathcal{B}_{i_2 j_1} & \mathcal{B}_{i_2 j_2} & \cdots & \mathcal{B}_{i_2 j_{N_\sigma}} \\
\vdots & \vdots & \ddots & \vdots \\
\mathcal{B}_{i_{N_\sigma} j_1} & \mathcal{B}_{i_{N_\sigma} j_2} & \cdots & \mathcal{B}_{i_{N_\sigma} j_{N_\sigma}}
\end{pmatrix}. \tag{H.4}
$$

To demystify this result let us remark that if the electrons were in orthonormal states, $\mathcal{B}_{ij} = \delta_{ij}$, the determinant would equal a product of δ-functions and would simply express the normalization of the many-electron basis.

We first note that (H.4) is trivially true for $N_\sigma = 1$. We next assume that it is true for $N_\sigma - 1$. Then, in the inner product on the left-hand side of (H.3), we commute electron operators until we move c_{jN_σ} all the way to the right, so it operates directly on the vacuum state $|0\rangle$ and produces 0. The anticommutation relation (H.3) says this sequence of commutations generates

$$
\langle 0 | c_{i_1} c_{i_2} \cdots c_{iN_\sigma} c_{jN_\sigma}^\dagger \cdots c_{j_2}^\dagger c_{j_1}^\dagger | 0 \rangle
$$

$$
= \mathcal{B}_{iN_\sigma jN_\sigma} \langle 0 | c_{i_1} c_{i_2} \cdots c_{iN_\sigma -1} c_{jN_\sigma -1}^\dagger \cdots c_{j_2}^\dagger c_{j_1}^\dagger | 0 \rangle
$$

$$
- \mathcal{B}_{iN_\sigma jN_\sigma -1} \langle 0 | c_{i_1} c_{i_2} \cdots c_{iN_\sigma -1} c_{jN_\sigma}^\dagger c_{jN_\sigma -2}^\dagger \cdots c_{j_2}^\dagger c_{j_1}^\dagger | 0 \rangle
$$

$$
+ \mathcal{B}_{iN_\sigma jN_\sigma -2} \langle 0 | c_{i_1} c_{i_2} \cdots c_{iN_\sigma -1} c_{jN_\sigma}^\dagger c_{jN_\sigma -1}^\dagger c_{jN_\sigma -3}^\dagger \cdots c_{j_2}^\dagger c_{j_1}^\dagger | 0 \rangle
$$

$$
- \cdots .
$$

By assumption, the inner products on the right-hand side of the above equation are determinants for $N_\sigma - 1$ electrons, and accordingly the right side is the cofactor expansion of the determinant we would obtain by expanding down the far right-hand column of the matrix for the N_σ electron system. Thus, we have proven the identity (H.4).

To start the proof of (H.1), we consider the imaginary-time propagation of an N_σ-state piecewise in $\Delta\tau$ steps:

$$
\mathcal{T} \exp\left(-\int_0^\beta H(\tau) d\tau \right) \equiv e^{-\Delta\tau H(\beta)} \cdots e^{-\Delta\tau H(2\Delta\tau)} e^{-\Delta\tau H(\Delta\tau)}.
$$

Because of the Hubbard-Stratonovich transformation, the electron operator $H(\ell\Delta\tau)$ is a quadratic form in the electron creation and destruction operators. Designating the $N \times N$ matrix that defines this form as M_ℓ and defining vectors of creation and destruction operators as

$$
c = \begin{pmatrix} c_1 \\ c_2 \\ \vdots \\ c_N \end{pmatrix}, \quad c^\dagger = \begin{pmatrix} c_1^\dagger & c_2^\dagger & \cdots & c_N^\dagger \end{pmatrix}
$$

yields

$$
\mathcal{T} \exp\left(-\int_0^\beta H(\tau) d\tau \right) \propto e^{-\Delta\tau c^\dagger M_{N_\tau} c} \cdots e^{-\Delta\tau c^\dagger M_2 c} e^{-\Delta\tau c^\dagger M_1 c}. \tag{H.5}
$$

Our task now is propagating the right-hand N_σ-body state $|i_{N_\sigma}, \ldots, i_2, i_1\rangle = c_{i_{N_\sigma}}^\dagger \cdots c_{i_2}^\dagger c_{i_1}^\dagger |0\rangle$ in (H.1) by the product of these exponentials. We first prove that

$$e^{-\Delta\tau c^\dagger Mc} c^\dagger = c^\dagger e^{-\Delta\tau M} e^{-\Delta\tau c^\dagger Mc}, \tag{H.6}$$

where $c^\dagger Mc = \sum_{ij} c_i^\dagger M_{ij} c_j$. For this, we note that

$$\left(\sum_{ij} c_i^\dagger M_{ij} c_j\right) c_k^\dagger = \sum_i c_i^\dagger \left[M_{ik} + \delta_{ik} \sum_j c_i^\dagger M_{ij} c_j\right],$$

or more compactly that $\left(c^\dagger Mc\right) c^\dagger = c^\dagger \left[M + I\left(c^\dagger Mc\right)\right]$. Using this commutation relation n times, we obtain

$$\left(c^\dagger Mc\right)^n c^\dagger = c^\dagger \left[M + I\left(c^\dagger Mc\right)\right]^n.$$

Then, using the power series representation of the exponential of an operator, $\exp(A) = I + A + \frac{1}{2!}A^2 + \cdots$, we find $e^{-\Delta\tau c^\dagger Mc} c^\dagger = c^\dagger e^{-\Delta\tau M - \Delta\tau (c^\dagger Mc)I}$ and, noting that M and $c^\dagger Mc$ commute, the relation (H.6).

To get (H.1), we make a proof by induction similar to the one used to go from (H.3) to (H.4). We first note that after a Trotter breakup (H.5) and an application of Thouless's theorem (Appendix F)

$$\mathcal{T} \exp\left(-\int_{\tau_1}^{\tau_2} H(\tau)\, d\tau\right) c_{j_{N_\sigma}}^\dagger \cdots c_{j_2}^\dagger c_{j_1}^\dagger |0\rangle$$

equals (F.1) and

$$b_i^\dagger = \sum_j c_j^\dagger \left[e^{-\Delta\tau M_{N_\tau}} \cdots e^{-\Delta\tau M_2} e^{-\Delta\tau M_1}\right]_{ji}. \tag{H.7}$$

We now define

$$B_{ij}(\tau_2, \tau_1) = \left[e^{-\Delta\tau M_{N_\tau}} \cdots e^{-\Delta\tau M_2} e^{-\Delta\tau M_1}\right]_{ij}.$$

With this definition, we see that the inner product in (H.1) is trivially true for $N_\sigma = 1$:

$$B_{i_1 j_1}(\tau_2, \tau_2) = \left\langle 0 \left| c_{i_1} \mathcal{T} \exp\left(-\int_{\tau_1}^{\tau_2} H(\tau)\, d\tau\right) c_{j_1}^\dagger \right| 0 \right\rangle$$

$$= \langle 0 | c_{i_1} b_{j_1}^\dagger | 0\rangle = B_{i_1 j_1}(\tau_2, \tau_1) \langle 0 | c_{i_1} c_{j_1}^\dagger | 0\rangle.$$

We next assume that it is true for $N_\sigma - 1$. Then, in the inner product on the left-hand side, we commute electron operators until we move $c_{j_{N_\sigma}}$ all the way to the right so it operates directly on the vacuum state $|0\rangle$ and produces 0. The relation (H.7) says that this sequence of commutations generates

$$\langle 0| \, c_{i_1} c_{i_2} \cdots c_{i_{N_\sigma}} \, c_{j_{N_\sigma}}^\dagger \cdots c_{j_2}^\dagger c_{j_1}^\dagger \, |0\rangle$$

$$= B_{i_{N_\sigma} j_{N_\sigma}} \, \langle 0| \, c_{i_1} c_{i_2} \cdots c_{i_{N_\sigma}-1} c_{j_{N_\sigma}-1}^\dagger \cdots c_{j_2}^\dagger c_{j_1}^\dagger \, |0\rangle$$

$$- B_{i_{N_\sigma} j_{N_\sigma}-1} \, \langle 0| \, c_{i_1} c_{i_2} \cdots c_{i_{N_\sigma}-1} c_{j_{N_\sigma}}^\dagger c_{j_{N_\sigma}-2}^\dagger \cdots c_{j_2}^\dagger c_{j_1}^\dagger \, |0\rangle$$

$$+ B_{i_{N_\sigma} j_{N_\sigma}-2} \, \langle 0| \, c_{i_1} c_{i_2} \cdots c_{i_{N_\sigma}-1} c_{j_{N_\sigma}}^\dagger c_{j_{N_\sigma}-1}^\dagger c_{j_{N_\sigma}-3}^\dagger \cdots c_{j_2}^\dagger c_{j_1}^\dagger \, |0\rangle$$

$$- \cdots,$$

where we left out the τ dependence of the B matrix elements for notational simplicity. By assumption, the inner products on the right-hand side are determinants for $N_\sigma - 1$ electrons, and accordingly, the right side is the cofactor expansion of the determinant we would obtain by expanding down the far right-hand column of the matrix for the N_σ-electron system. Thus, we have proven the identity (H.1).

Appendix I

Zero-temperature determinant method

While the zero-temperature determinant algorithm projects to the ground state from a trial state, it does so in a way different from that of the power methods discussed in Chapter 10. It focuses on (7.1) for a large but fixed value of β and is a procedure to estimate expectation values

$$\langle A \rangle = \frac{\langle \psi_L | e^{-(\beta-\tau)H} A e^{-\tau H} | \psi_R \rangle}{\langle \psi_L | e^{-(\beta-\tau)H} e^{-\tau H} | \psi_R \rangle}. \tag{I.1}$$

Usually, $\tau - \frac{1}{2}\beta$. One may also calculate

$$\langle A \rangle = \frac{\int_{\beta/2-\tau_1}^{\beta/2+\tau_1} d\tau \, \langle \psi_L | e^{-(\beta-\tau)H} A e^{-\tau H} | \psi_R \rangle}{\int_{\beta/2-\tau_1}^{\beta/2+\tau_1} d\tau \, \langle \psi_L | e^{-(\beta-\tau)H} e^{-\tau H} | \psi_R \rangle},$$

that is, average the expectation values over a small interval around $\frac{1}{2}\beta$ to reduce the variance. It is also possible to extend the integration so it ranges from 0 to β. Our purposes are served if we take $\tau = \frac{1}{2}\beta$.

The simulations are done for N_σ fixed. Almost always, $|\psi_L\rangle = |\psi_R\rangle$. These states are usually chosen to be a superposition of basis states for fixed N^σ, for example, the noninteracting or the Hartree-Fock solution for the system being simulated.

Finite-sized systems of electrons display shell effects caused by degenerate states being separated by energy gaps. For *closed shells*, electrons occupy all states below some gap and hence are in a nondegenerate state. In this case, the $|\psi_L\rangle$ and $|\psi_R\rangle$ are often adequately chosen to be the noninteracting state. For *open shells*, not all states below the gap are occupied, and hence the state is degenerate. In this case it is better to choose $|\psi_L\rangle$ and $|\psi_R\rangle$ as a symmetry-adapted linear combination $\sum_\phi c_\phi |\phi\rangle$ of the noninteracting degenerate states. Then, even if $|\psi_L\rangle = |\psi_R\rangle$, we need the expectation values of operators between a $|\phi'\rangle$ and $|\phi\rangle$ that differ. In the following, we assume that the $|\psi_L\rangle = |\psi_L^\uparrow\rangle|\psi_L^\downarrow\rangle$ and $|\psi_R\rangle = |\psi_R^\uparrow\rangle|\psi_R^\downarrow\rangle$ are represented by single but different Slater determinants defined by $N_\sigma \times N$ and $N \times N_\sigma$ matrices Φ_L^σ and Φ_R^σ. For example,

$$|\psi_R^\sigma\rangle = \left(\Phi_{11}c_{1\sigma}^\dagger + \Phi_{21}c_{2\sigma}^\dagger + \cdots + \Phi_{N1}c_{N\sigma}^\dagger\right)$$
$$\left(\Phi_{12}c_{1\sigma}^\dagger + \Phi_{22}c_{2\sigma}^\dagger + \cdots + \Phi_{N2}c_{N\sigma}^\dagger\right)$$
$$\cdots$$
$$\left(\Phi_{1N_\sigma}c_{1\sigma}^\dagger + \Phi_{2N_\sigma}c_{2\sigma}^\dagger + \cdots + \Phi_{NN_\sigma}c_{N\sigma}^\dagger\right)|0\rangle,$$

where

$$\Phi_R^\sigma = \begin{pmatrix} \Phi_{11} & \Phi_{12} & \cdots & \Phi_{1N_\sigma} \\ \Phi_{21} & \Phi_{22} & \cdots & \Phi_{2N_\sigma} \\ \vdots & \vdots & \ddots & \vdots \\ \Phi_{N1} & \Phi_{N2} & \cdots & \Phi_{NN_\sigma} \end{pmatrix}.$$

As is standard, we express the denominator of (I.1) as a path-integral (7.1), make the Trotter approximation and the Hubbard-Stratonovich transformation (7.9), and then recycle the analysis of Section 7.1.2 to find

$$w_C = \prod_\sigma \det\left[\Phi_L^\sigma B^\sigma(\beta,0)\Phi_R^\sigma\right].$$

With the definitions

$$L^\sigma(\tau) = \Phi_L^\sigma B^\sigma(\beta,\tau), \quad R^\sigma(\tau) = B^\sigma(\tau,0)\Phi_R^\sigma,$$

we can show that

$$R^\sigma = \det\left[I + \Delta^\sigma R^\sigma(L^\sigma R^\sigma)^{-1}L^\sigma\right] = \det\left[I + \Delta^\sigma(I - G^\sigma)\right],$$

which reduces to (7.43). In zero-temperature fixed N^σ calculations, L^σ and R^σ are $N^\sigma \times N$ and $N \times N^\sigma$ rectangular matrices. We still update the Green's function matrix by (7.45).

For the measurement of the equal-time Green's function, an analysis similar to that in Section 7.2.2 yields the formula[1]

$$G_{ij}^\sigma(\tau,\tau) = \delta_{ji} - \left\langle c_{j\sigma}^\dagger(\tau) c_{i\sigma}(\tau)\right\rangle = \delta_{ji} - \left[R^\sigma(L^\sigma R^\sigma)^{-1}L^\sigma\right]_{ji}. \qquad (I.2)$$

Here, in contrast to the finite-temperature algorithm, τ should be interpreted not as an imaginary time, but rather as a Trotter index. For "equal time" measurements nothing changes, while unequal-time measurements are unphysical. If we move from one Trotter step to a later one, we can use relations such as

---

[1] Note the transposition of the indices.

$$L^\sigma(\tau') = L^\sigma(\tau)B^\sigma(\tau',\tau)^{-1}, \quad R^\sigma(\tau') = B^\sigma(\tau',\tau)R^\sigma(\tau).$$

At any Trotter index, the Green's function matrix is primarily composed of two strings of B^σ matrices, one whose number of factors is decreasing from the left and another whose number of factors is increasing from the right as imaginary time advances. Creating and storing the sequences of partial matrix products (and their factorizations) accelerates the simulation. We can store the partial products for each spin σ as indicated in the following equation:

$$\begin{aligned}
L^\sigma(\beta) = P_L^\sigma &\rightarrow L^\sigma(\beta) = P_L^\sigma, \\
L^\sigma(\beta - \Delta\tau) &\rightarrow L^\sigma(\beta - \Delta\tau), \\
&\vdots \qquad\qquad \vdots \\
L^\sigma(\tau + \Delta\tau) &\rightarrow L^\sigma(\tau + \Delta\tau), \\
L^\sigma(\tau) &\rightarrow R^\sigma(\tau + \Delta\tau), \\
R^\sigma(\tau) &\rightarrow R^\sigma(\tau), \\
&\vdots \qquad\qquad \vdots \\
R^\sigma(\Delta\tau) &\rightarrow R^\sigma(\Delta\tau), \\
R^\sigma(0) = P_R^\sigma &\rightarrow R^\sigma(0) = P_R^\sigma.
\end{aligned}$$

The partial products generated in the computation of $L^\sigma(\tau)$ and $R^\sigma(\tau)$ are stored. To move from time τ to time $\tau + \Delta\tau$ we need the factors $L^\sigma(\tau + \Delta\tau)$ and $R^\sigma(\tau + \Delta\tau)$. The left factor $L^\sigma(\tau + \Delta\tau)$ is already available. The right factor $R^\sigma(\tau + \Delta\tau) = B^\sigma(\tau + \Delta\tau, \tau)R^\sigma(\tau)$ is easily computed. $R(\tau)$ may be stored in lieu of $L^\sigma(\tau)$, which is no longer needed. With this storage scheme, we can create the matrix needing inversion just with a few matrix multiplications. The cost of repeatedly stabilizing the matrix inversion becomes minor. Storing partial products in this manner has the added advantage of reducing the accumulation of round-off errors. Instead of storing the partial products for each Trotter index, doing the storage every so many steps is still advantageous. For example, in the above we replace $\Delta\tau \rightarrow m\Delta\tau$.

The matrix factorization procedure for doing the matrix inversion is a bit different. Using the factored forms

$$L^\sigma(\tau) = V_L^\sigma D_L^\sigma U_L^\sigma, \quad R^\sigma(\tau) = U_R^\sigma D_R^\sigma V_R^\sigma$$

in the expression for the Green's function

$$G^\sigma(\tau,\tau) = I - R^\sigma(\tau)\left[L^\sigma(\tau)R^\sigma(\tau)\right]^{-1}L^\sigma(\tau) \tag{I.3}$$

we find that

$$\begin{aligned}
G^\sigma &= I - \left(U_R^\sigma D_R^\sigma V_R^\sigma\right)\left[\left(V_L^\sigma D_L^\sigma U_L^\sigma\right)\left(U_R^\sigma D_R^\sigma V_R^\sigma\right)\right]^{-1}\left(V_L^\sigma D_L^\sigma U_L^\sigma\right) \\
&= I - U_R^\sigma\left(U_L^\sigma U_R^\sigma\right)^{-1}U_L^\sigma. \tag{I.4}
\end{aligned}$$

Algorithm 50 Modified Gram-Schmidt method: matrix column orthonormalization.

Input: $M \times N$ matrix U.

  for $k = 1$ to N **do**

    $d = 0$;

    for $i = 1$ to M **do**

      $d \leftarrow d + U(i,k)U(i,k)$;

    end for

    $d \leftarrow \sqrt{d}$;

    for $i = 1$ to M **do**

      $U(i,k) \leftarrow U(i,k)/d$;

    end for

    for $j = k + 1$ to N **do**

      $v = 0$;

      for $i = 1$ to N **do**

        $v \leftarrow v + U(i,k)U(i,j)$;

      end for

      for $i = 1$ to M **do**

        $U(i,j) \leftarrow U(i,j) - vU(i,k)$;

      end for

    end for

  end for

  return the column-orthonormalized matrix U.

The U_L^σ and U_R^σ are $N^\sigma \times N$ and $N \times N^\sigma$ rectangular matrices, while the D_L^σ, V_L^σ, D_L^σ, and V_L^σ matrices are $N^\sigma \times N^\sigma$. The Green's function, however, depends only on the rectangular U_L^σ and U_R^σ matrices, so we can stabilize the calculation (Algorithm 50) just by orthonormalizing the rows of L^σ and columns of R^σ. Sugiyama and Koonin (1986) and Sorella et al. (1989) were among the first to note the importance of matrix column orthonormalization.

The zero-temperature algorithm is an elegant complement to the finite-temperature one. Unfortunately, it too has a sign problem. The more recent lattice fixed-node, constrained-path, and constrained-phase methods (Chapter 11) are useful in taming it.

Appendix J

Anderson impurity model: chain representation

This appendix explains the mapping of the Anderson impurity model (8.8)–(8.11) onto a semi-infinite chain. For simplicity, we may assume that the hybridization parameters V_p are real. The goal is to transform the impurity and bath annihilation operators $\{d, c_{p_1}, c_{p_2}, \ldots\}$ to new operators $\{d, c_1, c_2, \ldots\}$ such that $H_0 + H_{\text{bath}} + H_{\text{mix}}$ becomes tridiagonal:

$$
\begin{pmatrix} d^\dagger & c_{p_1}^\dagger & c_{p_2}^\dagger & c_{p_3}^\dagger & \cdots \end{pmatrix} UU^T
\begin{pmatrix}
-\mu & V_{p_1} & V_{p_2} & V_{p_3} & \cdots \\
V_{p_1} & \varepsilon_{p_1} & & & \\
V_{p_2} & & \varepsilon_{p_2} & & \\
V_{p_3} & & & \varepsilon_{p_3} & \\
\vdots & & & & \ddots
\end{pmatrix}
UU^T
\begin{pmatrix} d \\ c_{p_1} \\ c_{p_2} \\ c_{p_3} \\ \vdots \end{pmatrix}
$$

$$
= \begin{pmatrix} d^\dagger & c_1^\dagger & c_2^\dagger & c_3^\dagger & \cdots \end{pmatrix}
\begin{pmatrix}
-\mu & -V & & & \\
-V & \tilde{\varepsilon}_1 & -t_1 & & \\
& -t_1 & \tilde{\varepsilon}_2 & -t_2 & \\
& & -t_2 & \tilde{\varepsilon}_3 & \ddots \\
& & & \ddots & \ddots
\end{pmatrix}
\begin{pmatrix} d \\ c_1 \\ c_2 \\ c_3 \\ \vdots \end{pmatrix}.
$$

The symmetric and orthogonal matrix U associated with this tranformation can be constructed as a product of *Householder transformations*

$$
h_v = 1 - \frac{2\vec{v}\vec{v}^T}{|\vec{v}|^2}, \tag{J.1}
$$

which describe mirror operations. Let us assume that the bath has N sites. In the first step, we choose v such that the vector $\vec{V} = (V_{p_1} \ V_{p_2} \ V_{p_3} \cdots V_{p_N})^T$ is mapped onto the first element $\vec{e}_1$: $\vec{v}_1 = \vec{V} \mp |V|\vec{e}_1 \Rightarrow h_{v_1}\vec{V} = \pm|V|\vec{e}_1$, with $|V|^2 = \sum_{p=1}^{N} |V_p|^2$. Defining the block matrix

$$U_1 = \left(\begin{array}{c|c} 1 & \\ \hline & h_{v_1} \end{array} \right) \qquad (\text{J.2})$$

with diagonal blocks of dimension 1 and N, respectively, we obtain the transformed Hamiltonian

$$U_1 \left(\begin{array}{c|ccc} -\mu & V_{p1} & V_{p2} & \cdots \\ \hline V_{p1} & \varepsilon_{p1} & & \\ V_{p2} & & \ddots & \\ \vdots & & & \end{array} \right) U_1 = \left(\begin{array}{c|ccc} -\mu & \pm|V| & 0 & \cdots \\ \hline \pm|V| & & & \\ 0 & & A_2 & \\ \vdots & & & \end{array} \right). \qquad (\text{J.3})$$

We can now repeat the procedure, and define a transformation h_{v_2} that maps the vector defined by the off-diagonal elements in the first column of the $N \times N$ matrix A_2 onto $\vec{e}_2$. The Householder matrix

$$U_2 = \left(\begin{array}{cc|c} 1 & 0 & \\ 0 & 1 & \\ \hline & & h_{v_2} \end{array} \right) \qquad (\text{J.4})$$

with diagonal blocks of dimension 2 and $N-1$ may then be used to bring the Hamiltonian into tridiagonal form up to the second row and column. The transformation U for an impurity model with N bath sites eventually becomes

$$U = U_1 U_2 \cdots U_{N-1}. \qquad (\text{J.5})$$

In the case of multi-orbital systems with diagonal baths, similar transformations map the system onto a star-shaped geometry, with the impurity site at the center.

Appendix K

Anderson impurity model: action formulation

In this appendix, we derive the effective action for the Anderson impurity model (8.8)–(8.11). We denote the impurity annihilation operators by d, the bath annihilation operators by c, and an arbitrary type of operator by a. The partition function of the impurity model is

$$Z = \text{Tr}_d \text{Tr}_c \left[e^{-\beta H} \right].$$

To derive the action, we introduce the *coherent states*

$$|\xi\rangle = \exp\left[-\sum_\alpha \xi_\alpha a_\alpha^\dagger \right] |0\rangle,$$

which are eigenstates of the destruction operators: $a_\alpha |\xi\rangle = \xi_\alpha |\xi\rangle$. The ξ_α are elements of a Grassmann algebra, which allows us to account for the anticommutation relations of the Fermionic operators in a path-integral formulation. Specifically, the ξ_α satisfy

$$\{\xi_\alpha, \xi_\beta\} = \{\xi_\alpha^*, \xi_\beta\} = \{\xi_\alpha^*, \xi_\beta^*\} = 0,$$

$$\{\xi_\alpha, a_\beta\} = \{\xi_\alpha, a_\beta^\dagger\} = \{\xi_\alpha^*, a_\beta\} = \{\xi_\alpha^*, a_\beta^\dagger\} = 0,$$

$$\partial_{\xi_\alpha} \xi_\alpha = \int d\xi_\alpha \xi_\alpha = 1,$$

$$\partial_{\xi_\alpha} 1 = \int d\xi_\alpha = 0.$$

Three important relations of a Grassman algebra, which we will need in the following, are the scalar product, the closure relation, and the formula for Gaussian integrals:

$$\langle \xi | \psi \rangle = e^{\sum_\alpha \xi_\alpha^* \psi_\alpha}, \tag{K.1}$$

$$\int \prod_\alpha d\xi_\alpha^* d\xi_\alpha e^{-\sum_\alpha \xi_\alpha^* \xi_\alpha} |\xi\rangle \langle \xi| = 1, \tag{K.2}$$

$$\int \prod_\alpha d\xi_\alpha^* d\xi_\alpha e^{-\sum_{\alpha\beta} \xi_\alpha^* M_{\alpha\beta} \xi_\beta + J_\alpha^* \xi_\alpha + \xi_\alpha^* J_\alpha} = (\det M) e^{\sum_{\alpha\beta} J_\alpha^* [M^{-1}]_{\alpha\beta} J_\beta}. \tag{K.3}$$

Using the coherent states and the anticommutation of Grassmann numbers, we express the partition function as

$$Z = \sum_n \langle n | e^{-\beta H} | n \rangle = \int \prod_\alpha d\xi_\alpha^* d\xi_\alpha e^{-\sum_\alpha \xi_\alpha^* \xi_\alpha} \sum_n \langle n | \xi \rangle \langle \xi | e^{-\beta H} | n \rangle$$

$$= \int \prod_\alpha d\xi_\alpha^* d\xi_\alpha e^{-\sum_\alpha \xi_\alpha^* \xi_\alpha} \langle -\xi | e^{-\beta H} | \xi \rangle = \int d\vec{\xi}^* d\vec{\xi} e^{-\vec{\xi}^* \cdot \vec{\xi}} \langle -\xi | e^{-\beta H} | \xi \rangle,$$

where $\{|n\rangle\}$ is some basis of the impurity model Hilbert space. In the last step, we introduced vectors of Grassmann variables to simplify the notation.

By repeatedly inserting the closure relation (K.2) and introducing antiperiodic boundary conditions, we find

$$Z = \int \prod_{m=0}^{M-1} d\vec{\xi}_m^* d\vec{\xi}_m e^{-\sum_{m=1}^{M-1} \vec{\xi}_m^* \cdot \vec{\xi}_m} e^{-\vec{\xi}_0 \cdot \vec{\xi}_0} \langle -\xi_0 | e^{-\frac{\beta}{M} H} | \xi_{M-1} \rangle \cdots \langle \xi_1 | e^{-\frac{\beta}{M} H} | \xi_0 \rangle$$

$$\approx \int \prod_{m=1}^{M} d\vec{\xi}_m^* d\vec{\xi}_m e^{-\sum_{m=1}^{M} \vec{\xi}_m^* \cdot \vec{\xi}_m} \prod_{m=1}^{M} \langle \xi_m | 1 - \Delta\tau H(a^\dagger, a) | \xi_{m-1} \rangle$$

$$= \int \prod_{m=1}^{M} d\vec{\xi}_m^* d\vec{\xi}_m e^{-\sum_{m=1}^{M} \vec{\xi}_m^* \cdot \vec{\xi}_m} \prod_{m=1}^{M} (1 - \Delta\tau H(\vec{\xi}_m^*, \vec{\xi}_{m-1})) \langle \xi_m | \xi_{m-1} \rangle$$

$$\approx \int \prod_{m=1}^{M} d\vec{\xi}_m^* d\vec{\xi}_m e^{-\sum_{m=1}^{M} \vec{\xi}_m^* \cdot \vec{\xi}_m} e^{\sum_{m=1}^{M} [-\Delta\tau H(\vec{\xi}_m^*, \vec{\xi}_{m-1}) + \vec{\xi}_m^* \cdot \vec{\xi}_{m-1}]},$$

with $\Delta\tau = \beta/M$. In the limit $\Delta\tau \to 0$ the exponent becomes

$$-\sum_{m=1}^{M} \Delta\tau \left[\vec{\xi}_m^* \cdot \frac{\vec{\xi}_m - \vec{\xi}_{m-1}}{\Delta\tau} + H(\vec{\xi}_m^*, \vec{\xi}_{m-1}) \right]$$

$$\to -\int_0^\beta d\tau \left[\vec{\xi}^*(\tau) \cdot \partial_\tau \vec{\xi}(\tau) + H(\vec{\xi}^*, \vec{\xi}) \right].$$

Hence, we obtain the expression for the partition function as a path integral over Grassmann variables:

$$Z = \int D\vec{\xi}^* D\vec{\xi}\, e^{-S(\vec{\xi}^*,\vec{\xi})},$$

$$S(\vec{\xi}^*,\vec{\xi}) = \int_0^\beta d\tau \left[\vec{\xi}^*(\tau)\cdot\partial_\tau\vec{\xi}(\tau) + H(\vec{\xi}^*,\vec{\xi}) \right].$$

To formally define the effective action for the impurity, we distinguish between the Grassmann variables ξ_d associated with the impurity and ξ_c associated with the bath and write

$$Z = \int D\vec{\xi}_d^* D\vec{\xi}_d D\vec{\xi}_c^* D\vec{\xi}_c\, e^{-S(\vec{\xi}_d^*,\vec{\xi}_d,\vec{\xi}_c^*,\vec{\xi}_c)} \equiv \int D\vec{\xi}_d^* D\vec{\xi}_d\, e^{-S_{\text{eff}}(\vec{\xi}_d^*,\vec{\xi}_d)}.$$

The Hamiltonian of our impurity model is of the form

$$H = H_{\text{loc}} + \sum_{p\sigma} \varepsilon_p c_{p\sigma}^\dagger c_{p\sigma} + \sum_{p\sigma}(V_{p\sigma} d_\sigma^\dagger c_{p\sigma} + V_{p\sigma}^* c_{p\sigma}^\dagger d_\sigma),$$

so the action explicitly reads

$$S = \int_0^\beta d\tau \left[\sum_\sigma \xi_{d\sigma}^*(\tau)\partial_\tau\xi_{d\sigma}(\tau) + H_{\text{loc}}(\vec{\xi}_d^*(\tau),\vec{\xi}_d(\tau)) + \sum_\sigma \left(\vec{\xi}_{c\sigma}^*(\tau)\cdot\partial_\tau\vec{\xi}_{c\sigma}(\tau) \right. \right.$$

$$\left. \left. + \vec{\xi}_{c\sigma}^*(\tau)\cdot\varepsilon\vec{\xi}_{c\sigma}(\tau) + \xi_{d\sigma}^*(\tau)\vec{V}_\sigma\cdot\vec{\xi}_{c\sigma}(\tau) + \vec{\xi}_{c\sigma}^*(\tau)\cdot\vec{V}_\sigma^*\xi_{d\sigma}(\tau) \right) \right]$$

$$= \int_0^\beta d\tau \left[\sum_\sigma \xi_{d\sigma}^*(\tau)\partial_\tau\xi_{d\sigma}(\tau) + H_{\text{loc}}(\vec{\xi}_d^*(\tau),\vec{\xi}_d(\tau)) + \sum_\sigma \left(-\vec{\xi}_{c\sigma}^*(\tau)\cdot M\vec{\xi}_{c\sigma}(\tau) \right. \right.$$

$$\left. \left. + \vec{J}_\sigma^*(\tau)\cdot\vec{\xi}_{c\sigma}(\tau) + \vec{\xi}_{c\sigma}^*(\tau)\cdot\vec{J}_\sigma(\tau) \right) \right],$$

where in the second line we have introduced the matrix $M_{ij} = \delta_{ij}(-\partial_\tau - \varepsilon_i)$ and the vectors $\vec{J}_\sigma^*(\tau) = \xi_{d\sigma}^*(\tau)\vec{V}_\sigma$ and $\vec{J}_\sigma(\tau) = \vec{V}_\sigma^*\xi_{d\sigma}(\tau)$.

We now use (K.3) to integrate out the bath degrees of freedom:

$$S_{\text{eff}} = \text{const} + \int_0^\beta d\tau \left[\sum_\sigma \xi_{d\sigma}^*(\tau)\partial_\tau\xi_{d\sigma}(\tau) + H_{\text{loc}}(\vec{\xi}_d^*(\tau),\vec{\xi}_d(\tau)) \right]$$

$$+ \int_0^\beta d\tau d\tau' \sum_\sigma \xi_{d\sigma}^*(\tau) \underbrace{\vec{V}_\sigma M^{-1}\vec{V}_\sigma^*}_{\equiv \Delta_\sigma(\tau-\tau')} \xi_{d\sigma}(\tau').$$

The coupling to the bath leads to a retarded term in the effective action, which is given by the *hybridization function* $\Delta_\sigma(\tau)$ defined in (8.14). In fact, the Matsubara coefficients of this function are[1]

$$\Delta_\sigma(i\omega_n) = \sum_p \frac{|V_{p\sigma}|^2}{i\omega_n - \varepsilon_p}.$$

Inserting the definition of $H_{\text{loc}} = Un_\uparrow n_\downarrow - \mu(n_\uparrow + n_\downarrow)$, we finally obtain

$$S_{\text{eff}} = \int_0^\beta d\tau\, Un_\uparrow(\tau)n_\downarrow(\tau)$$

$$+ \int_0^\beta d\tau d\tau' \sum_\sigma \left[\xi_{d\sigma}^*(\tau)\delta(\tau - \tau')\partial_\tau \xi_{d\sigma}(\tau') - \mu\xi_{d\sigma}^*(\tau)\delta(\tau - \tau')\xi_{d\sigma}(\tau')\right.$$

$$\left. + \xi_{d\sigma}^*(\tau)\Delta_\sigma(\tau - \tau')\xi_{d\sigma}(\tau')\right]$$

$$= \int_0^\beta d\tau\, Un_\uparrow(\tau)n_\downarrow(\tau) - \int_0^\beta d\tau d\tau' \sum_\sigma \xi_{d\sigma}^*(\tau)\mathcal{G}_{0\sigma}^{-1}(\tau - \tau')\xi_{d\sigma}(\tau').$$

$\mathcal{G}_{0\sigma}$ is the "Weiss Green's function" (8.15), which is related to the hybridization function by

$$\mathcal{G}_{0\sigma}^{-1}(i\omega_n) = i\omega_n + \mu - \Delta_\sigma(i\omega_n).$$

The Grassmann path integral expression for the partition function is therefore

$$Z = \int D\vec{\xi}_d^* D\vec{\xi}_d e^{-\int_0^\beta d\tau\, Un_\uparrow(\tau)n_\downarrow(\tau) + \int_0^\beta d\tau d\tau' \sum_\sigma \xi_{d\sigma}^*(\tau)\mathcal{G}_{0\sigma}^{-1}(\tau - \tau')\xi_{d\sigma}(\tau')}.$$

Alternatively, we can express the partition function in terms of the operators d, $d^\dagger$, and the hybridization function Δ:

$$Z = \text{Tr}_d\left[\mathcal{T}e^{-\int_0^\beta d\tau[Un_\uparrow(\tau)n_\downarrow(\tau) - \mu(n_\uparrow(\tau) + n_\downarrow(\tau))] - \int_0^\beta d\tau d\tau' \sum_\sigma d_\sigma^\dagger(\tau)\Delta_\sigma(\tau - \tau')d_\sigma(\tau')}\right].$$

---

[1] The Fourier transform of ∂_τ is $-i\omega_n$.

Appendix L

Continuous-time auxiliary-field algorithm

This appendix presents a continuous-time impurity solver that combines the weak-coupling expansion with an auxiliary-field decomposition of the interaction term (Gull et al., 2008). This *continuous-time auxiliary-field method* (CTAUX) is an adaptation of an algorithm by Rombouts et al. (1999b) for Fermion lattice models. It is suitable for models with density-density interactions and in this case is equivalent to the weak-coupling continuous-time method of Section 8.4.

We present the algorithm for the Anderson impurity model (8.8)–(8.11). Following Rombouts and collaborators, we define $H_2 = H_U - \frac{1}{2}U(n_\uparrow + n_\downarrow) - K/\beta$ and $H_1 = H - H_2 = H_\mu + \frac{1}{2}U(n_\uparrow + n_\downarrow) + H_{\text{bath}} + H_{\text{mix}} + K/\beta$, with K being some arbitrary nonzero constant. After the expansion of the partition function in H_2, (8.30) gives the statistical weight of a configuration of n "interaction vertices." We now extend our configuration space by decoupling each interaction vertex using the discrete Hubbard-Stratonovich decoupling formula (Hirsch, 1983)

$$-H_2 = K/\beta - U(n_\uparrow n_\downarrow - \tfrac{1}{2}(n_\uparrow + n_\downarrow)) = \frac{K}{2\beta} \sum_{s=-1,1} e^{\gamma s(n_\uparrow - n_\downarrow)},$$

with

$$\cosh(\gamma) = 1 + (\beta U)/(2K). \tag{L.1}$$

The configuration space becomes the collection of all possible auxiliary spin configurations $C = \{(\tau_1, s_1), \ldots, (\tau_n, s_n)\}$, $n = 0, 1, \ldots$, $\tau_i \in [0, \beta)$, $s_i = \pm 1$ on the imaginary-time interval $[0, \beta)$. These configurations have the weight

$$w_C = \text{Tr}\left[e^{-(\beta - \tau_n)H_1} e^{\gamma s_n(n_\uparrow - n_\downarrow)} \cdots e^{-(\tau_2 - \tau_1)H_1} e^{\gamma s_1(n_\uparrow - n_\downarrow)} e^{-\tau_1 H_1} \right] \left(\frac{K d\tau}{2\beta} \right)^n. \tag{L.2}$$

We first separate the electron spin components and then proceed to the analytical calculation of the trace. Introducing $H_1^\sigma = -(\mu - \frac{1}{2}U)n_\sigma + \sum_p \varepsilon_p c_{p\sigma}^\dagger c_{p\sigma} + \sum_p (V_{p\sigma} d_\sigma^\dagger c_{p\sigma} + V_{p\sigma}^* c_{p\sigma}^\dagger d_\sigma)$, we factorize the trace in (L.2) as

$$\mathrm{Tr}\,[\ldots] \propto e^{-K} \prod_\sigma \mathrm{Tr} \left[e^{-(\beta - \tau_n)H_1^\sigma} e^{\gamma s_n \sigma n_\sigma} \cdots e^{-(\tau_2 - \tau_1)H_1^\sigma} e^{\gamma s_1 \sigma n_\sigma} e^{-\tau_1 H_1^\sigma} \right]. \qquad (\text{L.3})$$

Then, using the identity $e^{\gamma s \sigma n_\sigma} = e^{\gamma s \sigma} d_\sigma^\dagger d_\sigma + d_\sigma d_\sigma^\dagger = e^{\gamma s \sigma} - (e^{\gamma s \sigma} - 1)d_\sigma d_\sigma^\dagger$, we express the trace factors in terms of noninteracting impurity Green's functions $\tilde{\mathcal{G}}_0$ (with chemical potential shifted as $\mu \to \tilde{\mu} = \mu - \frac{1}{2}U$):

$$\mathrm{Tr} \left[e^{-(\beta - \tau_n)H_1^\sigma} e^{\gamma s_n \sigma n_\sigma} \cdots e^{-(\tau_2 - \tau_1)H_1^\sigma} e^{\gamma s_1 \sigma n_\sigma} e^{-\tau_1 H_1^\sigma} \right] = \tilde{Z}_0^\sigma \det N_\sigma^{-1}(\{s_i, \tau_i\}), \quad (\text{L.4})$$

where $\tilde{Z}_0^\sigma = \mathrm{Tr}\,[e^{-\beta H_1^\sigma}]$. Here, N_σ represents an $n \times n$ matrix defined by the location of the decoupled interaction vertices, the spin orientations, and the noninteracting Green's functions:

$$N_\sigma^{-1}(\{s_i, \tau_i\}) = e^{\Gamma_\sigma} - [\tilde{\mathcal{G}}_0^\sigma]\left(e^{\Gamma_\sigma} - I\right),$$

where we used the notations $e^{\Gamma_\sigma} \equiv \mathrm{diag}(e^{\gamma \sigma s_1}, \ldots, e^{\gamma \sigma s_n})$, $[\tilde{\mathcal{G}}_0^\sigma]_{ij} = \tilde{\mathcal{G}}_0^\sigma(\tau_i - \tau_j)$ for $i \neq j$, $[\tilde{\mathcal{G}}_0^\sigma]_{ii} = \tilde{\mathcal{G}}_0^\sigma(0^+)$.[1] Combining (L.2), (L.3), and (L.4) yields

$$w_C = e^{-K} \left(\frac{K d\tau}{2\beta} \right)^n \prod_\sigma \tilde{Z}_0^\sigma \det N_\sigma^{-1}(\{s_i, \tau_i\})$$

for the weight of configuration $C = \{(\tau_1, s_1), \ldots, (\tau_n, s_n)\}$.

For ergodicity, it is sufficient to insert and remove spins with random orientations at random times. The sampling procedure with fast matrix updates is thus completely analogous to that of the weak-coupling algorithm discussed in Section 8.4.

To compute the contribution of a configuration C to the Green's function, $G_C^\sigma(\tau)$, we insert in (L.2) a creation operator $d_\sigma^\dagger$ at time 0 and an annihilation operator d_σ at time τ and divide by w_C. Wick's theorem then leads to

$$G_C^\sigma(\tau) = \tilde{\mathcal{G}}_0^\sigma(\tau) + \sum_{k=1}^n \sum_{l=1}^n \tilde{\mathcal{G}}_0^\sigma(\tau - \tau_k)(e^{\gamma \sigma s_k} - 1)[N_\sigma^{(n)}]_{kl}\tilde{\mathcal{G}}_0^\sigma(\tau_l).$$

As in the weak-coupling algorithm, it is advantageous to accumulate the quantity

$$S_\sigma(\tilde{\tau}) \equiv \sum_{k=1}^n \delta(\tilde{\tau} - \tau_k) \sum_{l=1}^n \left[(e^{\Gamma_\sigma} - I)N_\sigma^{(n)}\right]_{kl}\tilde{\mathcal{G}}_0^\sigma(\tau_l),$$

---

[1] We use here the sign convention of Chapter 7, that is, $\tilde{\mathcal{G}}_0(\tau) > 0$ for $0 \leq \tau < \beta$.

by binning the time points $\tilde{\tau}$ on a fine grid. Then, after the simulation, the Green's function is computed from the convolution integral

$$G^\sigma(\tau) = \tilde{\mathcal{G}}_0^\sigma(\tau) + \int_0^\beta d\tilde{\tau} \tilde{\mathcal{G}}_0^\sigma(\tau - \tilde{\tau}) \langle S_\sigma(\tilde{\tau}) \rangle_{MC}.$$

Let us briefly address the role of the parameter K and its effect on the expansion order. It follows from (8.69) that $\langle -H_2 \rangle = \frac{1}{\beta} \int_0^\beta d\tau \langle -H(\tau) \rangle = \frac{1}{\beta} \langle n \rangle$, and because $\langle -H_2 \rangle = K/\beta - U \langle n_\uparrow n_\downarrow - \frac{1}{2}(n_\uparrow + n_\downarrow) \rangle$, the average perturbation order $\langle n \rangle$ is related to the parameter K and the potential energy $E_{\text{pot}} = U \langle n_\uparrow n_\downarrow \rangle$ by

$$\langle n \rangle = K - \beta E_{\text{pot}} + \tfrac{1}{2} \beta U \langle (n_\uparrow + n_\downarrow) \rangle.$$

Increasing K leads to a higher perturbation order (and thus slower matrix updates), but through (L.1) also to a smaller value of γ and thus to less polarization of the auxiliary spins. The average perturbation order grows proportionally to U and the inverse temperature.

For models with density-density interaction, the weak-coupling algorithm discussed in Section 8.4 is equivalent to the continuous-time auxiliary-field method (Mikelsons et al., 2009). The weak-coupling algorithm is based on the following representation of the interaction term (see (8.31)):

$$U n_\uparrow n_\downarrow = \frac{U}{2} \sum_s \left[n_\uparrow - \frac{1}{2} - \left(\frac{1}{2} + \delta \right) s \right] \left[n_\downarrow - \frac{1}{2} + \left(\frac{1}{2} + \delta \right) s \right]$$

$$+ \frac{U}{2} (n_\uparrow + n_\downarrow) + U \left[\left(\frac{1}{2} + \delta \right)^2 - \frac{1}{4} \right].$$

The second term on the right-hand side can be absorbed into a shift of the chemical potential $\mu \to \tilde{\mu} = \mu - \frac{1}{2} U$ (with corresponding redefinition of the noninteracting Green's function $\mathcal{G}_0 \to \tilde{\mathcal{G}}_0$ and of the noninteracting partition function $Z_0 \to \tilde{Z}_0$), and for future convenience we write the third term as K^*/β, so that

$$K^* = \beta U \left[\left(\frac{1}{2} + \delta \right)^2 - \frac{1}{4} \right].$$

With this definition, the weight of a weak-coupling configuration becomes

$$w_C^{\text{weak-coupling}}((\tau_1, s_1), \dots, (\tau_n, s_n)) = \tilde{Z}_0 e^{-K^*} \left(-\frac{U d\tau}{2} \right)^n \prod_\sigma \det M_\sigma^{-1},$$

with $[M_\sigma^{-1}]_{ij} = \tilde{\mathcal{G}}_0^\sigma(\tau_i - \tau_j) - \frac{1}{2} \delta_{ij} - (\frac{1}{2} + \delta) s_i \sigma \delta_{ij}$. This expression should be compared with the weight of a corresponding configuration in the CTAUX formalism, which reads

$$w_C^{\text{CTAUX}}((\tau_1, s_1), \ldots, (\tau_n, s_n)) = \tilde{Z}_0 e^{-K} \left(\frac{K d\tau}{2\beta}\right)^n \prod_\sigma \det N_\sigma^{-1},$$

with $[N_\sigma^{-1}]_{ij} = [e^{\Gamma_\sigma} - \tilde{G}_0^\sigma (e^{\Gamma_\sigma} - I)]_{ij} = e^{\gamma \sigma s_i} \delta_{ij} - \tilde{G}_0^\sigma (\tau_i - \tau_j)(e^{\gamma \sigma s_j} - 1)$. The parameters γ and K are related by $\cosh(\gamma) = 1 + \beta U / 2K$.

To derive the relationship between the weak-coupling parameter δ and the CTAUX parameter γ, which makes the two methods identical, we manipulate the matrix N_σ^{-1} with the goal of bringing it into a form analogous to that of M_σ^{-1}. The first step is to rewrite

$$e^{\Gamma_\sigma} - \tilde{G}_0^\sigma (e^{\Gamma_\sigma} - I) = \left(\tilde{G}_0^\sigma - \frac{1}{2} I - \left\{ \frac{1}{2} I - [I - e^{\Gamma_\sigma}]^{-1} \right\} \right)(I - e^{\Gamma_\sigma}). \tag{L.5}$$

The expression in the curly brackets is a diagonal matrix with elements $1/(2 \tanh (\gamma \sigma s/2)) = \sigma s/(2 \tanh(\gamma/2))$. The last identity is true because $\sigma s = \pm 1$. We furthermore note that $\prod_\sigma (I - e^{\Gamma_\sigma})$ is a diagonal matrix with elements $2 - 2\cosh(\gamma) = -\beta U / K$. This means that the last factor in (L.5) contributes a factor $(-\frac{\beta U}{K})^n$ to the product of determinants. Hence, we obtain

$$\prod_\sigma \det \left(e^{\Gamma_\sigma} - \tilde{G}_0^\sigma \left(e^{\Gamma_\sigma} - I\right)\right) = \left(-\frac{\beta U}{K}\right)^n \prod_\sigma \det \left(\tilde{G}_0^\sigma - \frac{1}{2} I - \frac{1}{2 \tanh(\gamma/2)} s \sigma I\right).$$

By comparing the right-hand side with the expression for $w_C^{\text{weak-coupling}}$, we conclude that the two algorithms produce configurations $C = \{(\tau_1, s_1), \ldots (\tau_n, s_n)\}$ with identical weights if the parameters δ and γ satisfy the condition

$$\frac{1}{2} + \delta = \frac{1}{2} \frac{1}{\tanh(\gamma/2)},$$

or, equivalently, if $K = \beta U[(\frac{1}{2} + \delta)^2 - \frac{1}{4}] = K^*$. Note that $\delta \geq 0$ implies $K \geq 0$.

Appendix M

Continuous-time determinant algorithm

In this appendix, we describe the continuous-time version of the determinant method, which has been developed by Iazzi and Troyer (2014). This algorithm combines the efficient linear-in-β scaling of the determinant algorithm (Chapter 7) with a continuous-time sampling of interaction vertices analogous to the weak-coupling impurity solver described in Section 8.4 of Chapter 8.

First, let us recall that in the discrete-time determinant method we transformed the partition function via a Trotter approximation and Hubbard-Stratonovich transformation into

$$Z = \sum_{C} \prod_{\sigma} w_C^{\sigma},$$

with

$$w_C^{\sigma} = \det \left[I + B_C^{\sigma}(\beta, 0) \right].$$

and

$$B_C^{\sigma}(\beta, 0) \equiv B^{\sigma}(\beta, \beta - \Delta\tau) \cdots B^{\sigma}(2\Delta\tau, \Delta\tau) B^{\sigma}(\Delta\tau, 0),$$

a string of multielectron propagators for some noninteracting Hamiltonian. The continuous-time formulation is based on a similar determinant, but avoids a fixed time-discretization and Hubbard-Stratonovich decoupling. Instead the times in the multi-electron propagators are generated by a weak-coupling expansion and are sampled stochastically. We now describe the method for finite-temperature simulations, following the presentation in the original paper (Iazzi and Troyer, 2014).

As in the case of the continuous-time impurity solvers, we switch to an interaction representation. The Hamiltonian of the lattice model is split into $H = H_0 + H_1$, where H_0 is the noninteracting part of the Hamiltonian and H_1 is the interacting part. In this representation,

$$H_1(\tau) = e^{H_0\tau} H_1 e^{-H_0\tau},$$

where $H_1 = \sum_i V_i$ and V_i is some site-dependent two-body interaction in a lattice of N sites. To be specific, let us consider the Hubbard model, for which $V_i = U(n_{i\uparrow} - \frac{1}{2})(n_{i\downarrow} - \frac{1}{2})$. For future convenience, we also introduce the dimensionless vertex $v_i = 4(n_{i\uparrow} - \frac{1}{2})(n_{i\downarrow} - \frac{1}{2})$. As in the case of the weak-coupling continuous-time solver, we now expand the partition function in powers of the interaction term

$$Z = \sum_k \int_0^\beta d\tau_1 \cdots \int_{\tau_{k-1}}^\beta d\tau_k \sum_{i_1,\ldots,i_k} w(C),$$

where

$$w(C) = \mathrm{Tr}\left[e^{-\beta H_0} \prod_{\ell=1}^k (-V_{i_\ell}(\tau_\ell)) \right] = (-U/4)^k \mathrm{Tr}\left[e^{-\beta H_0} \prod_{\ell=1}^k (v_{i_\ell}(\tau_\ell)) \right]$$

and $C = \{(i_1, \tau_1), \ldots, (i_k, \tau_k)\}$ is a configuration of k interaction vertices at space-time coordinates (i_ℓ, τ_ℓ).

Since the propagation from one vertex to the next is given by the noninteracting part of the Hamiltonian, the weight can (in a manner analogous to the method described in Chapter 7) be expressed as a determinant,

$$w(C) = (-U/4)^k \det[I + B(C, \beta)], \quad B(C, \beta) = e^{-\beta H_0} \prod_{\ell=1}^k v(i_\ell, \tau_\ell), \qquad (\text{M.1})$$

where $v(i_\ell, \tau) = e^{H_0 \tau} v_{i_\ell} e^{-H_0 \tau}$, and the 2×2 matrices H_0 and $v_{i_\ell} = e^{i\pi(n_{i_\ell\uparrow} + n_{i_\ell\downarrow})}$ have the elements

$$[H_0]_{i\sigma,j\sigma'} = -t_{ij}\delta_{\sigma\sigma'},$$

$$[v_{i_\ell}]_{i\sigma,j\sigma'} = -2\delta_{i,i_\ell}\delta_{i_\ell j}\delta_{\sigma\uparrow}\delta_{\sigma'\uparrow} - 2\delta_{i,i_\ell}\delta_{i_\ell j}\delta_{\sigma\downarrow}\delta_{\sigma'\downarrow} + \delta_{ij}\delta_{\sigma\sigma'}.$$

The fact that the weights are such determinants allows us to use measurement formulas for Green's functions and other observables that are similar to those used in the discrete-time determinant methods. The Monte Carlo sampling and updating, however, are analogous to the procedures used in the weak-coupling impurity solver.

The factor $(-U/4)^k$ in (M.1) introduces a sign problem, which, as in the case of the weak-coupling impurity solver, we can avoid by introducing auxiliary fields that shift the chemical potential for spin up and down or by expanding in a shifted interaction term, as in the CTAUX method (Appendix L). Following Iazzi and Troyer (2014), we choose the latter approach and expand the partition function in powers of $V_i = -U(1 - n_{i\uparrow}n_{i\downarrow}) \equiv -\frac{1}{2}Uv_i$. Introducing the auxiliary Ising variable α (Appendix G), we then write

$$v_i = 2(1 - n_{i\uparrow}n_{i\downarrow}) = \sum_{\alpha=\pm1} (1 + \alpha n_{i\uparrow})(1 - \alpha n_{i\downarrow}).$$

The configuration $C = \{(i_1, \tau_1, \alpha_1), \ldots, (i_k, \tau_k, \alpha_k)\}$ becomes a collection of k auxiliary spin variables, and the matrix for the auxiliary spin dependent vertex becomes

$$[v_{i_\ell}^{\alpha_\ell}]_{i\sigma, j\sigma'} = \delta_{ij}\delta_{\sigma\sigma'} + \alpha_\ell \delta_{i,i_\ell}\delta_{i_\ell j}\delta_{\sigma\uparrow}\delta_{\sigma'\uparrow} - \alpha_\ell \delta_{i,i_\ell}\delta_{i_\ell j}\delta_{\sigma\downarrow}\delta_{\sigma'\downarrow}.$$

In complete analogy to the sampling discussed in Section 8.4, the Monte Carlo sampling of the configurations C proceeds by random insertion and removal of vertices. For example, if the initial configuration C has k vertices and the new configuration with $k + 1$ vertices is C', then

$$\mathcal{R}(C' \leftarrow C) = \frac{\beta N |U|}{k+1} \frac{\det[I + B(C', \beta)]}{\det[I + B(C, \beta)]}.$$

The computational bottleneck is the updating of the matrix $B(C, \beta)$. To do this calculation in a time $\mathcal{O}(\beta N^3)$, Iazzi and Troyer (2014) propose to switch to the eigenbasis of H_0: $H_0 = OEO^\dagger$, with E a diagonal matrix and O being an orthonormal transformation. In this basis, the time-evolution operators become diagonal, while the vertex operators become an identity matrix plus an outer product of two vectors: $\delta_{kk'} + \gamma O^\dagger_{ki_\ell} O_{i_\ell k'}$. As in the case of the finite-temperature determinant algorithm (Chapter 7), the calculation of the matrix products requires a stabilization procedure.

Equal-time observables are expressible as functions of the matrix B. For example, the equal-time Green's function $\langle c_{i\sigma} c^\dagger_{j\sigma'} \rangle$ is obtained by computing the matrix elements of

$$G = B(C, \beta)[I + B(C, \beta)]^{-1}.$$

The time-dependent Green's function at time τ is measured using the partial propagator $B(C, \tau) = e^{-\tau H_0} \prod_{\tau_i < \tau} v(i_\ell, \tau_\ell)$. In this case, we accumulate the matrix elements of

$$G(\tau) = B(C, \tau)[I + B(C, \beta)]^{-1}.$$

The continuous-time version of the zero-temperature determinant algorithm (Appendix I) has been discussed by Wang et al. (2015).

Appendix N
Correlated sampling

Correlated sampling is a technique for reducing variance. It is sometimes called *differential Monte Carlo* because its principal use is for reducing the variance of the average difference between two quantities. A related use is estimating derivatives. The method exploits the fact that by promoting a statistical dependence between variables, often simply by having estimates share all or part of a sequence of random numbers, the total variance is less than the sum of the variances.

To explain the basic idea, suppose we seek a Monte Carlo estimate of the integral

$$F = \int dx f(x) p_X(x),$$

where $p_X(x)$ is the probability distribution of a random variable X. In other words, $F = \langle f(x) \rangle$. We now consider another random variable Y, which is not necessarily independent of X, and construct a third random variable

$$\xi = f(X) - g(Y),$$

where for the time being $g(Y)$ is arbitrary. If $G = \langle g(y) \rangle$, then we can express the variance of ξ as

$$\text{var}(\xi) = \text{var}(f) + \text{var}(g) - 2\rho \sqrt{\text{var}(f)\,\text{var}(g)}, \qquad (\text{N.1})$$

where ρ is the correlation coefficient

$$\rho = \frac{\langle (f - F)(g - G) \rangle}{\sqrt{\text{var}(f)\text{var}(g)}}.$$

As noted in Section 3.6, $-1 \le \rho \le 1$, and if X and Y are statistically independent, that is, uncorrelated, $\rho = 0$ and the variances add.

If G is known, then since

$$F = \langle \xi \rangle + G,$$

we could estimate F by estimating $\langle \xi \rangle$. From (N.1), we see that the variance of ξ is small if ρ is close to unity. In fact, if $X = Y$ and $g = f$, that is, the expectation values are perfectly correlated, then $\rho = 1$ and the variance of ζ equals zero. Consequently, when $X = Y$ and $g \approx f$, it is more advantageous to sample ξ than X.

We comment that correlated sampling is often used in the following way: Suppose the random variable Y is correlated with the random variable X and we can compute $\text{cov}(X, Y)$ (3.23). If we want to estimate $\langle X \rangle$, know $\langle Y \rangle$, and compute $\text{var}(Y)$, then we construct the random variable $X(a) = X + a(Y - \langle Y \rangle)$. Noting that $X(a)$ and X have the same mean, we write

$$\text{var}(X\,(a)) = \text{var}(X) - 2a\,\text{cov}(X, Y) + a^2 \text{var}(Y).$$

If we take $a = \text{cov}(X, Y)/\text{var}(Y)$, then

$$\text{var}(X(a)) = \left[1 - \rho_{XY}^2\right]\text{var}(X) < \text{var}(X).$$

Finally, we note that if $\langle X \rangle = \langle Y \rangle$, then we can always choose an a such that the variance of $X(a) = aX + (1 - a)Y$ is smaller that the variance of X.

In both the energy and energy-variance minimization variational Monte Carlo methods (Section 9.3), the basic step is generating a sequence of trial states $|\psi_T^1\rangle, \psi_T^2\rangle, \ldots, |\psi_T^i\rangle, \ldots$ from a base state $|\psi_T^0\rangle$, where the parameters of each $|\psi_T^i\rangle$ differ only slightly from those of $|\psi_T^0\rangle$. If we were to compute the energy of each state by drawing independent configurations, the statistical errors of the differences between any pair of estimates would add and their confidence intervals could overlap. By drawing configurations from the base trial state and computing all energies with this common set of configurations, the statistical error of the difference between any pair is less than the sum because the pair's estimate is correlated.

Appendix O

The Bryan algorithm

There are multiple ways to maximize a nonlinear function of N variables. Here we describe the Levenberg-Marquardt method (Golub and Loan, 1989; Press et al., 2007), which is a variation of Newton's method. After its description, we discuss how to adjust it for the specific problem of maximizing Q as stated in (12.39).

Our problem here is to maximize some general function

$$Q(A) = Q(A_1, A_2, \ldots, A_N)$$

with respect to A. Starting from some initial guess for A, we iteratively improve it until the procedure converges. Suppose our current guess is A and we want to find a better solution A'. We start by considering the Taylor series expansion of Q about some point A,

$$Q\left(A'\right) \approx Q\left(A\right) + \delta A^T \cdot g - \tfrac{1}{2}\delta A^T \cdot H \cdot \delta A, \tag{O.1}$$

where the vector δA is the deviation $\delta A = A' - A$,

$$g_i = \left.\frac{\partial Q}{\partial A_i}\right|_A \tag{O.2}$$

are the components of the gradient of Q evaluated at A, and

$$H_{ij} = -\left.\frac{\partial^2 Q}{\partial A_i \partial A_j}\right|_A \tag{O.3}$$

are the elements of the Hessian matrix of Q evaluated at A. Near the maximum, the Hessian is positive definite.

We can regard (O.1) as defining a local approximation for the gradient

$$\nabla Q \approx g - H \cdot \delta A.$$

Because we want to make the gradient at A' vanish, within the second-order approximation (O.1), we should make δA satisfy

$$g \approx H \cdot \delta A,$$

so the relation between A' and A is

$$A' \approx A + H^{-1} \cdot g.$$

If the function Q were in fact quadratic with a positive-definite Hessian, then the above analysis shows that we could move from A to the maximum $\hat{A}$ in a single step, that is, $\hat{A} = A + H^{-1}g$. When (O.1) is not exact, we can execute this stepping procedure iteratively,

$$(H + \mu I) \cdot \delta A = g,$$
$$A' \leftarrow A + \delta A, \qquad (O.4)$$

until convergence (as measured, for example, by $0 < \sqrt{g^T \cdot g} \ll 1$). The method will converge provided we can keep stepping uphill. Adding the matrix μI to the Hessian shifts its eigenvalues, and adjusting μ at each step to maintain the local positive definiteness of $\mu I + H$ keeps us stepping uphill. This is the Levenberg-Marquardt strategy, summarized in Algorithm 51.

In Algorithm 51, what do we mean by solving for δA? Clearly, $[\mu I + H] \cdot \delta A = g$ is a simultaneous system of linear equations for δA. For a given value of μ, we use the standard Gaussian elimination method with partial pivoting to solve them for δA, and check whether the new position $A + \delta A$ moves the new value of Q above the old. If not, we could then repeat the process with new values of μ until it does. When we finally step uphill, we accept the new position and continue iterating.

Algorithm 51 Levenberg-Marquardt algorithm.

Input: Starting point A and a convergence parameter ϵ.

  Set $v = 10$, $\mu' = 0.01$, $A' = A$, and $Q' = Q(A)$;

  repeat

    $\mu \leftarrow \mu'/v$;

    repeat

      Solve $[\mu I + H(A')] \cdot \delta A = g(A')$ for δA ;

      if $Q(A + \delta A) < Q'$ **then**

        $\mu \leftarrow \mu v$;

      end if

    until $Q(A + \delta A) > Q'$.

    $\mu' \leftarrow \mu$;

    $A' \leftarrow A + \delta A$;

    $Q' \leftarrow Q(A + \delta A)$;

  until $\left\| g(A') \right\|_2 < \epsilon$.

  return A'.

Because H is symmetric, we alternatively could just as easily diagonalize it, writing the diagonalization as $H = Z \cdot \Xi \cdot Z^T$, where Z is an orthonormal matrix and Ξ is a diagonal matrix whose diagonal elements are the eigenvalues ξ_i of H. We now transform the system of linear equations

$$(H + \mu I) \cdot \delta A = g \rightarrow (\mu + \xi_i) \, \delta A'_i = g'_i$$

and thereby transform the coupled system of linear equations into a set of independent linear equations. Ensuring we step uphill becomes equivalent to adjusting μ so that $\mu + \xi_i > 0$ for all i. Because an orthonormal transformation preserves norms, our stopping criterion

$$\|g\|_2 = \sqrt{g^T \cdot g} = \sqrt{[g']^T \cdot g'} < \epsilon$$

remains unaltered by the coordinate transformation.

For our specific problem discussed in Section 12.3, the gradient is

$$\nabla Q = \alpha \nabla S - \nabla \mathcal{L},$$

with

$$\nabla S = - \ln (A_i/m_i), \quad \nabla \mathcal{L} = K^T \cdot C^{-1} \cdot \delta G,$$

where δG is a vector whose components are $G_i - \bar{G}_i$. The gradient of S follows from (12.43).

When $\nabla Q = 0$,

$$\alpha \ln (A_i/m_i) = -K^T \cdot \left[C^{-1} \cdot \delta G \right].$$

The expression $C^{-1} \cdot \delta G$ is just a vector. What the above says is that the solution is a linear combination of the vectors representing the columns of K^T. With the singular value decomposition $K^T = U \cdot D \cdot V^T$, these vectors are actually the orthogonal columns of U. Based on this observation, Bryan (1990) suggested parameterizing the solution as

$$A_i = m_i \exp \left(\sum_j U_{ij} u_j \right). \tag{O.5}$$

The condition for the maximum then becomes

$$\alpha u = -D \cdot V^T \cdot C^{-1} \cdot \delta G.$$

Hence,

$$g = \alpha u + D \cdot V^T \cdot C^{-1} \cdot \delta G$$

is the gradient in the space defined by the parameters u. The Hessian in this space is

$$H = \frac{\partial g}{\partial u} = \alpha I + D \cdot V^T \cdot C^{-1} \cdot K \cdot \frac{\partial A}{\partial u}.$$

We now proceed to the maximization procedure, invoking $H\delta u = g$ for the Newton step using the just-defined expressions for g and H.

At each step of the iteration, we must compute g and H for the current estimate of u. We now recall, however, that before the start of the maximization procedure, we transformed the Green's function and the kernel into a space where the covariance matrix C is diagonal, so C^{-1} is some diagonal matrix Σ (12.37). We also have $K = V \cdot D \cdot U^T$. Accordingly,

$$H = \frac{\partial g}{\partial u} = \alpha I + \left[D \cdot V^T \cdot \Sigma \cdot V \cdot D\right] \cdot U^T \cdot \frac{\partial A}{\partial u}.$$

We note that

$$\frac{\partial A}{\partial u} = \mathcal{A} \cdot U,$$

where $\mathcal{A}$ is a diagonal matrix whose diagonal elements are A_i. Finally, we obtain

$$H = \alpha I + \left[D \cdot V^T \cdot \Sigma \cdot V \cdot D\right] \cdot \left[U^T \cdot \mathcal{A} \cdot \mathcal{U}\right] \equiv \alpha I + X \cdot Y.$$

When we restrict the search space to the s-dimensional sub-dominant space of the kernel K, each of these matrices and their product becomes a symmetric $s \times s$ matrix. The solution to this reduced system of linear equations becomes a numerical task that we could handle by Gaussian elimination with partial pivoting. We could also proceed as before by diagonalizing the symmetric matrix $X \cdot Y$. However, another, more preferable strategy is available.

The preferred strategy transforms H into a form similar to one we would get with the diagonalization. First, we note that X depends on the matrices we obtain from singular-value decomposition of the kernel K and the diagonalization of the covariance C. Because these matrix factorizations are done prior to starting the maximization procedure, they are done only once, and accordingly X is invariant during the maximization procedure. On the other hand, Y changes at each step as the values of u_i and hence A_i change. We also note that the matrix U, completing the definition of Y, does not change.

To transform H, we first diagonalize Y and write $Y = W \cdot \Omega \cdot W^T$, where W is an orthonormal matrix and Ω is a diagonal matrix whose diagonal elements are the eigenvalues of Y. Next, we define the symmetric matrix $P = \sqrt{\Omega} \cdot W \cdot X \cdot W^T \cdot \sqrt{\Omega}$ and then diagonalize it, writing $P = Z \cdot \Xi \cdot Z^T$. If we define $F = W \cdot \sqrt{\Omega^{-1}} \cdot Z$, then direct substitution shows that $F^{-T} \cdot F^{-1} = Y$. Now, we perform the series of transformations

$[\alpha I + X \cdot Y] \cdot \delta u = g \rightarrow$

$\qquad [(\alpha + \mu) I + \Xi] (F \cdot \delta u) = F \cdot g \rightarrow$

$\qquad\qquad [(\alpha + \mu) I + \Xi] \delta u' = g' \quad (O.6)$

and again reduce the simultaneous system of linear equations to a set of independent linear equations but in this case for the step $\delta u'$. The Levenberg-Marquardt phase of the procedure chooses μ such that $\alpha + \xi_i + \mu > 0$ for all i.

The transformation just made is not norm preserving. The stopping condition $\sqrt{[\delta u']^T \cdot \delta u'} < \epsilon$ is, however, equivalent to $\sqrt{\delta u^T \cdot Y \cdot \delta u} < \epsilon$. Before parametrizing A by u and being mindful of the metric for flat curvature of the entropy, we would have used the stopping criterion $\sqrt{\sum_i (\delta A_i)^2 / A_i} < \epsilon$. Remarkably, $\sum_i (\delta A_i)^2 / A_i = \sum_{ij} \delta u_i Y_{ij} \delta u_j$; that is, Y is the metric in u-space.

The readily available LAPACK and BLAS software facilitates the numerical implementation of this algorithm, which is due to Bryan (1990).

References

Acton, F. S. 1970. *Numerical Methods That Work*. New York: Harper and Row.

Affleck, I., Kennedy, T., Lieb, E. L., and Tasaki, H. 1987. Rigorous results on valence bond ground states in antiferromagnets. *Phys. Rev. Lett.*, **59**, 799.

Affleck, I., Kennedy, T., Lieb, E. L., and Tasaki, H. 1988. Valence bond ground states in isotropic quantum antiferromagnets. *Comm. Math. Phys.*, **115**, 477.

Aho, A. V., Ullman, J. D., and Hopcroft, J. E. 1983. *Data Structures and Algorithms*. Lebanon, IN: Addison Wesley.

Aichhorn, M., Pourovskii, L., and Georges, A. 2011. Importance of electronic correlations for structural and magnetic properties of the iron pnictide superconductor LaFeAsO. *Phys. Rev. B*, **84**, 054529.

Alet, F., Wessel, S., and Troyer, M. 2005. Generalized directed loop method for quantum Monte Carlo simulations. *Phys. Rev. E*, **71**, 036706.

Allen, M. P., and Tildesley, D. J. 1987. *Computer Simulations of Liquids*. Oxford University Press.

Anderson, J. B. 1975. Random walk simulation of the Schrödinger equation: He_3^+. *J. Chem. Phys.*, **63**, 1499.

Anderson, J. B. 1976. Quantum chemistry by random walk. *J. Chem. Phys.*, **65**, 4121.

Anderson, M. H., Ensher, J. R., Matthews, M. R., Wieman, C. E., and Cornell, E. A. 1995. Observation of Bose-Einstein condensation in a dilute atomic vapor. *Science*, **269**, 198.

Anderson, P. W. 1973. Resonating valence bonds: a new kind of insulator? *Mater. Res. Bull.*, **8**, 153.

Anderson, P. W. 1987. The resonating valence bond state in La_2CuO_4 and superconductivity. *Science*, **235**, 1196.

Andreev, A. F., and Lifshitz, I. M. 1969. Quantum theory of defects in crystals. *Sov. Phys. JETP*, **29**, 1107.

Anisimov, V. I., and Gunnarsson, O. 1991. Density-functional calculation of effective Coulomb interactions in metals. *Phys. Rev. B*, **43**, 7570.

Anisimov, V. I., Zaanen, J., and Andersen, O. K. 1991. Band theory and Mott insulators: Hubbard-U instead of Stoner-I. *Phys. Rev. B*, **44**, 943.

Arnow, D. M., Kalos, M. H., Lee, M. A., and Schmidt, K. E. 1982. Green's function Monte Carlo for few fermion problems. *J. Chem. Phys.*, **77**, 5562.

Aryasetiawan, F., Imada, M., Georges, A., Kotliar, G., Biermann, S., and Lichtenstein, A. I. 2004. Frequency-dependent local interactions and low-energy effective models from electronic structure calculations. *Phys. Rev. B*, **70**, 195104.

Assaad, F. F. 1998. SU(2) invariant auxiliary field quantum Monte Carlo algorithm for Hubbard models. In: Krause, E., and Jager, W. (eds.), *High Performance Computing in Science and Engineering*. New York: Springer-Verlag.

Assaad, F. F., and Lang, T. C. 2007. Diagrammatic determinantal quantum Monte Carlo methods: projective schemes and applications to the Hubbard-Holstein model. *Phys. Rev. B*, **76**, 035116.

Barker, A. A. 1965. Monte Carlo calculations of the radial distribution functions for a proton-electron plasma. *Aust. J. Phys.*, **18**, 119.

Bartlett, J. H., Gibbons, J. J., and Dunn, C. G. 1935. The normal helium atom. *Phys. Rev.*, **47**, 679.

Batrouni, G., and Scalettar, R. T. 1990. Anomalous decouplings and the fermion sign problem. *Phys. Rev. B*, **42**, 2282.

Baxter, R. J. 1982. *Exactly Solved Models in Statistical Mechanics*. London: Academic Press.

Baym, G., and Mermin, N. D. 1961. Determination of thermodynamic Green's functions. *J. Math. Phys.*, **2**, 236.

Beach, K. S. D., Alet, F., Mambrini, M., and Capponi, S. 2009. SU(N) Heisenberg model on the square lattice: a continuous-N quantum Monte Carlo study. *Phys. Rev. B*, **80**, 184401.

Bemmel, H. J. M., ten Haaf, D. F. B., van Saarlos, W., van Leeuwen, J. M. J., and An, G. 1994. Fixed-node quantum Monte Carlo method for lattice fermions. *Phys. Rev. Lett.*, **72**, 2442.

Berg, B. A., and Neuhaus, T. 1991. Multicanonical algorithms for first order phase transitions. *Phys. Lett. B*, **267**, 249.

Berg, B. A., and Neuhaus, T. 1992. Multicanonical ensemble: a new approach to simulate first-order phase transitions. *Phys. Rev. Lett.*, **68**, 9.

Bethe, Hans A. 1935. Statistical theory of superlattices. *Proc. Roy. Soc. London A*, **150**, 552.

Blaizot, J. P., and Ripka, G. 1986. *Quantum Theory of Finite Systems*. Cambridge, MA: MIT Press.

Blankenbecler, R., Scalapino, D. J., and Sugar, R. L. 1981. Monte Carlo calculations of coupled boson-fermion systems. I. *Phys. Rev. D*, **24**, 2278.

Blöte, H. W. J., and Nightingale, M. P. 1982. Critical behavior of the two-dimensional Potts model with a continuous number of states: a finite size scaling analysis. *Physica A*, **112**, 405.

Blume, D., Lewerenz, M., Niyaz, P., and Whaley, K. B. 1997. Excited states by quantum Monte Carlo methods: imaginary time evolution with projection operators. *Phys. Rev. E*, **55**, 3664.

Blume, D., Lewerenz, M., and Whaley, K. B. 1998. Excited states by quantum Monte Carlo method. *Math. Comp. Simul.*, **47**, 133.

Boehnke, L., Hafermann, H., Ferrero, M., Lechermann, F., and Parcollet, O. 2011. Orthogonal polynomial representation of imaginary-time Green's functions. *Phys. Rev. B*, **84**, 075145.

Boguslawski, K., Marti, K. H., and Reiher, M. 2011. Construction of CASCI-type wavefunctions for very large matrices. *J. Chem. Phys.*, **134**, 224101.

Bonča, J., and Gubernatis, J. E. 1993a. Quantum Monte Carlo simulations of the degenerate single-impurity Anderson model. *Phys. Rev. B*, **47**, 13137.

Bonča, J., and Gubernatis, J. E. 1993b. Real-time dynamics from imaginary-time quantum Monte Carlo simulations: tests on oscillator chains. *Phys. Rev. B*, **47**, 13137.

Bonča, J., and Gubernatis, J. E. 1994. Degenerate Anderson impurity model in the presence of spin-orbit and crystal field splitting. *Phys. Rev B*, **50**, 10427.

Boninsegni, M., and Prokof'ev, N. 2005. Supersolid phase of hard-core bosons on a triangular lattice. *Phys. Rev. Lett.*, **95**, 237204.

Booth, T. E. 2003a. Computing the higher k-eigenfunctions by Monte Carlo power iteration: a conjecture, corrigendum. *Nucl. Sci. Eng.*, **144**, 113.

Booth, T. E. 2003b. Computing the higher k-eigenfunctions by Monte Carlo power iteration: a conjecture. *Nucl. Sci. Eng.*, **143**, 291.

Booth, T. E. 2003c. *Improvements to the Monte Carlo second eigenfunction estimation.* Tech. rept. LA-UR-03-4100. Los Alamos National Laboratory.

Booth, T. E. 2009. Particle transport applications. Page 215 of: Rubino, G., and Tuffin, B. (eds.), *Rare Event Simulation using Monte Carlo Methods.* Chichester: Wiley.

Booth, T. E. 2010. *Examples of superfast power iteration.* Tech. rept. LA-UR-10-06663. Los Alamos National Laboratory.

Booth, T. E. 2011a. *A multiple eigenvalue power iteration convergence metric.* Tech. rept. LA-UR-11-02223. Los Alamos National Laboratory.

Booth, T. E. 2011b. A simple eigenfunction convergence acceleration method for Monte Carlo. In: *International Conference on Mathematics and Computational Methods Applied to Nuclear Science and Engineering (MC2011).*

Booth, T. E. 2011c. *Test of the multiple k-eigenfunction convergence acceleration method in MCNP.* Tech. rept. LA-UR-11-02222. Los Alamos National Laboratory.

Booth, T. E., and Gubernatis, J. E. 2009a. Monte Carlo determination of multiple extremal eigenpairs. *Phys. Rev. E.*, **80**, 46704.

Booth, T. E., and Gubernatis, J. E. 2009b. The sewing algorithm. *Comp. Phys. Commun.*, **180**, 509.

Bouchoud, J. P., Georges, A., and Lihuillier, C. 1988. Pair wave functions for strongly correlated fermions and the determinantal representation. *J. de Phys.*, **49**, 553.

Brissenden, R. J., and Garlick, A. R. 1986. Biases in the estimations of k_{eff} and its errors by Monte Carlo estimation. *Annals of Nuclear Energy*, **13**, 63.

Brower, R., Chandrasekharan, S., and Wiese, U.-W. 1998. Green's functions from quantum cluster algorithms. *Physica A.*, **261**, 520.

Bryan, R. K. 1990. Maximum entropy analysis of oversampled data problems. *Eur. Biophys. J.*, **18**, 165.

Cafferel, M., and Ceperley, D. M. 1992. A Bayesian analysis of Green's function Monte Carlo correlation functions. *J. Chem Phys.*, **97**, 8415.

Camp, W. J., and Fisher, M. E. 1972. Decay of order in classical many-body systems I: introduction and formal theory. *Phys. Rev. B.*, **6**, 946.

Cardy, J. 1996. *Scaling and Renormalization in Statistical Physics.* Cambridge University Press.

Carlson, J., Gubernatis, J. E., Ortiz, G., and Zhang, S. 1999. Issues and observations on applications of the constrained-path Monte Carlo method to many-fermion systems. *Phys. Rev. B.*, **59**, 12788.

Ceperley, D. M., and Bernu, B. 1988. The calculation of excited state properties with quantum Monte Carlo. *J. Chem. Phys.*, **89**, 6316.

Chandrasekharan, S., Cox, J., and Wiesse, U.-J. 1999. Meron cluster solution of fermion sign problems. *Phys. Rev. Lett.*, **83**, 3116.

Chandrasekharan, S., Cox, J., Osborn, J. C., and Wiesse, U.-J. 2003. Meron cluster approach to systems of strongly correlated electrons. *Nucl. Phys. B*, **673**, 405.

Changlani, H. J., Kinder, J. M., Umrigar, C. J., and Chan, G. K.-L. 2009. Approximating strongly correlated wave functions with correlator product states. *Phys. Rev. B*, **80**, 245116.

Cirac, J. I., and Verstraete, F. 2009. Renormalization and tensor product states in spin chains and lattices. *J. Phys. A: Math. Theor.*, **42**, 504004.

Coldwell, R. L. 1977. Zero Monte Carlo calculations or quantum mechanics is easier. *Int. J. Quant. Chem. Symp.*, **11**, 215.

Corboz, P., Jordan, J., and Vidal, G. 2010a. Simulation of fermionic lattice models in two spatial dimensions with fermionic projected entangled pairs: next-nearest neighbor Hamiltonians. *Phys. Rev. B*, **82**, 245119.

Corboz, P., Orus, R., Bauer, B., and Vidal, G. 2010b. Simulation of strongly correlated fermions in two spatial dimensions with fermionic projected entangled pairs. *Phys. Rev. B*, **81**, 165104.

Costi, T. A. 2000. Kondo effect in a magnetic field and the magnetoresistivity of Kondo alloys. *Phys. Rev. Lett.*, **85**, 1504–1507.

Courant, R., and Hilbert, D. 1965. *Methods of Mathematical Physics*. Vol. 1. New York: Interscience.

Creutz, M. 1980. Monte Carlo study of quantized SU(2) gauge theory. *Phys. Rev. D*, **21**, 2308.

de Oliveira, P. M. C., Penna, T. J. P., and Herrmann, H. J. 1998a. Broad histogram Monte Carlo. *Braz. J. Phys.*, **26**, 677.

de Oliveira, P. M. C., Penna, T. J. P., and Herrmann, H. J. 1998b. Broad histogram Monte Carlo. *Eur. Phys. J. B*, **1**, 205.

Dukelsky, J., Martin-Delgado, M. A., Nishino, T., and Sierra, G. 1998. Equivalence of the variational matrix product method and the density-matrix renormalization group applied to spin chains. *Europhys. Lett.*, **43**, 457.

Edegger, B., Muthukumar, V. N., and Gros, C. 2007. Gutzwiller-RVB theory of high temperature superconductivity: results from renormalized mean field theory and variational Monte Carlo calculations. *Adv. Phys.*, **56**, 927.

Enz, C. P. 1992. *A Course on Many-Body Theory Applied to Solid State Physics*. Lecture Notes in Physics, vol. 11. Singapore: World Scientific.

Everett, C. J., and Cashwell, E. D. 1983. *A third Monte Carlo sampler*. Tech. rept. LA-9721-MS. Los Alamos National Laboratory.

Evertz, H. G., and Marcu, M. 1994. Page 65 in: Suzuki, M. (ed.), *Quantum Monte Carlo Methods in Condensed Matter Physics*. Singapore: World Scientific.

Evertz, H. G., Lana, G., and Marcu, M. 1993. Cluster algorithm for vertex models. *Phys. Rev. Lett.*, **70**, 875.

Fahy, S. B., and Hamann, D. R. 1990. Positive-projection Monte Carlo simulation: a new variational approach to strongly interacting fermion systems. *Phys. Rev. Lett.*, **65**, 3437.

Fahy, S. B., and Hamann, D. R. 1991. Diffusive behavior of states in the Hubbard-Stratonovich transformation. *Phys. Rev. B*, **43**, 765.

Fazekas, P. 1999. *Lecture Notes on Electron Correlation and Magnetism*. Singapore: World Scientific.

Fazekas, P., and Anderson, P. W. 1974. Ground state properties of anisotropic triangular antiferromagnets. *Phil. Mag.*, **30**, 423.

Ferrenberg, A. M., and Swendsen, R. H. 1988. New Monte Carlo technique for studying phase transitions. *Phys. Rev. Lett.*, **61**, 2635.

Fetter, A. L., and Walecka, J. D. 1971. *Quantum Theory of Many-Particle Systems*. New York: McGraw-Hill.

Feynman, R. P., and Hibbs, A. R. 1965. *Quantum Mechanics and Path Integrals*. New York: McGraw-Hill.

Fisher, M. E., and Barber, M. N. 1972. Scaling theory for finite-size effects in critical region. *Phys. Rev. Lett.*, **28**, 1516.

Fishman, G. S. 1996. *Monte Carlo: Concepts, Algorithms, and Applications*. New York: Springer-Verlag.

Fortuin, C. M., and Kasteleyn, P. W. 1972. Random cluster model. 1. Introduction and relation to other models. *Physica*, **57**, 536.

Friel, N., and Wyse, J. 2012. Estimating the evidence: a review. *Statistica Neerlandica*, **66**, 288.

Fuchs, S., Pruschke, T., and Jarrell, M. 2010. Analytic continuation of quantum Monte Carlo data by stochastic analytic inference. *Phys. Rev. E*, **81**, 56701.

Fuchs, S., Gull, E., Pollet, L., Burovski, E., Kozik, E., Pruschke, T., and Troyer, M. 2011. Thermodynamics of the 3D Hubbard model on approaching the Néel transition. *Phys. Rev. Lett.*, **106**, 030401.

Fukui, K., and Todo, S. 2009. Order-N cluster Monte Carlo method for spin systems with long-range interactions. *J. Comp. Phys.*, **228**, 2629.

Fulde, P. 1991. *Electron Correlations in Molecules and Solids*. Solid-State Sciences, vol. 100. Berlin: Springer-Verlag.

Gammel, J. T., Campbell, D. K., and Loh, E. Y. 1993. Extracting infinite system properties from finite-size clusters: phase randomization boundary condition averaging. *Synthetic Metals*, **57**, 4437.

Gelbard, E., and Gu, A. G. 1994. Biases in Monte Carlo eigenvalue calculations. *Nuclear Science and Engineering*, **117**, 1.

Georges, A., Kotliar, G., Krauth, W., and Rozenberg, M. J. 1996. Dynamical mean-field theory of strongly correlated fermion systems and the limit of infinite dimensions. *Rev. Mod. Phys.*, **68**, 13.

Geyer, C. J. 1991. *Computing Science and Statistics: Proceedings of the 23rd Symposium on the Interface*. American Statistical Association, vol. 156, New York: American Statistical Association.

Giamarchi, T., and Lhuillier, C. 1990. Variational Monte Carlo study of incommensurate antiferromagnetic phases in the two-dimensional Hubbard model. *Phys. Rev. B*, **42**, 10641.

Giamarchi, T., and Lhuillier, C. 1991. Phase diagrams of the two-dimensional Hubbard and tJ models by a variational Monte Carlo method. *Phys. Rev. B*, **43**, 12943.

Goldstone, J. 1961. Field theories with superconductor solutions. *Nuovo Cimento*, **19**, 154.

Golub, G. H., and Loan, C. F. Van. 1989. *Matrix Computations*. Baltimore, MD: Johns Hopkins University Press.

Greiner, M., Mandel, O., Esslinger, T., Hänsch, T. W., and Bloch, I. 2002. Quantum phase transition from a superfluid to a Mott insulator in a gas of ultracold atoms. *Nature*, **415**, 39.

Gros, C. 1988. Superconductivity in correlated wave functions. *Phys. Rev. B*, **38**, 931.

Gubernatis, J. E. Editor's note to "Proof of validity of Monte Carlo method for canonical averaging" by M. Rosenbluth. *The Monte Carlo Method in the Physical Sciences: Celebrating the 50th Anniversary of the Metropolis Algorithm*. ed. J. E. Gubernatis, AIP Conference Proceedings, **690**, 31 (2003). New York: American Institute of Physics.

Gubernatis, J. E. 2012. Unpublished manuscript.

Gubernatis, J. E., and Booth, T. E. 2008. Multiple extremal eigenvalues by the power method. *J. Comp. Phys.*, **227**, 8508.

Gubernatis, J. E., and Zhang, X. Y. 1994. Negative weights in quantum Monte Carlo simulations at finite temperatures using the auxiliary field method. *Intl. J. Mod. Phys. C*, **8**, 590.

Gubernatis, J. E., Hirsch, J. E., and Scalapino, D. J. 1987. Spin and charge corelations around an Anderson magnetic impurity. *Phys. Rev. B*, **35**, 8478.

Gull, E., Werner, P., Millis, A. J., and Troyer, M. 2007. Performance analysis of continuous-time solvers for quantum impurity models. *Phys. Rev. B*, **76**, 235123.

Gull, E., Werner, P., Parcollet, O., and Troyer, M. 2008. Continuous-time auxiliary-field Monte Carlo for quantum impurity models. *Europhys. Lett.*, **82**, 57003.

Gull, E., Millis, A. J., Lichtenstein, A. I., Rubtsov, A. N., Troyer, M., and Werner, P. 2011. Continuous-time Monte Carlo methods for quantum impurity models. *Rev. Mod. Phys.*, **83**, 349.

Gull, S. F., and Skilling, J. 1984. Maximum entropy image reconstruction. *IEE Proc.*, **131F**, 646.

Gutzwiller, M. C. 1965. Effect of correlation on the ferromagnetism of transition metals. *Phys. Rev. Lett.*, **10**, 159.

Hamann, D. R., and Fahy, S. B. 1990. Energy measurement in auxiliary-field many-electron calculations. *Phys. Rev. B*, **41**, 11352.

Hammond, B. L., W. A. Lester, Jr., and Reynolds, P. J. 1994. *Monte Carlo Methods in ab Initio Quantum Chemistry*. Singapore: World Scientific.

Handscomb, D. C. 1962a. A Monte Carlo method applied to the Heisenberg ferromagnet. *Proc. Cambridge Philos. Soc.*, **58**, 594.

Handscomb, D. C. 1962b. The Monte Carlo method in quantum statistical mechanics. *Proc. Cambridge Philos. Soc.*, **58**, 594.

Harada, K., and Kawashima, N. 2001. Loop algorithm for Heisenberg models with biquadratic interaction and phase transitions in two dimensions. *J. Phys. Soc. Jpn.*, **70**, 13.

Harada, K., and Kawashima, N. 2002. Coarse-grained loop algorithms for Monte Carlo simulation of quantum spin systems. *Phys. Rev. E.*, **66**, 056705.

Harada, K., Troyer, M., and Kawashima, N. 1998. The two-dimensional $S = 1$ quamtum Heisenberg antiferromagnet at finite temperatures. *J. Phys. Soc. Jpn.*, **67**, 1130.

Hastings, W. K. 1970. Monte Carlo sampling methods using Markov chains and their applications. *Biometrika*, **57**, 97.

Hatano, N., and Suzuki, M. 1992. Representation basis in quantum Monte Carlo calculations and the negative-sign problem. *Phys. Lett. A*, **163**, 246.

Hatano, N., and Suzuki, M. 2005. Finding exponential product formulas of higher orders. *Lect. Notes Phys.*, **679**, 37.

Haule, K. 2007. Quantum Monte Carlo impurity solver for cluster dynamical mean-field theory and electronic structure calculations with adjustable cluster base. *Phys. Rev. B*, **75**, 155113.

Haule, K. 2015. Exact Double Counting in Combining the Dynamical Mean Field Theory and the Density Functional Theory. *Phys. Rev. Lett.*, **115**, 196403.

Hettler, M. H., Tahvildar-Zadeh, A. N., Jarrell, M., Pruschke, T., and Krishnamurthy, H. R. 1998. Nonlocal dynamical correlations of strongly interacting electron systems. *Phys. Rev. B*, **58**, R7475.

Himeda, A., and Ogata, M. 2000. Spontaneous deformation of the Fermi surface due to strong correlation in the two-dimensional *tJ* model. *Phys. Rev. Lett.*, **85**, 4345.

Himeda, A., Kato, T., and Ogata, M. 2002. Stripe states with spatially oscillating *d*-wave superconductivity in the two-dimensional $tt'J$ model. *Phys. Rev. Lett.*, **88**, 117001.

Hirsch, J. E. 1983. Discrete Hubbard-Stratonovich transformation for fermion lattice models. *Phys. Rev. B*, **28**, 1983.

Hirsch, J. E. 1985. Two-dimensional Hubbard model. *Phys. Rev. B*, **31**, 4403.

Hirsch, J. E. 1987. Simulations of the three-dimensional Hubbard model: half-filled band sector. *Phys. Rev. B*, **35**, 1851.

Hirsch, J. E., and Fye, R. M. 1986. Monte Carlo method for magnetic impurities in metals. *Phys. Rev. Lett.*, **56**, 2521.

Hoshen, J., and Kopelman, R. 1976. Percolation and cluster distribution. I. Cluster multiple labeling technique and critical concentration algorithm. *Phys. Rev. B*, **14**, 3438.

Householder, A. S. 2006. *The Theory of Matrices in Numerical Analysis*. New York: Dover.

Hukushima, K., and Nemoto, K. 1996. Exchange Monte Carlo method and application to spin glass simulations. *J. Phys. Soc. Jpn.*, **65**, 1604.

Huscroft, C., Gross, R., and Jarrell, M. 2000. Maximum entropy method of obtaining thermodynamic properties from quantum Monte Carlo simulations. *Phys. Rev. B*, **61**, 9300.

Iazzi, M., and Troyer, M. 2014. Efficient continuous-time quantum Monte Carlo algorithm for fermionic lattice models. *arXiv:1411.0683*.

Janke, W., and Kappler, S. 1995. Multibondic cluster algorithm for Monte Carlo simulations of first-order phase transitions. *Phys. Rev. Lett.*, **74**, 212.

Jarrell, M., and Gubernatis, J. E. 1996. Bayesian inference and the analytic continuation of quantum Monte Carlo data. *Phys. Rept.*, **269**, 133.

Jarrell, M., Macridin, A., Mikelsons, K., Doluweera, D. G. S. P., and Gubernatis, J. E. "The dynamical cluster approximation with quantum Monte Carlo cluster solvers." AIP Conference Proceedings, **1014**, 24 (2008).

Jordan, P., and Wigner, E. 1928. Über das Paulische Äquivalenzverbot. *Zeit. Phys.*, **47**, 631.

Kalos, M. H., and Pederiva, F. 2000. Exact Monte Carlo for continuum fermion systems. *Phys. Rev. Lett.*, **85**, 3547.

Kalos, M. H., and Whitlock, P. A. 1986. *Monte Carlo Methods I: Basics*. New York: Wiley-Interscience.

Karlin, S., and Taylor, H. W. 1975. *A First Course in Stochastic Processes*. Academic Press.

Kashurnikov, V. A., Prokof'ev, N. V., Svistunov, B. V., and Troyer, M. 1999. Quantum spin chains in a magnetic field. *Phys. Rev. B*, **59**, 1162.

Kasteleyn, P. W., and Fortuin, C. M. 1969. Phase transitions in lattice systems with random local properties. *J. Phys. Soc. Jpn. Suppl.*, **26**, 11.

Kato, Y., and Kawashima, N. 2009. Quantum Monte Carlo method for the Bose-Hubbard model with harmonic confining potential. *Phys. Rev. E*, **79**, 021104.

Kato, Y., and Kawashima, N. 2010. Finite-size scaling for quantum criticality above the upper critical dimension: superfluid–Mott-insulator transition in three dimensions. *Phys. Rev. E*, **81**, 011123.

Kaul, R. 2007. Private communication.

Kawashima, N. 1996. Cluster algorithms for anisotropic quantum spin models. *J. Stat. Phys.*, **82**, 131.

Kawashima, N. 2007. Unpublished manuscript.

Kawashima, N., and Gubernatis, J. E. 1994. Loop algorithms for Monte Carlo simulations of quantum spin systems. *Phys. Rev. Lett.*, **73**, 1295.

Kawashima, N., Evertz, H., and Gubernatis, J. E. 1994. Loop algorithms for quantum simulations of fermions on lattices. *Phys. Rev. B*, **50**, 136.

Kawashima, N., Jarrell, M., and Gubernatis, J. E. 1996. Cluster Monte Carlo study of the quantum XY model in two dimensions. *Int. J. Mod. Phys. C*, **7**, 433.

Khinchin, A. I. 1957. *Mathematical Foundations of Information Theory*. New York: Dover.

Kim, D. Y., and Chan, M. H. W. 2012. Absence of supersolidity in solid helium in porous Vycor glass. *Phys. Rev. Lett.*, **109**, 155301.

Knuth, D. E. 1997. *The Art of Computer Programming*, vol. 2: *Seminumerical Algorithms*, third ed. Reading: Addison Wesley. Page 119.

Kotliar, G., Savrasov, S. Y., Haule, K., Oudovenko, V. S., Parcollet, O., and Marianetti, C. A. 2006. Electronic structure calculations with dynamical mean-field theory. *Rev. Mod. Phys.*, **78**, 865.

Kuklov, A. B., and Svistunov, B. V. 2003. Counterflow superfluidity of two-species ultracold atoms in a commensurate optical lattice. *Phys. Rev. Lett.*, **90**, 100401.

Kuklov, A. B., Prokof'ev, N., and Svistunov, B. V. 2004. Commensurate two-component bosons in an optical lattice: ground state phase diagram. *Phys. Rev. Lett.*, **92**, 050402.

Landau, D. P., and Binder, K. 2000. *A Guide to Monte Carlo Simulations in Statistical Physics*. Cambridge University Press.

Läuchli, A. M., and Werner, P. 2009. Krylov implementation of the hybridization expansion impurity solver and application to 5-orbital models. *Phys. Rev. B*, **80**, 235117.

Li, Z.-X., Jiang, Y.-F., and Yao, H. 2014. Solving fermion sign problem in quantum Monte Carlo by Majorana representation. *arXiv:1408.2269*.

Liang, S., Doucot, D., and Anderson, P. W. 1988. Some new variational resonating valence bond type wave functions for the spin-$\frac{1}{2}$ antiferromagnetic Heisenberg model on a square lattice. *Phys. Rev. Lett.*, **61**, 365.

Lichtenstein, A. I., and Katsnelson, M. I. 2000. Antiferromagnetism and d-wave superconductivity in cuprates: a cluster dynamical mean-field theory. *Phys. Rev. B*, **62**, R9283.

Lin, C., Zong, F. H., and Ceperley, D. M. 2001. Twist-averaged boundary conditions in continuum quantum Monte Carlo algorithms. *Phys. Rev. E*, **64**, 016702.

Lin, X., Zhang, H., and Rappe, A. M. 2000. Optimization of quantum Monte Carlo wave functions using analytical energy derivatives. *J. Chem. Phys.*, **112**, 2650.

Liu, J. S. 2001. *Monte Carlo Strategies for Scientific Computing*. New York: Springer-Verlag.

Loh, E. Y. Jr., and Gubernatis, J. E. 1992. Stable numerical simulations of models of interacting electrons in condensed-matter physics. Chap. 4 of: Hanke, W., and Kopaev, Yu. V. (eds.), *Electronic Phase Transitions*. Modern Problems in Condensed-Matter Physics, vol. 32. Amsterdam: North-Holland.

Loh, E. Y. Jr., Gubernatis, J. E., Scalettar, R. T., Sugar, R. L., and White, S. R. 1989. Stable matrix-multiplication algorithms for the low-temperature numerical simulation of fermions. Page 55 of: Baeriswyl, D., and Campbell, D. K. (eds.), *Interacting Electrons in Reduced Dimensions*. NATO ASI Series, vol. 213. New York: Plenum.

Loh, E. Y. Jr., Gubernatis, J. E., Scalettar, R. T., White, S. R., Scalapino, D. J., and Sugar, R. L. 1990. Sign problem in the numerical simulation of many-electron systems. *Phys. Rev. B*, **41**, 9301.

Loh, E. Y. Jr., Gubernatis, J. E., Scalettar, R. T., White, S. R., Scalapino, D. J., and Sugar, R. L. 2005. Numerical stability and the sign problem in the determinant quantum Monte Carlo method. *Int. J. Mod. Phys. C*, **16**, 1319.

Luijten, E., and Blöte, H. W. J. 1995. Monte Carlo method for spin models with long-range interactions. *Int. J. Mod. Phys. C*, **6**, 359.

Ma, S.-K. 1985. *Modern Theory of Critical Phenomena*. Singapore: World Scientific.

Maier, T., Jarrell, M., Pruschke, T., and Hettler, M. H. 2005. Quantum cluster theories. *Rev. Mod. Phys.*, **77**, 1027.

Majumdar, D. K., and Ghosh, C. K. 1969a. On nearest-neighbor interaction in linear chain: I. *J. Math. Phys.*, **10**, 1388.

Majumdar, D. K., and Ghosh, C. K. 1969b. On nearest-neighbor interaction in linear chain: II. *J. Math. Phys.*, **10**, 1399.

Marinari, E., and Parisi, G. 1992. Simulated tempering: a new Monte Carlo scheme. *Europhys. Lett.*, **19**, 451.

Marshall, W. 1955. Antiferromagetism. *Proc. Roy. Soc. (London) A*, **232**, 48.

Marti, K., Bauer, B., Reiher, M., Troyer, M., and Verstraete, F. 2010. Complete-graph tensor network states: a new fermionic wavefunction for molecules. *New J. Phys.*, **12**, 103008.

Marzari, N., and Vanderbilt, D. 1997. Maximally localized generalized Wannier functions for composite energy bands. *Phys. Rev. B*, **56**, 12847.

Masaki, A., Suzuki, T., Harada, K., Todo, S., and Kawashima, N. 2013. Parallelized quantum Monte Carlo algorithm with non-local worm update. Phys. Rev. Lett. 112, 140603 (2014).

Masaki-Kato, A., Suzuki, T., Harada, K., Todo, S., and Kawashima, N. 2014. Parallelized quantum Monte Carlo algorithm with nonlocal worm updates. *Phys. Rev. Lett.*, **112**, 140603.

Matsubara, T., and Masuda, H. 1956. A lattice model of liquid helium. *Prog. Theor. Phys.*, **16**, 416.

McCulloch, I. 2007. From density-matrix renormalization group to matrix product states. *J. Stat. Mech.*, P10014.

McQueen, P. G., and Wang, C. S. 1991. Variational Monte Carlo evaluation of Gutzwiller states for the Anderson lattice model. *Phys. Rev. B*, **44**, 10021.

Metropolis, N. 1985. Monte Carlo: in the beginning and some great expectations. Page 62 of: Alcouffe, R., et al. (eds.), *Monte Carlo Calculations and Applications in Neutronics, Photonics, and Statistical Physics*. Berlin: Springer-Verlag.

Metropolis, N. 1987. The beginning of the Monte Carlo method. *Los Alamos Science*, Special Issue, 125.

Metropolis, N., and Ulam, S. 1949. The Monte Carlo method. *J. Am. Stat. Assoc.*, **44**, 335.

Metropolis, N., Rosenbluth, A., Rosenbluth, M., Teller, A., and Teller, A. 1953. Equations of state by fast computing machines. *J. Chem. Phys.*, **21**, 1087.

Metzner, W., and Vollhardt, D. 1989. Correlated lattice fermions in d-dimensions. *Phys. Rev. Lett.*, **62**, 324.

Meyer, C. D. 2000. *Matrix Analysis and Applied Linear Algebra Research*. Philadelphia: SIAM.

Mezzacapo, F. 2011. Variational study of a mobile hole in a two-dimensional quantum antiferromagnet using entangled-plaquette states. *Phys. Rev. B*, **83**, 115111.

Mezzacapo, F., Schuch, N., Boninsegni, M., and Cirac, J. I. 2009. Ground-state properties of quantum many-body systems: entangled-plaquette states and variational Monte Carlo. *New J. Phys.*, **11**, 83026.

Mikelsons, K., Macridin, A., and Jarrell, M. 2009. Relationship between Hirsch-Fye and weak-coupling diagrammatic quantum Monte Carlo methods. *Phys. Rev. E*, **79**, 057701.

Moskowitz, J. W., Schmidt, K. E., Lee, M. A., and Kalos, M. H. 1982. A new look at correlation energy in atomic and molecular systems. II: the application of the Green's function Monte Carlo method to LiH. *J. Chem. Phys.*, **77**, 349.

Muramatsu, A., Zumbach, G., and Zotos, X. 1992. A geometric view of the minus-sign problem. *Int. J. Modern Phys. C*, **3**, 185.

Nambu, Y. 1960. Quasiparticles and gauge invariance in the theory of superconductivity. *Phys. Rev.*, **117**, 648–663.

Nave, C. P., Ivanov, D. A., and Lee, P. A. 2006. Variational Monte Carlo study of the current carried by a quasiparticle. *Phys. Rev. B*, **73**, 104502.

Negele, J. W., and Orland, H. 1988. *Quantum Many-Particle Systems*. New York: Addison-Wesley.

Neuscamman, E., Changlani, H., Kinder, J. M., and Chan, G. K.-L. 2011. Nonstochastic algorithms for Jastrow-Slater and correlator product state wavefunctions. *Phys. Rev. B*, **84**, 205132.

Neuscamman, E., Umrigar, C. J., and Chan, G. K.-L. 2012. Optimizing large parameter set in variational quantum Monte Carlo. *Phys. Rev. B*, **85**, 045103.

Newman, M. E. J., and Barkema, G. T. 1999. *Monte Carlo Methods in Statistical Physics*. New York: Oxford University Press.

Nielsen, M. A., and Chaung, I. L. 2000. *Quantum Computation and Quantum Information*. Cambridge University Press.

Nightingale, M. P. 1999. Basics, quantum Monte Carlo, and statistical mechanics. In: Nightingale, M. P., and Umrigar, C. J. (eds.), *Quantum Monte Carlo Methods in Physics and Chemistry*. NATO Science Series, vol. 525. Dordrecht: Kluwer.

Nightingale, M. P., and Melik-Alaverdian, V. 2001. Optimization of ground and excited states wave functions and van der Waals clusters. *Phys. Rev. Lett.*, **87**, 43401.

Nishimori, H., and Ortiz, G. 2010. *Elements of Phase Transitions and Critical Phenomena*. Oxford University Press.

Oguri, A., and Asahata, T. 1992. Brinkman-Rice transition in the three-band Hubbard model. *Phys. Rev. B*, **46**, 14073.

Oguri, A., Asahata, T., and Maekawa, S. 1994. Gutzwiller wave function in the three-band Hubbard: a variational Monte Carlo study. *Phys. Rev. B*, **49**, 6880.

Ohgoe, T. Suzuki, T. and Kawashima, N. 2012. Commensurate supersolid of three-dimensional lattice bosons. *Phys. Rev. Lett.*, **108**, 185302.

Ortiz, G., Ceperley, D. M., and Martin, R. M. 1993. New stochastic method for systems with broken time-reversal symmetry: 2D fermions in a magnetic field. *Phys. Rev. Lett.*, **71**, 2777.

Ortiz, G., Gubernatis, J. E., and Carlson, J. A. 1997. Stochastic approach to lattice fermions in a magnetic field. *Bull. Am. Phys. Soc.*, March Meeting, M11.10.

Ostlund, S., and Rommer, S. 1995. Thermodynamic limit of density-matrix renormalization. *Phys. Rev. Lett.*, **75**, 3557.

Otsuki, J., Kusunose, H., Werner, P., and Kuramoto, Y. 2007. Continuous-time quantum Monte Carlo method for the Coqblin–Schrieffer model. *J. Phys. Soc. Japan*, **76**, 114707.

Paramekanti, A., Randeria, M., and Trivedi, N. 2001. Projected wave functions and high temperature superconductivity. *Phys. Rev. Lett.*, **87**, 217002.

Parisi, G. 1988. *Statistical Field Theory*. Frontiers in Physics, vol. 66. New York: Addison Wesley.

Peskun, P. H. 1973. Optimum Monte Carlo sampling using Markov chains. *Biometrika*, **60**, 607.

Pollock, E. L., and Ceperley, D. M. 1987. Path-integral computation of superfluid densities. *Phys. Rev. B*, **36**, 8343.

Press, W. H., Teukolsky, S. A., Vetterling, W. T., and Flannery, B. P. 2007. *Numerical Recipes: The Art of Scientific Computing*, third edition. Cambridge University Press.

Prokof'ev, N., and Svistunov, B. 2001. Worm algorithms for classical statistical models. *Phys. Rev. Lett.*, **87**, 160601.

Prokov'ev, N. V., Svistunov, B. V., and Tupitsyn, I. S. 1998. Exact, complete, and universal continuous-time worldline Monte Carlo approach to the statistics of discrete quantum systems. *Sov. Phys. JETP*, **87**, 310.

Purwanto, W., and Zhang, S. 2004. Quantum Monte Carlo for the ground state of many bosons. *Phys. Rev. B*, **70**, 056702.

Renyi, A. 1960. On measures of entropy and information. Page 547 of: *Proceedings of the Fourth Berkeley Symposium on Mathematical Statistics and Probability*. Berkeley: University of California Press.

Reynolds, P. J., Ceperley, D. M., Alder, B. J., and Lester, W. A. 1982. Fixed-node Monte Carlo for molecules. *J. Chem. Phys.*, **77**, 5593.

Rieger, H., and Kawashima, N. 1999. Application of a continuous time cluster algorithm to the two-dimensional random quantum Ising ferromagnet. *Eur. Phys. J. B*, **9**, 233.

Rombouts, S., Heyde, K., and Jachowicz, N. 1999a. A discrete Hubbard-Stratonovich decomposition for general, fermionic two-body interactions. *Phys. Lett.*, **82**, 4155.

Rombouts, S. M. A., Heyde, K., and Jachowicz, N. 1999b. Quantum Monte Carlo method for fermions, free of discretization errors. *Phys. Rev. Lett.*, **82**, 4155.

Rosenbluth, M. 1953. *Proof of validity of Monte Carlo method for canonical averaging*. Tech. rept. LADC-1567 (AECU-2773). Los Alamos National Laboratory.

Rosenbluth, M. N. "Genesis of the Monte Carlo algorithm for statisitcal mechanics," in *The Monte Carlo Method in the Physical Sciences:Celebrating the 50th Anniversary of the Metropolis Algorithm*, ed. J. E. Gubernatis, AIP Conference Proceedings, **690**, 22 (2003). New York: American Institute of Physics.

Rubenstein, B. M., Gubernatis, J. E., and Doll, J. D. 2010. Comparative Monte Carlo efficiency by Monte Carlo analysis. *Phys. Rev. E*, **82**, 36701.

Rubtsov, A. N., Savkin, V. V., and Lichtenstein, A. I. 2005. Continuous-time quantum Monte Carlo method for fermions. *Phys. Rev. B*, **72**, 035122.

Sachdev, S. 1999. *Quantum Phase Transitions*. New York: Cambridge University Press.

Samson, J. H. 1993. Quantum Monte Carlo computation: the sign problem as a Berry phase. *Phys. Rev. Lett.*, **47**, 3408.

Samson, J. H. 1995. Auxiliary fields and the sign problem. *Int. J. Modern Phys. C*, **6**, 427.

Sandvik, A. 2008. Scale-renormalized matrix-product states for correlated quantum systems. *Phys. Rev. Lett.*, **101**, 140603.

Sandvik, A. W. 1992. A generalization of Handscomb's quantum Monte Carlo scheme-application to the 1D Hubbard mode. *J. Phys. A: Math. Gen.*, **25**, 3667.

Sandvik, A. W. 1999. Stochastic series expansion method with operator-loop update. *Phys. Rev. B*, **59**, 14157.

Sandvik, A. W., and Evertz, H. G. 2010. Loop updates and projector quantum Monte Carlo in the valence bond basis. *Phys. Rev. B*, **82**, 024407.

Sandvik, A. W., and Kurkijärvi, J. 1991. Quantum Monte Carlo simulation method for spin systems. *Phys. Rev. B*, **43**, 5950.

Sandvik, A. W., and Vidal, G. 2007. Variational Monte Carlo simulations with tensor-network states. *Phys. Rev. Lett.*, **99**, 220602.

Sarsa, A., Fatoni, S., Schmidt, K. E., and Pederiva, F. 2003. Neutron matter at zero temperature with the auxiliary-field diffusion Monte Carlo method. *Phys. Rev. C*, **68**, 024308.

Scalapino, D. J., and Sugar, R. L. 1981. Method for performing Monte Carlo calculations for systems with fermions. *Phys. Rev. Lett.*, **46**, 519.

Scalettar, R. T., Scalapino, D. J., Sugar, R. L., and Toussaint, D. 1987. Hybrid Monte Carlo for the numerical simulation of many-electron systems. *Phys. Rev. B*, **36**, 8632.

Scalettar, R. T., Noack, R. M., and Singh, R. P. 1991. Ergodicity at large couplings with the determinant Monte Carlo algorithm. *Phys. Rev. B*, **44**, 10502.

Schautz, F., and Filippi, C. 2004. Optimized Jastrow-Slater wave functions for ground and excited states: application to the lowest states of ethene. *J. Chem. Phys.*, **120**, 10931.

Schmidt, K. E., and Fantoni, S. 1999. A quantum Monte Carlo method for nucleon systems. *Phys. Lett.*, **B446**, 99.

Schmidt, K. E., Sarsa, A., and Fatoni, S. 2001. A constrained path Monte Carlo for nucleon systems. *Int. J. Mod. Phys. B*, **15**, 1510.

Schmidt, K. E., Fatoni, S., and Sarsa, A. 2003. Constrained path calculation of the ^{4}He and ^{16}O nuclei. *Eur. J. Phys.*, **A17**, 469.

Schollwöck, U. 2005. The density matrix renormalization group. *Rev. Mod. Phys.*, **77**, 259.

Schrödinger, E. 1952. *Statistical Thermodynamics*. Cambridge University Press.

Schuch, N., Wolf, M. M., Verstraete, F., and Cirac, J. I. 2008. Simulation of quantum many-body systems with strings of operators and Monte Carlo tensor contractions. *Phys. Rev. Lett.*, **100**, 40501.

Sfondrini, A., Cerrillo, J., Schuch, N., and Cirac, J. I. 2010. Simulating two- and three-dimensional frustrated quantum systems with string-bond states. *Phys. Rev. B*, **81**, 214426.

Shi, H., and Zhang, S. 2013. Symmetry in auxiliary-field quantum Monte Carlo calculations. *Phys. Rev. B*, **88**, 125132.

Shiba, H. 1986. Properties of strongly correlated Fermi liquid in valence fluctuation system – a variational Monte Carlo study. *J. Phys. Soc. Japan*, **55**, 2765.

Shiba, H., and Fazekas, P. 1990. Correlated Fermi-liquid state formed with overlapping Kondo clouds. *Prog. Theor. Phys. Suppl.*, **101**, 403.

Shore, J. E., and Johnson, R. W. 1980. Axiomatic derivation of the principle of maximum entropy and of minimum cross-entropy. *IEEE Trans. Inform. Theory*, **IT-26**, 26.

Shore, J. E., and Johnson, R. W. 1983. Comments on and corrections to "Axiomatic derivation of the principle of maximum entropy and of minimum cross-entropy." *IEEE Trans. Inform. Theory*, **IT-29**, 942.

Sivia, D. S., and Skilling, J. 2006. *Data Analysis: A Bayesian Tutorial*. Oxford University Press.

Skilling, J. 1998. Massive inference and maximum entropy. In: Fischer, R., Preuss, R., and von Toussaint, U. (eds.), *Maximum Entropy and Bayesian Methods*. Dordrecht: Kluwer.

Sorella, S. 1998. Green function Monte Carlo with stochastic reconfiguration. *Phys. Rev. Lett.*, **80**, 4558.

Sorella, S. 2000. Green function Monte Carlo with stochastic reconfiguration: an effective remedy for the sign problem. *Phys. Rev. B*, **61**, 2599.

Sorella, S. 2005. Wave function optimization in the variational Monte Carlo method. *Phys. Rev. B*, **71**, 241103.

Sorella, S., Baroni, S., Car, R., and Parrinello, M. 1989. A novel technique for the simulation of interacting fermion systems. *Eurphys. Lett.*, **8**, 663.

Sorella, S., Martins, G. B., Becca, F., Gazza, C., Capriotti, L., Parola, A., and Dagotto, E. 2002. Superconductivity in the two-dimensional tJ model. *Phys. Rev. Lett.*, **88**, 117002.

Spanier, J., and Gelbard, E. M. 1969. *Monte Carlo Principles and Neutron Transport Problems*. Reading, MA: Addison-Wesley.

Stewart, G. W. 2001a. *Matrix Algorithms I: Basic Decompositions*. Philadelphia: SIAM.

Stewart, G. W. 2001b. *Matrix Algorithms II: Eigensystems*. Philadelphia: SIAM.

Sugiyama, G., and Koonin, S. E. 1986. Auxiliary field Monte Carlo for quantum many-body ground states. *Ann. Phys.*, **168**, 1.

Surer, B., Troyer, M., Werner, P., Wehling, T. O., Läuchli, A. M., Wilhelm, A., and Lichtenstein, A. I. 2012. Multiorbital Kondo physics of Co in Cu hosts. *Phys. Rev. B*, **85**, 085114.

Suzuki, M. 1976a. Generalized Trotter's formula and systematic approximants of exponential operators and inner derivatives with applications to many-body problems. *Comm. Math. Phys.*, **51**, 183.

Suzuki, M. 1976b. Relationship between d-dimensional quantal spin systems and $(d + 1)$-dimensional Ising systems: equivalence, critical exponents and systematic approximants of the partition function and spin correlations. *Prog. of Theor. Phys.*, **56**, 1454.

Swendsen, R. H., and Wang, J.-S. 1987. Nonuniversal critical dynamics in Monte Carlo simulations. *Phys. Rev. Lett.*, **58**, 86.

Syljuåsen, O. F. 2008. Using the average spectrum method to extract dynamics from quantum Monte Carlo simulations. *Phys. Rev. B*, **78**, 174429.

Syljuåsen, O. F., and Sandvik, A. W. 2002. Quantum Monte Carlo with directed loops. *Phys. Rev. E*, **66**, 046701.

Tahara, D., and Imada, M. 2008. Variational Monte Carlo method with quantum number projection and multi-variable optimization. *J. Phys. Soc. Japan*, **77**, 114701.

Takasaki, H., Hikihara, T., and Nishino, T. 1999. Fixed point of the finite system DMRG. *J. Phys. Soc. Japan*, **68**, 1537.

ten Haaf, D. F. B., Bemmel, H. J. M., van Leeuwen, J. M. J., van Saarlos, W., and Ceperley, D. 1995. Proof of upper bound in fixed-node Monte Carlo for lattice fermions. *Phys. Rev. B*, **51**, 13039.

Thompson, C. P. 1972. *Mathematical Statistical Mechanics*. Princeton, NJ: Princeton University Press.

Thouless, D. J. 1961. *The Quantum Mechanics of Many-Body Systems*. New York: Academic.

Todo, S., and Kato, K. 2001. Cluster algorithms for general-S quantum spin systems. *Phys. Rev. Lett.*, **87**, 047203.

Todo, S., Matsuo, H., and Shitara, H. 2012. Private communication.

Toulouse, J., and Umrigar, C. J. 2007. Optimization of quantum many-body wave function by energy minimization. *J. Chem. Phys.*, **126**, 84102.

Toulouse, J., and Umrigar, C. J. 2008. Full optimization of Jastrow-Slater wave functions with application to the first-row atoms and homonuclear diatomic atoms. *J. Chem. Phys.*, **128**, 174101.

Trotzky, S., Pollet, L., Gerbier, F., Schnorrberger, U., Bloch, I., Prokof'ev, N. V., Svistunov, B., and Troyer, M. 2010. Suppression of the critical temperature for superfluidity near the Mott transition. *Nature Physics*, **6**, 998.

Troyer, M., Wessel, S., and Alet, F. 2003. Flat histogram methods for quantum systems: algorithms to overcome tunneling problems and calculate the free energy. *Phys. Rev. Lett.*, **90**, 120201.

Umrigar, C. J., and Filippi, C. 2005. Energy and variance optimization of many-body wave functions. *Phys. Rev. Lett.*, **94**, 150201.

Umrigar, C. J., Wilson, K. G., and Wilkins, J. W. 1988. Optimized trial wave functions for quantum Monte Carlo calculations. *Phys. Rev. Lett.*, **60**, 1719.

Umrigar, C. J., Toulouse, J., Filippi, C., Sorella, S., and Hennig, R. G. 2007. Alleviation of the fermion-sign problem by optimization of many-body wave functions. *Phys. Rev. Lett.*, **98**, 110202.

van Hove, L. 1950. Sur l'integrale de configuration pour les systemémes de particules á une dimension. *Physica*, **XVI**, 137.

Verstraete, F., Murg, V., and Cirac, J. I. 2008. Matrix product states, projected entangled pair state, and variational renormalization group methods for quantum spin systems. *Adv. Phys.*, **37**, 143.

Walker, A. J. 1977. An efficient method for generating discrete random variables with general distributions. *ACM Trans. Math. Softw.*, **3**, 253.

Wang, F., and Landau, D. P. 2001. Efficient, multiple-range random walk algorithm to calculate the density of states. *Phys. Rev. Lett.*, **86**, 2050.

Wang, J.-S., Tay, T. K., and Swendsen, R. H. 1999. Transition matrix Monte Carlo reweighting and dynamics. *Phys. Rev. Lett.*, **82**, 476.

Wang, L., Pizorn, I., and Verstraete, F. 2011a. Monte Carlo simulation with tensor network states. *Phys. Rev. B*, **2011**, 134421.

Wang, L., Kao, Y.-J., and Sandvik, A. W. 2011b. Plaquette renormalization scheme for tensor network states. Phys. Rev. E 83, 056703 (2011).

Wang, L., Iazzi, M., Corboz, P., and Troyer, M. 2015. Efficient continuous-time quantum Monte Carlo method for the ground state of correlated fermions. Phys. Rev. B 91, 235151 (2015).

Watanabe, H., and Ogata, M. 2009. Fermi surface reconstruction in the periodic Anderson model. *J. Phys. Soc. Japan*, **78**, 024715.

Watanabe, T., Yokoyama, H., Tanaka, Y., and Inoue, J. I. 2006. Superconductivity and a Mott transition in a Hubbard model on an anisotropic triangular lattice. *J. Phys. Soc. Japan*, **75**, 074707.

Weber, C., Läuchli, A., Mila, F., and Giamarchi, T. 2006. Magnetism and superconductivity of strongly correlated electrons on the triangular lattice. *Phys. Rev. B*, **73**, 014519.

Werner, P., and Millis, A. J. 2006. Hybridization expansion impurity solver: general formulation and application to Kondo lattice and two-orbital models. *Phys. Rev. B*, **74**, 155107.

Werner, P., and Millis, A. J. 2007. High-spin to low-spin and orbital polarization transitions in multiorbital Mott systems. *Phys. Rev. Lett.*, **99**, 126405.

Werner, P., Comanac, A., de' Medici, L., Troyer, M., and Millis, A. J. 2006. Continuous-time solver for quantum impurity models. *Phys. Rev. Lett.*, **97**, 076405.

White, S. R. 1992. Density-matrix formulation for quantum renormalization groups. *Phys. Rev. Lett.*, **69**, 2863.

Wilson, K. G. 1975. The renormalization group: critical phenomena and the Kondo problem. *Rev. Mod. Phys.*, **47**, 773.

Wolff, U. 1989. Collective Monte Carlo updating for spin systems. *Phys. Rev. Lett.*, **62**, 361.

Wood, W. W., and Parker, F. R. 1957. Monte Carlo equation of state of molecules interacting with the Lennard-Jones potential. I. A supercritical isotherm at about twice the critical temperature. *J. Chem. Phys.*, **27**, 720.

Wu, C. J., and Zhang, S. C. 2005. Sufficient condition for absence of the sign problem in the fermionic quantum Monte Carlo algorithm. *Phys. Rev. B*, **71**, 155115.

Yamaguchi, C., and Kawashima, N. 2002. Combination of improved multibondic method and the Wang-Landau method. *Phys. Rev. E.*, **65**, 056710.

Yamaguchi, C., Kawashima, N., and Okabe, Y. 2002. Broad histogram relation for the bond number and its applications. *Phys. Rev. E.*, **66**, 036704.

Yanagisawa, T. 1999. Wave functions of correlated electron state in the periodic Anderson model. *J. Phys. Soc. Japan*, **1999**, 893.

Yang, H.-Y., Yang, F., Jiang, Y.-J., and Li, T. 2007. On the origin of the tunneling asymmetry in the cuprate superconductors: a variational perspective. *J. Phys.: Condens. Matter*, **19**, 016217.

Yokoyama, H., and Shiba, H. 1987a. Variational Monte Carlo studies of the Hubbard model. I. *J. Phys. Soc. Japan*, **56**, 1490.

Yokoyama, H., and Shiba, H. 1987b. Variational Monte Carlo studies of the Hubbard model. II. *J. Phys. Soc. Japan*, **56**, 3582.

Yokoyama, H., and Shiba, H. 1988. Variational Monte Carlo studies of superconductivity in strongly correlated electron systems. *H. Yokoyama and H. Shiba*, **57**, 2482.

Yoo, J., Chandrasekharan, S., Kaul, R. K., Ullmo, D., and Baranger, H. U. 2005. On the sign problem in the Hirsch-Fye algorithm for impurity problems. *J. Phys. A: Math. Gen.*, **38**, 10307.

Yunoki, S., Dagotto, E., and Sorella, S. 2005. Role of strong correlation in the recent angle-resolved photoemission spectroscopy experiments on cuprate superconductors. *Phys. Rev. Lett.*, **94**, 037001.

Zhang, S. 2013. Auxiliary-field quantum Monte Carlo for correlated electron systems. In: Pavarini, E., Kock, E., and Schollwöck, U. (eds.), *Emergent Phenomena in Correlated Matter Modeling and Simulation*, vol. 3. http://hdl.handle.net/2128/5389: Open source.

Zhang, S., and Kalos, M. H. 1991. Exact Monte Carlo for few electron systems. *Phys. Rev. Lett.*, **67**, 3074.

Zhang, S., and Krakauer, H. 2003. Quantum Monte Carlo using phase-free random walks with Slater determinants. *Phys. Rev. Lett.*, **90**, 136401.

Zhang, S., Carlson, J., and Gubernatis, J. E. 1995. Constrained path Monte Carlo for fermion ground states. *Phys. Rev. Lett.*, **74**, 3652.

Zhang, S., Carlson, J., and Gubernatis, J. E. 1997. Constrained path Monte Carlo method for fermion ground states. *Phys. Rev. B*, **55**, 7464.

Index

acceptance probability, 29
algorithm pseudocode
  analytic continuation, 393
  comb reconfiguration, 319
  constrained-path method, 356
    B_V propagation, 358
  continuous-time cluster algorithm, 95
  directed loop algorithm, 152
    "on the fly," 165
    "on-the-fly" $S = \frac{1}{2}XY$ model, 166
  discrete probability sampling
    elegant, 19
    standard, 18
  DMFT self-consistency loop, 222
  finite T determinant method, 195
  heat-bath algorithm, 34
  LDA+DMFT self-consistency loop, 226
  Levenberg-Marquardt algorithm, 465
  long-range interactions, 430
  loop algorithm
    directed, 109
    quantum, 106
  Metropolis algorithm, 30
    single-spin flip, 69
  modified Gram-Schmidt method, 207
    column orthonormalization, 448
    UDV factorization, 207
  Monte Carlo program structure, 61
  Newton's method, 297
  power method
    Monte Carlo, 309
    shifted, 303
    shifted inverse, 305
  quantum loop and cluster algorithm, 129
  rejection method, 418
  sampling time-ordered sequence of imaginary times
    faster, 99
    simple, 97
  single-spin update, 400
    distributed-memory computer, 402
    shared-memory computer, 401

  stochastic series expansion method, 143
    loops and clusters, 144
  Swendsen-Wang algorithm, 71
  union-find algorithm
    serial, 406
    serial (improved), 407
    shared-memory computer, 410
  wave function optimization methods
    linearized, 299
    Newton, 300
  weight window reconfiguration, 317
  Wolff algorithm, 74
  world-line method, 103
  worm algorithm, 78
    simplified, 79
alias method, 18, 416
Anderson impurity model, 189, 215, 220
  action, 218
  chain representation, 217
Anderson lattice model, 189
antiferromagnetic structure factor, 197
area law, 289
autocorrelation time, 46, 49, 52
auxiliary field method, 180
  BSS algorithm (determinant algorithm), 189
  continuous-time auxiliary field algorithm, 455
  continuous-time determinant algorithm, 459
  Hirsch transformation, 183
  Hirsch-Fye algorithm, 198
  Hubbard-Stratonovich transformation, 182
auxiliary fields
  continuous, 182
  discrete, 183
average spectrum method, 382

back propagation, 328
bath Green's function, 218
  Kondo model, 251
Bayesian constrained fit, 380
Bayesian statistical inference, 367
  average spectrum method, 382

484